ORGANIC CHEMISTRY
concepts and applications

ORGANIC CHEMISTRY

concepts and applications

James O. Schreck

DEPARTMENT OF CHEMISTRY
UNIVERSITY OF NORTHERN COLORADO
GREELEY, COLORADO

with 93 illustrations

The C. V. Mosby Company

SAINT LOUIS   1975

Library of Congress Cataloging in Publication Data

Schreck, James O   1937-
  Organic chemistry: concepts and applications.

  Includes index.
  1. Chemistry, Organic.   I. Title.   [DNLM:
1. Chemistry, Organic.   QD251.2 S377o]
QD251.2.S37   547   74-20866
ISBN 0-8016-4358-9

TS/CB/B   9   8   7   6   5   4   3   2   1

Preface

This textbook is written with a functional group approach and incorporates a mechanistic viewpoint to explain the reactions of organic compounds. Written for a one-semester or two-quarter course, it is intended for nonmajors in the paramedical curriculum, home economics, nursing, and industrial arts, and for other nonscience majors who want or need to take a course in introductory organic chemistry.

This book treats the International Union of Pure and Applied Chemistry and common systems of nomenclature for a particular functional group, the common laboratory preparations of that group, and the characteristic class reactions. Mechanisms are used throughout. Energy considerations and profiles are included to illustrate a mechanism. Practical organic chemistry is stressed where appropriate. Some applications include discussions on insect pheromones, analysis of alcoholic beverages, the use and misuse of drugs, commercial materials derived from benzene, food additives, a stereospecific world, and the free radical theory of human aging. Chapter 14 is an important application of organic chemistry; in fact, it is a discussion of the organic chemistry of the living system. In addition, the fourteen chapters include a review of fundamental concepts, aliphatic and aromatic hydrocarbons, stereochemistry, organic halides, alcohols and phenols, ethers and epoxides, aldehydes and ketones, carboxylic acids and their derivatives, organic nitrogen compounds, structure determination through the use of classical and spectral methods, and biomolecules.

Each chapter consists of core material, a summary of the important concepts, a list of important terms for the student's vocabulary of organic chemistry, and a set of problems and questions that review the concepts presented in the chapter. Furthermore, several parts of these problems and questions are answered or explained so that students can check their progress. The answers to all the questions within the chapter are given in the back of the text; the answers to the questions at the end of the chapter are in the instructor's manual. Finally, a self-test is provided at the end of each chapter containing material that I consider of most importance.

Chapter 1 introduces the student to the functional groups that will be encountered throughout the text. After a discussion on the distinction between ionic and covalent bonding, the student is introduced to the concept of the hybridized carbon atom and the energy requirements of organic reactions.

Instead of using the traditional approach of teaching the chemistry of alkanes, alkenes, and alkynes in three separate chapters, I have combined the chemistry of aliphatic hydrocarbons in two chapters. Chapter 2 deals with the structure, nomenclature, and laboratory preparations of the aliphatic hydrocarbons. The concept of the hybridized carbon atom for alkanes, alkenes, and alkynes is reviewed first. The nomenclature of alkanes is presented in detail and the nomenclature of alkenes and alkynes builds upon the IUPAC rules for alkanes. Although methods are available, none are presented for the laboratory synthesis of alkanes. Instead a discussion of petroleum as a source of alkanes is given. The laboratory methods for preparing alkenes are illustrated by dehydrohalogenation and dehydration. Mechanisms are discussed for each method and the presence of carbonium ions in the biochemical synthesis of cholesterol is illustrated. For alkynes, dehydrohalogenation and the reaction of sodium acetylides with primary alkyl halides are presented as methods of forming alkynes. Other applications found in this chapter are geometrical isomers among insect pheromones and the relationship of the carbon-carbon triple bond to the birth control pill. Chapter 3, on the other hand, considers the reactions of aliphatic hydrocarbons. For alkanes, the reactions discussed are free radical substitution (predominantly for methane and propane) and combustion. Also discussed are combustion of alkanes with respect to gasoline and pollution; the typical reactions of alkenes dealing with electrophilic addition, polymerization and polymers as an application of free radical addition in alkenes; and tautomerization and acidity of alkynes.

Chapter 4 deals with the chemistry of aromatic hydrocarbons, particularly benzene. The structure, molecular orbital picture, and nomenclature of benzene and its derivatives are presented. These concepts are then followed by a discussion of electrophilic aromatic substitution. The concept of resonance is used to explain orientation of substitution in benzene derivatives but can be omitted in a one-quarter course. The students are taught how to synthesize using the orientation concept. The chemistry of some other benzene derivatives (for example, naphthalene) and the free radical substitution in arenes is included but usually omitted in the short course. The chapter concludes with a discussion of some commercially important products derived from benzene.

The outline of Chapter 5, on isomerism, is not traditional and is the result of an idea developed by Dr. Gordon Tomasi and myself. Its purpose is to present in a uniform manner the most important concepts of isomerism and usage of terminology, particularly since the organic student of one quarter is the biochemistry student of the next quarter at our university.

Chapters 6 through 11 are concerned with the chemistry of the functional groups. Each chapter is written in basically the same manner: structure, nomenclature, laboratory methods, of synthesizing, reactions involved, and uses or occurrences of compounds containing the functional group.

In Chapter 6 the laboratory methods discussed for introducing a hydroxy group into a molecule include hydrolysis of alkyl halides, Grignard synthesis, and hydroboration-oxidation. The reactions of alcohols and phenols include the formation of alkyl halides (alcohols), dehydration (alcohols), acidity (alcohols and phenols), ester formation (alcohols and phenols), oxidation (alcohols), and electrophilic aromatic substitution (phenols). The reaction dealing with oxidation of alcohols contains a discussion of the roadside breath test. The chapter concludes with an introduction to gas chromatography and its use in the analysis of alcoholic beverages.

Chapter 7 deals with the four types of reactions that an alkyl halide can undergo: S_N1, S_N2, E1, and E2. A section on organochlorine insecticides is included. Topics that are included in the text but are of less importance and not covered in a short course include carbene and benzyne formation and nucleophilic and electrophilic substitution reactions of aromatic halides.

The applications in Chapter 8 are the uses of ethers and products derived from epoxides. There is also a section on ethyl ether as an anesthetic and another on crown ethers and their potential use to the biochemist.

Chapter 9 is on the chemistry of the carboxyl group. Oxidation of alcohols and aromatic hydrocarbons, carbonation of Grignard reagents, and hydrolysis of nitriles are the preparative methods discussed for carboxylic acids. The reactions of carboxylic acids include reduction, salt formation and acidity, formation of functional derivatives, and substitution. The reactions of the functional derivatives are presented as well. Mechanistically, ester formation and nucleophilic acyl substitution are discussed. The organophosphate insecticides are mentioned briefly. The reactivity of the functional derivatives compared to carboxylic acids is emphasized because a similar selectivity occurs in biochemical systems. The usefulness of malonic acid in the synthesis of carboxylic acids and barbiturates is included but can be omitted in a short course. This chapter concludes with a discussion of the chemicals we eat (food additives).

In Chapter 10 aldehydes and ketones are discussed. The preparative procedures for aldehydes include oxidation of primary alcohols, hydrolysis of geminal dihalides, and reduction of acid chlorides, while those for ketones include oxidation of secondary alcohols and Friedel-Crafts acylation. The nucleophilic addition reactions of carbonyl compounds are discussed, but generally the aldol condensation is omitted in a short course.

Chapter 11 discusses the chemistry of amines. Amines are synthesized by reduction of nitro compounds, reductive amination, reduction of nitriles, and the Hofmann degradation of amides. Reductive amination includes a theoretical implication in chemical evolution. The reactions of amines include salt formation and basicity, formation of diazonium compounds and their subsequent reactions, conversion into amides, including sulfonamide. There is an application section on sulfanilamide and sulfa drugs. The section on heterocyclic amines can be omitted in a short course. The chapter includes a discussion on drugs such as hallucinogens, amphetamines, and narcotics.

A brief introduction to the chemical and physical methods used by organic chemists to deduce the structure of a molecule is presented in Chapter 12. The chemical tests for functional groups is summarized and outlined and the student is shown how to use infrared techniques for functional group identification and nuclear magnetic resonance for a more refined structure identification. A discussion on the use of classical and spectral techniques in the identification of a pheromone concludes the chapter.

Chapters 13 and 14 deal with biomolecules and biochemistry. The organic chemistry of biomolecules is divided into lipids, carbohydrates, and proteins. The topics of interest regarding lipids are fats, saponification, and the method by which a soap cleans. Brief introductions to waxes, phosphoglycerides, cerebrosides, and steroids are included but can be omitted in a short course. The role of vitamin A in the visual process and a proposal of the role of vitamin E as an antioxidant are described. The chemistry of carbohydrates is discussed in terms of the monosaccharide glucose, but sucrose, cellulose, and starch are also discussed. The last section on proteins includes the kinds and structures of amino acids, some of the reactions of amino acids, the Merrifield approach to the laboratory synthesis of peptides, and the primary and secondary structures of proteins (DNA and RNA).

Chapter 14 is an application of organic chemistry to the living system. It includes a discussion of enzyme catalyzed reactions, the function of the coenzyme, biochemical energetics, the Embden-Meyerhof pathway, Krebs' cycle, oxidation of fatty acids, and protein biosynthesis. The latter is developed in terms of the biosynthesis of normal and sickle-cell hemoglobin.

A number of people have given their time generously during the writing of this text. I would like to thank Clive Grant, David Pringle, and Charles DePuy for their comments and suggestions on early portions of this text. Gordon Tomasi was immensely helpful with the biochemistry chapters. John Idoux read the entire manuscript. His comments were valuable, appreciated, and always encouraging. I am grateful to Faye Magneson, who typed the entire manuscript, and also to many of my students who showed an interest in the text during the writing and rewriting stages. Finally, I am most grateful to my family, Jean, Steven, and Michael, who were so patient with me during the writing of this text.

James O. Schreck

Contents

4   Aromatic hydrocarbons, 82

5   Stereochemistry, 114

14 Organic chemistry of living systems, 428

Selected answers, 473

ORGANIC CHEMISTRY
concepts and applications

chapter 1

Some fundamental concepts

Up until the early 1800's organic chemistry was considered to be a study of those compounds isolated from living organisms, such as plants and animals. Sugar was considered to be an organic compound because it could be obtained from sugar cane or the sugar beet. Similarly, urea, which is the main end product of protein metabolism, formed in the liver of animals and excreted in the urine, was classified as an organic compound.

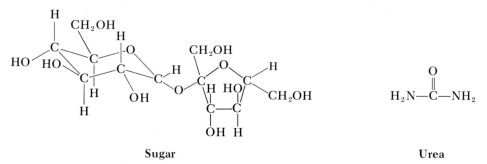

Sugar   Urea

In 1828 the first organic compound was prepared in the laboratory without the use of a living organism. The organic compound synthesized was urea. Although organic compounds are often obtained from living organisms, chemists have now learned how to prepare a number of compounds found in nature. These include such diverse compounds as lysergic acid and *cis*-9-tricosene, the structures of which are as follows:

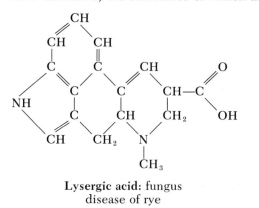

$$CH_3(CH_2)_7CH{=}CH(CH_2)_{12}CH_3$$

Lysergic acid: fungus
disease of rye

cis-9-**Tricosene:** sex attractant
of housefly

1

In addition, chemists have synthesized many organic compounds not found in nature. Some of these compounds, such as carbon tetrachloride, acetylsalicylic acid, and amyl acetate, find application in our everyday life:

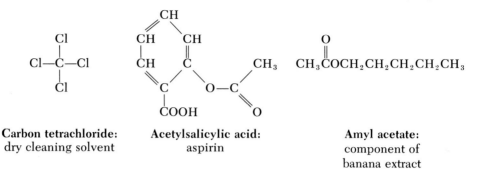

Carbon tetrachloride:
dry cleaning solvent

Acetylsalicylic acid:
aspirin

Amyl acetate:
component of
banana extract

Other interesting compounds not found in nature include cubane, asterane, and radialene, which have structures resembling their names. The synthesis of these compounds is a challenge to the organic chemist because of their unique structures:

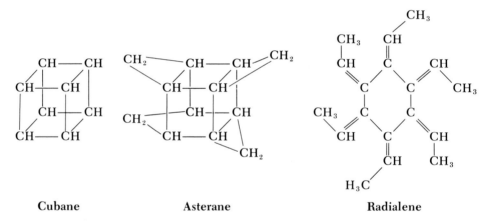

Cubane   **Asterane**   **Radialene**

Although millions of organic compounds are known, many millions more remain to be synthesized and studied. Following are two examples. The compound on the right is an example of a single chain of atoms that is knotted.

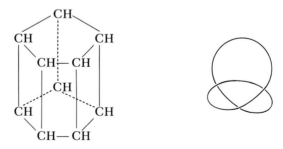

All these compounds have one feature in common. They contain the element carbon. Generally today, organic chemistry is considered to be a study of the chemistry of the compounds of carbon. However, some carbon

Table 1-1
Classes of organic compounds

Class name	Class structure	Example	Common use or source
Acid halide	—COX or —C (=O)(\X)	CH_3COCl	Acetylating agent—used to introduce an acetyl group into a molecule
Alcohol	—OH°	CH_2OHCH_2OH	Antifreeze
Aldehyde	—CHO or —C (=O)(\H)	C_6H_5CHO	Oil of bitter almond
Amide	—CONH$_2$ or —C (=O)(\NH$_2$)	H_2NCONH_2	Barbiturate precursor
Amine	—NH$_2$	$H_2N(CH_2)_6NH_2$	Used to synthesize nylon
Carboxylic acid	—COOH or —C (=O)(\OH)	CH_3COOH	Vinegar
Ester	—COOR† or C (=O)(\OR)	$CH_3COOC_5H_{11}$	Banana flavoring
Ether	—OR	$C_2H_5OC_2H_5$	Anesthetic
Halogen compound	—X	Cl—⬡—Cl	Mothballs
Hydrocarbons Alkane	—C—C— (single bonds only)	$CH_3(CH_2)_6CH_3$	Component of gasoline
Alkene	—CH=CH$_2$	$CH_2=CCl_2$	Substance from which plastic wrap is made
Alkyne	—C≡CH	CH≡CH‡	Oxyacetylene torch
Ketone	—COR or —C (=O)(\R)	$CH_3COC_2H_5$	Fingernail polish remover
Phenol	—OH°	C_6H_5OH	Antiseptic

°The —OH group of an alcohol is chemically different from that of a phenol.
†The symbol R attached to the functional group represents an alkyl group, such as methyl (CH_3) (see Chapter 2).
‡Acetylene (CH≡CH) is a remarkably reactive substance, and almost every simple organic compound can be prepared from it. During World War II, Germany prepared large amounts of the simple organic compounds, such as ethanol, ethyl chloride, butadiene, and acetic acid, from acetylene and, as a result, became chemically independent of the rest of the world.

compounds, such as alkali carbonates (for example, Na_2CO_3) and ferri-cyanides (for example, $K_3Fe(CN)_6$), are not considered to be organic compounds.

A study of organic chemistry is of interest to us because of the relationship of carbon compounds to life and life processes. Organic chemistry is the chemistry of plastics, gasoline, drugs, food, and clothing. The living system is made up of water and organic compounds. The study of biological processes, then, involves a study of organic chemistry. We will approach our study of organic chemistry from a study of the properties, reactions, and interrelationships of relatively simple organic compounds in relatively simple systems. We will then use these ideas to make predictions regarding more complex molecules in complex systems such as biological processes.

Organic chemists correlate the large amount of information about individual molecules by systematically dividing organic compounds into *classes*. Each class of organic compounds is characterized by a particular *functional group*. A functional group can be defined as a unique collection of chemically bonded atoms that displays a distinctive set of properties. All alkenes, for example, contain the carbon–carbon double bond, $C=C$. The reactions of alkenes are the reactions of the carbon–carbon double bond. Knowing the properties of the carbon–carbon double bond, one can make fairly good predictions regarding the physical and chemical properties of a new compound also known to be an alkene. For example, the carbon–carbon double bond in *cis*-9-tricosene would be expected to be very similar in properties to the carbon–carbon double bond in a simple alkene such as 1-butene $(CH_3CH_2CH=CH_2)$.

The most common classes of organic compounds and the structure common to each class encountered in the study of organic chemistry are listed in Table 1-1.

Types of bonding: ionic and covalent

The molecules illustrated in the previous section consist of atoms held together by chemical bonds. It is important to understand how these bonds are formed. The nucleus of an atom is surrounded by electrons arranged in energy levels. The number of electrons in each energy level is limited. The first energy level is limited to two electrons, the second, to eight, and the third, to eighteen. The distribution of electrons in the energy levels for the first ten elements is shown in Table 1-2. The greatest stability for hydrogen is reached when the first energy level is full (two electrons) as in helium. In the case of the elements lithium through fluorine, the greatest stability is reached when the outermost energy level (second energy level in these elements) contains eight electrons as in neon. These atoms attain a stable electron configuration in one of two ways: by transfer of electrons from one atom to another or by sharing of electrons between atoms.

The transfer of electrons from one atom to another results in the formation of an ionic bond, as in the formation of beryllium oxide:

Table 1-2
Electrons in energy levels for first ten elements

| Element | Electron configuration | | Outermost level electrons° |
	First energy level	Second energy level	
H	1		H·
He	2		$\overset{\cdot}{He}$·
Li	2	1	Li·
Be	2	2	$\overset{\cdot}{Be}$·
B	2	3	$\overset{\cdot}{B}$·
C	2	4	·$\overset{\cdot}{\underset{\cdot}{C}}$·
N	2	5	·$\overset{\cdot\cdot}{N}$·
O	2	6	·$\overset{\cdot\cdot}{\underset{\cdot}{O}}$:
F	2	7	·$\overset{\cdot\cdot}{\underset{\cdot\cdot}{F}}$:
Ne	2	8	:$\overset{\cdot\cdot}{\underset{\cdot\cdot}{Ne}}$:

°Only those electrons in the outermost energy level are represented.

$$\overset{\cdot}{Be}\cdot + \cdot\overset{\cdot\cdot}{\underset{\cdot\cdot}{O}}: \longrightarrow Be^{2+} :\overset{\cdot\cdot}{\underset{\cdot\cdot}{O}}:^{2-}$$

The loss of the two electrons from the outer energy level of beryllium leaves two electrons in the first energy level (see Table 1-2). Beryllium has attained stability, since it has two electrons in the first energy level. An oxygen atom gains both electrons lost by beryllium. This gain of two electrons gives oxygen a second energy level of eight electrons. As a result of losing two electrons, the beryllium atom becomes positively charged. Oxygen, which has gained two electrons, has a negative charge. The *ionic bond* is the electrostatic attraction between the oppositely charged ions.

The sharing of electrons between atoms results in the formation of a covalent bond, as in the formation of hydrogen fluoride:

$$H\cdot + \cdot\overset{\cdot\cdot}{\underset{\cdot\cdot}{F}}: \longrightarrow H:\overset{\cdot\cdot}{\underset{\cdot\cdot}{F}}:$$

A hydrogen atom completes its energy level of two by sharing two electrons. Fluorine completes its energy level of eight by sharing two electrons. The *covalent bond* is the electrostatic attraction between each electron and both atoms. The covalent bond is the typical bond found in organic compounds. Following are some examples of the formation of a covalent bond:

$$H\cdot + H\cdot \longrightarrow H:H$$

$$2H\cdot + \cdot\overset{\cdot\cdot}{\underset{\cdot}{O}}: \longrightarrow \begin{array}{l} H:\overset{\cdot\cdot}{\underset{\cdot\cdot}{O}}: \\ H \end{array}$$

$$4H\cdot + \cdot\overset{\cdot}{\underset{\cdot}{C}}\cdot \longrightarrow \begin{array}{c} H \\ H:\overset{\cdot\cdot}{\underset{\cdot\cdot}{C}}:H \\ H \end{array}$$

In formulas in which two dots represent a covalent bond, a single line can be substituted for the two dots; for example:

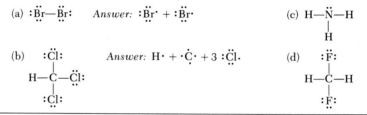

1-1. Using electronic notation, show the formation of the covalent bonds in the following molecules:

(a) $:\ddot{B}r{-}\ddot{B}r:$ *Answer:* $:\ddot{B}r\cdot + :\ddot{B}r\cdot$

(c) $H{-}\ddot{N}{-}H$
       $\quad\ \ |$
       $\quad\ \ H$

(b) $\quad :\ddot{C}l:$
$\quad\quad |$
$\ H{-}\overset{\textstyle |}{C}{-}\ddot{C}l:$
$\quad\quad |$
$\quad :\ddot{C}l:$

Answer: $H\cdot + \cdot\dot{C}\cdot + 3 :\ddot{C}l\cdot$

(d) $\quad :\ddot{F}:$
$\quad\quad |$
$\ H{-}\overset{\textstyle |}{C}{-}H$
$\quad\quad |$
$\quad :\ddot{F}:$

Atomic orbitals

Table 1-2 shows that electrons exist in different energy levels. Furthermore, within each energy level there are subdivisions called *sublevels.* These sublevels are designated as *s, p, d,* and *f.* (The *d* and *f* sublevels will not be encountered in our study of organic chemistry.) The sublevel is a volume wherein there is a good probability of finding an electron of a given energy level. The region wherein the electron with this energy operates is called the *atomic orbital* for that electron.

For electrons in *s* sublevels, the orbitals are spherical in shape and symmetrical about the nucleus. The *s* atomic orbital is shown in Fig. 1-1.

The *p* atomic orbitals are dumbbell shaped with the nucleus between the two lobes. The probability of finding the electron is at a minimum near the nucleus. There are three *p* atomic orbitals available to electrons in all ele-

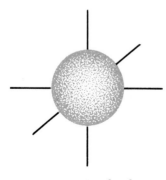

s atomic orbital

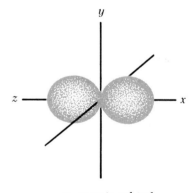

p_x atomic orbital

Fig. 1-1
Atomic orbitals *s* and p_x. Nucleus is at the center.

ments except hydrogen and helium (see Table 1-2.) All three orbitals are at right angles to one another. They are usually referred to as p_x, p_y, and p_z. The p_x atomic orbital is shown in Fig. 1-1. Each p atomic orbital can hold only two electrons. Furthermore, these electrons must be paired; that is, they must have opposite spins.

In filling orbitals, electrons will occupy the level of lowest energy according to the following sequence: $1s$, $2s$, $2p$. The electron configurations of the first ten elements are as follows:

H	$1s^1$
He	$1s^2$
Li	$1s^2 2s^1$
Be	$1s^2 2s^2$
B	$1s^2 2s^2 2p^1 2p^0 2p^0$
C	$1s^2 2s^2 2p^1 2p^1 2p^0$
N	$1s^2 2s^2 2p^1 2p^1 2p^1$
O	$1s^2 2s^2 2p^2 2p^1 2p^1$
F	$1s^2 2s^2 2p^2 2p^2 2p^1$
Ne	$1s^2 2s^2 2p^2 2p^2 2p^2$

Beginning with carbon, the configurations are more complicated. In the case of carbon, two electrons are placed in each of the $1s$ and $2s$ orbitals; the remaining two electrons are placed in different p orbitals in the same energy level because more energy is required for two electrons to pair and occupy the same orbital than for them to remain unpaired and occupy different orbitals. Carbon still has one p orbital that is vacant.

Molecular orbitals: covalent bonds

If the electron of one atomic orbital combines or overlaps with the electron of another atomic orbital so that the two electrons occupy a common orbital holding the two nuclei together, a new orbital is formed, which is called a *molecular orbital*. The formation of a molecular orbital is the formation of a covalent bond. A molecular orbital always encompasses two or more nuclei.

The shape of a molecular orbital is usually deduced from the shapes of the atomic orbitals involved. Consider the formation of the molecular orbital from two hydrogen atoms as shown in Fig. 1-2. The hydrogen molecule is formed by the overlap of two s atomic orbitals. The molecular orbital has the shape expected from the overlapping of two s atomic orbitals. Fig. 1-3 shows

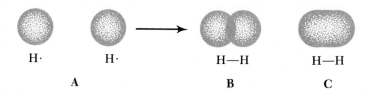

Fig. 1-2
Formation of molecular orbital of hydrogen molecule. **A,** Individual s atomic orbitals for two hydrogen atoms. **B,** Overlap of s atomic orbitals. **C,** Shape of hydrogen molecule molecular orbital.

H·   ·F̈:   H—F̈:   H—F̈:

          A   **B**   **C**

Fig. 1-3
Formation of molecular orbital of hydrogen fluoride. **A,** The s atomic orbital of hydrogen and the p atomic orbital of fluorine. **B,** Overlap of an s atomic orbital and a p atomic orbital. **C,** Shape of hydrogen fluoride molecular orbital.

the shape of the hydrogen fluoride molecular orbital. This molecular orbital is formed by the overlap of an s atomic orbital of hydrogen and a p atomic orbital of fluorine. The molecular orbital has the shape expected from the overlapping of an s and a p atomic orbital.

In the hydrogen molecule and the hydrogen fluoride molecule, the molecular orbital has an electron cloud that is symmetrical about the axis joining the two nuclei. Such a molecular orbital is called a sigma (σ) molecular orbital. The covalent bond is called a σ bond.

Hybrid atomic orbitals

The arrangement of the electrons in the outermost level of carbon when carbon is covalently bonded is not the same as when carbon is in the atomic state. The electron configuration of carbon in the atomic state is $1s^2 2s^2 2p^1 2p^1$. From this configuration, it would be expected that carbon would form bonds by combining or overlapping of the two partially filled p orbitals with s or p atomic orbitals of other atoms and thus that the combination with two hydrogen s atomic orbitals would lead to CH_2. However, the stable hydride of carbon is methane (CH_4). Furthermore, the four carbon–hydrogen bonds in CH_4 are equivalent. The equivalent bonding is achieved by hybridization of the one s and three p atomic orbitals in the second energy level to give four orbitals of a new type, which are called sp^3 orbitals. These orbitals are equivalent, and each contains one electron. The process of hybridization results in the following change in the electron configuration of carbon:

$$1s^2 2s^2 2p^1 2p^1 2p^0 \xrightarrow{\text{hybridization}} 1s^2 2[(sp^3)^1 (sp^3)^1 (sp^3)^1 (sp^3)^1]$$

Atomic state of carbon   **sp^3 hybridized state of carbon**

In addition, the four sp^3 hybrid orbitals are directed to the corners of a tetrahedron. The angle between any two of these orbitals is 109.5°. The formation of these orbitals is illustrated in Fig. 1-4.

The shape of methane is tetrahedral. With the use of the concepts of the covalent bond, the formation of methane is illustrated in Fig. 1-5. The carbon–hydrogen bonds in methane are symmetrical about a line joining the atomic nuclei. These bonds, formed by overlap of atomic orbitals along an axis, are σ bonds.

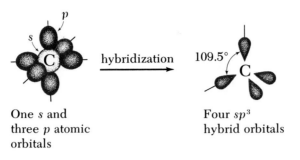

Fig. 1-4
Hybridization of an s and three p orbitals to form four sp^3 hybrid orbitals directed toward the corners of a tetrahedron.

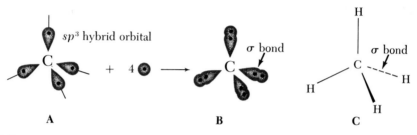

Fig. 1-5
Formation of methane molecular orbitals (covalent bonds). **A,** Four s atomic orbitals of hydrogen approach an sp^3 hybrid atomic orbital of carbon. The number of electrons in each orbital is indicated by dots. **B,** Overlap of four s atomic orbitals with four sp^3 hybrid orbitals to form four molecular orbitals or four σ bonds. **C,** Perspective line drawing for methane. Solid lines indicate bonds in the plane of the paper, dashed line indicates a bond behind the plane, and wedged line indicates a bond directed out of the plane of the paper toward the reader.

Carbon uses sp^3 hybridized orbitals when it forms bonds with four other atoms. However, organic compounds do not contain carbons with sp^3 hybridized orbitals only. In addition, carbons with sp^2 and sp hybridized orbitals are possible.

The combination of one s orbital and two p orbitals forms three sp^2 hybrid orbitals. The hybridization process results in the following change in the electron configuration:

$$1s^2 2s^2 2p^1 2p^1 2p^0 \xrightarrow{\text{hybridization}} 1s^2 2[(sp^2)^1(sp^2)^1(sp^2)^1]2p^1$$

Atomic state of carbon sp^2 **hybridized state of carbon**

The three sp^2 orbitals are coplanar, and the angles between them are 120°. Each orbital contains one electron. The remaining p orbital contains one electron and lies perpendicular to the plane of the sp^2 orbitals. The sp^2 orbitals with the lobes of the p orbital above and below the plane are shown in Fig. 1-6.

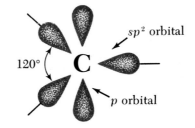

Fig. 1-6
An sp^2 hybrid orbital. The p orbital lies perpendicular to the plane of the sp^2 orbital.

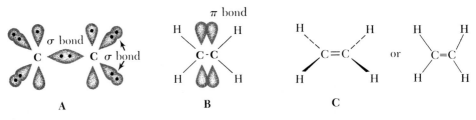

Fig. 1-7
Formation of ethene. **A,** Overlap of two sp^2 orbitals of carbon and four s orbitals of hydrogen with two sp^2 orbitals of carbon. Overlap described here forms five σ bonds. The number of electrons in each bond is indicated by dots. **B,** Side-by-side overlap of p orbitals to form a π bond. **C,** Perspective line drawing for ethene.

Carbon uses sp^2 hybrid orbitals when it is bonded to three other atoms. Consider the bonding in ethene ($CH_2{=}CH_2$), in which a carbon is bonded to three other atoms (one carbon and two hydrogens). In this case, the carbon is using sp^2 hybrid orbitals in forming bonds to the other carbon and the two hydrogens. The four hydrogens in ethene use s atomic orbitals to form bonds to carbon. Maximum overlap of the sp^2 and s orbitals leads to the partial structure of ethene shown in Fig. 1-7, A. Five bonds are formed as a result of this overlap. From Fig. 1-7, A, it can be seen that the p orbitals on the two carbons are parallel. These orbitals with one electron each can overlap side by side as shown in Fig. 1-7, B. The result of this overlap is a new molecular orbital, which is called a pi (π) molecular orbital or π bond. Fig. 1-7, C, is a perspective line drawing for ethene. The σ and π bonds in ethene are represented by a carbon–carbon double bond.

The third possible hybrid orbital used by carbon is the sp hybrid orbital. The combination of one s orbital and one p orbital produces two sp hybrid orbitals. The hybridization process results in the following change in the electron configuration:

$$1s^2 2s^2 2p^1 2p^1 2p^0 \xrightarrow{\text{hybridization}} 1s^2 2[(sp)^1(sp)^1]2p^1 2p^1$$

Atomic state of carbon $\qquad\qquad$ **sp hybridized state of carbon**

The two sp orbitals are 180° apart and lie along the same line. Each orbital contains one electron. The remaining two p orbitals each contain an electron

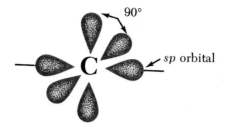

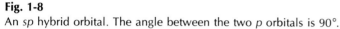

90°

sp orbital

Fig. 1-8
An *sp* hybrid orbital. The angle between the two *p* orbitals is 90°.

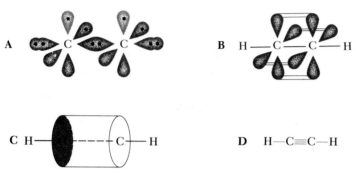

A   B   H——C——C——H

C H——————C——H   D   H—C≡C—H

Fig. 1-9
Formation of acetylene. **A,** Overlap of *sp* orbitals and *s* orbitals. Overlap forms three σ bonds. The number of electrons in each bond is indicated by dots. **B,** Side-by-side overlap of *p* orbitals to form two bonds. **C,** Barrel-shaped electron cloud. **D,** Perspective line drawing for acetylene.

and are perpendicular to one another. The *sp* orbitals with the lobes of one *p* orbital above and below the plane of the line joining the two *sp* orbitals and the other *p* orbital in front of and behind the line joining the two *sp* orbitals are depicted in Fig. 1-8.

Carbon uses *sp* hybrid orbitals when it is bonded to two other atoms. Consider the bonding in acetylene (HC≡CH), in which a carbon atom is bonded to one other carbon atom and a hydrogen atom. Carbon uses *sp* hybrid orbitals in forming bonds to the other carbon and to one of the hydrogens. Each of the hydrogens uses an *s* atomic orbital to bond to carbon. Maximum overlap of an *sp* hybrid orbital with another *sp* hybrid orbital and of an *sp* hybrid orbital with an *s* orbital leads to the formation of three σ bonds and the partial structure of acetylene shown in Fig. 1-9, *A*. Overlap of the two pairs of *p* orbitals remaining on each carbon in Fig. 1-9, *A*, results in two bonds at right angles to one another as shown in Fig. 1-9, *B*. These π bonds merge, and when viewed from the end (Fig. 1-9, *C*), the symmetrical π electron cloud appears to be barrel shaped around the axis joining the nuclei. The σ and two π bonds in acetylene are represented as a triple bond.

In summary, carbon uses sp^3 hybrid orbitals when it forms bonds with

Table 1-3
Characteristics of hybrid orbitals

Characteristics	Hybrid orbital		
	sp^3	sp^2	sp
Used when carbon is bonded to this number of other (not necessarily carbon) atoms	4	3	2
Number of equivalent hybrid orbitals	4	3	2
Shape			
Geometry associated with	Tetrahedral	Planar or trigonal	Linear
Bond angles	109.5°	120°	180°
Number of π bonds associated with	0	1	2

four other atoms, as in methane; sp^2 hybrid orbitals when it forms bonds with three other atoms, as in ethene; and sp hybrid orbitals when it is attached to two other atoms, as in acetylene. The characteristic features of these orbitals are summarized in Table 1-3.

1-2. What is the hybridization of the orbitals of each circled carbon atom?

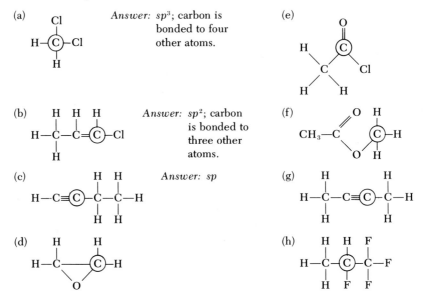

(a)   *Answer:* sp^3; carbon is bonded to four other atoms.

(b)   *Answer:* sp^2; carbon is bonded to three other atoms.

(c)   *Answer:* sp

1-3. What is the bond angle between the indicated atoms?

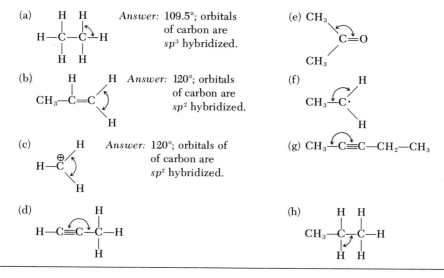

(a)   H   H   *Answer:* 109.5°; orbitals
      H—C—C—H   of carbon are
      H   H sp^3 hybridized.

(b)   H   H   *Answer:* 120°; orbitals
      CH₃—C=C   of carbon are
              H sp^2 hybridized.

(c)   H   *Answer:* 120°; orbitals of
      H—C⊕   of carbon are
          H sp^2 hybridized.

(d)   H
      H—C≡C—C—H
              H

(e) CH₃
          C=O
    CH₃

(f)   H
      CH₃—C·
             H

(g) CH₃—C≡C—CH₂—CH₃

(h)   H   H
      CH₃—C—C—H
           H   H

Energy requirements for organic reactions

Generally, in organic chemistry we are concerned with the preparation of a compound, such as AB, from available starting materials, such as A and B, which can be represented by the following equation:

$$A + B \xrightarrow{\text{special condition}} AB$$

In this chemical equation, A and B are the reactants, AB is the product, and the special condition might be one of the following: temperature (−5° C or 200° C), pressure (1 atm or 2.5 atm), inert atmosphere (nitrogen rather than air), or solvent (alcohol, dry ether, or water).

The rate at which A and B react to form AB is proportional to the concentrations of one or both of the reacting species A and B. There are several possible rate relationships, and one of these is as follows:

$$\text{rate of formation of AB} \propto [\text{A}][\text{B}]$$

According to this expression, the rate of formation of AB depends on the concentration of the reactants A and B. Such a reaction is a second-order reaction. When the proportionality is removed, the rate expression takes the following form:

$$\text{rate of formation of AB} = k[\text{A}][\text{B}]$$

where k is the *rate constant*. If the concentration of one or both reactants is changed, this change will be reflected in the rate of formation of AB.

In other situations, the rate of formation can be proportional to either the concentration of A alone or the concentration of B alone. In either of these cases, the reaction is said to be a first-order reaction.

The preparation or synthesis of AB from A and B involves bond breaking and bond forming. An energy diagram or energy profile, which represents

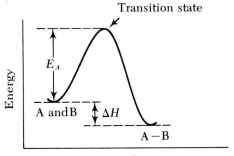
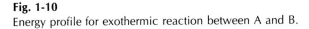

Fig. 1-10

Energy profile for exothermic reaction between A and B.

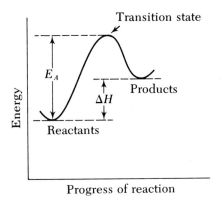
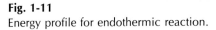

Fig. 1-11

Energy profile for endothermic reaction.

the progress of a reaction, is shown in Fig. 1-10 for the previously discussed second-order reaction. The energy, which changes during the reaction, increases along the vertical axis from bottom to top. The progress of the reaction indicates the distance between molecules of A and B, which are forming a new bond; these changes occur along the horizontal axis.

Several requirements must be met in order for A and B to form AB. First, A and B must collide. Second, the collision between A and B must be energetic; that is, it must supply a minimum amount of energy to get the molecules over the barrier. This minimum amount of energy is called the *energy of activation* (E_A), and it determines the rate of the reaction. The larger the energy of activation is, the slower the reaction is; conversely, the smaller the energy of activation is, the faster the reaction is. Third, the reactants A and B must be properly oriented.

In the reaction shown in Fig. 1-10, the energy of the product AB is less than that of the reactants A and B. Such a reaction is *exothermic*, and heat, *H*, is released during the formation of products. On the other hand, if the

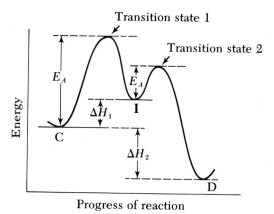

Fig. 1-12
Energy profile for reaction involving intermediate.

energy of the product(s) is more than that of the reactant(s), heat is absorbed in the process, and the reaction is *endothermic*. The energy profile of an endothermic reaction is shown in Fig. 1-11.

The *transition state*, which lies along the reaction path, is characterized by the breaking and formation of bonds. The formation or breaking of bonds is represented by dashed lines. In the reaction between A and B, the transition state would be represented by the formation of a bond between A and B:

$$A + B \longrightarrow [A\text{-}\text{-}\text{-}B] \longrightarrow AB$$

The reverse reaction would have a transition state in which bonds are being broken:

$$AB \longrightarrow [A\text{-}\text{-}\text{-}B] \longrightarrow A + B$$

A reaction in which one bond is being broken and another is being made would be represented as follows:

$$\overset{\text{breaking}}{} \quad \overset{\text{forming}}{}$$
$$AB + C \longrightarrow [A\text{-}\text{-}\text{-}B\text{-}\text{-}\text{-}C] \longrightarrow A + BC$$

The structure of a transition state cannot be known with certainty, since a transition state is unstable and has too short a lifetime. In general, a reaction that is highly exothermic has a low activation energy and has a transition state that greatly resembles the reactants, whereas an endothermic reaction usually has a transition state that resembles the products.

Many organic reactions have more than one step. Usually an intermediate is involved. An *intermediate* is a high-energy species that is formed sometime during a reaction but has all its bonds intact. Consider the reaction in which C goes to D (C $\longrightarrow$ D) that involves the single intermediate I; that is, C $\longrightarrow$ I $\longrightarrow$ D. The energy profile for this two-step reaction can be represented as in Fig. 1-12. In general, the lower the energy of the intermediate is, the greater its stability is. Some intermediates can actually be isolated and their stability thereby demonstrated.

The slowest step in a reaction, which is the one with the largest energy of activation, is the *rate-determining step*. The formation of the intermediate (C $\longrightarrow$ I) is the rate-determining step, since this step has a larger energy of activation than the step I $\longrightarrow$ D. This means that the rate of formation of the product cannot be faster than the rate of conversion of C to the intermediate. The overall reaction is exothermic, since D possesses less energy than does C. The formation of the intermediate is an endothermic reaction.

In this course, you will be introduced to four carbon-containing intermediates: the carbonium ion (Chapter 2), the carbanion (Chapters 9 and 10), the free radical (Chapter 3), and carbene (Chapter 7).

Summary

Organic chemistry is the study of the properties, reactions, and interrelationships of organic compounds. These compounds are divided into classes, each of which is characterized by a particular functional group. A functional group is a unique collection of chemically bonded atoms that displays a distinctive set of properties.

The covalent bond is the typical bond found in organic compounds. The covalent bond, or molecular orbital, is formed by overlapping of atomic orbitals. Carbon forms covalent bonds by using three hybrid atomic orbitals: sp^3, sp^2, and sp.

In order for organic molecules to react with one another, they must collide, have a minimum energy to make the collision effective, and be properly oriented. The progress of a reaction is represented in an energy profile. Organic reactions can be classified as exothermic or endothermic, depending on whether energy is released or absorbed during the formation of the products.

Important terms

Atomic orbital	π bond
Covalent bond	p orbital
Electron configuration	Rate constant
Endothermic reaction	Rate-determining step
Energy of activation	Reaction rate
Exothermic reaction	σ bond
Functional group	s orbital
Hybridization	sp hybrid orbital
Intermediate	sp^2 hybrid orbital
Ionic bond	sp^3 hybrid orbital
Molecular orbital	Transition state
Organic classes	

Problems

1-4. Name the class to which each of the following compounds belong:

(a) $CH_3CH_2CH_2COOH$
(b) $CH_3CH_2CH_2CH_2CH_3$
(c) $CH_3CH_2NH_2$
(d) CH_3Br
(e) $C_6H_5OCH_3$

(f) $HC{\equiv}CCH_2CH_3$
(g) $HCONH_2$
(h) $HOCH_2CH_3$
(i) CH_3CH_2CHO
(j) $HCOOCH(CH_3)_2$

1-5.   Identify the functional group(s) present in each structure:

(a)

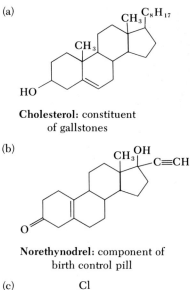

Cholesterol: constituent
of gallstones

(b)

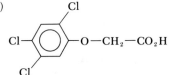

Norethynodrel: component of
birth control pill

(c)

Cl

$Cl-$⟨benzene ring⟩$-O-CH_2-CO_2H$

Cl   Cl

**2,4,5-Trichlorophenoxyacetic
acid:** herbicide

(d) $HC{\equiv}CC{\equiv}CCH{=}C{=}CHCH{=}CHCH{=}CHCH_2COOH$

Mycomycin: antibiotic

1-6.   An organic reaction proceeds in such a manner that it absorbs heat from the surroundings. Draw an energy diagram and show the relative positions of the reactants and products, the energy of activation, and the heat of the reaction.

1-7.   In the following energy diagram, which step is the rate-determining step?

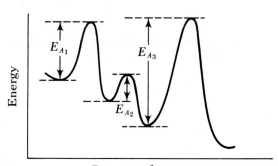

Progress of reaction

1-8. The following mechanism has been proposed for the addition of hydrogen chloride to ethylene. Assume that the reaction is exothermic.

(a) $CH_2=CH_2 + H^+ \xrightarrow{\text{slow}} CH_3CH_2^+$

(b) $CH_3CH_2^+ + Cl^- \xrightarrow{\text{fast}} CH_3CH_2Cl$

Write the rate expression for the rate-determining step. Assume that the carbonium ion $(CH_3CH_2^+)$ is an intermediate, and draw the energy diagram for the two-step reaction.

1-9. What feature of sp, sp^2, and sp^3 hybrid orbitals allows them to be recognized in a molecule?

1-10. Nitrogen and oxygen can also form four sp^3 hybrid orbitals. Change the atomic state configuration of each of these atoms to illustrate the hybridized state. Remember that the number of electrons in the outermost energy level of the atom in its hybridized state is the same as the number of electrons in the outermost energy level of the atomic state.

Carbon: $1s^2 2s^2 2p^1 2p^1 2p^0 \xrightarrow{\text{hybridization}} 1s^2 2[(sp^3)^1(sp^3)^1(sp^3)^1(sp^3)^1]$

        Atomic state   **Hybridized state**

Nitrogen: $1s^2 2s^2 2p^1 2p^1 2p^1 \xrightarrow{\text{hybridization}}$

        Atomic state   **Hybridized state**

Oxygen: $1s^2 2s^2 2p^2 2p^1 2p^1 \xrightarrow{\text{hybridization}}$

        Atomic state   **Hybridized state**

1-11. What is the hybridization of the orbitals of each circled carbon atom?

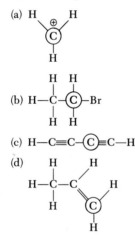

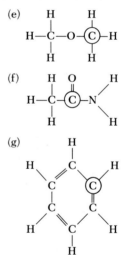

Self-test

1. Match the functional group found in the structure in column A with the name of the functional group found in column B.

A	B
$CH_3CH_2\overset{\displaystyle O}{\overset{\displaystyle \|}{C}}CH_3$	Alkene
$CH_3CH_2CH_2OH$	Ether
CH_3NH_2	Ketone
$CH_3CH=CHCH_3$	Alcohol
$CH_3CH_2CH_2COOH$	Amine
	Carboxylic acid

2. Consider the following energy profile for an organic reaction:

(a) Label the positions of the reactant, the product, and the transition state.
(b) Indicate the energy of activation and the heat of the reaction.
(c) Is the reaction endothermic or exothermic?

3. Two atomic p orbitals can overlap as shown in A and B to form either a σ or a π bond. Which is which, and how can the two bonds be distinguished?

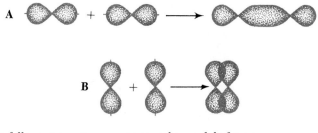

4. Answer the following questions concerning this model of propene:

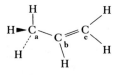

(a) Which carbon(s) is(are) sp^3 hybridized?
(b) Which carbon(s) is(are) sp^2 hybridized?
(c) Which carbon–carbon bond is the shorter: C_a–C_b or C_b–C_c?
(d) What is the C_a–C_b–C_c bond angle?

5. Give the correct answer for each of the following questions:
(a) What is the total number of electrons between two carbon atoms bonded by a triple bond?
(b) What is the value of the bond angle associated with an sp hybrid orbital of the carbon atom?
(c) In an endothermic reaction, which species – reactant or product – possesses less energy?
(d) How many electrons are in the secondary energy level of carbon?
(e) What name is given to the class of compounds formed through the transfer of electrons from one atom to another?
(f) Which atomic orbitals have shapes resembling dumbbells?
(g) What name is given to the four covalent bonds in methane?
(h) What geometric shape is associated with an sp^3 hybrid orbital?
(i) What is the hybridization of the orbitals of the carbon atom in formaldehyde, shown as follows?

(j) What name is given to the slowest step in a reaction?

Answers to self-test

1. The functional groups found in the compounds in column A are ketone, alcohol, amine, alkene, and carboxylic acid.

2.

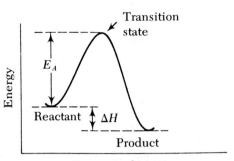

Progress of reaction

The reaction is exothermic, since the products contain less energy than do the reactants.

3. (a) σ bond. (b) π bond. A σ bond is formed by the overlap of two atomic orbitals along their axis. A π bond is formed by the side-by-side overlap of orbitals.

4. (a) C_a (b) C_b and C_c (c) C_b–C_c (d) 120°

5. (a) Six   (e) Ionic   (h) Tetrahedral
   (b) 180°   (f) p orbitals   (i) sp^2
   (c) Reactant   (g) σ bonds   (j) Rate-determining step
   (d) Four

chapter 2

Hydrocarbons: structure, nomenclature, and preparation

Compounds containing only carbon and hydrogen are called *hydrocarbons*. They include alkanes, alkenes, alkynes, and aromatics. Aromatic hydrocarbons are discussed in Chapter 4. Alkanes are often referred to as *saturated hydrocarbons*, since only single covalent bonds are present in the molecule. Alkenes and alkynes are called *unsaturated hydrocarbons*, since there is one or more multiple bonds between the carbon atoms. Following are examples of each kind of hydrocarbon:

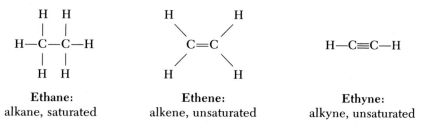

| Ethane: | Ethene: | Ethyne: |
| alkane, saturated | alkene, unsaturated | alkyne, unsaturated |

Structure and bonding in alkanes

The general formula of an alkane is $C_n H_{2n+2}$ where n is the number of carbon atoms. If $n = 2$, then the formula of the alkane is $C_2 H_6$, which is ethane. Both carbons in ethane are bonded to four other atoms. As a general rule, whenever carbon is bonded to four other atoms, the orbitals of the carbon atom are sp^3 hybridized (Chapter 1). The carbon–carbon bond in ethane results from the overlapping of sp^3 hybrid orbitals. The six carbon–hydrogen bonds are formed by the overlap of an sp^3 hybrid orbital of carbon and an s orbital of hydrogen. All the bonds in ethane are symmetrical about a line joining the atomic nuclei. Such bonds, formed by overlap of atomic orbitals along an axis, are σ bonds. The carbon–carbon bond distance in ethane is 1.54 angstrom units (Å); the carbon–hydrogen distance is 1.09 Å.

Since each carbon is bonded to four other atoms, the bonding orbitals are directed toward the corners of a tetrahedron. A three-dimensional perspective formula for ethane is as follows:

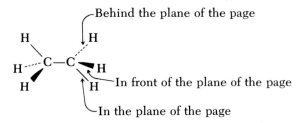

The structural formula of ethane is more conveniently written in a three-dimensional perspective formula (I), a two-dimensional projection or expanded structural formula (II), or a condensed structural formula (III):

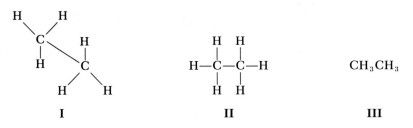

With the use of perspective formula I, all the information regarding the structure and bonding in ethane can be summarized as follows:

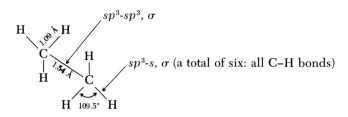

where sp^3-sp^3, σ indicates that the particular bond is formed by the overlap of two sp^3 hybrid orbitals and that the overlap of these two orbitals results in the formation of a σ bond.

2-1. Using the condensed structural formula for propane, $CH_3CH_2CH_3$, answer the following questions:

(a) Rewrite the condensed structural formula as an expanded structural formula.
(b) How many σ bonds does the molecule contain? *Answer:* 10(8 C–H and 2 C–C)
(c) Which σ bond(s) is (are) formed by the overlap of two sp^3 hybrid orbitals? *Answer:* (2 C–C)
(d) Which σ bonds are sp^3-s?
(e) What is the carbon–carbon bond distance?
(f) What is the bond angle between the three carbons?
(g) What is the angle of any H–C–H bond?

In an alkane, all the carbons are either in a *continuous* chain or in a *branched* chain. Every alkane that has a branched chain also has at least one continuous chain of carbon atoms. Butane, a continuous-chain alkane, contains a continuous chain of four carbon atoms; 2-methylpropane, a branched-chain alkane, contains a continuous chain of three carbon atoms:

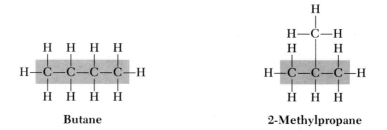

Butane   **2-Methylpropane**

In a continuous chain, the carbon atoms do not lie along the same straight line. In an alkane, the angles between the carbon atoms are 109.5° because of the sp^3 hybrid orbitals. Thus the continuous chains in butane and 2-methylpropane have the following appearance:

Instead of being written in the expanded form shown above, the formula for butane is more commonly written in the condensed form as $CH_3CH_2CH_2CH_3$ or $CH_3(CH_2)_2CH_3$. In the condensed form $CH_3(CH_2)_2CH_3$, the parentheses indicate that the group within them is bonded to the two carbons immediately to the left and to the right. The number written as a subscript after the parentheses indicates the number of such groups. This is the case for any continuous-chain alkane. Instead of being written in the expanded form shown above, the formula for 2-methylpropane is more commonly written in the condensed form as $CH_3—\overset{\overset{\displaystyle CH_3}{|}}{CH}—CH_3$ or $CH_3CH(CH_3)CH_3$. In the condensed form $CH_3CH(CH_3)CH_3$, the parentheses indicate that the group within them is bonded to the atom before the parentheses. This is true for any branched-chain alkane. Following are examples of formulas written in the condensed and expanded forms:

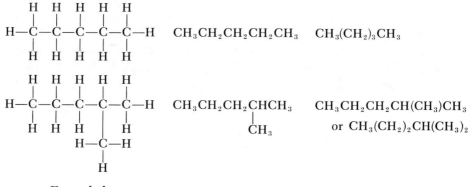

Expanded   **Condensed**

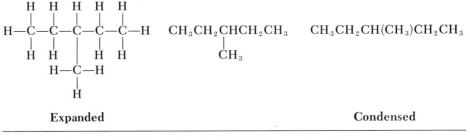

Expanded	Condensed

2-2. Write the expanded form of each of the following formulas:

(a) $C(CH_3)_4$ *Answer:*

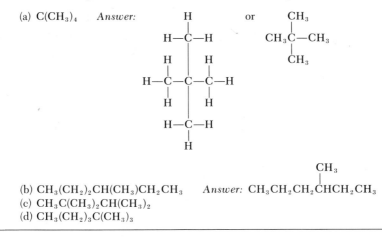

(b) $CH_3(CH_2)_2CH(CH_3)CH_2CH_3$ *Answer:* $CH_3CH_2CH_2\overset{\displaystyle CH_3}{\overset{\displaystyle |}{C}}HCH_2CH_3$

(c) $CH_3C(CH_3)_2CH(CH_3)_2$

(d) $CH_3(CH_2)_3C(CH_3)_3$

Nomenclature of alkanes

Some organic compounds have common names. However, all organic compounds can be named according to the system of the International Union of Pure and Applied Chemistry (IUPAC). In this system, each series of compounds has a characteristic ending usually related to and derived from the class. The common suffix for alkanes is *-ane*. To name a continuous-chain alkane, a combination of a prefix and a suffix is used. The prefix refers to the number of carbon atoms in the continuous chain. The IUPAC names of some continuous-chain alkanes are listed in Table 2-1.

Each member of the alkane series in Table 2-1 differs from the next highest or the next lowest member by a constant amount: one carbon and two hydrogen atoms. A series of compounds in which each member differs from the next member by a constant amount is called a *homologous series*, and the members of such a series are called *homologs*. Thus butane is a homolog of propane and of pentane.

An *alkyl group* is derived from the corresponding alkane by removal of a hydrogen. For example, a methyl group (CH_3—) is derived from methane (CH_4). Alkyl groups are named by dropping *-ane* from the name of the corresponding alkane and replacing it by the suffix *-yl*. Since an alkyl group has one less hydrogen than the corresponding alkane, an alkyl group has the general formula C_nH_{2n+1}.

The ethyl group (CH_3CH_2—) is derived from ethane (CH_3CH_3).

Table 2-1
IUPAC names of alkanes

Number of carbon atoms	Prefix	Formula	Name
1	Meth-	CH_4	Methane
2	Eth-	CH_3CH_3	Ethane
3	Prop-	$CH_3CH_2CH_3$	Propane
4	But-	$CH_3CH_2CH_2CH_3$	Butane
5	Pent-	$CH_3CH_2CH_2CH_2CH_3$	Pentane
6	Hex-	$CH_3(CH_2)_4CH_3$	Hexane
7	Hept-	$CH_3(CH_2)_5CH_3$	Heptane
8	Oct-	$CH_3(CH_2)_6CH_3$	Octane
9	Non-	$CH_3(CH_2)_7CH_3$	Nonane
10	Dec-	$CH_3(CH_2)_8CH_3$	Decane
11	Undec-	$CH_3(CH_2)_9CH_3$	Undecane
12	Dodec-	$CH_3(CH_2)_{10}CH_3$	Dodecane
13	Tridec-	$CH_3(CH_2)_{11}CH_3$	Tridecane
14	Tetradec-	$CH_3(CH_2)_{12}CH_3$	Tetradecane

In propane, six of the hydrogen atoms are identical and different from the two hydrogen atoms circled in the following structure:

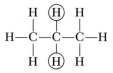

Removal of any one of these six hydrogens gives a normal propyl (*n*-propyl) group ($CH_3CH_2CH_2$—). Removal of one of the other two identical circled hydrogen atoms gives an isopropyl group (CH_3CHCH_3 or CH_3CH—).

There are four butyl groups: two derived from the continuous-chain alkane butane and two derived from its isomer (discussed in next section), the branched-chain alkane 2-methylpropane:

$$CH_3CH_2CH_2CH_3 \qquad\qquad CH_3\!-\!\overset{\displaystyle CH_3}{\underset{\displaystyle H}{\overset{|}{\underset{|}{C}}}}\!-\!CH_3$$

Butane   **2-Methylpropane**

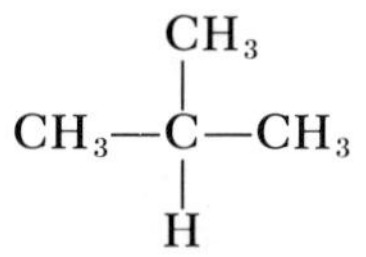

These four groups and their structures are as follows:

$CH_3CH_2CH_2CH_2$——   **Normal butyl (*n*-butyl)**

$CH_3CH_2\underset{\displaystyle |}{\underset{\displaystyle CH_3}{C}}HCH_3$ or CH_3CH_2CH——   **Secondary butyl (*sec*-butyl)**

$CH_3\underset{\displaystyle |}{\underset{\displaystyle CH_3}{C}}HCH_2$——   **Isobutyl**

$$CH_3\!-\!\underset{\underset{\displaystyle |}{|}}{\overset{\overset{\displaystyle CH_3}{|}}{C}}\!-\!CH_3$$

Tertiary butyl (*tert*-butyl)

The nomenclature of branched-chain alkanes is based on the *longest continuous* chain of carbon atoms, which is called the *parent structure* and is given the name of the alkane corresponding to the number of carbon atoms it contains. For example, the longest continuous chain in the following structure contains three carbon atoms and thus is named propane:

$$\underset{\underset{\displaystyle 1\quad\ 2\quad\ 3}{}}{CH_3\!-\!CH\!-\!CH_3}$$ $\overset{CH_3}{|}$ ⟵ Branch
⟵ Continuous chain

Next, the branch is named and located. The branch, which is an alkyl group, is specifically a methyl group. The position of the methyl group is determined by numbering the parent structure from the end of the longest continuous chain that has the largest number of branches. In the example, the branch is located on carbon 2 whether the parent structure is numbered from one end or the other. Therefore, the IUPAC name of the compound is 2-methyl-propane.

Branched-chain alkanes are named in accordance with the following IUPAC rules:

1. Select as the parent structure the longest continuous chain, and then consider the compound derived from this structure by the replacement of hydrogen with various alkyl groups.
2. To locate the alkyl-groups, number the parent structure, starting at the end of the chain that has the most substituted groups.
3. Name the alkyl groups. If the same alkyl group appears more than once, indicate this by using the prefixes *di-*, *tri-*, *tetra-*, and the like to show how many of these alkyl groups there are. The position of each group must be indicated.
4. The name of the compound is written as one word. Numbers are separated from other numbers by commas, and numbers are separated from names of alkyl groups by hyphens.

Following are examples of the use of the rules in naming branched-chain alkanes:

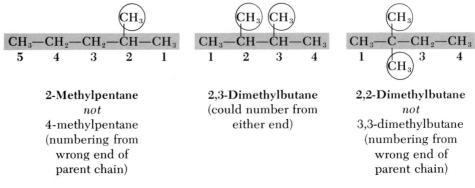

2-Methylpentane
not
4-methylpentane
(numbering from
wrong end of
parent chain)

2,3-Dimethylbutane
(could number from
either end)

2,2-Dimethylbutane
not
3,3-dimethylbutane
(numbering from
wrong end of
parent chain)

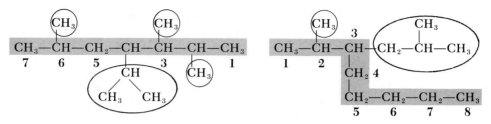

2,3,6-Trimethyl-4-isopropylheptane

2-Methyl-3-isobutyloctane

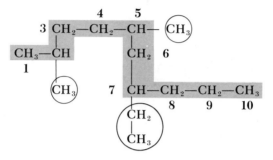

2,5-Dimethyl-7-ethyldecane

Isomers of alkanes

Isomers are compounds that have the same numbers and kinds of atoms but have different structures. For example, butane and 2-methylpropane are isomers. These two alkanes have the same molecular formula, C_4H_{10}. However, the bonding of the four carbons and ten hydrogens in one compound is not the same as that in the other compound:

Butane   2-Methylpropane

Table 2-2
Number of isomers

Number of carbon atoms	Number of isomers
5	3
6	5
7	9
8	18
9	35
10	75
15	4347
20	366,319
30	4×10^9
40	6×10^{13}

As the number of carbons increases in a molecule, the number of isomers increases exponentially, as shown in Table 2-2.

2-3. Draw the structures and give the IUPAC names for all the compounds of formula C_7H_{16}.

> *Answer:* One possibility is heptane ($CH_3CH_2CH_2CH_2CH_2CH_2CH_3$). The eight other possibilities include two hexanes, five pentanes, and one butane as the parent hydrocarbons.

Types of carbons and hydrogens

The carbon atoms in an alkane can be classified according to the number of other carbon atoms to which it is attached. A *primary* (1°) *carbon atom* is attached to one other carbon atom; a *secondary* (2°) *carbon atom* is attached to two other carbon atoms; and a *tertiary* (3°) *carbon atom* is attached to three other carbon atoms. A hydrogen is similarly classified according to the type of carbon atom to which it is attached. This concept is illustrated as follows:

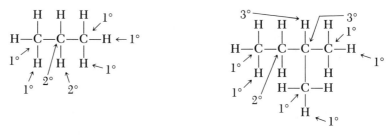

2-4. Classify the carbon atoms and hydrogen atoms in the following examples as primary, secondary, or tertiary:

(a) 2-Methylpentane   *Answer:* There are three primary carbons (nine primary hydrogens), two secondary carbons (four secondary hydrogens), and one tertiary carbon (one tertiary hydrogen).

(b) 3,3-Dimethylhexane

(c) 2-Methyl-3-ethylheptane

Structure and bonding in alkenes

As an example of an alkene, we will consider ethene ($CH_2=CH_2$). Both carbons in ethene use sp^2 hybrid orbitals, since each carbon is bonded to three other atoms. The molecule is planar and all the bond angles are 120°. There are five σ bonds, four of which are identical and are formed by the overlap of an sp^2 hybrid orbital of carbon and an s orbital of hydrogen. The other σ bond is formed by the overlap of two carbon sp^2 hybrid orbitals. The π bond is formed through a side-by-side overlap of two p orbitals. Thus the orbital makeup of the bond between the two carbons is σ and π. The σ and π bonds of ethene are represented by a carbon–carbon double bond, which is the characteristic feature of alkenes. As a result of formation of the π bond in ethene, the two carbons are brought closer together. Thus the carbon–carbon double bond distance is 1.34 Å as compared to the carbon–carbon single bond distance in ethane of 1.54 Å. The structure and bonding in ethene are summarized as follows:

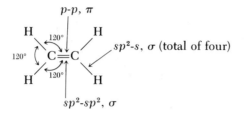

2-5. Answer the following questions in regard to the molecule propene (CH_3CH=CH_2):

    (a) How many σ bonds are there?   *Answer:* 8
    (b) How many π bonds are there?   *Answer:* 1
    (c) How many sp^3-s bonds are there?
        Answer: 3 (the three C_{sp3}–H bonds of the methyl group)
    (d) How many bonds are sp^2-s?
        Answer: 3 (the three C–H bonds of —CH=CH_2)
    (e) What is the orbital makeup of the π bond?
    (f) What is the angle of the =CH_2 bond?
    (g) Which atoms lie in the same plane?

Structure and bonding in alkynes

    As an example of an alkyne, we will consider ethyne, or acetylene (CH≡CH). Each carbon is bonded to two other atoms. Thus the two carbons use sp hybrid orbitals. The two hydrogens and the two carbons lie along the same straight line. Three of the bonds are σ bonds; the other two bonds are π bonds. Two of the σ bonds are formed by the overlap of an sp hybrid orbital of carbon with an s orbital of hydrogen; the other σ bond is formed by the overlap of two carbon sp hybrid orbitals. Each π bond is formed through the side-by-side overlap of two p orbitals. The orbital makeup of the bond between the two carbons is σ and two π bonds. The σ and two π bonds of acetylene are represented by a carbon–carbon triple bond, which is the characteristic feature of alkynes. In acetylene, the two carbons are closer together than in ethene, since two π bonds are present. Thus the carbon–carbon triple bond distance is 1.21 Å. The structure and bonding in acetylene are summarized as follows:

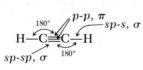

2-6. Answer the following questions in regard to the molecule propyne (CH_3C≡CH):

    (a) How many σ bonds are there?   *Answer:* 6

(b) How many π bonds are there?   *Answer:* 2
(c) How many $sp^3\text{-}s$ bonds are there?
   Answer: 3 (the three C–H bonds of the methyl group)
(d) How many bonds are $sp\text{-}s$?   *Answer:* 1
(e) Are there any $sp^2\text{-}s$ bonds? Why or why not?
(f) What is the makeup of the triple bond?
(g) What is the angle of the C—C≡C?
(h) Which atoms lie along the same straight line?

Nomenclature of alkenes and alkynes

Alkenes and alkynes are referred to as unsaturated hydrocarbons because they contain less than the maximum quantity of hydrogen atoms. The maximum number of bonds for a carbon atom is four. The characteristic feature of alkenes and alkynes is multiple bonds between carbon atoms, which are due to the presence of π bonds. Because of these multiple bonds between carbon atoms, alkenes and alkynes do not need as many hydrogens to meet the maximum number of bonds for a carbon atom:

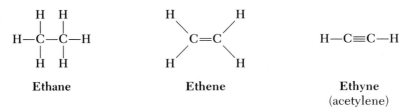

Ethane	**Ethene**	**Ethyne** (acetylene)

An alkene contains two fewer hydrogens than the corresponding alkane and thus has the general formula C_nH_{2n}. An alkyne has four fewer hydrogens than the corresponding alkane and thus has the general formula C_nH_{2n-2}.

The same prefixes are used in naming alkenes and alkynes as are used in naming alkanes (Table 2-1). The suffix for an alkene is *-ene* and that for an alkyne is *-yne*.

The rules of the IUPAC system for naming alkenes and alkynes are as follows:

1. Select as the parent structure the longest continuous chain that contains the carbon–carbon double bond or the carbon–carbon triple bond. The name of the parent structure is obtained with the prefix corresponding to the number of carbon atoms (Table 2-1) and the ending -ene or -yne. For example, the following two continuous chains can be found in the same structure, one containing eight carbon atoms and the carbon–carbon double bond and the other containing nine carbon atoms but no carbon–carbon double bond:

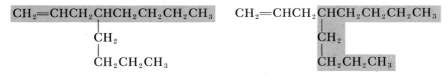

The eight-carbon continuous chain is chosen as the parent structure, since it contains the carbon–carbon double bond. Hence the parent structure has the name octene.

2. Indicate by a number the position of the double bond or triple bond in the parent chain. The position of the bond is designated by the number of the first doubly or triply bonded carbon encountered when the parent chain is numbered from the end nearest the double or triple bond:

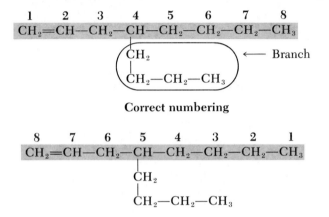

Correct numbering

$$\overset{8}{CH_2}=\overset{7}{CH}-\overset{6}{CH_2}-\overset{5}{CH}-\overset{4}{CH_2}-\overset{3}{CH_2}-\overset{2}{CH_2}-\overset{1}{CH_3}$$
$$CH_2$$
$$CH_2-CH_2-CH_3$$

Incorrect numbering

Thus the name of the parent structure is 1-octene.

3. Indicate by number(s) the position(s) of the alkyl group(s) attached to the parent chain. The complete name of the alkene is then 4-*n*-butyl-1-octene.

Following are examples of the naming of alkenes and alkynes:

4,4-Dimethyl-2-pentene	3-Heptyne

not 2,2-dimethyl-3-pentene
(parent chain numbered
from wrong end)

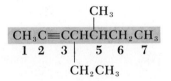

4-Ethyl-5-methyl-2-heptyne

2-7. Draw the structures and give the IUPAC names for the isomeric hexenes (C_6H_{12}) and the isomeric hexynes (C_6H_{10}).

Answer: There is one 1-hexene, one 2-hexene, one 3-hexene, three 1-pentenes, three 2-pentenes, two 1-butenes, and one 2-butene for C_6H_{12}.

Geometric isomers of alkenes

In alkenes, unlike in alkanes, the four atoms attached to the carbon–carbon double bond must lie in the same plane. Rotation about the double bond is restricted, since it takes considerable energy to break a π bond. Thus geometric isomerism is possible. The isomer with the like groups on the same side of the molecule is called the *cis* isomer, and the isomer with the like groups on opposite sides of the molecule is called the *trans* isomer. No geometric isomers are possible when one of the carbons of the carbon–carbon double bond has two like groups. These isomers are discussed in more detail in Chapter 5. Following are an example of a compound with geometric isomers and an example of a compound with no isomers:

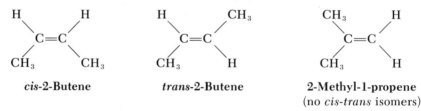

cis-2-Butene   *trans*-2-Butene   2-Methyl-1-propene
                                                  (no *cis-trans* isomers)

The concept of geometric isomerism manifests itself in naturally occurring substances such as insect pheromones, which are sex attractants. Usually, one isomer is biologically active and the other is completely inactive. Following are some examples of active sex attractants:

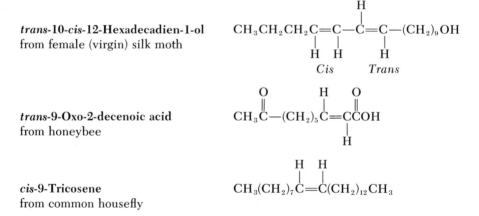

trans-10-*cis*-12-Hexadecadien-1-ol
from female (virgin) silk moth

$CH_3CH_2CH_2C{=}C{-}C{=}C{-}(CH_2)_9OH$

trans-9-Oxo-2-decenoic acid
from honeybee

cis-9-Tricosene
from common housefly

Insect sex attractants are fantastically powerful. A female moth contains about 0.01 μg (10^{-8} g) of attractant. The threshold concentration for obtaining a characteristic sexual response in the male is only several hundred molecules. Thus 0.01 μg of active attractant is capable of exciting over a billion male moths. There are economic benefits to be reaped from these facts. One of the major difficulties in the use of insecticides is their accumulation in plants and animals, notably, fish and birds. Insect pheromones possibly can be used as insecticides. Much research has been done on synthesizing these compounds in the laboratory with the idea of using them in place of DDT and other insecticides. They are highly specific and insects may be incapable of building an immunity to them. This idea has been applied successfully to the boll weevil, an insect that causes over $200 million of damage per year to cotton crops. The male boll weevil secretes the attractant to which the female responds. Traps are baited with the attractant, and the females are lured in and destroyed and thus eliminated from the reproductive cycle.

Multiple double bonds

Sometimes a molecule contains more than one double bond. A molecule that contains two double bonds is called a *diene;* one containing three double bonds, a *triene;* and so on. There are three types of dienes. *Conjugated dienes,* such as 1,3-butadiene, contain alternate double and single bonds. *Isolated dienes,* such as 1,4-pentadiene, contain double bonds separated by more than one single bond. *Cumulenes,* or *allenes,* are dienes with double bonds joining adjacent carbon atoms; propadiene, or allene, is the simplest hydrocarbon containing such cumulative double bonds.

$$CH_2{=}CH{-}CH{=}CH_2 \qquad CH_2{=}CH{-}CH_2{-}CH{=}CH_2 \qquad CH_2{=}C{=}CH_2$$

1,3-Butadiene   **1,4-Pentadiene**   **Propadiene**

The IUPAC rules for naming alkenes can be used for naming compounds with more than one double bond. Following are some examples:

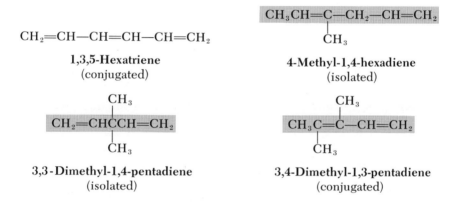

$$CH_2{=}CH{-}CH{=}CH{-}CH{=}CH_2$$

1,3,5-Hexatriene
(conjugated)

4-Methyl-1,4-hexadiene
(isolated)

3,3-Dimethyl-1,4-pentadiene
(isolated)

3,4-Dimethyl-1,3-pentadiene
(conjugated)

Structure and nomenclature of cyclohydrocarbons

The compounds studied so far are open-chain compounds. There is also a class of compounds in which the carbon atoms are arranged in rings. These are called *cyclohydrocarbons* or *cyclic compounds* and include cycloalkanes and cycloalkenes. The names of cyclohydrocarbons are formed by the prefix cyclo- added to the name of the corresponding open-chain hydrocarbon. Following are some examples. For convenience, rings are often represented by simple geometric figures:

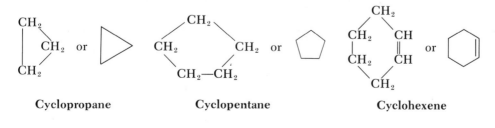

Cyclopropane   **Cyclopentane**   **Cyclohexene**

If there are substituents on the ring, they are named and their positions are indicated by use of the lowest combination of numbers. In cycloalkenes, the unsaturated bond is always located at positions 1 and 2. For example:

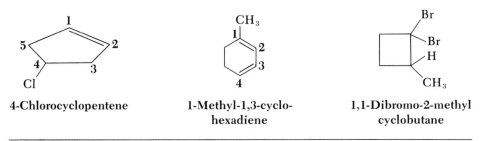

4-Chlorocyclopentene   **1-Methyl-1,3-cyclo-hexadiene**   **1,1-Dibromo-2-methyl cyclobutane**

2-8. The general formula for a cycloalkane is $C_n H_{2n}$. With what class of compounds are cycloalkanes isomeric? With what class of compounds are cycloalkenes isomeric?

2-9. Draw the structures and give IUPAC names for the isomeric cyclohexanes.

Answer: Two possibilities are cyclohexane 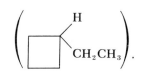 and ethylcyclobutane

.

As an alkyl group is derived from the corresponding alkane by removal of one hydrogen, so a cycloalkyl group is formed by removal of a hydrogen from the corresponding cycloalkane. Following are examples of cycloalkyl groups:

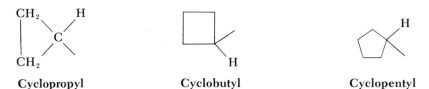

Cyclopropyl   **Cyclobutyl**   **Cyclopentyl**

These groups can be used as substituents, as shown in the following examples:

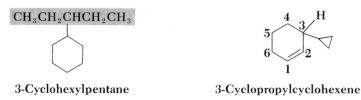

3-Cyclohexylpentane   **3-Cyclopropylcyclohexene**

Some naturally occurring compounds contain rings. The cyclopropane ring occurs in chrysanthemums, pear juice, and caraway seeds, and the cyclobutane ring is found in pine oil:

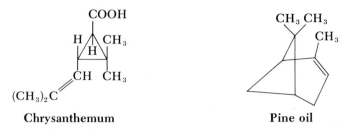

Chrysanthemum   **Pine oil**

Cyclopentane and cyclohexane rings are found abundantly in nature. For example, deoxyribose, which is the sugar in DNA, contains five atoms in a ring; and the cyclohexane ring is found in cholesterol, which is the chief constituent of gallstones:

Deoxyribose   **Cholesterol**

Petroleum: source of alkanes

Crude petroleum as delivered by oil wells is a liquid mixture of hydrocarbons, which are mainly alkanes. The decomposition of the remains of living plants and animals over the ages is believed to be the principal method of formation of petroleum, which exists under high pressures in porous rock as a dark viscous oil. Natural gas (chiefly methane) is often found in association with petroleum.

When petroleum is refined, it is separated into fractions by distillation. Each fraction differs in boiling range. Following is a schematic diagram of the refining process:

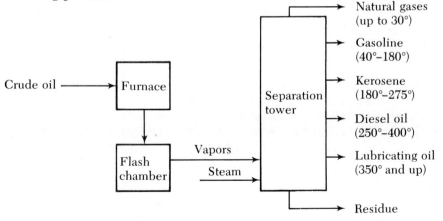

Crude oil is fed continuously through a hot furnace. The hot liquid is passed into a flash chamber, where it is vaporized. The vapors then pass into a sepa-

Table 2-3
Composition of natural gas

Component	Kansas-Nebraska Natural Gas Company	Public Service of Colorado
Methane	83.9	92.9
Nitrogen	9.8	0.62
Ethane	4.1	4.53
Propane	1.73	0.47
Butane	0.21	0.05
2-Methylpropane	0.19	0.05
C_5 to C_7 hydrocarbons	0.04	Trace to 0.05
Carbon dioxide		1.30
Helium	0.05	0.01
Hydrogen	0.002	

ration tower: the more volatile components collect in the upper portion, and the less volatile (high boiling point) components collect in the lower portion. Each fraction is a complex mixture of hydrocarbons. The main petroleum fractions are natural gas, gasoline, kerosene, diesel oil, lubricating oil, and the residue.

Natural gas is used as a fuel. It is chiefly methane with smaller amounts of ethane, propane, butane, and 2-methylpropane, but the composition may vary, as shown in Table 2-3. Generally, these composition differences are necessitated by different local weather conditions.

The gasoline fraction, which is composed of C_5 to C_{10} hydrocarbons, is used mainly as a motor fuel. The fraction of lower hydrocarbons (isomeric pentanes, hexanes, and heptanes) with a boiling range of 30° to 100° C, which is called petroleum ether, is used as a solvent and in dry cleaning.

Kerosene, which is mainly C_{11} and C_{12} hydrocarbons, is used as an illuminant and as fuel in jet engines but not in automobile engines.

Diesel oil, which is the fraction composed of C_{13} to C_{25} hydrocarbons, is also known as fuel oil. Large quantities of it are used in oil-burning furnaces and in diesel engines.

The fraction boiling above 350° C, which is composed of C_{26} to C_{28} hydrocarbons, is solid at room temperature. It is used as paraffin wax for candles; petrolatum (Vaseline), a mixture of paraffin wax and compounds with low melting points; and lubricating oil, which is a liquid because it is a mixture.

The residue, the black, tarry material remaining after all the volatile components are removed, is used as asphalt.

Preparation of alkenes

Industrially, alkenes (mainly ethylene) are obtained by pyrolysis of the petroleum fraction, which is cleavage by heat and is often referred to as *cracking*. In this process, smaller alkanes as well as ethylene are obtained:

$$C_{12}H_{26} \xrightarrow{\text{heat}} CH_2{=}CH_2 + C_{10}H_{22}$$

Ethylene

In the laboratory, the introduction of a carbon–carbon double bond into a molecule containing only single bonds involves the elimination of atoms or groups from two adjacent carbons:

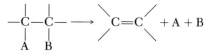

The two principal methods of introducing the double bond are dehydrohalogenation of an alkyl halide and dehydration of an alcohol. Another useful method is the Diels-Alder reaction.

Dehydrohalogenation of alkyl halides

This method involves removal from an alkyl halide of a halogen atom and a hydrogen atom on adjacent carbon atoms. An alkyl halide has the general formula RX where R is an alkyl group and X is a halogen — fluorine, chlorine, bromine, or iodine. An alkyl bromide is often represented as RBr, an alkyl chloride as RCl, and an alkyl iodide as RI. The IUPAC nomenclature of alkyl halides is an extension of the IUPAC nomenclature of alkanes. Chlorine, bromine, and iodine are called chloro, bromo, and iodo substituents or groups, respectively. Following are examples of the naming of alkyl halides by the IUPAC system:

$$CH_3CH_2CH_2Br \qquad\qquad CH_3\underset{\underset{Cl}{|}}{C}HCH_3 \qquad\qquad CH_3\underset{\underset{I}{|}}{C}H\underset{\overset{CH_3}{|}}{C}HCH_2CH_2CH_3$$

1-Bromopropane   **2-Chloropropane**   **2-Methyl-3-iodohexane**

The reagent necessary to cause the elimination is a base, usually an alcohol solution of potassium hydroxide (KOH). Following are examples of the dehydrohalogenation process:

$$CH_3{-}\underset{\underset{H}{|}}{C}H{-}\underset{\underset{Cl}{|}}{C}H_2 \xrightarrow[\text{CH}_3\text{CH}_2\text{OH}]{\text{KOH}} CH_3CH{=}CH_2$$

1-Chloropropane   **1-Propene**

$$CH_3CH_2\underset{\underset{H}{|}}{C}H\underset{\underset{Cl}{|}}{C}H_2 \xrightarrow[\text{CH}_3\text{CH}_2\text{OH}]{\text{KOH}} CH_3CH_2CH{=}CH_2$$

1-Chlorobutane   **1-Butene**

When an isomeric mixture of alkenes is formed, the preferred or major product is the alkene with the greater number of alkyl groups attached to the doubly bonded carbon atoms, which is more easily formed. The following reactions illustrate this:

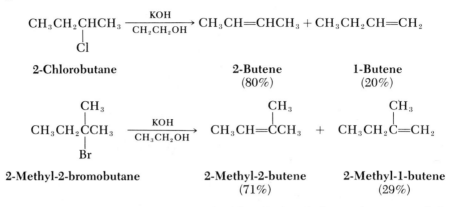

	2-Butene	1-Butene
2-Chlorobutane	**(80%)**	**(20%)**

2-Methyl-2-bromobutane	**2-Methyl-2-butene**	**2-Methyl-1-butene**
	(71%)	(29%)

We can see that 2-butene is more highly alkylated than 1-butene and that 2-methyl-2-butene is more highly alkylated than 2-methyl-1-butene by drawing the expanded structural formula for each compound, boxing in the carbon–carbon double bond, and counting the number of alkyl groups bonded to each carbon of the double bond:

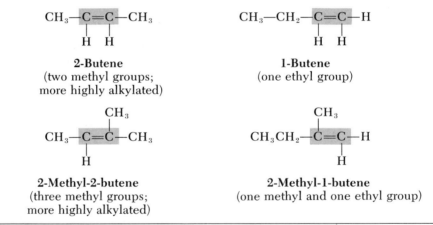

2-Butene
(two methyl groups;
more highly alkylated)

1-Butene
(one ethyl group)

2-Methyl-2-butene
(three methyl groups;
more highly alkylated)

2-Methyl-1-butene
(one methyl and one ethyl group)

2-10. Predict the major or preferred product in the dehydrohalogenation of each of the following compounds:

(a) 2-Bromopropane   *Answer:* $CH_3CH{=}CH_2$
(b) $(CH_3)_2CHCH_2I$
(c) $CH_3CHClCH(CH_3)CH_3$ *Answer:* Two alkenes are formed: 3-methyl-1-butene and 3-methyl-2-butene (preferred product).

To see how the hydroxide ion (OH^-) causes the dehydrohalogenation, we can study the *mechanism* of the reaction, which is the detailed, step-by-step description of the path through which the reactants pass to the products. In dehydrohalogenation, the function of the hydroxide ion is to pull a hydrogen ion away from a carbon atom as a halide ion simultaneously separates and the double bond forms. This *simultaneous* mechanism, which is also referred to as a *concerted* mechanism, is as follows:

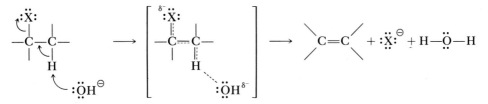

The arrows show the direction of electron shift. The hydroxide ion removes a hydrogen ion, which relinquishes both electrons to the carbon atom. These electrons are available for formation of the π bond. The halogen atom retains both electrons. Then since the carbon atom has "five" bonds around it, the halogen atom departs with its pair of electrons as a halide ion. In the transition state, which is shown in brackets in the mechanism, broken lines indicate either formation or breaking of bonds. In the transition state of dehydrohalogenation, bonds are being formed between hydrogen and oxygen atoms and also between two carbon atoms; bonds are being broken between carbon and hydrogen atoms and between carbon and halogen atoms. The partial negative charge (δ^-) indicates that the oxygen atom, which is negatively charged in the reactant OH^-, is losing its negative charge. The halogen atom, which carries no negative charge in the alkyl halide, begins to become negative in the transition state and is a negative halide ion in the product. The structure of the transition state is somewhere between that of the reactants and that of the products.

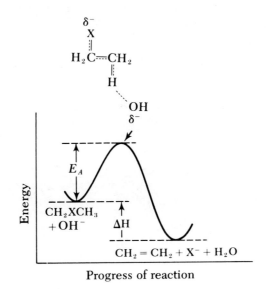

Progress of reaction

Fig. 2-1
Energy profile for the dehydrohalogenation of alkyl halide. The reactants, transition state, and products are shown. The energy of activation and the heat of the reaction are indicated.

The energy changes occuring during dehydrohalogenation, which is an exothermic reaction, are shown in the energy profile in Fig. 2-1. In an exothermic reaction, the products possess less energy than the reactants.

2-11. The mechanism for the dehydrohalogenation of 2-methyl-2-bromobutane follows the course outlined in the following steps:

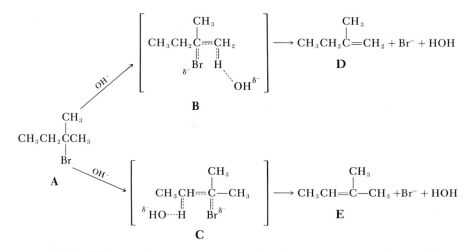

If the following diagram represents the changes occurring during this transformation, where would you place the letters A through E so that each species is identified at its correct position in the energy profile? Assume that the preferred product is the more easily formed product.

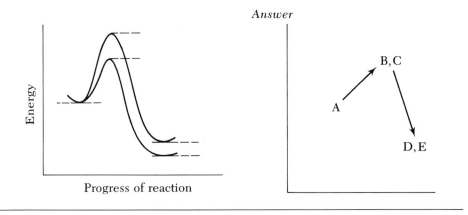

Dehydration

This process by which alcohols are converted to alkenes involves the elimination of water. An alcohol has the general formula ROH where symbol R is an alkyl group and OH is the characteristic functional group of an alcohol. Nomenclature is discussed in Chapter 6.

Dehydration requires the presence of an acid, usually sulfuric acid (H_2SO_4), and the application of heat. Following are examples of the dehydration process:

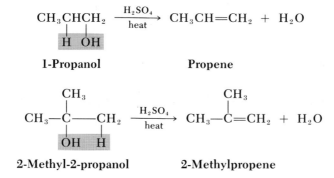

1-Propanol   **Propene**

2-Methyl-2-propanol   **2-Methylpropene**

When an isomeric mixture of alkenes is formed, one isomer tends to pre-dominate; unlike in dehydrohalogenation, the double bond can be formed at a position remote from the carbon originally holding the —OH group. The following reactions illustrate this:

$$CH_3CH_2CH_2CH_2Cl \xrightarrow[CH_3CH_2OH]{KOH} CH_3CH_2CH{=}CH_2$$

1-Chlorobutane   **1-Butene**
                                 (only product)

$$CH_3CH_2CH_2CH_2OH \xrightarrow[heat]{H_2SO_4} CH_3CH_2CH{=}CH_2 + CH_3CH{=}CHCH_3$$

1-Butanol   **1-Butene**   **2-Butene**
                                                    (major product)

The mechanism of dehydration of *tert*-butyl alcohol $[(CH_3)_3COH]$ is as follows:

(1) $(CH_3)_3C$ + $H^{\oplus}$ $\longrightarrow$ $(CH_3)_3{-}C$
          | |
         :ÖH   :ÖH$_2$
                                                 $\oplus$

Protonated alcohol

(2) $(CH_3)_3{-}C$ $\longrightarrow$ $(CH_3)_3C^{\oplus}$ + $H{-}\ddot{O}{-}H$
          |
        :ÖH$_2$

Carbonium ion

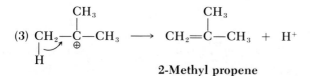

(3) $CH_2{-}\underset{\oplus}{C}{-}CH_3$ $\longrightarrow$ $CH_2{=}C{-}CH_3$ + H^+

2-Methyl propene

In step 1, the alcohol combines with a proton from the sulfuric acid catalyst to form a protonated alcohol, which in step 2 dissociates into a carbonium ion and water. The carbonium ion then in step 3 loses a proton (acid catalyst is regenerated) to form the alkene. Unlike dehydrohalogenation, in which a hydrogen atom and a halogen atom are simultaneously lost, dehydration is a stepwise process.

The *carbonium ion*, the intermediate formed in step 2, is a group of atoms that contains a carbon atom bearing only six electrons. Carbonium ions are classified as primary, secondary, or tertiary according to the type of carbon atom bearing the positive charge. For example, in a primary carbonium ion, the carbon bearing the positive charge is attached to one other carbon atom. Carbonium ions are named *cations*. Following are examples of specific names:

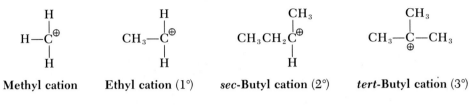

Methyl cation   Ethyl cation (1°)   *sec*-Butyl cation (2°)   *tert*-Butyl cation (3°)

2-12. Draw the structures for the following ions:

    (a) A three-carbon primary carbonium ion   *Answer:* $CH_3CH_2CH_2^{\oplus}$
    (b) Isopropyl cation   *Answer:* $(CH_3)_2CH^{\oplus}$
    (c) Isobutyl cation
    (d) Cyclopropyl cation

An alkyl group, being an electron-releasing group, is capable of dispersing the positive charge over a large area. The more alkyl groups there are attached to the positive carbon atom, the more dispersed is the positive charge. In *tert*-butyl cation, the positive charge is dispersed over all three methyl groups; whereas in ethyl cation, the positive charge is dispersed over only one methyl group:

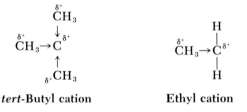

tert-Butyl cation   Ethyl cation

Thus the more alkyl groups there are, the greater is the stability of the ion. Also, the more stable a carbonium ion is, the more easily it is formed. The order of stability of carbonium ions then is tertiary > secondary > primary > methyl ion. Hence the ease of formation of carbonium ions is in the same order as the stability. For example, the energy profile in Fig. 2-2 shows that it is easier (smaller E_A) to form a tertiary carbonium ion from *tert*-butyl alcohol than it is to form a methyl carbonium ion from methyl alcohol.

Dehydration sometimes gives alkenes that, at first observation, do not seem to be formed from the same carbonium ion initially formed from the alcohol. For example, the formation of 2-butene from 1-butanol cannot be explained by the concepts just presented. However, it can be explained if it is assumed that a carbonium ion can rearrange to form a more stable carbo-

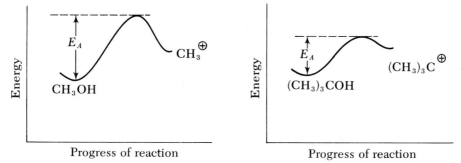

Fig. 2-2
Energy profile for the formation of *tert*-butyl cation and methyl cation from *tert*-butyl alcohol and methyl alcohol, respectively.

nium ion. The *n*-butyl cation from 1-butanol rearranges by a *hydride shift* to the more stable carbonium ion, *sec*-butyl cation, as follows. The hydride shift, which is a 1,2 shift, is the migration of a hydrogen atom with its pair of electrons from one atom to the very next atom.

(1) $CH_3CH_2CH_2CH_2OH + H^\oplus \longrightarrow CH_3CH_2CH_2CH_2OH_2^\oplus$

       1-Butanol

(2) $CH_3CH_2CH_2CH_2OH_2^\oplus \longrightarrow CH_3CH_2CH_2CH_2^\oplus + H_2O$

                         n-Butyl cation

(3) $CH_3CH_2CHCH_2^\oplus \xrightarrow[\text{hydride shift}]{\text{rearranges by}} CH_3CH_2CHCH_3$

             H   **sec-Butyl cation**

(4a) $CH_3CH{-}CHCH_3 \longrightarrow CH_3CH{=}CHCH_3 + H_3O^+$

       H   **2-Butene**

             H_2O

(4b) $CH_3CH_2CH{-}CH_2 \quad\longrightarrow\quad CH_3CH_2CH{=}CH_2 \ + \ H_3O^+$

            H   **1-Butene**

            H_2O

    The energy profile for the dehydration of 1-butanol to form 1-butene and 2-butene is shown in Fig. 2-3. The formation of the protonated alcohol and *n*-butyl cation and the rearrangement of the *n*-butyl cation to the more stable cation are not shown on the energy profile. The important part of the mechanism, as illustrated on the energy profile, is the formation of 1-butene and 2-butene from the *sec*-butyl cation. 2-Butene is formed in preference to 1-butene, since the energy of activation for 2-butene formation is less than that for 1-butene formation. As shown in the energy profile, dehydration is an endothermic reaction, since the reactants lie at a lower energy level than the products.

    In dehydration, the more easily formed alkene, which is formed from the

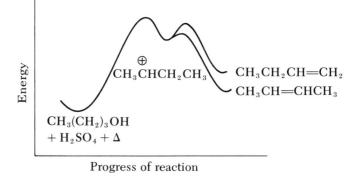

Progress of reaction

Fig. 2-3
Formation of 2-butene and 1-butene from *sec*-butyl cation in dehydration of 1-butanol.

more stable carbonium ion, is the preferred (major) product, as shown in the following examples:

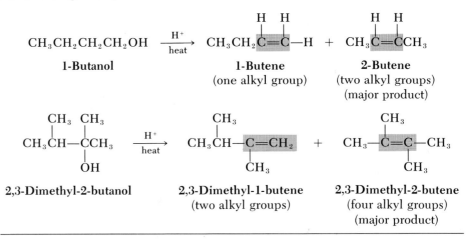

2-13. Predict the product(s) in the dehydration of each of the following compounds. Indicate the major product:

(a) $CH_3CH(OH)CH_3$ *Answer:* Propene
(b) $(CH_3)_2CHCHOHCH_3$ *Answer:* $(CH_3)_2CHCH=CH_2$ from
$(CH_3)_2CH\overset{\oplus}{C}HCH_3$; $(CH_3)_2C=CHCH_3$
and $CH_2=C(CH_3)CH_2CH_3$ from
$(CH_3)_2\overset{\oplus}{C}CH_2CH_3$. Major product is
$(CH_3)_2C=CHCH_3$
(c) $(CH_3)_2CHCH_2CH_2OH$
(d) $(CH_3)_2COHCH_2CH_3$

Carbonium ions are important intermediates in other organic reactions besides the preparation of an alkene by the dehydration of an alcohol. These reactive intermediates can undergo substitution reactions (Chapter 7), and the formation of certain molecules in the living system is thought to proceed through the intermediate formation of carbonium ions. For example, in the following reaction, the bromo group of an alkyl halide undergoes substitution by a methoxyl group via a carbonium ion intermediate:

$$CH_3CH_2CHBr \longrightarrow \left[CH_3CH_2CH^{\oplus} \right] \xrightarrow{NaOCH_3} CH_3CH_2CHOCH_3 + NaBr$$
$$\quad\;\; | \qquad\qquad\qquad\quad | \qquad\qquad\qquad\qquad\quad |$$
$$\quad\;\; CH_3 \qquad\qquad\qquad CH_3 \qquad\qquad\qquad\qquad\;\; CH_3$$

The formation of squalene, an important intermediate in the formation of cholesterol, a component of gallstones, and the formation of limonene, a constituent of lemon and orange rinds, are examples of biochemical syntheses involving carbonium ions. The formation of squalene is shown as follows. Nerolidol pyrophosphate (I) condenses with farnesyl pyrophosphate (II) to form a carbonium ion (step 1); the carbonium ion loses a proton to form an alkene (step 2); attack on the doubly bonded carbon atom by a hydride ion (H:⁻) with the subsequent loss of a pyrophosphate anion gives squalene (step 3); and then squalene is converted in several steps to cholesterol:

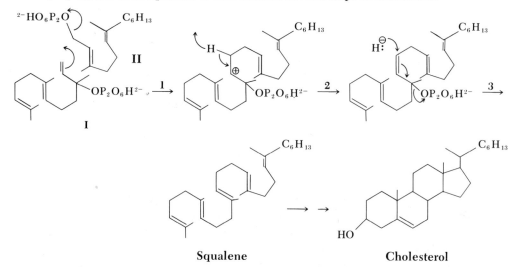

Squalene   Cholesterol

The formation of limonene is as follows. Geranyl pyrophosphate (I) is converted to limonene via loss of pyrophosphate (step 1); the initially formed carbonium ion undergoes a ring-forming reaction resulting in a new carbonium ion (step 2); and then loss of a proton forms limonene (step 3):

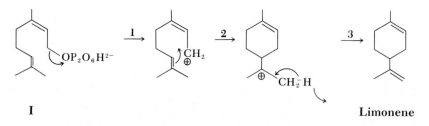

I   Limonene

All the carbon and hydrogen atoms of these substances are not shown. The structures are represented in a shorthand notation in which the carbon–carbon bonds are represented by lines and the carbon atoms are understood to be located at the junction of or the ends of lines.

Diels-Alder reaction

This reaction, which is useful for synthesizing a six-membered cyclo-alkene, involves a conjugated diene and a dienophile. The dienophile is an unsaturated compound containing an electron-withdrawing group, which is a carbonyl group (—CHO), a carboxyl group (—COOH) or a nitrile group (—CN). Some useful Diels-Alder reactions that form cycloalkenes are as follows:

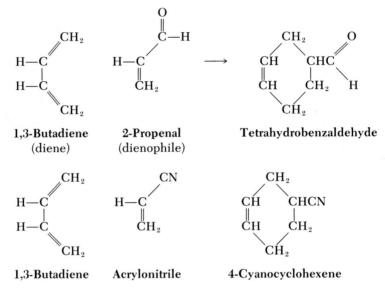

1,3-Butadiene (diene)	2-Propenal (dienophile)	Tetrahydrobenzaldehyde

1,3-Butadiene	Acrylonitrile	4-Cyanocyclohexene

Preparation of alkynes

A carbon–carbon triple bond is introduced into a molecule by elimination of atoms or groups from adjacent carbons. The groups eliminated and the reagents used in forming these bonds are essentially the same as those used in the formation of carbon–carbon double bonds. The formation of the triple bond can be represented as follows:

$$\underset{\substack{| \\ D}}{\overset{\substack{A \\ |}}{-C}}\!-\!\underset{\substack{| \\ E}}{\overset{\substack{B \\ |}}{C}}- \xrightarrow{-DE} \overset{\substack{A \;\; B \\ | \;\; |}}{-C\!=\!C-} \xrightarrow{-AB} -C\!\equiv\!C-$$

Dehydrohalogenation of vicinal dihalides

This is the most useful way of introducing a carbon–carbon triple bond into a molecule, since vicinal dihalides are readily prepared from alkenes by the addition of halogen (Chapter 3). Following is an example of the dehydrohalogenation reaction:

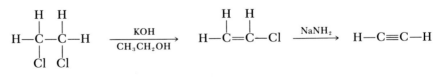

1,2-Dichloroethane (vicinal dihalide)	1-Chloroethene (vinyl chloride)	Acetylene

This reaction involves two steps. It is a useful method for preparing vinyl halides, in which a halogen atom is directly attached to a doubly bonded carbon. Vinyl halides are very unreactive, and dehydrohalogenation will stop at this stage unless a stronger base than an alcoholic solution of potassium hydroxide is used, such as sodium amide ($NaNH_2$).

From sodium acetylides and primary alkyl halides

This method involves the reaction of sodium acetylides with primary alkyl halides. The main advantage of this reaction is that it permits conversion of smaller alkynes into larger ones. Sodium acetylides can be formed from 1-alkynes only. The formation of alkynes by this method can be represented as follows:

$$R—C{\equiv}C—H \xrightarrow{Na} R—C{\equiv}C{:}^-Na^+ \xrightarrow{R'X} R—C{\equiv}C—R' + NaX$$

Following is an example of this reaction:

$$CH_3CH_2C{\equiv}CH \xrightarrow{Na} CH_3CH_2C{\equiv}C{:}^-Na^+ \xrightarrow{CH_3CH_2Br}$$

1-Butyne

$$CH_3CH_2C{\equiv}CCH_2CH_3 + NaBr$$

3-Hexyne

This reaction is limited to primary alkyl halides, since tertiary and secondary halides undergo an elimination rather than a substitution reaction with sodium acetylides. This is discussed in Chapter 7.

Application of triple bond: birth control pill

The two important sex hormones of the woman are estradiol and progesterone:

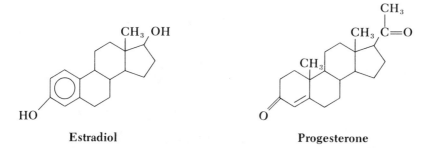

Estradiol   **Progesterone**

These hormones are routinely produced in pregnancy and aid the maintenance of pregnancy and prevent ovulation, that is, release of an ovum from the ovaries. Pregnancy results if fertilization of an ovum by a sperm occurs followed by implantation of the fertilized egg in the uterus. Once pregnancy begins, no more ova are released until the pregnancy is terminated either mechanically or naturally. It was found that if the natural hormones, estradiol and progesterone, were taken orally, they induced a pregnancy-like state as far as the ovaries were concerned and, as a result, ovulation was prevented and conception was impossible. However, it was soon discovered that the natural sex hormones were very weak agents when taken orally and furthermore had to be taken in large doses.

The oral contraceptives, or birth control pills, represent a major breakthrough in this area. They are synthetic hormones that are related to the natural hormones found in the body. Chemically, they are alkyne derivatives of the natural hormones. Their physiological effect is similar but is not well understood. Their effectiveness may in part be due to the ease with which they are passed from the intestines into the bloodstream. The addition of the acetylenic group imparts a greater activity to these compounds when taken orally, so that they can be used in lower dosages. The more common and commercially available birth control pills are combinations of mestranol and norethindrone (Ortho-Novum) and of mestranol and norethynodrel (Enovid). The combination of mestranol with either norethindrone or norethynodrel forms the estradiol-progesterone combination occuring naturally. Following are the structures of these compounds:

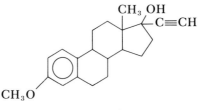

Mestranol

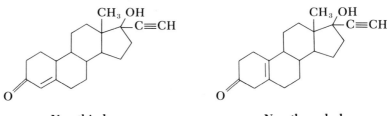

Norethindrone   **Norethynodrel**

Unfortunately, the problem of birth control via the pill has not been completely solved for the woman (or the man). Perhaps the biggest question is the safety of the pill. The use of the pill results in numerous side effects, such as acne, breast cancer, blood vessel clots, hypertension, and metabolic changes.

Summary

In an alkane, the orbitals of the carbon atoms are sp^3 hybridized. The orbitals of the carbon atoms of the double bond in an alkene are sp^2 hybridized. In an alkyne, the orbitals of the carbon atoms of the triple bond are sp hybridized. The carbon–carbon distance of the single bond in alkanes is 1.54 Å, the distance of the double bond in alkenes is 1.34 Å, and the distance of the triple bond in alkynes is 1.21 Å. Alkanes, alkenes and alkynes are named by the IUPAC system. Many alkanes are obtained from petroleum refining. Alkenes can be prepared by the dehydrohalogenation of alkyl halides. This reaction requires the aid of potassium hydroxide in alcohol. When an isomeric mixture is formed, the product with more attached alkyl groups is preferred. Alkenes can also be prepared by the dehydration of alcohols. This reaction requires the aid of sulfuric acid and heat. Unlike dehydrohalogenation, dehydration is a stepwise reaction. The intermediate formed in dehydration is the carbonium ion. The order of stability and ease of formation of carbonium ions are tertiary > secondary > primary > methyl ion. The alkene with more attached alkyl groups is preferred. Rearrangement products can occur in dehydration. Alkynes are formed either by the dehydrohalogenation of vicinal dihalides or by the reaction of sodium salts of 1-alkynes with primary alkyl halides.

Important terms

Alkane	Isomer
Alkene	IUPAC system
Alkyl group	Mechanism
Alkyne	Parent structure
Branched-chain alkane	Petroleum
Carbonium ion	Petroleum fraction
Concerted mechanism	π bond
Conjugated diene	Primary carbon atom
Continuous-chain alkane	Primary hydrogen atom
Cumulated diene	Saturated hydrocarbon
Dehydration	Secondary carbon atom
Dehydrohalogenation	Secondary hydrogen atom
Diene	σ bond
Geometric isomerism	sp hybridization
Homologs	sp^2 hybridization
Homologous series	sp^3 hybridization
Hydride shift	Tertiary carbon atom
Hydrocarbon	Tertiary hydrogen atom
Isolated diene	Unsaturated hydrocarbon

Problems

2-14. Give the correct IUPAC name for each of the following compounds:

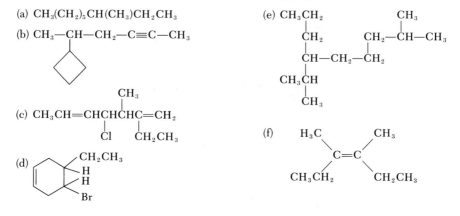

(a) $CH_3(CH_2)_5CH(CH_3)CH_2CH_3$

(b) $CH_3{-}CH{-}CH_2{-}C{\equiv}C{-}CH_3$

(c) $CH_3CH{=}CHCHCHC{=}CH_2$

(d)

(e)

(f)

2-15. Draw the structure and give the IUPAC name of the compound(s) that fit(s) each of the following descriptions:

(a) Isomer of 2-hexyne
(b) Homolog of 3-octene
(c) C_5H_{12} compound that has only primary hydrogen atoms
(d) C_6H_{10} compound that has conjugated bonds
(e) Most stable carbonium ion of $C_4H_9^+$
(f) $C_5H_{10}Br_2$ compound in which the halogens are vicinal
(g) Most stable alkene with the molecular formula C_6H_{12}
(h) *Trans* isomer of C_5H_{10}

2-16. Draw all the structures for the compounds with the formula C_6H_{14}. How many primary, secondary, and tertiary hydrogens are there in each isomer? Give the correct IUPAC name for each isomer.

2-17. Answer the following questions concerning this molecule:

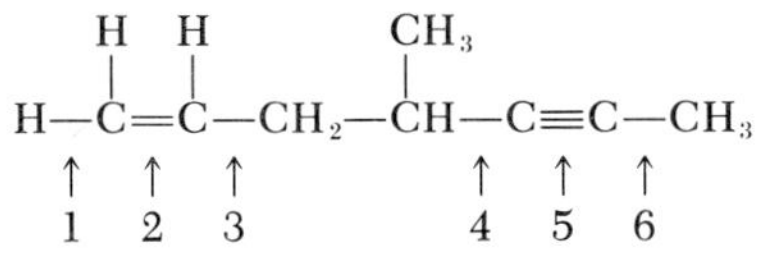

$$\underset{1}{\overset{H}{\underset{\uparrow}{H-C}}}=\underset{2}{\overset{H}{\underset{\uparrow}{C}}}-\underset{3}{\overset{}{\underset{\uparrow}{CH_2}}}-\underset{4}{\overset{CH_3}{\underset{\uparrow}{CH}}}-\underset{5}{\overset{}{\underset{\uparrow}{C}}}\equiv\underset{6}{\overset{}{\underset{\uparrow}{C}}}-CH_3$$

(a) How many π bonds are there?
(b) What is the orbital makeup of bonds 2, 3, 4, 5, and 6?
(c) How many tertiary hydrogen atoms are there?
(d) Which atoms lie in the same plane? Which lie along the same straight line?
(e) What is the bond angle of the $H-\overset{|}{\underset{|}{C}}=$ portion of the molecule? What is the bond angle
H
of the $C\equiv C-C$ portion?

2-18. Predict the product(s) of the following reactions. Indicate the major product if there is more than one product:

(a) 1-Bromopentane + alcoholic potassium hydroxide $\longrightarrow$

(b) $CH_3CH_2CH_2CH_2CH_2OH + H_2SO_4 + heat \longrightarrow$

(c) 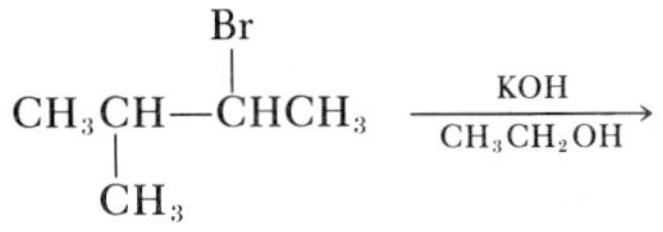 $CH_3\underset{\overset{|}{CH_3}}{CH}-\overset{\overset{Br}{|}}{C}HCH_3 \xrightarrow[CH_3CH_2OH]{KOH}$

(d) 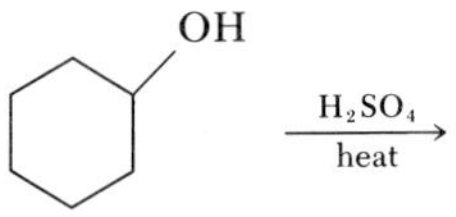 $\underset{}{\overset{OH}{\bigcirc}} \xrightarrow[heat]{H_2SO_4}$

(e) 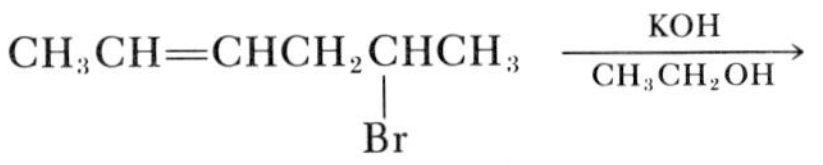 $CH_3CH=CHCH_2\underset{\overset{|}{Br}}{C}HCH_3 \xrightarrow[CH_3CH_2OH]{KOH}$ (2 products)

(f) 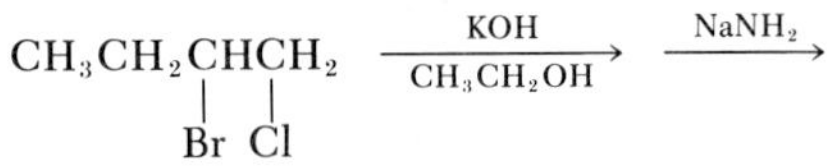 $CH_3CH_2\underset{\overset{|}{Br}\ \overset{|}{Cl}}{C}HCH_2 \xrightarrow[CH_3CH_2OH]{KOH} \xrightarrow{NaNH_2}$

(g) 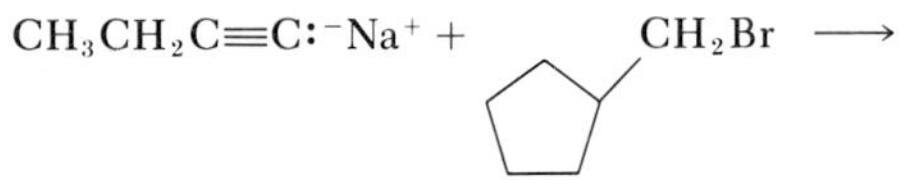 $CH_3CH_2C\equiv C:^- Na^+ + \underset{}{\bigtriangleup} CH_2Br \longrightarrow$

2-19. Draw the structural formula for each of the following compounds:

(a) 1-Chloro-2-butyne
(b) *cis*-4-Methyl-2-hexene
(c) Cyclohexylcyclohexane
(d) 3,5-Octadiene
(e) 3-Bromo-2,3-dimethylpentane

(f) 2-Methyl-4-ethyl-3-hexene
(g) 2-Cyclobutyl-4-methylheptane
(h) *trans*-3-Hexene
(i) 2,4,4-Trichloro-2-pentene
(j) 4-Methyl-1-pentyne

2-20. Indicate which compounds are isomeric, identical, or homologs?

(a) 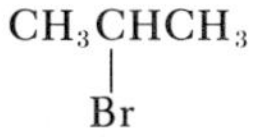 CH_3CHCH_3
$\quad\;\;|$
$\quad\;\;Br$

(b) $CH_3CH=CH_2$

(c) $CH_3CH_2CH_3$

(d) $(CH_3)_2CHCH(CH_3)_2$

(e) 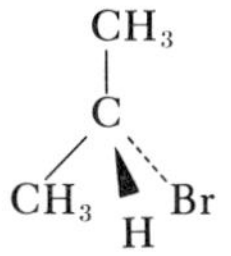 $\quad CH_3$
$\quad\;\;|$
$\quad\;\;C$
$CH_3 \; \blacktriangleright \; Br$
$\quad\;\; H$

(f) 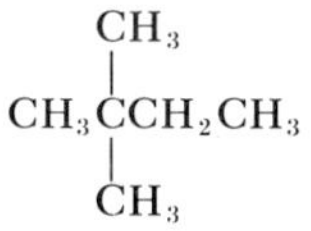 $\quad CH_3$
$\quad\;\;|$
$CH_3CCH_2CH_3$
$\quad\;\;|$
$\quad CH_3$

(g) 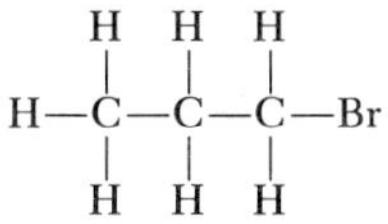 $\quad H \;\; H \;\; H$
$\quad\;\;|\;\;\;|\;\;\;|$
$H-C-C-C-Br$
$\quad\;\;|\;\;\;|\;\;\;|$
$\quad H \;\; H \;\; H$

(h) $CH_3CH_2CH_2CH_3$

(i) 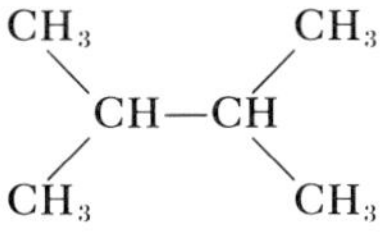 $CH_3 \qquad\qquad CH_3$
$\quad\;\searrow \qquad\qquad \swarrow$
$\qquad\quad CH-CH$
$\quad\swarrow \qquad\qquad\;\; \searrow$
$CH_3 \qquad\qquad\;\; CH_3$

(j) $(CH_3)_2CHBr$

(k) 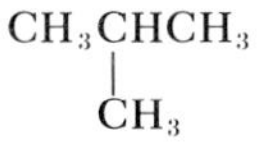 CH_3CHCH_3
$\quad\;\;|$
$\quad\;\;CH_3$

(l) $CH_3CH_2CH_2Br$

(m) $CH_2=CHCH_3$

(n) 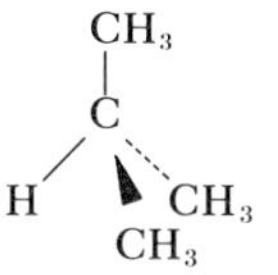 $\quad CH_3$
$\quad\;\;|$
$\quad\;\;C$
$H \; \blacktriangleright \; CH_3$
$\quad\;\;CH_3$

2-21. Answer the following questions concerning alkane $CH_3CH_2CH_2CH_2CH_2CH_2CH_3$:

(a) What is the IUPAC name?
(b) Draw the condensed formula.
(c) Draw the structure of an isomer and give its IUPAC name.
(d) Draw the condensed formula for the isomer in (c).
(e) Draw the structure of a homolog and give its IUPAC name.

2-22. The dehydration of an alcohol produced the following isomeric mixture of alkenes. Assuming that the most stable alkene is the most easily formed alkene, list them in the order of ease of formation:

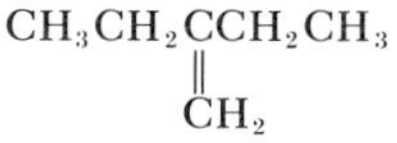

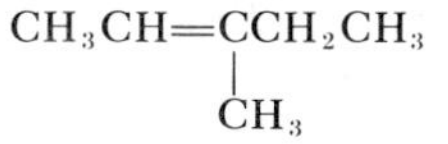

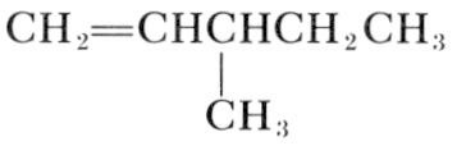

$CH_3CH_2CCH_2CH_3$
$\qquad\;\; \|$
$\qquad\;\; CH_2$

$CH_3CH=CCH_2CH_3$
$\qquad\qquad\;\; |$
$\qquad\qquad\;\; CH_3$

$CH_2=CHCHCH_2CH_3$
$\qquad\qquad\;\; |$
$\qquad\qquad\;\; CH_3$

Self-test

1. Give a correct name for each of the following structures:

(a) $CH_3CHCH_2CH_2CH_3$
$\quad\;\;\;|$
$\quad\;\;\;CH_3$

(b) $ClCH_2CH_2Cl$

(c) 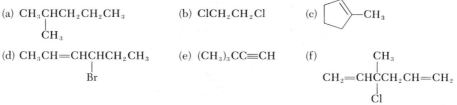 $-CH_3$

(d) $CH_3CH=CHCHCH_2CH_3$
$\qquad\qquad\quad |$
$\qquad\qquad\quad Br$

(e) $(CH_3)_3CC\equiv CH$

(f) $\qquad\qquad\qquad CH_3$
$\qquad\qquad\qquad\;\; |$
$CH_2=CHCCH_2CH=CH_2$
$\qquad\qquad\qquad\;\; |$
$\qquad\qquad\qquad\;\; Cl$

2. Draw a complete structure corresponding to each of the following compounds:

(a) 3-Methyl-2-pentene
(b) 1,2-Dibromocyclohexane

(c) 3-Heptyne
(d) 1,4-Hexadiene

(e) 3-Chloro-2-methylhexane
(f) Octane

3. Dehydrohalogenation and dehydration are two methods of alkene formation. Give the conditions, intermediate or transition state, and product(s) (indicate the preferred product) for (a) dehydrohalogenation with the reagent $CH_3(CH_2)_3Br$ and (b) dehydration with the reagent $CH_3(CH_2)_3OH$.

4. Give the correct answer for each of the following questions:
   (a) How many π bonds are contained in acetylene?
   (b) Which carbonium ion is more stable:

$$CH_3-\overset{\overset{\textstyle CH_3}{|}}{\underset{\underset{\textstyle CH_3}{|}}{C}}{}^{\oplus} \quad \text{or} \quad CH_3CH_2CH_2\overset{\oplus}{C}H_2$$

   (c) Which diene contains conjugated double bonds: 1,3-butadiene or 1,4-pentadiene?
   (d) What is the hybridization of the orbitals of the indicated carbon atom in the following molecule?

   (e) What alkene is formed if n-butyl cation eliminates a proton?
   (f) What carbonium ion is formed if n-butyl cation rearranges to a more stable carbonium ion?
   (g) To what class of organic compounds does the organic product of the following reaction belong?

$$CH_3C\equiv C:^-Na^+ + CH_3CH_2Br \longrightarrow$$

   (h) Draw the structure of an isomer of 2-methylpentane.
   (i) Draw the structure of a homolog of heptane.
   (j) What is the major source of alkanes?
   (k) Write the expanded formula of $(CH_3)_3CCH_2CH(CH_3)CH_2CH_2CH(CH_3)_2$.
   (l) Give the correct IUPAC name for the compound in (k).
   (m) Draw the structure of *tert*-butyl chloride.

Answers to self-test

1. (a) 2-Methylpentane
   (b) 1,2-Dichloroethane
   (c) Methylcyclopentene

   (d) 4-Bromo-2-hexene
   (e) 3,3-Dimethyl-1-butyne
   (f) 3-Methyl-3-chloro-1,5-hexadiene

2. (a)
$$CH_3CH_2\overset{\overset{\textstyle CH_3}{|}}{C}=CHCH_3$$

   (b)

Br

Br

   (d) $CH_2{=}CHCH_2CH{=}CHCH_3$

   (e)
$$CH_3CH_2CH_2\overset{\overset{\textstyle Cl}{|}}{C}H\overset{}{\underset{\underset{\textstyle CH_3}{|}}{C}}HCH_3$$

   (f) $CH_3CH_2CH_2CH_2CH_2CH_2CH_2CH_3$

   (c) $CH_3CH_2CH_2C{\equiv}CCH_2CH_3$

3.

Conditions	Intermediate or transition state	Product(s)
(a) KOH, CH_3CH_2OH	$$\overset{\delta^-}{Br}\\ {\|}{\vdots}\\ CH_3CH_2CH\text{---}CH_2\\ {\vdots}{\vdots}\\ \delta^-\,HO\text{----}H$$	$CH_3CH_2CH{=}CH_2$
(b) H_2SO_4, heat	$CH_3CH_2\underset{\oplus}{C}HCH_3$	$CH_3CH_2CH{=}CH_2$, $CH_3CH{=}CHCH_3$ (preferred)

4. (a) Two

(b) $(CH_3)_3C^{\oplus}$, tertiary

(c) 1,3-Butadiene $(CH_2{=}CHCH{=}CH_2)$

(d) sp^2

(e) 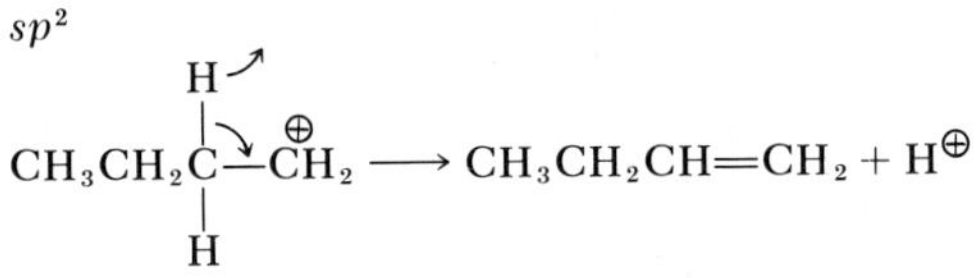 $CH_3CH_2\overset{\displaystyle H\nearrow}{\underset{\displaystyle H}{\overset{}{C}}}\!\!\underset{}{\overset{\oplus}{\rightharpoondown}}CH_2 \longrightarrow CH_3CH_2CH{=}CH_2 + H^{\oplus}$

(f) 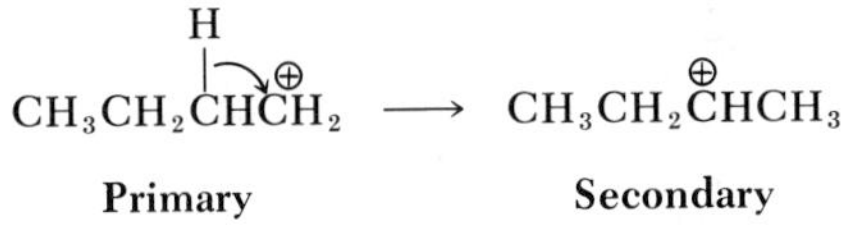 $CH_3CH_2\overset{\displaystyle H}{\underset{}{\overset{}{C}}}\!\!\overset{\oplus}{\rightharpoondown}HCH_2 \longrightarrow CH_3CH_2\overset{\oplus}{C}HCH_3$

       Primary   **Secondary**

(g) Alkyne, $CH_3C{\equiv}CCH_2CH_3$

(h) 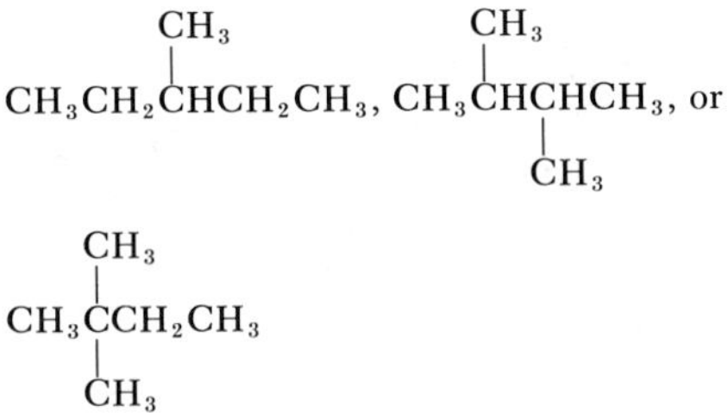 $CH_3CH_2\overset{\displaystyle CH_3}{\underset{}{\overset{|}{C}}}HCH_2CH_3$, $CH_3\overset{\displaystyle CH_3}{\underset{\displaystyle CH_3}{\overset{|}{C}}}\!H\underset{|}{C}HCH_3$, or

$CH_3\overset{\displaystyle CH_3}{\underset{\displaystyle CH_3}{\overset{|}{\underset{|}{C}}}}CH_2CH_3$

(i) $CH_3CH_2CH_2CH_2CH_2CH_3$ or
$CH_3CH_2CH_2CH_2CH_2CH_2CH_2CH_3$

(j) Petroleum

(k) 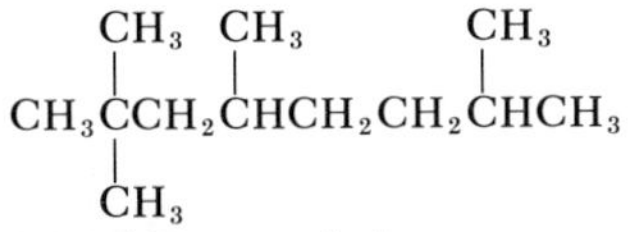 $CH_3\overset{\displaystyle CH_3}{\underset{\displaystyle CH_3}{\overset{|}{\underset{|}{C}}}}CH_2\overset{\displaystyle CH_3}{\overset{|}{C}}HCH_2CH_2\overset{\displaystyle CH_3}{\overset{|}{C}}HCH_3$

(l) 2,2,4,7-Tetramethyloctane

(m) 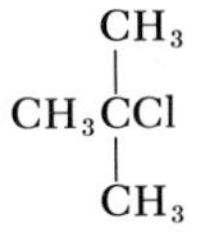 $CH_3\overset{\displaystyle CH_3}{\underset{\displaystyle CH_3}{\overset{|}{\underset{|}{C}}}}Cl$

chapter 3

Hydrocarbons: reactions

Alkanes are inert compounds. They are stable toward acids, bases, and oxidizing and reducing agents. Under vigorous conditions, they undergo a free-radical reaction; the resulting product is usually a mixture. Alkanes can be burned, and the combustion process leads to problems of pollution. Alkenes and alkynes, on the other hand, are much more reactive because of the presence of the carbon–carbon double and triple bonds, respectively, and they undergo a number of addition reactions.

Reactions of alkanes

Halogenation

In this reaction, a halogen atom is substituted for a hydrogen atom of the alkane, and the hydrogen atom thus replaced combines with a second atom of halogen. This occurs under the influence of ultraviolet (UV) light or at a temperature of about 250° C. Alkanes react with chlorine or bromine to form alkyl chlorides (chloroalkanes) or alkyl bromides (bromoalkanes) and equivalent amounts of hydrogen chloride or hydrogen bromide. Fluorine is so reactive with alkanes that it causes other types of reactions to occur, and iodine is so unreactive that iodination does not occur. The alkyl halide formed can itself undergo further substitution to form a new dihalogen substitution product. In theory, this process of substitution of halogen for hydrogen can continue as long as there are hydrogen atoms in the molecule. The reaction can be limited to monosubstitution if a large excess of alkane is used. The halogenation reaction can be illustrated by the chlorination of methane, which can yield any one of four products:

$$CH_4 \ + \ Cl_2 \ \xrightarrow[\text{or UV light}]{\text{heat} (\sim 250°C)} \ CH_3Cl + HCl$$

Methane $\qquad\qquad\qquad\qquad$ Methyl chloride

$$CH_3Cl + Cl_2 \ \longrightarrow \ CH_2Cl_2 + HCl$$

$\qquad\qquad\qquad\qquad\qquad\qquad\qquad$ Methylene chloride

$$CH_2Cl_2 + Cl_2 \longrightarrow CHCl_3 + HCl$$

Chloroform

$$CHCl_3 + Cl_2 \longrightarrow CCl_4 + HCl$$

**Carbon
tetrachloride**

Halogenation of other alkanes is essentially the same as that of methane, but it sometimes results in a mixture of isomers.

The chlorination of methane is a *chain reaction*, which involves a series of steps, each of which generates a reactive substance (in this case a free radical) that brings about the next step. There are three characteristics of a chain reaction: (1) *chain initiation*, in which energy is absorbed and a reactive particle is generated; (2) *chain propagation*, in which a reactive particle is consumed but not generated; and (3) *chain termination*, in which the reactive particles are consumed but not generated. There is always one chain-initiating step and one or more chain-propagating and chain-terminating steps. The mechanism for the chlorination of methane can be summarized as follows:

Chain initiation   (a) Cl_2 $\xrightarrow{\text{light or heat}}$ $2Cl\cdot$

Chain propagation $\begin{cases} \text{(b) } CH_4 + Cl\cdot \longrightarrow HCl + CH_3\cdot \\ \text{(c) } CH_3\cdot + Cl_2 \longrightarrow CH_3Cl + Cl\cdot \end{cases}$

(b and c repeat until chain termination occurs)

Chain termination $\begin{cases} \text{(d) } CH_3\cdot + CH_3\cdot \longrightarrow CH_3CH_3 \\ \text{(e) } CH_3\cdot + Cl\cdot \longrightarrow CH_3Cl \\ \text{(f) } Cl\cdot + Cl\cdot \longrightarrow Cl_2 \end{cases}$

Step (a) involves the breaking of a chlorine–chlorine bond in a symmetrical way so that each chlorine atom formed retains one electron of the pair from the original bond. This is commonly referred to as *homolytic cleavage*. The energy required to break this bond is supplied by the heat or light.

In step (b), the chlorine atom abstracts a hydrogen atom from a methane molecule to form a methyl free radical and a molecule of hydrogen chloride. Once formed, the methyl free radical reacts with a chlorine molecule in step c to form 1 mole of product and to regenerate a chlorine atom. Steps b and c are repeated over and over, but this sequence cannot continue forever.

Eventually, one of the reactive particles collides with itself or another reactive particle. Such a collision (steps d, e, or f) causes the formation of a covalent bond, consumes the reactive particles, generates no new reactive particles, and as a result, terminates the reaction.

The energy changes occurring in the two chain-propagating steps, b and c, are shown in Fig. 3-1. The rate-determining step is b because of its higher E_A. Once formed, methyl free radicals easily react with chlorine molecules to form the product CH_3Cl and chlorine atoms. The overall reaction of course is exothermic.

The chlorination of methane is a *free-radical substitution* reaction. The

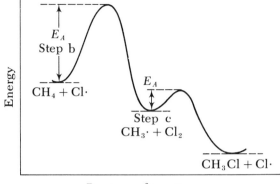

Progress of reaction

Fig. 3-1
Energy changes for the two chain-propagating steps in chlorination of methane.

free radical is an atom with an unpaired electron and is another type of reactive carbon-containing intermediate.

The methyl free radical contains a carbon atom bonded to three other atoms. The carbon atom has sp^2 hybrid orbitals associated with a p orbital. The unpaired electron of the methyl free radical occupies the p orbital. All four atoms of the methyl free radical lie in the same plane, and the p orbital is perpendicular to the plane. All the bonds in the methyl free radical, which is commonly written as $CH_3 \cdot$, are formed by overlapping of an sp^2 orbital of the carbon atom and an s orbital of a hydrogen atom about a line joining the nuclei. Thus there are three σ bonds in the methyl free radical, the structure of which is as follows:

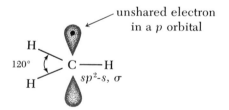

The ethyl free radical, unlike the methyl free radical, is not planar. It contains one carbon atom with sp^3 hybrid orbitals. Its structure and bonding are shown in Fig. 3-2.

The free-radical chlorination of methane has commercial application. The substitution of all the hydrogens in methane forms carbon tetrachloride (CCl_4), which is a colorless, heavy liquid. It is very volatile, and its vapor is toxic. However, it is nonflammable. It has been used in industry and in the dry cleaning of clothes as a solvent for fats, oils, and greases. It has also been used as a fire extinguisher, although it has largely been replaced by carbon dioxide because it can be hydrolyzed under high temperatures to phosgene, which is a very poisonous gas:

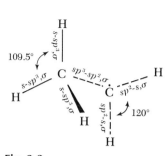

Fig. 3-2
Structure of the ethyl free radical.

$$CCl_4 + H_2O \xrightarrow{\text{heat}} Cl—\overset{\displaystyle \|}{\underset{\displaystyle O}{C}}—Cl + 2HCl$$

Phosgene

The chlorination of higher alkanes yields isomeric mixtures. For example, in propane there are two types of hydrogen atoms, six equivalent primary hydrogen atoms and two equivalent secondary hydrogen atoms. Thus it might predicted on a statistical basis that the ratio of *n*-propyl chloride to isopropyl chloride would be 6:2, or 3:1. Furthermore, it might be predicted that *n*-propyl chloride would be the major product, since there are three times as many hydrogen atoms. However, isopropyl chloride is the major product. This can be explained by a consideration of the reactive intermediate in halogenation, the free radical. The stability of free radicals is tertiary > secondary > primary > methyl radical. The more stable the free radical is, the more easily it will be formed. Both a secondary free radical and a primary free radical arise from the abstraction of a hydrogen atom of propane by a chlorine atom. However, the more stable free radical, the isopropyl free radical, is formed more easily than the *n*-propyl free radical. In other words, a secondary hydrogen is more easily abstracted than a primary hydrogen. The ease of abstraction of hydrogen atoms then is in the same order as the stability. Since preferred product arises from the most easily formed free radical, isopropyl chloride is the preferred product:

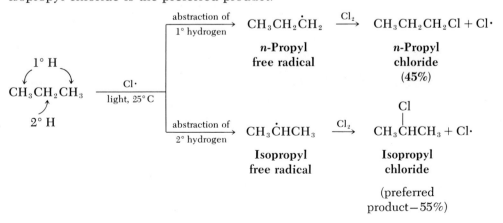

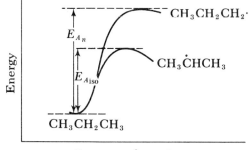

Progress of reaction

Fig. 3-3
Formation of *n*-propyl and isopropyl free radicals from propane. Energy relationships show that it takes more energy to form *n*-propyl (primary) free radical than it does to form isopropyl (secondary) free radical. The smaller the energy required to form a radical is, the more stable the free radical is.

The energy profile in Fig. 3-3 illustrates the ease of formation of the *n*-propyl and isopropyl free radicals from propane.

3-1. Predict the product(s) formed from the chlorination of each of the following compounds if only monosubstitution occurs. Indicate the preferred product:

(a) 2-Methylpropane   *Answer:* $(CH_3)_3CCl$ (preferred), formed from *tert*-butyl free radical, and $(CH_3)_2CHCH_2Cl$, formed from isobutyl free radical

(b) 2,3-Dimethylbutane   *Answer:* $(CH_3)_2CClCH(CH_3)_2$ (preferred), formed from $(CH_3)_2\dot{C}CH(CH_3)_2$, and $(CH_3)_2CHCH(CH_3)CH_2Cl$, formed from $(CH_3)_2CHCH(CH_3)CH_2\cdot$

(c) Pentane
(d) 3,3-Dimethylpentane
(e) 2,2-Dimethylbutane
(f) Cyclohexane

3-2. Draw the structure of the alkane with a molecular weight of 72 that forms only one monochlorinated product.

Answer: The alkane must contain less than six carbon atoms.

Combustion: gasoline and pollution

The complete combustion (oxidation by oxygen) of an alkane produces carbon dioxide, water, and energy:

$$alkane + O_2 \longrightarrow CO_2 + H_2O + energy$$

For example, the complete combustion of methane yields carbon dioxide, water, and heat:

$$CH_4 + 2O_2 \longrightarrow CO_2 + 2H_2O$$

Methane

The heat of combustion is large; that is, the reaction is highly exothermic. Thus alkanes can be used as fuels. More than 95% of the energy used in the United States is supplied by chemical processes involving petroleum and natural gas. Hydroelectric power and nuclear power account for the remainder.

In the operation of the automobile engine, the oxygen supply is usually not sufficient enough for there to be complete combustion of the fuel. Thus some of the alkane is oxidized to carbon monoxide or even carbon, the black sooty deposit often seen around the end of an exhaust pipe:

$$2CH_4 + 3O_2 \longrightarrow 2CO + 4H_2O$$

Methane   Carbon
                           monoxide

$$CH_4 + O_2 \longrightarrow C + 2H_2O$$

Methane   Carbon

Unfortunately, this incomplete combustion is necessary, since the automobile engine operates most efficiently when there is a slight deficiency in the ratio of oxygen to hydrocarbon. However, some automobile manufacturers have devised a method of introducing air to complete the combustion process in the exhaust manifold.

The efficiency of the modern-day gasoline used as fuel depends on the smoothness with which the alkane burns. Branched-chain alkanes perform better than continuous-chain alkanes. The heats of combustion of these two types of alkanes are, for all practical purposes, the same. However, branched-chain alkanes burn more smoothly, that is, more slowly and less explosively. When the combustion process is too rapid, "knocking" results because the heat of combustion is released to the engine block rather than causing the piston to move.

The amount of knocking depends on the octane rating of gasoline, which is defined in terms of mixtures of heptane (a continuous-chain alkane) and 2,2,4-trimethylpentane (a branched-chain alkane sometimes called isooctane). Heptane is given a rating of 0, since it is prone to knocking; 2,2,4-trimethylpentane is given a rating of 100. The octane ratings of some hydrocarbons are given in Table 3-1. The rating increases as the molecular weight decreases and the branching increases. A 90-octane fuel would produce the same amount of knocking as a mixture that is 10% heptane and 90% 2,2,4-trimethylpentane.

Octane ratings can be improved by the addition of small amounts of tetraethyl lead (TEL) $[(C_2H_5)_4Pb]$ to the gasoline. However, when gasoline containing this additive is burned, the lead is oxidized to lead (II) oxide:

$$(C_2H_5)_4Pb + \tfrac{27}{2}O_2 \longrightarrow PbO + 8CO_2 + 10H_2O$$

For prevention of harmful deposits of lead in cylinder walls and valves, 1,2-dibromoethane $(C_2H_4Br_2)$ is added with TEL. 1,2-Dibromoethane and lead (II) oxide react to form lead tetrabromide $(PbBr_4)$, which is volatile and

Table 3-1
Octane rating of hydrocarbons

Hydrocarbon	Molecular weight	Octane rating
Nonane	128	−45
Octane	114	−17
Heptane	100	0
3-Methylheptane	114	35
2,3-Dimethylhexane	114	78.9
2,2,3-Trimethylpentane	114	99.9
2,2,4-Trimethylpentane	114	100
2,2,3,3-Tetramethylbutane	114	103

toxic and is swept out with the exhaust gases, polluting the atmosphere further. A current solution is the use of nonleaded gasolines. Another additive currently on the market is tricresyl phosphate (TCP):

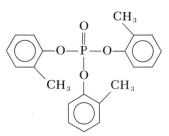

Some gasoline manufacturers also use *platforming,* a process involving a platinum catalyst, to make high-octane gasoline.

Two other by-products of combustion also contribute to the pollution problem. One of these is sulfur trioxide, which is oxidized in the combustion reaction from the sulfur compounds present in gasoline. Sulfuric acid is then formed from the sulfur trioxide and water vapor present in the atmosphere. It is damaging to animal and plant life, since it reacts readily with tissue. Large amounts in rivers and streams can change the pH of the water and alter or kill the biological systems in them.

The other pollution problem does not originate in the fuel. The oxygen used in the combustion process is taken from the air, which is a mixture primarily of oxygen and nitrogen. Because of the intense heat of the flame in the combustion process, some of the nitrogen is oxidized to the poisonous nitrogen oxides, nitric oxide (NO) and nitrogen dioxide (NO_2). Nitrogen dioxide is a brown gas that can undergo a reaction with sunlight (ultraviolet light) to form oxygen atoms (O), which are so reactive that they react with the unburned hydrocarbons from automobile exhausts to form acyl radicals. Acyl radicals are precursors of peroxyacyl nitrates, which are lacrimators often associated with smog. Another component of smog, ozone (O_3), oxidizes rubber in automobile tires by reacting with the carbon–carbon double bonds of the rubber, so that it becomes brittle and, in essence, has a shorter life.

The chemical reactions occurring in the smog-forming process are summarized as follows:

$$N_2 + O_2 \longrightarrow 2NO$$
$$2NO + O_2 \longrightarrow 2NO_2$$
$$NO_2 + \text{sunlight} \longrightarrow NO + O$$
$$O + O_2 \longrightarrow O_3$$

$$O \;+\; \text{hydrocarbons} \;\longrightarrow\; \underset{\text{Aldehyde}}{R\overset{\displaystyle O}{\overset{\|}{C}}H} \;+\; R\!-\!\overset{\displaystyle O}{\overset{\|}{C}}\!-\!O\cdot$$

$$RCO_2\cdot \;+\; NO \longrightarrow NO_2 + R\!-\!\underset{\displaystyle O}{\underset{\|}{C}}\cdot$$

$$RCO\cdot \;+\; NO_2 + O_2 \longrightarrow R\!-\!\underset{\displaystyle O}{\underset{\|}{C}}\!-\!O\!-\!O\!-\!NO_2$$

Peroxyacyl nitrate

Reactions of alkenes

The reactions of alkenes occur at the carbon–carbon double bond, which consists of a strong σ bond and a weak π bond. These reactions involve the breaking of the weaker π bond and the resultant formation of two strong σ bonds:

$$\underset{\diagup \;\; \underset{\sigma}{} \;\; \diagdown}{\overset{\diagdown \;\; \overset{\pi}{} \;\; \diagup}{C\!=\!C}} \;+\; A-B \longrightarrow \; \overset{\sigma}{-\!C\!-\!\!-\!\!-\!C\!-}$$

Since two molecules combine to form one molecule, this characteristic reaction of an alkene is called an *addition reaction.*

A π bond consists of a pair of electrons and thus acts as a source of electrons (a Lewis base). It reacts with a compound that is deficient in electrons (a Lewis acid). An acidic compound that is seeking a pair of electrons is called an *electrophilic reagent* (Greek: electron loving). A basic compound that is capable of supplying a pair of electrons is called a *nucleophile* (Greek: nucleus loving). Since the carbon–carbon double bond reacts with electrophilic reagents, the typical reaction of an alkene is *electrophilic addition.*

Electrophilic addition reactions

The addition of hydrogen halide, halogen, sulfuric acid, water, or hypohalous acid proceeds through an electrophilic addition mechanism.

Addition of hydrogen halide. The reaction between an alkene and a hydrogen halide (HX) forms an alkyl halide:

$$R\!-\!CH\!=\!CH_2 \;+\; HX \;\longrightarrow\; \underset{X \;\; H}{RCHCH_2}$$

Alkene   **Alkyl halide**

Following are some examples:

$$CH_2\!\!=\!\!CH_2 \;+\; HCl \;\longrightarrow\; \underset{\underset{\displaystyle H \quad\; Cl}{|\quad\;\; |}}{CH_2\!\!-\!\!CH_2}$$

Ethene	**Ethyl chloride**
	(1-chloroethane)

$$CH_3CH\!\!=\!\!CHCH_3 + HBr \longrightarrow \underset{\underset{\displaystyle H \quad\; Br}{|\quad\;\; |}}{CH_3CH\!\!-\!\!CHCH_3}$$

2-Butene	***sec*-Butyl bromide**
	(2-bromobutane)

$$CH_3CH\!\!=\!\!CH_2 + HI \longrightarrow \underset{\underset{\displaystyle H \quad\; I}{|\quad\; |}}{CH_3CH\!\!-\!\!CH_2} + \underset{\underset{\displaystyle I \quad\; H}{|\quad\; |}}{CH_3CH\!\!-\!\!CH_2}$$

Propene	***n*-Propyl iodide**	**Isopropyl iodide**
	(1-iodopropane)	(2-iodopropane)
		(major product)

$$CH_3CH_2CH\!\!=\!\!CH_2 + HCl \longrightarrow \underset{\underset{\displaystyle H \quad\; Cl}{|\quad\; |}}{CH_3CH_2CH\!\!-\!\!CH_2} + \underset{\underset{\displaystyle Cl \quad\; H}{|\quad\; |}}{CH_3CH_2CH\!\!-\!\!CH_2}$$

1-Butene	***n*-Butyl chloride**	***sec*-Butyl chloride**
	(1-chlorobutane)	(2-chlorobutane)
		(major product)

The addition of hydrogen halide to ethene or 2-butene forms a single product. This is the case with all symmetrical alkenes, in which the alkyl groups on either side of the carbon–carbon double bond are the same.

In the addition reaction with propene or 1-butene, two isomeric products are possible. The product that predominates can be predicted from a generalization known as *Markovnikov's rule*, which states that in the electrophilic addition of a hydrogen halide to the carbon–carbon double bond of an alkene, the hydrogen atom of the hydrogen halide attaches itself to the carbon atom of the double bond that already holds the greater number of hydrogens. Therefore, in propene and 1-butene, the carbon of the double bond that has two hydrogens gets the hydrogen of the hydrogen halide.

The addition of the acid HX proceeds as follows:

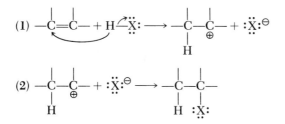

Step 1 involves the formation of a carbonium ion by transfer of a hydrogen from the hydrogen halide. The π bond supplies the electrons for the newly formed bond between carbon and hydrogen. Step 2 is the formation of a covalent bond between the positive carbon and the base X^-, which might be Cl^-, Br^-, or I^-.

With the use of this mechanism for electrophilic addition, it can be shown why the major product in the addition of HCl to propene is isopropyl chloride. The addition of H^+ can form either isopropyl or n-propyl cations:

$$
\begin{array}{c}
\underset{C_a}{\xrightarrow{H^+ \text{ to}}} \quad CH_3\underset{\oplus}{C}HCH_3 \quad \xrightarrow{Cl^-} \quad CH_3CHCH_3 \\
2° \text{ Carbonium ion} \qquad \overset{|}{Cl} \\[1em]
\underset{C_b}{\xrightarrow{H^+ \text{ to}}} \quad CH_3CH_2\underset{\oplus}{C}H_2 \quad \xrightarrow{Cl^-} \quad CH_3CH_2CH_2Cl \\
1° \text{ Carbonium ion}
\end{array}
$$

$$\overset{b}{}\quad\overset{a}{}\qquad CH_3CH{=}CH_2$$

Using the order of stability of carbonium ions, tertiary > secondary > primary > methyl ion, as a basis of Markovnikov's rule, we can state the rule more generally: electrophilic addition to a carbon–carbon double bond involves the intermediate formation of the more stable carbonium ion. Since the ease of formation of a carbonium ion parallels its stability, isopropyl cation, a secondary carbonium ion, is formed faster than n-propyl cation, a primary carbonium ion. Hence, isopropyl chloride is the major product in the electrophilic addition to the carbon–carbon double bond of propene, since it is formed from the more stable isopropyl cation.

3-3. (a) Write the mechanism for the formation of *sec*-butyl chloride from the addition of HCl to 1-butene.

(b) In the addition of HI to 3-methyl-1-butene, the major product formed is 2-methyl-2-iodobutane. Explain this, using the carbonium ion mechanism and the concepts of carbonium ion rearrangement discussed in Chapter 2.

Answer: The initially formed secondary carbonium ion rearranges to a more stable tertiary carbonium ion.

Addition of halogen. The reaction between a carbon–carbon double bond and halogen, either chlorine or bromine, produces a vicinal dihalide and is usually carried out in carbon tetrachloride:

$$
R{-}CH{=}CH_2 + X_2 \longrightarrow R{-}\underset{X}{\overset{|}{C}}H{-}\underset{X}{\overset{|}{C}}H_2
$$

Alkene $\qquad\qquad$ Vicinal dihalide

Following are some examples:

$$
CH_2{=}CH_2 \quad + \quad Br_2 \quad \xrightarrow{CCl_4} \quad CH_2{-}CH_2
$$
$$
\qquad\qquad\qquad\qquad\qquad\qquad \overset{|}{Br}\quad\overset{|}{Br}
$$

Ethene $\qquad\qquad$ 1,2-Dibromoethane
(colorless) $\quad$ (reddish) $\qquad\qquad$ (colorless)

$$CH_3CH{=}CH_2 \quad + \quad Cl_2 \quad \xrightarrow{CCl_4} \quad CH_3CH{-}CH_2$$
$$\underset{Cl}{\overset{|}{}} \quad \underset{Cl}{\overset{|}{}}$$

Propene   **1,2-Dichloropropane**

Carbon tetrachloride serves as a solvent for the gaseous chlorine or bromine. The addition of bromine actually constitutes a test for the presence of a carbon–carbon double bond: the alkene is colorless, and the bromine solution is reddish; once the addition is complete, the solution turns colorless again.

In the case of bromine, it is believed that the reaction proceeds through a cyclic intermediate, thought be to a cyclic bromonium ion:

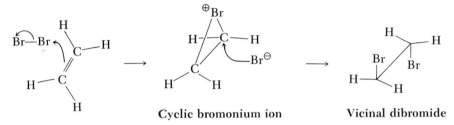

Cyclic bromonium ion   **Vicinal dibromide**

A positive bromine ion attaches itself to both carbons of the carbon–carbon double bond, so that a cyclic bromonium ion is formed. The bromide ion then attacks the bromonium ion, yielding a vicinal dibromide.

Addition of sulfuric acid. This reaction proceeds according to Markovnikov's rule. The addition of sulfuric acid ($HOSO_3H$) to a carbon–carbon double bond forms an alkyl hydrogen sulfate:

$$R{-}CH{=}CH_2 \quad + \quad HOSO_3H \quad \longrightarrow \quad R{-}CH{-}CH_2$$
$$\underset{HO_3SO}{\overset{|}{}} \quad \underset{H}{\overset{|}{}}$$

Alkene   **Alkyl hydrogen sulfate**

Following is an example:

$$CH_3CH{=}CH_2 + HOSO_3H \longrightarrow CH_3{-}CH{-}CH_3$$
$$\underset{OSO_3H}{\overset{|}{}}$$

Propene   **Isopropyl hydrogen sulfate**

It is important to realize that in this addition reaction carbon is bonded to oxygen, not to sulfur. Thus alkyl hydrogen sulfates are easily hydrolyzed to form alcohols:

$$CH_3{-}CH{-}CH_3 \xrightarrow{HOH} CH_3{-}CH{-}CH_3$$
$$\underset{OSO_3H}{\overset{|}{}} \qquad\qquad \underset{OH}{\overset{|}{}}$$

Isopropyl alcohol

Addition of water. This reaction follows Markovnikov's rule. Alcohols are formed by the addition of water to alkenes in the presence of acid:

$$R—CH=CH_2 + HOH \xrightarrow{H^+} R—CH—CH_2$$
$$\overset{|}{OH} \quad \overset{|}{H}$$

Alkene   **Alcohol**

Following is an example:

$$CH_3\underset{\underset{CH_3}{|}}{C}=CH_2 \xrightarrow{H_2O, H^+} (CH_3)_3COH$$

2-Methylpropene   *tert*-**Butyl alcohol**

Hydration of alkenes is the principal industrial source of lower alcohols.

Addition of hypohalous acid. The addition of hypohalous acid (HOX) to alkenes forms *halohydrins*, compounds in which halogen and hydroxyl groups are positioned on adjacent carbon atoms:

$$R—CH=CH_2 + HOX \longrightarrow R—CH—CH_2$$
$$\overset{|}{OH} \quad \overset{|}{X}$$

Alkene   **Halohydrin**

Either hypobromous acid, HOBr ($H_2O + Br_2 \longrightarrow$ HOBr), or hypochlorous acid, HOCl ($H_2O + Cl_2 \longrightarrow$ HOCl), is used. Since oxygen is more electronegative than either —Br or —Cl, then —Br or —Cl is the positive portion of the adding group:

$$CH_3CH=CH_2 + HOBr \longrightarrow CH_3—CH—CH_2$$
$$\overset{|}{OH} \quad \overset{|}{Br}$$

Propene   **1-Bromo-2-propanol**

Other addition reactions

Some other addition reactions of alkenes include hydrogenation, free-radical addition, ozonolysis, and oxidation by a cold aqueous solution of potassium permanganate.

Addition of hydrogen. This is called *hydrogenation*. The addition of hydrogen to a carbon–carbon double bond requires the presence of a catalyst and usually heat or pressure. The product is an alkane:

$$R—CH=CH_2 + H_2 \xrightarrow{Ni} RCH—CH_2$$
$$\overset{|}{H} \quad \overset{|}{H}$$

Alkene   **Alkane**

Following are some examples:

$$CH_2=CH_2 + H_2 \xrightarrow[\text{heat}]{Ni} CH_2—CH_2$$
$$\overset{|}{H} \quad \overset{|}{H}$$

Ethene   **Ethane**

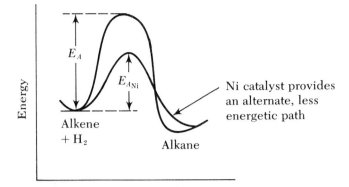

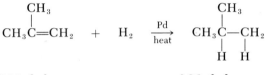

Progress of reaction

Fig. 3-4

Energy profile for hydrogenation of an alkene in the presence and absence of a catalyst. E_A is the energy of activation in the absence of a catalyst. $E_{A_{Ni}}$ is the energy of activation in the presence of a nickel catalyst.

$$
\underset{\textbf{2-Methylpropene}}{CH_3\overset{\displaystyle CH_3}{\underset{\displaystyle |}{C}}=CH_2} \quad + \quad H_2 \quad \xrightarrow[\text{heat}]{\text{Pd}} \quad \underset{\textbf{2-Methylpropane}}{CH_3\overset{\displaystyle CH_3}{\underset{\displaystyle \underset{H}{|}}{\underset{|}{C}}}\underset{H}{\overset{|}{-}}CH_2}
$$

The catalyst is usually nickel, platinum, or palladium. The purpose of the catalyst is to provide an alternate, less energetic path for the reacting molecules to attain the product state. This is shown in the energy profile in Fig. 3-4.

It is postulated that the mechanism of addition of hydrogen is as follows:

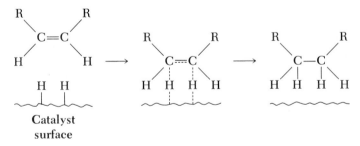

The hydrogen is adsorbed on the surface of the catalyst as hydrogen atoms. The alkene approaches the surface of the catalyst and encounters the hydrogen atoms adsorbed on the surface. The hydrogens then begin to form covalent bonds with the carbon atoms, and the carbon–carbon double bond begins to break. Finally, the formation of the alkane is complete, and the alkane moves away from the catalyst surface.

Addition involving free radicals. The reaction of the addition of HBr to an unsymmetrical alkene in the presence of peroxides is as follows:

$$R-CH=CH_2 + HBr \xrightarrow{\text{peroxides}} R-CH-CH_2$$
$$\qquad\qquad\qquad\qquad\qquad\quad \underset{H}{|} \quad \underset{Br}{|}$$

Alkene   **Alkyl halide**

The result is the reverse of that predicted by Markovnikov's rule:

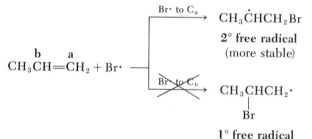

$$CH_3CH=CH_2 + HBr$$

absence of peroxides → $CH_3\underset{Br}{\underset{|}{C}}HCH_3$ **Isopropyl bromide**   Markovnikov addition

presence of peroxides → $CH_3CH_2CH_2Br$ ***n*-Propyl bromide**   Anti-Markovnikov addition

The addition in the presence of peroxides occurs not by a carbonium ion mechanism but rather by a free-radical mechanism. The peroxide causes the formation of the reactive bromine atom ($Br\cdot$), which then adds to the carbon–carbon double bond in such a manner that the more stable secondary free radical is formed:

$$\underset{b \qquad a}{CH_3CH=CH_2} + Br\cdot$$

$Br\cdot$ to C_a → $CH_3\overset{\cdot}{C}HCH_2Br$ **2° free radical (more stable)**

Br to C_b → $CH_3\underset{Br}{\underset{|}{C}}HCH_2\cdot$ **1° free radical**

This step is then followed by abstraction of a hydrogen atom from HBr, which leads to the product:

$$CH_3\overset{\cdot}{C}HCH_2Br \xrightarrow{HBr} CH_3CH_2CH_2Br + Br\cdot$$

Hence, *free-radical addition* of HBr in the presence of peroxides to a carbon–carbon double bond gives the anti-Markovnikov product.

The same product is obtained when HCl reacts with a carbon–carbon double bond in the presence or the absence of peroxides. The product is that predicted by Markovnikov's rule:

$$CH_3CH=CH_2 \ + \ HCl \xrightarrow{\text{peroxides}} CH_3CH-CH_2$$
$$\qquad\qquad\qquad\qquad\qquad\qquad\qquad \underset{Cl}{|} \quad \underset{H}{|}$$

Propene   **Isopropyl chloride**

Polymerization. Alkenes and dienes undergo a free-radical addition reaction that has commercial application. This reaction is polymerization, which is a process of joining together many small molecules called monomers to make very large molecules called polymers (Greek: *poly*, many, and *meros*, parts). Polymerization reactions take place under a variety of conditions:

Table 3-2
Commonly used polymers

Monomer		Polymer		
Formula	*Name*	*Name*	*Formula*	*Use*
$CH_2{=}CH_2$	Ethylene	Polyethylene	$-CH_2CH_2CH_2CH_2-$ or $+(CH_2CH_2)_n$	Plastic material used in packing film
$CH_2{=}CHCl$	Vinyl chloride	Polyvinyl chloride	$+(CH_2{-}CH)_n$, Cl	Phonograph records, raincoats, shower curtains
$CH_2{=}CCl_2$	1,1-Dichloroethene	Polydichloroethene	$+(CH_2{-}CCl_2)_n$	Saran wrap
$CF_2{=}CF_2$	Tetrafluoroethene	Polytetrafluoroethene	$+(CF_2{-}CF_2)_n$	Teflon (chemically resistant coating used in cookware)
$CH_2{=}CHCN$	Acrylonitrile	Polyacrylonitrile	$+(CH_2{-}CH)_n$, CN	Orlon and Acrilan (for fibers and fabrics)
(phenyl)$-CH{=}CH_2$	Styrene	Polystyrene	$+(CH{-}CH_2)_n$ (with two phenyl groups)	Plastics for combs and kitchen utensils; insulation for refrigerators and air conditioners
$CH_2{=}C{-}CO_2CH_3$, CH_3	Methyl methacrylate		$+(CH_2{-}C)_n$, CO_2CH_3, CH_3	Lucite and Plexiglas (safety glass)
$CH_2{=}C{-}CH{=}CH_2$, CH_3	Isoprene	cis-Polyisoprene	$+(CH_2{-}C{=}CH{-}CH_2)_n$, CH_3	Natural rubber (rubber made by polymerization of isoprene is a *cis-trans* mixture)
$CH_2{=}C{-}CH{=}CH_2$, Cl	Chloroprene	Polychloroprene	$+(CH_2{-}C{=}CH{-}CH_2)_n$, Cl	Neoprene, Duprene (first rubber substitute)

heat, pressure, and the presence of oxygen or peroxides. Following are some examples:

$$nCH_2{=}CH_2 \xrightarrow{\text{O}_2,\ \text{heat, pressure}} [-CH_2-CH_2-CH_2-CH_2-]_n$$

Ethylene (monomer) **Polyethylene** (polymer)

$$nCH_2{=}CH-CH{=}CH_2 \xrightarrow{\text{peroxides}} [-CH_2-CH{=}CH-CH_2-]_n$$

1,3-Butadiene (monomer) **Polybutadiene** (polymer)

One particular diene, isoprene (2-methyl-1,3-butadiene), is a source of natural rubber:

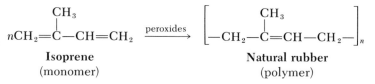

Isoprene (monomer) **Natural rubber** (polymer)

A diene with substituted groups, chloroprene (2-chloro-1,3-butadiene), on polymerization, yields polychloroprene, a rubber substitute:

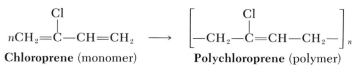

Chloroprene (monomer) **Polychloroprene** (polymer)

Each of these reactions proceeds through the formation of a free radical followed by addition of the radical to a monomer. This is illustrated as follows for the formation of polyethylene from ethylene:

Radicals· $^{\frown}$CH$_2{=}$CH$_2$ $^{\frown}$CH$_2{=}$CH$_2$ $^{\frown}$CH$_2{=}$CH$_2$ $^{\frown}$CH$_2{=}$CH$_2$ $^{\frown}$CH$_2{=}$CH$_2$ $\longrightarrow$
polyethylene

Table 3-2 lists some polymers in common use today that are formed through free-radical addition reactions.

To meet the needs of space-age civilization, a whole new generation of plastics has been developed. Space travel demands substances that will not shatter or even crack under intense pounding and that will not decompose even at temperatures several times the melting point of steel. Although these substances must withstand pressure changes and extremes of heat and cold unknown before the space ventures, their discoverers have been quick to devise more conventional uses for them as well. Some of these materials are described as follows:

Lexan, a polycarbonate resin, is transparent enough to be mistaken for glass, but it behaves more like steel. A General Electric chemist discovered it in 1955 when he attempted to remove his stirrer after a reaction and found that the residue in the flask refused to let go. He broke the flask and went to work on the clear mass surrounding his stirrer, but no amount of battering could induce the substance to let go of it. A sheet of Lexan can stop the bullet of a .38-caliber gun fired from a distance of 12 feet. It is used in visors to protect astronauts from the perils of space and in windows to protect home-owners from vandals. The structure of Lexan is shown as follows, where n is 80 to 100:

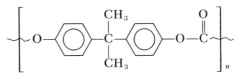

A new cyanoacrylic is a powerful plastic adhesive that handles the job of nails, screws, and even rivets better and more easily than these fasteners ever did themselves. While most adhesives set only in the presence of heat or under evaporation, the miniscule amounts of moisture and oxygen normally present on ordinary surfaces are sufficient to set off this plastic's unique poly-merization process. Furthermore, it takes only a single drop to create a chemical bond hardy enough to support a half ton. Its strength and adaptabil-ity allow it to be used in hard-to-reach corners of electronic equipment such as hearing aids, computers, and the tiny new calculators. Its structure is shown as follows, where n is undetermined:

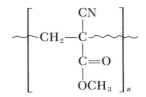

Most elementary school children learn that fish breathe through gills, thin membranes that keep water out of their bodies, but allow air dissolved in the water to pass through. A versatile chemical, silicone, may soon allow man to do the same thing. A skin diver in a suit made of rubber laminated with a new, thin, silicone film would be protected from the effects of the water, but could inhale oxygen and expel carbon dioxide through the suit just like a fish. Scientists are experimenting with nonaquatic uses for the film; one of the most promising is the manufacture of oxygen tents. The structural formula for this film is shown as follows, where n is 10,000 to 40,000:

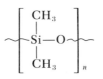

Glycol formation. A glycol, which is a dihydroxy alcohol, is formed when an oxidizing agent such as potassium permanganate ($KMnO_4$) is reacted with an alkene:

$$R-CH{=}CH_2 + KMnO_4 \longrightarrow \underset{\underset{\displaystyle OH \quad\; OH}{|\qquad |}}{RCH-CH_2}$$

              Alkene   **Glycol**

Following is an example:

$$CH_3CH=CH_2 \quad + \quad KMnO_4 \quad \longrightarrow \quad CH_3-\underset{OH}{\underset{|}{CH}}-\underset{OH}{\underset{|}{CH_2}} \quad + \quad MnO_2 \downarrow$$

| **Propene** | | **1,2-Dihydroxypropane** | (brown |
| (colorless) | (purple) | (colorless) | precipitate) |

Oxidation by $KMnO_4$ is the basis for the *Baeyer test for unsaturation.* The presence of a carbon–carbon double bond is indicated by conversion of the purple color of $KMnO_4$ to a brown precipitate of manganese dioxide (MnO_2) when a compound is added to the cold alkaline solution of $KMnO_4$. The observed reaction is not as clearly distinguishable to the unexperienced eye as is the decolorization of bromine, since the brown precipitate MnO_2 is obscured by the deep purple color of the unreduced $KMnO_4$.

Ozonolysis. This is the cleavage of an alkene molecule into two smaller molecules by the reagent ozone. The cleavage products are carbonyl compounds, aldehydes and ketones:

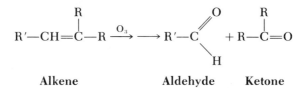

| **Alkene** | **Aldehyde** | **Ketone** |

Aldehydes and ketones are usually identified by being converted to compounds of known melting points. (The preparation and identification of compounds of known melting points is discussed in Chapter 9.) If the number and arrangement of the carbon atoms in the aldehydes and ketones are known, the structure of the original alkene and hence the position of the double bond can be determined. For example, from 1-pentene and 2-pentene, the following kinds of aldehydes are obtained:

$$H-\overset{\parallel}{\underset{O}{C}}-H + H-\overset{\parallel}{\underset{O}{C}}CH_2CH_2CH_3 \longleftarrow \overset{O_3}{\longleftarrow} CH_2=CHCH_2CH_2CH_3$$

| **Aldehyde** | **Aldehyde** | **1-Pentene** |

$$CH_3\overset{\parallel}{\underset{O}{C}}-H + H-\overset{\parallel}{\underset{O}{C}}CH_2CH_3 \longleftarrow \overset{O_3}{\longleftarrow} CH_3CH=CHCH_2CH_3$$

| **Aldehyde** | **Aldehyde** | **2-Pentene** |

3-4. What is the structure of the alkene that forms the following indicated carbonyl compounds?

(a) $\quad$ $CH_3CH_2\overset{\overset{O}{\parallel}}{C}-H + H\overset{\overset{O}{\parallel}}{C}H$ $\quad$ *Answer:* $CH_3CH_2CH=CH_2$

(b) $\quad\quad\quad\quad\quad\quad$ Two moles $CH_3CH_2CH_2\overset{\overset{O}{\parallel}}{C}-H$ $\quad$ *Answer:* $CH_3CH_2CH_2CH=CHCH_2CH_2CH_3$

(c) $CH_3C=O + CH_3CH_2C=O$
      | |
      CH_3 H

(d)
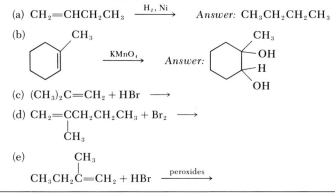

$$CH_3CH_2C=O + H-\overset{\displaystyle O}{\underset{\displaystyle}{\overset{\|}{C}}}-H$$
        |
        CH_3

3-5. Predict the addition products of the following reactions:

(a) $CH_2=CHCH_2CH_3 \xrightarrow{H_2,\ Ni}$ *Answer:* $CH_3CH_2CH_2CH_3$

(b)
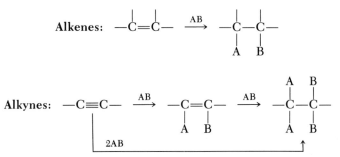

(c) $(CH_3)_2C=CH_2 + HBr \longrightarrow$

(d) $CH_2=CCH_2CH_2CH_3 + Br_2 \longrightarrow$
         |
         CH_3

(e)
              CH_3
              |
$CH_3CH_2C=CH_2 + HBr \xrightarrow{\text{peroxides}}$

Reaction of alkynes

Alkynes undergo addition reactions similar to those of alkenes. Whereas 1 mole of reagent is required to form a completely saturated molecule from an alkene, 2 moles of reagent is required to form such a molecule from an alkyne:

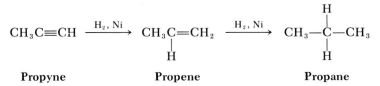

Addition reactions occur with the carbon–carbon triple bond for the same reason that they occur with the carbon–carbon double bond, namely, the availability of the loosely held π electrons.

Addition of hydrogen, halogen, and hydrogen halides

Alkynes react with hydrogen, halogens, and hydrogen halides as alkenes do. Following are some examples:

$$CH_3C\equiv CH \xrightarrow{H_2,\ Ni} CH_3C=CH_2 \xrightarrow{H_2,\ Ni} CH_3-\overset{\displaystyle H}{\underset{\displaystyle H}{\overset{|}{\underset{|}{C}}}}-CH_3$$
                                      |
                                 H

   Propyne   **Propene**   **Propane**

$$CH_3C{\equiv}CCH_3 \xrightarrow{\text{Br}_2/\text{CCl}_4}$$

CH₃—C=C—CH₃ with Br, Br $\xrightarrow{\text{Br}_2/\text{CCl}_4}$

2-Butyne   **2,3-Dibromo-2-butene**

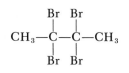

2,2,3,3-Tetrabromobutane

3-6. If you had two bottles of clear liquids and knew that one of them was 1-hexene and the other was 1-hexyne, could you distinguish them on the basis of the addition of 1 mole of bromine in carbon tetrachloride? Why or why not?

Answer: No. Each compound will decolorize at least 1 mole of bromine.

In the following example, the initial product after addition of 1 mole of HCl is a vinyl halide. The addition of each mole of HCl follows Markovnikov's rule.

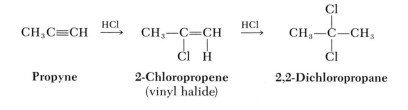

Propyne   **2-Chloropropene**   **2,2-Dichloropropane**
               (vinyl halide)

3-7. Predict the final product of the reaction when 1 mole of HI and then 1 mole of HBr is added to 1-pentyne.

Answer: 2-Bromo-2-iodopentane

Addition of water

The addition of water in the presence of mercuric sulfate and acid to the carbon–carbon triple bond forms compounds containing the carbonyl group (—C—). In the addition of water to acetylene, the product is acetaldehyde:
‖
O

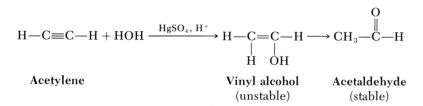

Acetylene   **Vinyl alcohol**   **Acetaldehyde**
                        (unstable)   **(stable)**

The vinyl alcohol formed initially is unstable, it cannot be isolated, and it rearranges to acetaldehyde. A structure with an —OH group directly bonded to a carbon–carbon double bond is called an *enol*, and a structure that contains a carbonyl group is called a *keto*. The equilibrium is very much in favor of the keto form. In simple terms, the enol form changes to the keto form by the migration of the hydroxyl hydrogen atom to the adjacent carbon atom of the carbon–carbon double bond. A pair of electrons from the double bond then shifts to the carbon–oxygen single bond, forming a carbon–oxygen double bond. This mechanism is represented as follows:

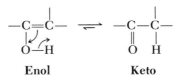

Enol   **Keto**

This kind of isomerism is called *keto-enol tautomerism*.

The addition of water to any other alkyne follows Markovnikov's rule and yields ketones. Following are some examples:

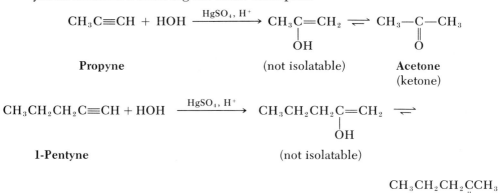

Reactions of terminal alkynes

In general, a hydrogen atom attached to a fluorine or chlorine atom is quite acidic. Thus both HF and HCl are strong acids. Both fluorine and chlorine possess the ability to attract electrons to them; as a result, the bond between hydrogen and fluorine and that between hydrogen and chlorine are polarized. The ability of an atom to attract electrons is called electronegativity. Thus fluorine and chlorine are electronegative atoms.

Carbon is only weakly electronegative. A hydrogen atom bonded to a carbon atom as in CH_4 is weakly acidic. On the other hand, a hydrogen atom attached to a triply bonded carbon atom is appreciably acidic. (A compound in which a hydrogen atom is bonded to one of the carbon atoms of the triple bond is a 1-alkyne, or terminal alkyne.) For example, acetylene reacts with sodium to form the compound sodium acetylide and liberates hydrogen:

$$H—C≡C—H + Na \longrightarrow H—C≡C:^- \ Na^+ + \tfrac{1}{2}H_2$$

Acetylene   **Sodium acetylide**

Moreover, acetylene is a stronger acid than ethene, which in turn is a stronger acid than ethane.

The acidity of acetylene or any 1-alkyne can be explained by the atomic orbital description of the carbon–hydrogen bonds. The more s character there is in a hybrid atomic orbital, the more that orbital approaches a spherical shape in which the electrons are relatively close to and more strongly held by the nucleus. The amount of s character in various hybrid atomic orbitals is as follows:

Orbital	Percent s character
sp^3	25
sp^2	33.3
sp	50

Since an sp hybrid orbital has more s character and thus the electrons are more strongly held by the nucleus, the carbon–hydrogen bond is weakened, and the hydrogen atom is relatively acidic.

Although acetylene is classified as an acid, it is a weaker acid than water. Water can liberate acetylene from salts of acetylene as follows:

$$H_2O + H—C\equiv C:^- Na^+ \longrightarrow H—C\equiv C—H + NaOH$$

Stronger acid   **Weaker acid**

1-Alkynes react readily with an alcoholic solution of silver nitrate ($AgNO_3$) to form a white precipitate. A nonterminal alkyne does not give this reaction. Therefore, the reaction constitutes a test for a terminal alkyne:

$$CH_3CH_2C\equiv CH + AgNO_3 \xrightarrow{\text{alcohol}} CH_3CH_2C\equiv C:^- Ag^+$$

1-Butyne   **White precipitate**
(terminal alkyne)

$$CH_3C\equiv CCH_3 \quad + \quad AgNO_3 \xrightarrow{\text{alcohol}} \quad \text{no reaction}$$

2-Butyne
(nonterminal alkyne)

3-8. Can you now distinguish between 1-hexene and 1-hexyne (see problem 3-6)?

The reaction between a sodium acetylide and a primary alkyl halide results in the formation of a higher alkyne:

$$CH_3C\equiv C:^- Na^+ + CH_3CH_2Br \longrightarrow CH_3C\equiv CCH_2CH_3$$

2-Pentyne

If the product alkyne is hydrogenated, an alkene or alkane can be obtained:

$$CH_3C\equiv CCH_2CH_3 \xrightarrow{\text{H}_2,\ \text{Ni}} CH_3CH=CHCH_2CH_3 \xrightarrow{\text{H}_2,\ \text{Ni}} CH_3CH_2CH_2CH_2CH_3$$

2-Pentyne   **2-Pentene**   **Pentane**

Summary

The halogenation of alkanes is a free-radical substitution reaction. The free radical, a carbon-containing reaction intermediate, contains a carbon

Table 3-3
Reactions of carbon-carbon double bond in $RCH{=}CH_2$

Reagent	Product
H_2, Ni	RCH_2CH_3, alkane
X_2 (Br_2 or Cl_2)	$\underset{\underset{X}{\|}}{R}CH\underset{\underset{X}{\|}}{C}H_2$, vicinal dihalide
HX (X = Cl, Br, or I)°	$\underset{\underset{X}{\|}}{R}CH\underset{\underset{H}{\|}}{C}H_2$, alkyl halide
$HOSO_3H$°	$\underset{\underset{OSO_3H}{\|}}{R}CH CH_3$, alkyl hydrogen sulfate
HOH, H^+°	$\underset{\underset{OH}{\|}}{R}CH\underset{\underset{H}{\|}}{C}H_2$, alcohol
HOX (X = Cl or Br)°	$\underset{\underset{OH}{\|}}{R}CH\underset{\underset{X}{\|}}{C}H_2$, halohydrin
$KMnO_4$	$\underset{\underset{OH}{\|}}{R}CH\underset{\underset{OH}{\|}}{C}H_2$, glycol
HBr, H_2O_2†	$\underset{\underset{H}{\|}}{R}CH\underset{\underset{Br}{\|}}{C}H_2$, alkyl bromide

°Markovnikov addition.
†Anti-Markovnikov addition.

atom with sp^2 hybrid orbitals. The order of stability and ease of formation of free radicals are the same as those of the carbonium ion: tertiary > secondary > primary > methyl radical.

Alkenes undergo electrophilic addition reactions. These reactions involve the intermediate formation of a carbonium ion. Alkanes also undergo the addition reactions of hydrogenation, glycol formation, polymerization, and ozonolysis, a method of determining the position of a carbon–carbon double bond in an alkene.

Alkynes undergo reactions similar to those of alkenes but consume 2 moles of hydrogen, halogen, or hydrogen halide. 1-Alkynes have appreciable acidity and are useful for preparing higher alkynes.

The characteristic reactions of the carbon–carbon double bond and the carbon–carbon triple bond are summarized in Tables 3-3 and 3-4.

Important terms

Addition reaction	Chain reaction
Baeyer test	Chain termination
Chain initiation	Chloroalkane
Chain propagation	Combustion

Table 3-4

Reactions of carbon-carbon triple bond in RC≡CH

Reagent	Product
H_2, Ni (1 mole)	$RCH{=}CH_2$, alkene
H_2, Ni (2 moles)	RCH_2CH_3, alkane
X_2 (Br_2 or Cl_2) (1 mole)	$RC{=}CH$ with X, X on the double bond carbons
X_2 (2 moles)	RCXCHX with X, X
HX (X = Cl, Br, or I) (1 mole)	$RC{=}CH_2$, vinyl halide, with X
HX (2 moles)°	$RCXCH_3$, with X
HOH ($HgSO_4$, H^+)°	$RC{=}CH_2$ (OH) $\rightleftharpoons$ $RCCH_3$ (O)
Na†	$RC{\equiv}C{:}^{\ominus}Na^{\oplus}$
$Ag^{\oplus}$†	$RC{\equiv}C{:}^{\ominus}Ag^{\oplus}$
Na,† then R′X‡	$RC{\equiv}CR′$

°Markovnikov addition.
†1-Alkynes only.
‡Primary alkyl halides only. R′ is some alkyl group other than R.

Important terms — cont'd

Electrophile
Electrophilic addition
Enol
Free radical
Free-radical addition
Glycol
Halogenation
Halohydrin
Homolytic cleavage
Hydrogenation

Markovnikov's rule
Nucleophile
Ozonolysis
Polymer
Polymerization
Substitution reaction
Symmetrical alkene
Tautomerism
Terminal alkyne
Vicinal dihalide

Problems

3-9. Give the organic product(s) of each of the following reactions. If there are several products, indicate the preferred product. Indicate if no reaction occurs.

(a) 3-Methylpentane + Br_2 + UV light $\longrightarrow$

(b) $CH_3CH{=}C(CH_3)_2 + Cl_2/CCl_4 \longrightarrow$

(c) $CH_3CH{=}CHCH_2CH_3 + H_2 \xrightarrow{Ni}$

(d)

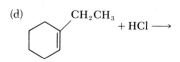

+ HCl $\longrightarrow$

(e) 2-Butyne + $AgNO_3$ $\xrightarrow{\text{alcohol}}$

(f) $CH_3CH_2CHCH_3$ + HOH $\longrightarrow$
    with OSO_3H on the CH

(g) $CH_3CH_2—C{\equiv}CH$ + HBr (1 mole) $\longrightarrow$

(h) Product in g + HI $\longrightarrow$

(i) 2-Methylpropene + $KMnO_4$ $\xrightarrow{\text{cold}}$

(j) 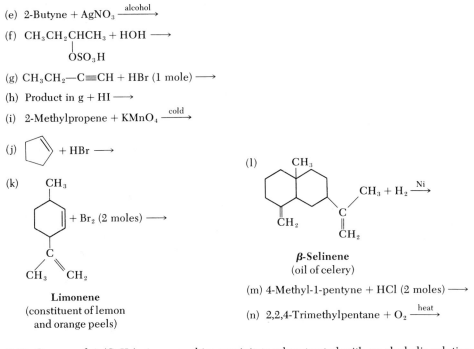 + HBr $\longrightarrow$

(k)

Limonene
(constituent of lemon
and orange peels)

+ Br_2 (2 moles) $\longrightarrow$

(l)

β-Selinene
(oil of celery)

CH_3 + H_2 $\xrightarrow{\text{Ni}}$

(m) 4-Methyl-1-pentyne + HCl (2 moles) $\longrightarrow$

(n) 2,2,4-Trimethylpentane + O_2 $\xrightarrow{\text{heat}}$

3-10. Compound A (C_4H_6) gives no white precipitate when treated with an alcoholic solution of silver nitrate. It consumes 2 moles of hydrogen to form butane. Identify A, and show the reactions involved.

3-11. Compound B (C_5H_{12}) does not decolorize Br_2 in CCl_4 and gives only one monochloro-substitution product, C. Identify B and C, and show the reactions involved.

3-12. An unknown naturally occurring compound is known to be composed of only carbon and hydrogen and consumes 3 moles of bromine. Which of the following structures best fits these observations?

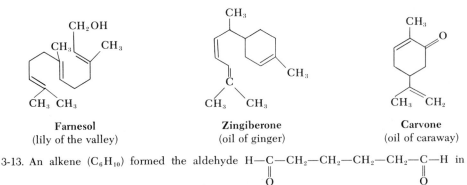

Farnesol
(lily of the valley)

Zingiberone
(oil of ginger)

Carvone
(oil of caraway)

3-13. An alkene (C_6H_{10}) formed the aldehyde $H—\underset{\underset{O}{\|}}{C}—CH_2—CH_2—CH_2—CH_2—\underset{\underset{O}{\|}}{C}—H$ in ozonolysis. What is the structure of the alkene?

3-14. Compare the reactions of 1-butene and 1-butyne with the following reagents:

(a) Br_2 (1 mole)
(b) Br_2 (2 moles)
(c) HCl (1 mole)
(d) HCl (2 moles)
(e) H_2 (1 mole) in the presence of Ni
(f) H_2 (2 moles) in the presence of Ni

3-15. Name the reagent that will cause each of the following transformations:

$$CH_3CH_3 \xrightarrow{(a)} CH_3CH_2Br \xrightarrow{(b)} CH_2{=}CH_2 \xrightarrow{(c)} \underset{\substack{| \quad\; | \\ Br \quad Br}}{CH_2{-}CH_2} \xrightarrow{(d)} HC{\equiv}CH$$

$$\downarrow (e)$$

$$\underset{\substack{| \\ Cl}}{CH_3CHBr} \xleftarrow{(f)} CH_2{=}CHBr$$

3-16. Name the reagent that will cause each of the following transformations:

$$H{-}C{\equiv}C{-}H \xrightarrow{(a)} H{-}C{\equiv}C{:}^-Na^+ \xrightarrow{(b)}$$

$$H{-}C{\equiv}C{-}CH_2CH_3 \xrightarrow{(c)} \underset{\substack{| \\ Cl}}{H_2C{=}C{-}CH_2CH_3}$$

$$\downarrow (d)$$

$$\underset{\substack{| \\ Cl}}{CH_3CHCH_2CH_3}$$

3-17. Show how the following conversions can be accomplished using the reactions of this chapter and the dehydration and dehydrohalogenation reactions in Chapter 2. The transformations outlined in problems 3-15 and 3-16 should be helpful.

(a) CH_3CH_3 to $CH_2{=}CH_2$

Answer: $CH_3CH_3 \xrightarrow[]{Br_2,\ UV\ light} CH_3CH_2Br \xrightarrow[CH_3CH_2OH]{KOH} CH_2{=}CH_2$

(b) Ethane to 1,2-dibromoethane

Answer: $CH_3CH_3 \xrightarrow[]{Br_2,\ UV\ light} CH_3CH_2Br \xrightarrow[CH_3CH_2OH]{KOH}$

$$CH_2{=}CH_2 \xrightarrow[CCl_4]{Br_2} CH_2BrCH_2Br$$

(c) Ethane to ethyl chloride
(d) Propane to isopropyl chloride
(e) 2-Butene to butane
(f) Cyclohexane to cyclohexyl iodide
(g) 1,2-Dibromopentane to $CH_3CH_2CH_2\underset{\substack{\| \\ O}}{C}CH_3$

(h) 1-Butyne to 2-pentyne
(i) $(CH_3)_3COH$ to 2-methylpropane
(j) 1-Butyne to 2-chloro-2-brombutane

3-18. The following phrases are used by organic chemists to designate the mechanisms of particular reactions: free-radical addition, free-radical substitution, electrophilic addition, electrophilic substitution, nucleophilic addition, and elimination. Use these to name the proper mechanism of each of the following reactions:

(a) Preparation of 2-bromobutane from 2-butene
(b) Formation of ethyl chloride from ethane
(c) Formation of 1-bromopropane from HBr in the presence of peroxides
(d) Preparation of cyclopentene from cyclopentyl bromide
(e) Decolorization of bromine by 1-hexene

Self-test

1. Draw the structure of the organic product of each of the following reactions:

(a) $CH_3CH{=}CH_2 + H_2$, Ni $\longrightarrow$

(b) $CH_3CH_3 + Cl_2$, ultraviolet light $\longrightarrow$

(c) $CH_3CH_2C{\equiv}CH + HI \longrightarrow$

(d) $CH_3CH=CHCH_3$ + cold $KMnO_4$ $\longrightarrow$

(e) $CH_3CH_2CH-CH_2$ + HBr, peroxides $\rangle$

(f) $CH_3C{\equiv}CH$ + 2 moles Br_2 $\longrightarrow$

(g)

CH_3 + HCl $\longrightarrow$

(h) CH_3C=CH_2 $\rightleftharpoons$
         |
        OH

2. Write the mechanism for the formation of *tert*-butyl chloride from 2-methylpropene and HCl.
3. Give the intermediate, product, and mechanism leading to the product for the halogenation of (a) ethene with bromine and carbon tetrachloride and (b) ethane with bromine and light.
4. Give the correct answer for each of the following questions.

   (a) What is the typical reaction of an alkene?
   (b) What is the common name of chloromethane?
   (c) What is the general name given to the class of compounds containing a halogen atom and a hydroxyl group positioned on adjacent carbon atoms?
   (d) How many monochlorinated products can be obtained from the free-radical chlorination of butane?
   (e) Of what value is ozonolysis of an alkene?
   (f) Draw the structure of the isopropyl free radical.
   (g) What is the structure of the monomeric unit of polystyrene?
   (h) What class of organic compounds is formed in the addition of hydrogen to an alkene?
   (i) Are 1-butene and 1-butyne isomers?
   (j) Draw the structure of the isobutyl cation.
   (k) Name one way in which the octane rating of gasoline can be improved.
   (l) What is the hybridization of carbon in the methyl free radical?

Answers to self-test

1. (a) $CH_3CH_2CH_3$ (hydrogenation)

   (b) CH_3CH_2Cl (free-radical substitution)

   (c) CH_3CH_2C=CH_2 (Markovnikov addition)
                  |
                  I

   (d) CH_3CH—CHCH_3 (glycol formation)
          | |
          OH   OH

   (e) $CH_3CH_2CH_2CH_2Br$ (free-radical addition)

   (f)   Br Br
            | |
        CH_3C—CH (Markovnikov addition)
            | |
            Br Br

   (g)   CH_3

                  Cl   (Markovnikov addition)

   (h) CH_3—C—CH_3 (keto-enol tautomerization)
            ‖
            O

          CH_3   CH_3
           | |
2. CH_3C=CH_2 + H$^\oplus$ $\longrightarrow$ CH_3C—CH_3
                                      $\oplus$

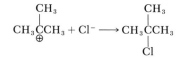

3.   *Intermediate*   *Product*   *Mechanism leading to product*

(a) 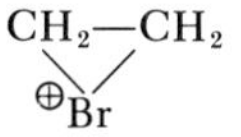 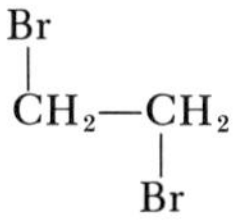 CH$_2$—CH$_2$ CH$_2$—CH$_2$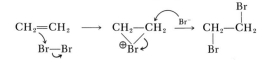

(b) CH$_3$CH$_2$·   CH$_3$CH$_2$Br

(1) Br$_2$ $\xrightarrow{\text{light}}$ 2Br·

(2) CH$_3$CH$_3$ + Br· ⟶ HBr + CH$_3$CH$_2$·

(3) CH$_3$CH$_2$· + Br$_2$ ⟶ CH$_3$CH$_2$Br + Br·

4. (a) Addition (electrophilic addition)
   (b) Methyl chloride
   (c) Halohydrin
   (d) Two: CH$_3$CH$_2$CH$_2$CH$_2$Cl and
        CH$_3$CH$_2$CHCH$_3$
                      |
                      Cl

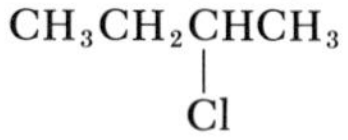

   (e) Location of the carbon–carbon double
        bond in an alkene

   (f) 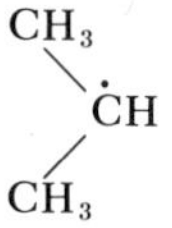 CH$_3$
          \
           ·CH
          /
        CH$_3$

   (g) 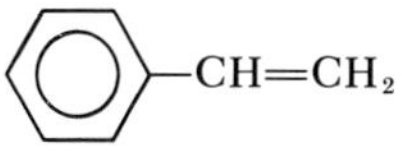

   ⬡—CH=CH$_2$

   (h) Alkane
   (i) No. They do not contain the same
        number of hydrogens.

   (j) 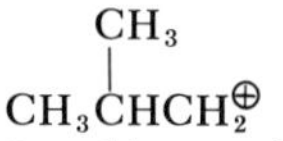   CH$_3$
               |
        CH$_3$CHCH$_2^{\oplus}$

   (k) By addition of tetraethyl lead or
        tricresyl phosphate or through
        platforming

   (l) sp^2

chapter 4

Aromatic hydrocarbons

Aromatic hydrocarbons are ring hydrocarbons. They are different from the other hydrocarbons that we have studied. They do not undergo the electrophilic addition or free-radical substitution and addition reactions characteristic of alkanes, alkenes, and alkynes. Historically, this class of compounds was referred to as aromatic because they possessed a pleasant fragrance. Today, the name aromatic hydrocarbons refers to the class of organic compounds with the common structural feature of one or more six-carbon structures related to benzene. Thus a study of the chemistry of aromatic compounds must begin with a study of benzene.

Structure of benzene

The molecular formula of benzene is C_6H_6. Benzene has fewer hydrogens than either a hexene (C_6H_{12}) or a hexyne (C_6H_{10}) and therefore is expected to be unsaturated. However, benzene does not appear to have the reactivity possessed by a hexene or hexyne. Benzene does not decolorize bromine as do hexene and hexyne. Furthermore, unlike hexene and hexyne, benzene is not easily oxidized by potassium permanganate.

The six carbon atoms of benzene are known to be in a ring, since benzene can be hydrogenated to form cyclohexane:

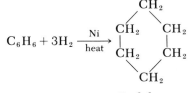

Cyclohexane

It can be seen that there is only one hydrogen atom attached to each carbon atom of benzene, since 1,3-cyclohexadiene can be dehydrogenated to form benzene:

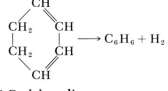

1,3-Cyclohexadiene

Furthermore, all the hydrogen atoms in benzene are equivalent, since the substitution of a bromine for one hydrogen in benzene forms only one bromobenzene:

$$C_6H_6 \quad + \quad Br_2 \quad \xrightarrow{\text{FeBr}_3} \quad C_6H_5Br \quad + \quad HBr$$

Bromobenzene

To account for these observations, it was proposed that benzene has a cyclic hexagonal structure of six carbon atoms with alternate double and single bonds with one hydrogen atom attached to each carbon atom. Furthermore, it was suggested that the double and single bonds are rapidly alternating positions so that each carbon–carbon bond is neither single nor double but something intermediate. In 1865 August Kekulé proposed the following structures, which are thus known as the Kekulé structures:

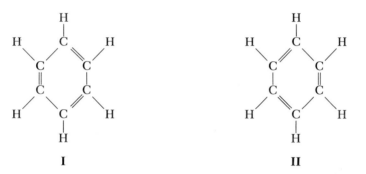

I   II

Structures I and II differ only in the positions of electrons. Structures that are similar to one another but differ from one another in the position of electrons are said to be in *resonance* and are referred to as resonance structures. Thus benzene, which is neither of the Kekulé structures, is a resonance hybrid of the two structures.

The resonance hybrid nature of benzene imparts to it the important characteristic of a high degree of stability. This unusual stability is manifested by the low heat of hydrogenation and the tendency to undergo substitution reactions rather than addition reactions. The heat of hydrogenation is the amount of heat evolved when an alkene reacts with 1 mole of hydrogen in the presence of a nickel catalyst. For an alkene with one carbon–carbon double bond, the heat of hydrogenation is about 29 kcal. For an alkene with two carbon–carbon double bonds (a diene), the heat of hydrogenation is about twice this value, or 58 kcal. If benzene is considered to have three carbon–carbon double bonds, the heat of hydrogenation would be expected to be about 87 kcal (3×29 kcal). In fact, the experimental heat of hydrogenation is approximately 51 kcal, 36 kcal less than the expected amount. This means that benzene evolves 36 kcal less energy than predicted and therefore contains 36 kcal less energy and is more stable by 36 kcal than would be expected for a cyclic compound with three carbon–carbon double bonds. This 36 kcal represents the extra stabilization of benzene due to resonance and is referred to as the resonance stabilization energy of benzene.

It is known that all the carbon–carbon bond lengths in benzene are identical. The bond distance of 1.40 Å is intermediate between the carbon–carbon single-bond distance (1.54 Å) and the carbon–carbon double-bond distance (1.34 Å).

In benzene, each carbon atom is bonded to three other atoms (two carbons and one hydrogen) and therefore has sp^2 hybrid orbitals. If the carbons are arranged to permit maximum overlap of the sp^2 hybrid orbitals, all the atoms will lie in the same plane. Thus benzene is a flat or planar molecule. Furthermore, each carbon in benzene has one p orbital with one electron in it. The p orbital of any one carbon overlaps equally well the p orbitals of both carbon atoms to which it is bonded. The result is two continuous doughnut-shaped electron clouds, one lying above and the other lying below the plane of the atoms. There are twelve σ bonds and one π cloud containing six electrons. Six of the σ bonds are formed by the overlap of sp^2-sp^2 orbitals. All the bond angles are 120°. Fig. 4-1 illustrates the molecular orbital representation of bonding in benzene.

Since the six π electrons of benzene are in an electron cloud covering all six carbon atoms, they are said to be delocalized. The result of delocalization is the formation of a stronger bond and a more stable molecule. The difference between delocalized electrons and localized electrons can be illustrated by benzene and 1,4-pentadiene:

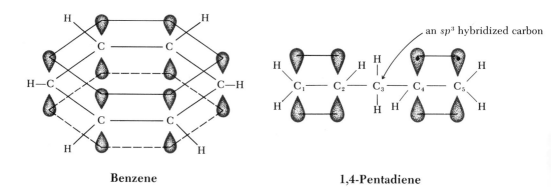

Benzene   **1,4-Pentadiene**

In benzene, a p orbital overlaps another p orbital of both carbon atoms to which it is bonded. However, in 1,4-pentadiene, the p orbitals overlap only the p orbitals of the carbon atoms with sp^2 hybrid orbitals, which are C1, C2, C4, and C5. The lack of a p orbital on the carbon with sp^3 orbitals prevents the p orbital of C2 from overlapping the p orbital of C4.

The reactions of benzene are those in which the stability of the ring is preserved, namely, substitution reactions. Addition destroys the stability of the ring because it destroys the complete delocalization of the π electrons of the ring. The π electrons are still delocalized but only over four carbon atoms:

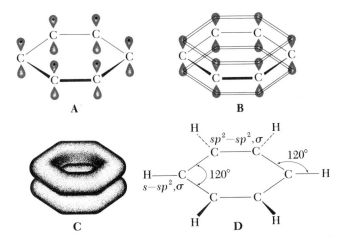

Fig. 4-1

Molecular orbital representation for benzene. **A,** Cyclic arrangment of six sp^2 hybrid orbitals of carbon atoms, each with a p orbital containing one electron. **B,** Overlap of p orbitals above and below plane of the carbon atoms. **C,** Electron distribution in doughnut-shaped clouds. **D,** Carbon–carbon and carbon–hydrogen σ bonds. All bond angles are 120°.

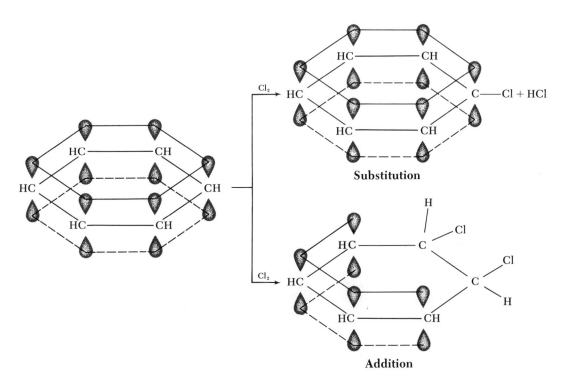

In summary, benzene and aromatic compounds in general possess a planar cyclic structure, have unusual stability because of delocalization of π electrons, and undergo substitution rather than addition reactions.

Formula representation of benzene

The benzene molecule is not generally represented in the form of the two Kekulé structures. Instead, it is represented as a regular hexagon containing a circle. It is understood that a hydrogen atom is attached to each angle of the hexagon unless another group or atom is indicated. Thus indicates a hybrid of and . The circle stands for the cloud of six delocalized π electrons.

4-1. Compare and contrast the following properties of benzene and the molecule 1,3-hexadiyne (H—C≡C—C≡C—CH$_2$—CH$_3$):

(a) Molecular formula   *Answer:* C$_6$H$_6$ for each molecule
(b) Positive or negative reaction with Br$_2$ in CCl$_4$ *Answer:* Positive with 1,3-hexadiyne
(c) Positive or negative reaction with cold KMnO$_4$
(d) Product of complete hydrogenation
(e) Predicted heat of hydrogenation   *Answer:* For 1,3-hexadiyne: four π bonds × 29 kcal/mole
(f) Number of monosubstitution products in bromination
(g) Number of π electrons
(h) Number of carbon atoms covered by π cloud of electrons
On the basis of your comparison, could the 1,3-hexadiyne molecule be benzene? Explain your answer.

Nomenclature of benzene derivatives

Derivatives of benzene are named according to the following rules:
1. For monosubstitution products of benzenes, the name of the substituent group is prefixed to the word *benzene.* Following are some examples:

Cl
Chlorobenzene

 NO$_2$
Nitrobenzene

Br

Bromobenzene

Some derivatives of benzene have special names:

CH$_3$	OH	COOH	NH$_2$	SO$_3$H
Toluene (methyl-benzene)	Phenol (hydroxy-benzene)	Benzoic acid (carboxy-benzene)	Aniline (amino-benzene)	Benzenesulfonic acid

2. A disubstitution product of benzene can exist in one of three isomeric forms and can be named by either of two methods. If two substituents are in the 1 and 2 positions, they are also *ortho (o-)* to each other; if they are in the 1 and 3 positions, they are *meta (m-)* to each other; if they are in the 1 and 4 positions, they are *para (p-)* to each other:

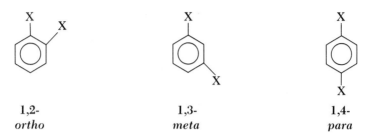

1,2-	1,3-	1,4-
ortho	*meta*	*para*

The two substituents do not have to be identical. If one of the substituents gives a special name to a derivative, the compound is named as a derivative of the compound with the special name, and the carbon bearing the group that gives the special name is carbon 1. Following are some examples:

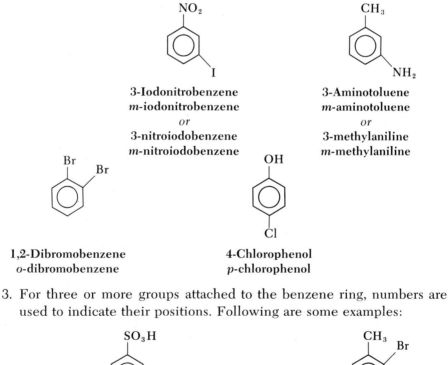

3-Iodonitrobenzene
m-iodonitrobenzene
or
3-nitroiodobenzene
m-nitroiodobenzene

3-Aminotoluene
m-aminotoluene
or
3-methylaniline
m-methylaniline

4-Chlorophenol
p-chlorophenol

1,2-Dibromobenzene
o-dibromobenzene

3. For three or more groups attached to the benzene ring, numbers are used to indicate their positions. Following are some examples:

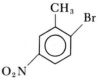

3,5-Dichlorobenzenesulfonic acid

2-Bromo-5-nitrotoluene
not
3-nitro-6-bromotoluene

2,3,6-Triiodoaniline

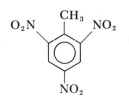

2,4,6-Trinitrotoluene (TNT)
(powerful explosive)

4-2. Draw structures corresponding to the following compounds:

(a) *p*-Iodoaniline   *Answer:* I—⬡—NH₂

(b) 2-Nitro-4-chlorobenzoic acid   *Answer:*

(c) 2,6-Dibromotoluene
(d) 3-Bromoiodobenzene
(e) *o*-Nitrophenol

4-3. What is objectionable about the name given to the following compound?

3-Nitrochlorotoluene

4-4. Give a correct name for each of the following compounds:

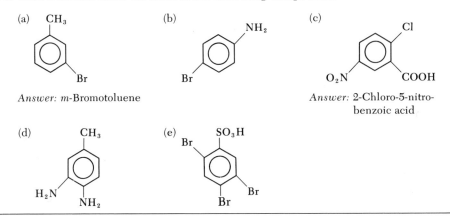

(a) *Answer:* m-Bromotoluene

(c) *Answer:* 2-Chloro-5-nitro-
benzoic acid

Reactions of benzene

The π electrons of benzene are available to a reagent that is seeking electrons. Thus the benzene ring serves as a source of electrons (a Lewis base), reacting with compounds that are deficient in electrons (Lewis acids or electrophilic reagents). Furthermore, benzene undergoes substitution rather than addition reactions, since substitution preserves the stability of the ring. Thus the characteristic reaction of substances containing the benzene ring is *electrophilic substitution,* some types of which are nitration, halogenation, sulfonation, and Friedel-Crafts alkylation.

Nitration. This reaction, which is generally carried out in a mixture of nitric acid and sulfuric acid, introduces a nitro group into the ring. Nitric acid, HNO_3, is sometimes written $HONO_2$. For example:

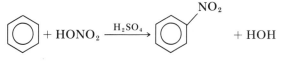

Nitrobenzene

Halogenation. This is the substitution of hydrogen by halogen, which occurs in the presence of Lewis acids such as ferric bromide ($FeBr_3$) and ferric chloride ($FeCl_3$). Fluorine and iodine do not act as reagents in this substitution reaction. Following are examples of this reaction:

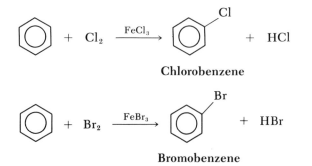

Chlorobenzene

Bromobenzene

Sulfonation. This reaction, for which sulfuric acid containing an excess of sulfur trioxide is the reagent, introduces the $-SO_3H$ group into the ring. For example:

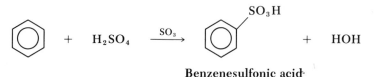

Benzenesulfonic acid

Friedel-Crafts alkylation. The product of this reaction, an alkylbenzene, is formed by the reaction of benzene and an alkyl halide (RX) in the presence of aluminum chloride ($AlCl_3$). RX is generally a chloride, bromide, or iodide. Following is an example of this reaction:

Alkylbenzene

Mechanism of substitution

The substitution reactions of benzene proceed by the same mechanism regardless of the reagent involved. If we assume that the reagent is ENu, the mechanism can be summarized as follows:

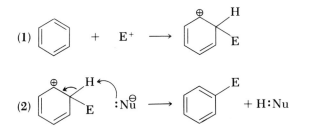

Step 1 involves the attack by the electrophile (E^+) to form a carbonium ion. The π electron cloud supplies the electrons by which the electrophile bonds to a ring carbon. In step 2, the base ($:N\overset{\ominus}{u}$) abstracts a proton from the carbonium ion.

Three structures can be written for the carbonium ion formed in step 1:

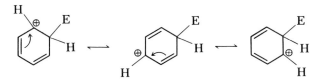

These structures differ from one another only in the position of the π electrons and the positive charge. They are resonance structures. The actual carbonium ion is a resonance hybrid of these three structures and is repre-

sented as ⬡. The positive charge is not localized on one carbon atom but is distributed over the molecule (delocalized). Delocalization of the positive charge makes the ion more stable than an ion with a localized positive charge.

The energy profile for electrophilic aromatic substitution is shown in Fig. 4-2. In step 1 of the mechanism, most of the resonance stabilization energy of benzene is lost in formation of the intermediate carbonium ion. The energy of the intermediate is high compared to that of either the reactants or products. The formation of the intermediate requires a relatively large activation energy. However, only a small amount of activation energy must be supplied in step 2, in which the nucleophile abstracts a proton from the intermediate. Consequently, the formation of the intermediate is the slow, or rate-determining, step of electrophilic aromatic substitution.

In nitration, the electrophilic reagent is the nitronium ion (NO_2^+) which is generated according to the following reaction:

$$HONO_2 + 2H_2SO_4 \longrightarrow H_3O^+ + 2HSO_4^- + NO_2^+$$

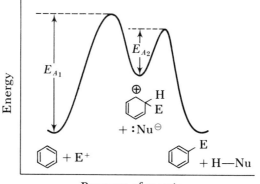

Fig. 4-2
Energy profile for electrophilic substitution in benzene.

The electrophilic reagent in halogenation is either the Lewis acid complex or a positive halogen ion. With chlorination as an example, these reagents are formed as follows:

$$Cl_2 \; + \; FeCl_3 \; \longrightarrow \; Cl{-}\overset{\oplus}{Cl}{-}\overset{\ominus}{Fe}Cl_3 \; \rightleftharpoons \; Cl^+ \; + \; FeCl_4^-$$

Lewis acid	Positive
complex	halogen

In the case of the Lewis acid complex, chlorine is transferred, without its electrons, to the ring.

The attacking electrophile in sulfonation is sulfur trioxide (SO_3), which is formed from two molecules of sulfuric acid as follows:

$$2H_2SO_4 \longrightarrow H_3O^+ + HSO_4^- + SO_3$$

The structure of sulfur trioxide is $\overset{\displaystyle :\ddot{O}:}{\underset{\displaystyle :\ddot{O}:}{S{-}\ddot{O}:}}$. Thus the sulfur atom can act as an electrophile, or Lewis acid, and accept a pair of electrons from the benzene ring.

In the Friedel-Crafts alkylation, the electrophilic reagent is a carbonium ion (R^+), which is formed as follows:

$$RX + AlX_3 \longrightarrow AlX_4^- + R^+$$

In this reaction, rearrangement to a more stable carbonium ion can occur. Thus the Friedel-Crafts alkylation can give rise to a product mixture when the initially formed carbonium ion is not the more stable ion. For example, the reaction between benzene and n-propyl chloride in the presence of $AlCl_3$ gives a mixture of n-propylbenzene and isopropylbenzene:

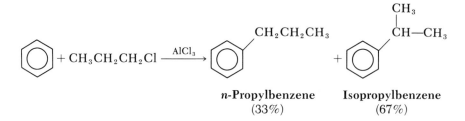

The isopropylbenzene arises from attack of the benzene ring by isopropyl cation, which is formed by rearrangement of the *n*-propyl cation. Furthermore, the amounts of products formed seem to indicate that more *n*-propyl cations rearrange than attack the benzene ring. This mechanism can be represented as follows:

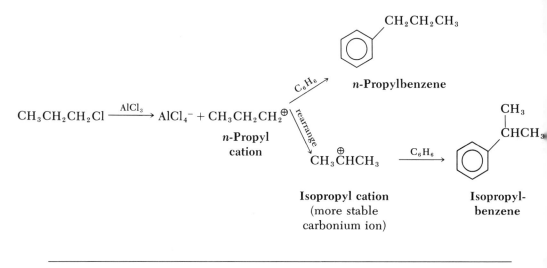

4-5. Predict the products of the reaction between benzene and *n*-butyl chloride in the presence of aluminum chloride. Which is the major product?

Reactions of the Friedel-Crafts type are also likely to occur between benzene and carbonium ion precursors such as the carbonium ions formed in the acid-catalyzed dehydrations of alcohols or the electrophilic addition reactions of hydrogen halides to alkenes (see problem 4-16).

Electrophilic substitution in benzene derivatives

Electrophilic aromatic substitution reactions are characteristic of benzene and the benzene ring wherever it is found. For example, both phenol and nitrobenzene undergo electrophilic substitution reactions. However, they differ in that they undergo electrophilic substitution at different rates than benzene does. Furthermore, phenol and nitrobenzene produce a mixture of isomeric monosubstitution products in electrophilic substitution, but usually one or two of the isomers are preferentially formed.

Table 4-1

Orientation of substituents on the benzene ring

Ortho,para directors, activating	Ortho,para directors, deactivating	Meta-directors, deactivating
—CH$_3$ (any alkyl group)	—B̈r:	—C—H or —CHO (aldehyde) ∥ O
—C$_6$H$_5$ (phenyl)	—C̈l:	—C≡N or —CN (cyano)
—ÖCH$_3$ (methoxy)	—F̈:	—C—OH or —COOH (carboxyl) ∥ O
—ÖH (hydroxyl)	—Ï:	—C—R or —COR (keto) ∥ O
—N̈H$_2$ (amino)		—N—O or —NO$_2$ (nitro) ∥ O
—N̈HCOCH$_3$ (acetamido)		—SO$_3$H (sulfo)

Orientation and reactivity

The group or substituent on the ring affects the rate at which substitution occurs and determines the position of substitution. The ability of a substituent or group already on the benzene ring to direct an incoming second group is called the *orienting effect* of the substituent or group on the ring. A substituent that directs an incoming second group to the *ortho* and *para* positions is called an *ortho,para director*. A substituent that directs an incoming second group to the *meta* position is called a *meta director*. A substituent that causes substitution to occur faster than the same substitution in benzene is called an *activating substituent*. On the other hand, a *deactivating substituent* causes substitution to occur more slowly than the same substitution in benzene. The orienting effects and reactivities of some common substituents are listed in Table 4-1. A characteristic feature of all *meta* directors is that the atom bonded to the ring carbon has a multiple bond to another atom. A characteristic of *most ortho, para* directors is that the atom bonded to the ring carbon has at least one unshared pair of electrons.

Following are reactions illustrating the orienting effect of substituents. The methyl group is activating and an *ortho,para* director. Thus sulfonation of toluene occurs faster than the sulfonation of benzene:

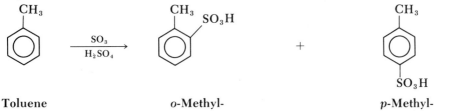

Toluene

o-**Methyl-benzenesulfonic acid**

p-**Methyl-benzenesulfonic acid**

The nitro group is deactivating and a *meta* director. Thus chlorination of nitrobenzene proceeds more slowly than the chlorination of benzene:

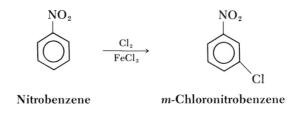

The bromo group is deactivating and an *ortho,para* director. Thus nitration of bromobenzene proceeds more slowly than the nitration of benzene:

4-6. What is the product of each of the following reactions? Which reactions go faster than the same reaction with benzene?

(a) Bromination of anisole ($C_6H_5OCH_3$)   *Answer:* o- and p-Bromoanisole; faster because —OCH_3 is an activating group

(b) Sulfonation of benzonitrile (C_6H_5CN)   *Answer:* m-Cyanobenzenesulfonic acid; slower because —CN is a deactivating group

(c) Chlorination of benzaldehyde (C_6H_5CHO)
(d) Nitration of iodobenzene
(e) Alkylation by ethyl bromide of toluene

4-7. Which of the following is (are) the correct product(s) of the given reaction? In other words, does —COOH or —Br determine the position of orientation?

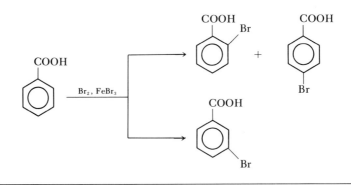

Consider the nitration reactions of phenol and nitrobenzene:

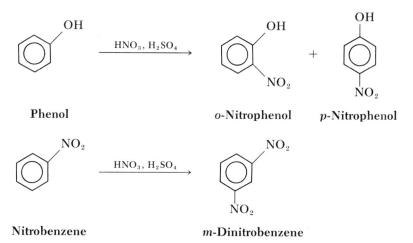

Phenol   o-Nitrophenol   p-Nitrophenol

Nitrobenzene   m-Dinitrobenzene

Phenol is nitrated faster than benzene; this means that the π electrons are more available for bonding to occur. Thus the hydroxyl group must release electrons to the ring. On the other hand, nitrobenzene undergoes nitration more slowly than benzene; this means that the π electrons are less available for bonding than those in benzene. The nitro group then must withdraw electrons from the ring. In other words, the hydroxyl group, an activating group, activates all the positions of the ring including the *meta* position. It directs *ortho* and *para* because it activates these positions more than it does the *meta* position. The nitro group, a deactivating group, deactivates *all* the positions in the ring, including the *meta* position. It directs *meta* because it deactivates the *ortho* and *para* positions more than it does the *meta* position.

For a better understanding of this concept, consider the ions formed in nitration of phenol by attack at all three positions and those formed in nitration of benzene:

Phenol (*ortho* attack)

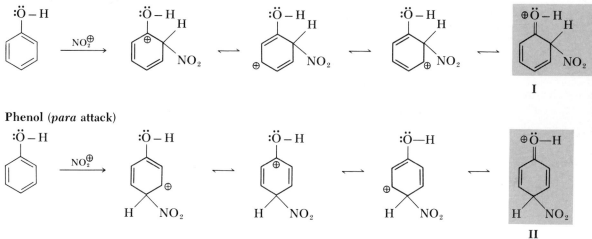

Phenol (*meta* attack)

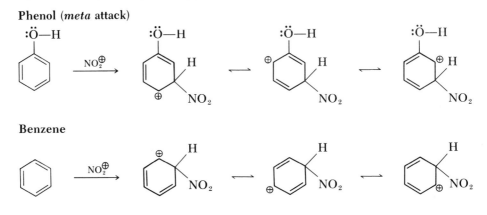

Benzene

Structures I and II, which result from attack at the *ortho* and *para* positions, are particularly stable. Their stability results from the complete octet of electrons of every atom (except hydrogen). They are more stable than any of the ions formed from *meta* attack on phenol or the carbonium ions formed from attack on benzene itself. Thus substitution in phenol occurs faster than substitution in benzene and occurs predominantly at the *ortho* and *para* positions to the hydroxyl group.

Now consider the ions formed in nitration of nitrobenzene by attack at all three positions. It is helpful to write the structure for the nitro group as

Nitrobenzene (*ortho* attack)

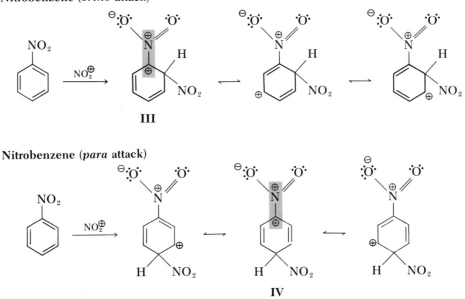

Nitrobenzene (*meta* attack)

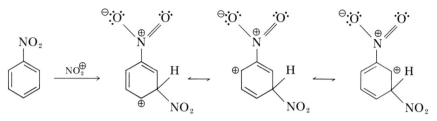

In structures III and IV, the positive charges are located on adjacent atoms. These structures are of high energy and are less stable than any of the other structures. In other words, the nitro group, which is an electron-withdrawing

group , withdraws electrons most effectively from the carbon atom

nearest it; since in structures III and IV this carbon is already positive, it has little tendency to accommodate the positive charge. Thus substitution in nitrobenzene occurs more slowly at the *ortho* and *para* positions than at the *meta* position.

In considering stabilities of the ionic structures of other *meta*-directing groups, we can use the following resonance structures in interpreting substitution in benzene compounds containing these *meta* directors:

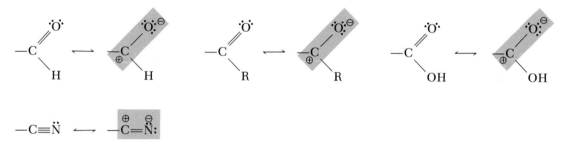

4-8. Using resonance structures, predict the point of substitution in the nitration of

anilinium chloride $\left(\begin{array}{c} \text{NH}_3^+\text{Cl}^- \\ \bigcirc \end{array} \right)$

Answer: The important resonance structures are as follows:

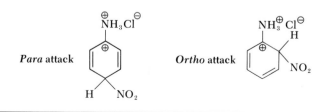

Orientation in disubstitution products of benzenes

The problem of orientation in a disubstitution product of benzene can be complicated, but the following rules can be successfully used for predicting the orientation:

1. The orientation can be easily predicted (see arrows in the following examples) if the two substituents are located so that the directive influence of one substituent is reinforced by the other. Following are some examples:

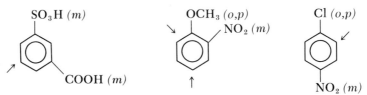

2. When the directive effect of one group opposes another, then activating groups win out over deactivating groups; that is, *ortho,para* activators predominate over *ortho,para* deactivators, and the later predominate over *meta* deactivators. Following are some examples:

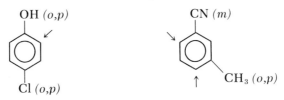

3. Often there is little substitution between two groups that are *meta* to each other. For example, in the nitration of *m*-bromochlorobenzene, there is less than 1% substitution at the indicated position:

This is undoubtedly due to the lack of room (steric factor) between the two *meta* substituents resulting from their size or the size of the substituting group, which in this case is the bulky nitro group.

4-9. Predict the point(s) of substitution in each of the following compounds:

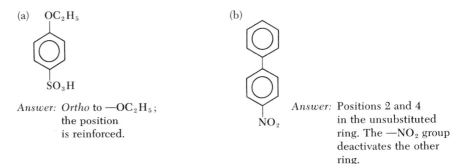

(a) OC_2H_5

SO_3H

Answer: Ortho to —OC_2H_5;
   the position
   is reinforced.

(b)

NO_2

Answer: Positions 2 and 4
   in the unsubstituted
   ring. The —NO_2 group
   deactivates the other
   ring.

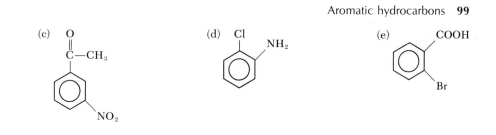

(c) benzene ring with C(=O)—CH₃ group and NO₂ substituent

(d) benzene ring with Cl and NH₂ substituents

(e) benzene ring with COOH and Br substituents

Orientation and synthesis

The concepts of orientation in disubstitution products of benzene can be useful in a laboratory synthesis of compounds. It is generally desirable to prepare a single pure compound. In order to do this, it is necessary to consider the order in which the various substituents are introduced into the ring. For example, it must be recalled in the preparation of *m*-nitrochlorobenzene that the chloro group is an *ortho,para* director and the nitro group is a *meta* director. Since it is desired that the substituents be *meta* to one another in the product, the correct order of introduction is nitro first and chloro second:

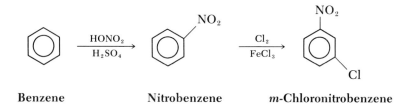

Benzene   **Nitrobenzene**   **m-Chloronitrobenzene**

If the substitutents were introduced in the reverse order, the product would be a mixture of *o*- and *p*-nitrochlorobenzene:

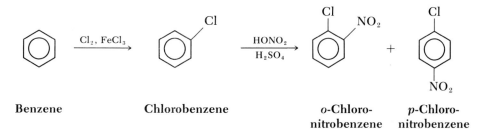

Benzene   **Chlorobenzene**   **o-Chloro-**   **p-Chloro-**
                                                       nitrobenzene   **nitrobenzene**

If a synthesis results in a product mixture and one isomer is desired, usually the desired isomer can be separated on the basis of boiling point (distillation) or solubility (fractional recrystallization). In the previous example, if *p*-nitrochlorobenzene were the desired product, it could be separated from the *ortho* isomer on the basis of being less soluble in the recrystallizing solvent.

Sometimes a synthesis involves conversion of one group into another, so that the proper time for the conversion must be considered. For example, a common reaction undergone by benzenes with alkyl substitution groups is oxidation by potassium permanganate or potassium dichromate ($K_2Cr_2O_7$):

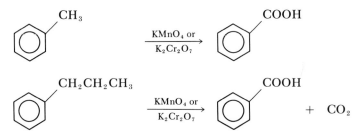

In the oxidation of an alkylbenzene, all the carbons of the alkyl group are lost as carbon dioxide except the one directly attached to the ring, which is converted to the carboxyl group (—COOH). The reaction in effect converts an *ortho,para* director to a *meta* director. Thus in the synthesis of *p*-chlorobenzoic acid from toluene, it is best to carry out oxidation after chlorination has been accomplished as follows:

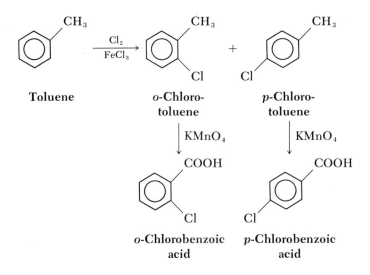

The desired *para* isomer can be separated from the *ortho* isomer on the basis of solubility.

4-10. Indicate the steps in the following synthetic reactions:

(a) Benzene ⟶ *o*-chlorotoluene

Answer: Friedel-Crafts alkylation with CH_3Cl and then chlorination with Cl_2 and $FeCl_3$

(b) Benzene ⟶ *m*-nitrobenzenesulfonic acid

Answer: Sulfonation of benzene followed by nitration of the benzenesulfonic acid (the substitution reactions can be reversed)

(c) Toluene ⟶ *m*-nitrobenzoic acid
(d) Benzene ⟶ *p*-nitrotoluene
(e) Toluene ⟶ 1,4-dicarboxybenzene [$C_6H_4(COOH)_2$], terephthalic acid

Other important derivatives of benzene

Arenes

Many important hydrocarbon compounds contain both an aliphatic and an aromatic unit. They are known as *arenes*. The aromatic unit can be attached to alkanes (alkylbenzenes), alkenes, or alkynes or to other aromatic units. Following are some examples:

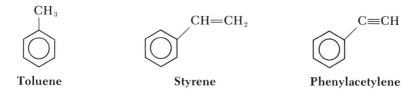

Toluene   Styrene   Phenylacetylene

In the names of alkylbenzenes, the name of the alkyl group is prefixed to the word *benzene*. Following are some examples:

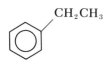   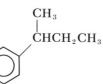

Ethylbenzene   *sec*-Butylbenzene   1,3-Dimethylbenzene
(*m*-xylene)

If the compound contains a complicated side chain, then the ring is named as a phenyl substituent or C_6H_5— :

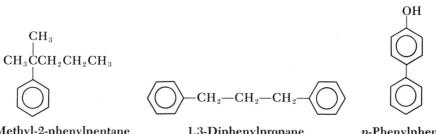

2-Methyl-2-phenylpentane   1,3-Diphenylpropane   *p*-Phenylphenol

Compounds in which the side chain contains a double or triple bond are named as alkenes or alkynes with substituent groups. The simplest aromatic-alkene compound is often called styrene. Following are some examples:

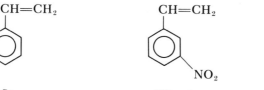 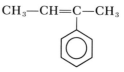

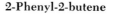

Styrene   *m*-Nitrostyrene   2-Phenyl-2-butene

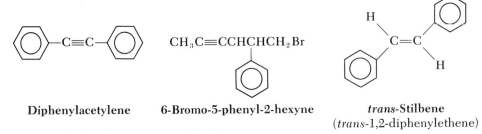

Diphenylacetylene   6-Bromo-5-phenyl-2-hexyne   *trans*-Stilbene
(*trans*-1,2-diphenylethene)

Free-radical substitution in alkylbenzenes

Alkylbenzenes undergo free-radical substitution in the aliphatic side chain in addition to electrophilic ring substitution and oxidation. This is not unexpected, since these alkylbenzenes can be considered to be substituted alkanes. The necessary conditions for free-radical substitution in the side chain are the reagent halogen and high temperature or ultraviolet light, whereas the necessary conditions for ring substitution are the reagent halogen and a Lewis acid catalyst, such as Fe or $FeBr_3$. These reactions are illustrated for toluene as follows:

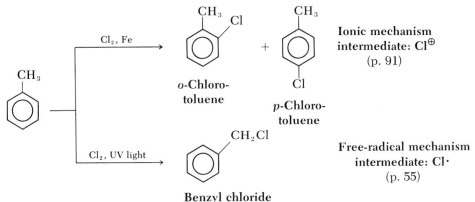

o-Chloro-
toluene

p-Chloro-
toluene

Ionic mechanism
intermediate: $Cl^{\oplus}$
(p. 91)

Benzyl chloride

Free-radical mechanism
intermediate: $Cl\cdot$
(p. 55)

With toluene, the position for attack is obvious; but with a more complicated side chain than methyl, it is possible to obtain a mixture of isomers. For example, if ethylbenzene is chlorinated with ultraviolet light, the two possible products are 1-chloro-1-phenylethane and 2-chloro-1-phenylethane:

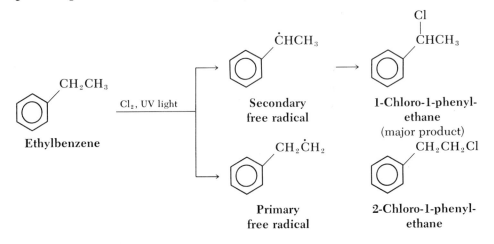

Ethylbenzene

Secondary
free radical

1-Chloro-1-phenyl-
ethane
(major product)

Primary
free radical

2-Chloro-1-phenyl-
ethane

Actually, only 1-chloro-1-phenylethane is formed. This must mean that the intermediate free radical from which it is formed is more stable and that the abstraction of hydrogen atoms that gives the radical is easier than that giving rise to the other free radical. The fact that secondary hydrogens are more easily abstracted than primary hydrogens in part explains why more 1-chloro-1-phenylethane is formed. However, we must look more closely at the free radical formed from abstraction of one of these secondary hydrogens to see what makes these secondary hydrogens so unique that they are so easily abstracted and no 2-chloro-1-phenylethane is formed. The secondary free radical is actually stabilized by resonance, that is, by delocalization of the odd electron over the ring carbons:

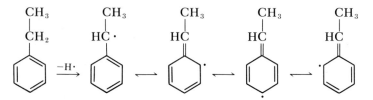

Such structures are not possible for the primary free radical. Thus the only product is 1-phenyl-1-chloroethane.

Hydrogen atoms attached to a carbon atom joined directly to an aromatic ring are called *benzylic hydrogens.* These hydrogens are easier to abstract than tertiary hydrogens because of the resulting stabilization of the radical through delocalization of the odd electron (resonance). The benzyl free radical has the following structure:

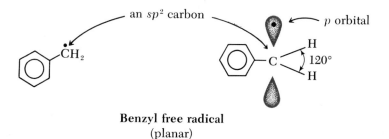

Benzyl free radical
(planar)

It should not be confused with the benzyl group, which has the following structure:

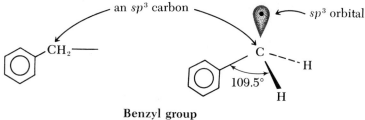

Benzyl group
(tetrahedral)

4-11. Give a correct IUPAC name for each of the following compounds:

(a) CH_3—CH_2—CH—CH_2—CH_3
　　　　　　　　　|
　　　　　　　　CH_2

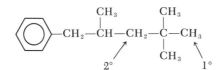

(c)
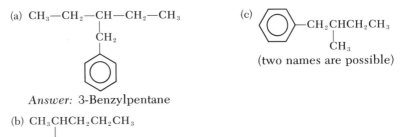

(two names are possible)

Answer: 3-Benzylpentane

(b) $CH_3CHCH_2CH_2CH_3$

4-12. Label the hydrogens in the following molecule in order of ease of abstraction:

Answer: Benzylic > tertiary > secondary > primary hydrogens

Polynuclear aromatic hydrocarbons

Some aromatic compounds contain two or more aromatic rings and share a pair of carbon atoms. Such rings are said to be fused, and fused-ring hydrocarbons are called *polynuclear aromatic hydrocarbons*. Some examples are as follows:

Naphthalene　　**1,4-Dinitronaphthalene**　　**Anthracene**　　**Phenanthrene**

Naphthalene is the simplest polynuclear aromatic compound. It contains two flat six-membered rings, each of which contains a delocalized cloud of six π electrons:

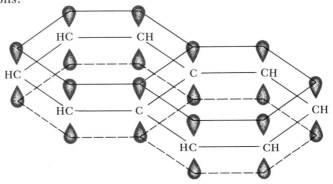

It is a white solid, sublimes readily, and is toxic to small insects and larvae. It is commonly used in mothballs.

Polynuclear aromatic hydrocarbons undergo the typical electrophilic substitution reactions. Some examples with the preferred position of substitution indicated in the compound are as follows:

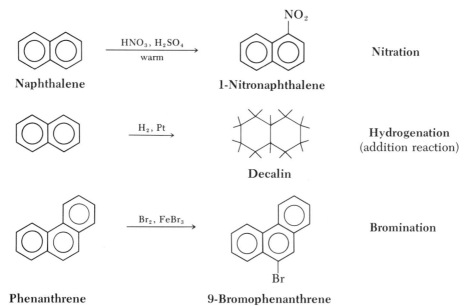

Naphthalene

1-Nitronaphthalene

Nitration

Decalin

Hydrogenation
(addition reaction)

Phenanthrene

9-Bromophenanthrene

Bromination

An interesting feature of these aromatic hydrocarbons is the fact that the color intensity increases as the number of rings increases. This is illustrated in Table 4-2. The color changes observed are related to the increase in conjugation.

Some of the complex polynuclear aromatic hydrocarbons are carcinogens, that is, have cancer-producing properties. Following are two examples:

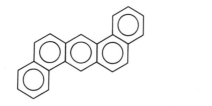

1,2,5,6-Dibenzanthracene

3,4-Benzpyrene

3,4-Benzpyrene is formed in the combustion of cigarettes and in the combustion of the fat that drops onto hot charcoal when a steak is cooked over hot coals.

Other aromatic compounds

An aromatic compound has a planar cyclic structure and has unusual stability (resonance stabilization energy) because of delocalization of electrons. Chemically, there are many compounds that do not resemble benzene

Table 4-2
Color of polynuclear aromatic hydrocarbons

Compound	Structure	Color
Benzene		Colorless
Naphthalene	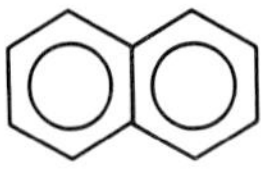	Colorless
Anthracene	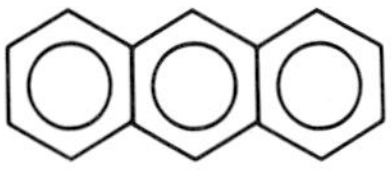	Buff
Naphthacene	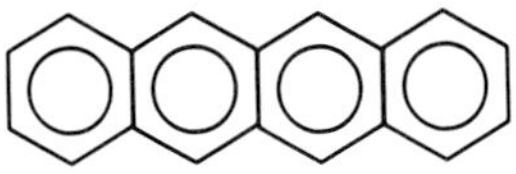	Yellow
Pentacene	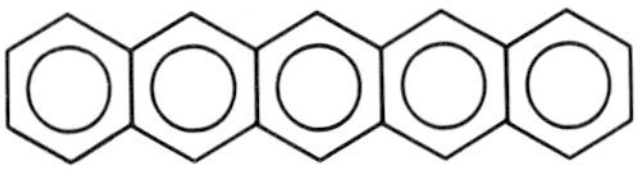	Blue
Hexacene	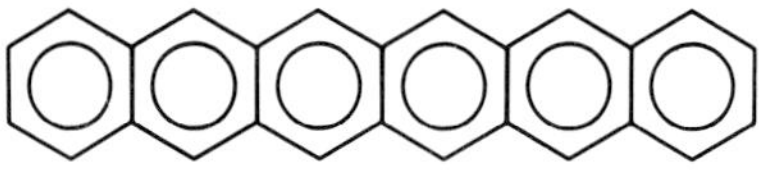	Green

but that have aromatic character. Some examples of these compounds are as follows:

Potassium cyclopentadienide

Tropylium bromide

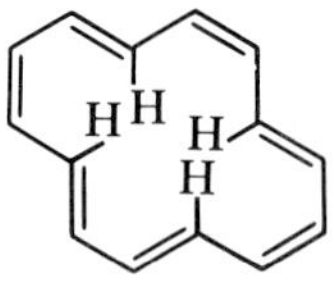

[14]-Annulene

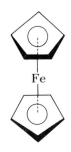

Ferrocene

Commercially important products derived from benzene

Approximately 100 chemicals are obtained from crude oil. The most useful are methane, ethene, propene, the butenes, and benzene. Although these

Table 4-3
Commercial products derived from benzene

Intermediate derivative or step(s)	*Commercial product*

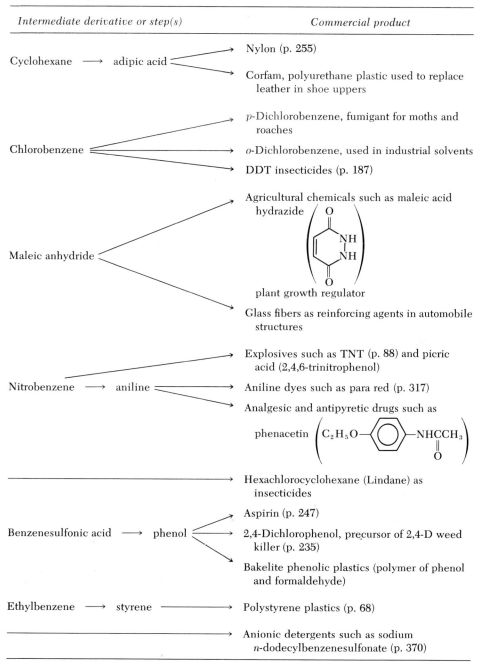

Cyclohexane ⟶ adipic acid

Nylon (p. 255)

Corfam, polyurethane plastic used to replace leather in shoe uppers

Chlorobenzene

p-Dichlorobenzene, fumigant for moths and roaches

o-Dichlorobenzene, used in industrial solvents

DDT insecticides (p. 187)

Maleic anhydride

Agricultural chemicals such as maleic acid hydrazide

plant growth regulator

Glass fibers as reinforcing agents in automobile structures

Nitrobenzene ⟶ aniline

Explosives such as TNT (p. 88) and picric acid (2,4,6-trinitrophenol)

Aniline dyes such as para red (p. 317)

Analgesic and antipyretic drugs such as

phenacetin $\left(C_2H_5O\!-\!\!\bigcirc\!\!-\!NHCCH_3 \atop \underset{O}{\|} \right)$

Hexachlorocyclohexane (Lindane) as insecticides

Benzenesulfonic acid ⟶ phenol

Aspirin (p. 247)

2,4-Dichlorophenol, precursor of 2,4-D weed killer (p. 235)

Bakelite phenolic plastics (polymer of phenol and formaldehyde)

Ethylbenzene ⟶ styrene

Polystyrene plastics (p. 68)

Anionic detergents such as sodium *n*-dodecylbenzenesulfonate (p. 370)

compounds represent less than 3% of the output of oil refining, the products derived from them account for more than two thirds of the organic chemicals used in the United States. These derivatives appear in plastics, fibers, synthetic rubber, soaps and detergents, cosmetics, pharmaceuticals, and insecticides.

When crude oil is refined (that is, separated into its components), the gasoline fraction obtained includes, among other chemicals, many continuous-chain hydrocarbons, which are often referred to as open-chain hydrocarbons. This mixture of open-chain hydrocarbons is heated in a reactor to convert some of the open-chain molecules into cyclic molecules such as benzene. The benzene structure shows up in a variety of commercially useful products. As shown in Table 4-3, some of these products are obtained directly from benzene, whereas in many cases benzene must first be converted to some intermediate, which then is converted to the important product.

Summary

Aromatic compounds have chemical behavior similar to that of benzene. Their aromaticity is manifested in their stability, and even though they are seemingly unsaturated, they undergo substitution rather than addition reactions.

Benzene is a planar molecule. All the carbons have sp^2 hybrid orbitals. There is a π cloud of six electrons encompassing all six carbon atoms. This delocalization of the π electrons accounts for the stability of the molecule.

Benzene undergoes the electrophilic substitution reactions, nitration, halogenation, sulfonation, and Friedel-Crafts alkylation. The rate of substitution and the orientation in a substitution product of benzene depend on the nature of the substituent already on the ring. Free-radical substitution occurs in the alkyl group substituted on a benzene ring. The reactive intermediate is the benzyl free radical, which is more stable than tertiary, secondary, primary, and methyl free radicals.

Important terms

Activating group	Naphthalene
Aniline	Nitration
Aromatic compound	Nitronium ion
Benzenesulfonic acid	Orbital picture
Benzoic acid	Orientation
Benzyl free radical	*Ortho* director
Benzyl group	*Ortho* position
Chlorination	*Para* director
Deactivating group	*Para* position
Electrophilic substitution	Phenol
Friedel-Crafts alkylation	Phenyl group
Heat of hydrogenation	Resonance
Hybrid structure	Styrene
Kekulé structure	Sulfonation
Meta director	Toluene
Meta position	

Problems

4-13. Give a correct name for each of the following structures:

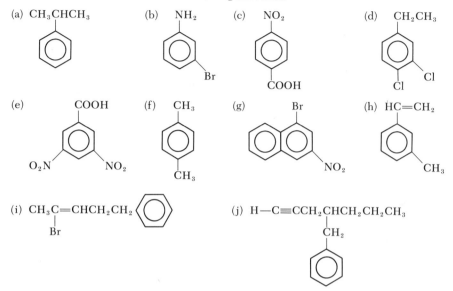

(a) CH_3CHCH_3

(b) NH_2 ... Br

(c) NO_2 ... $COOH$

(d) CH_2CH_3 ... Cl Cl

(e) $COOH$ ... O_2N ... NO_2

(f) CH_3 ... CH_3

(g) Br ... NO_2

(h) $HC{=}CH_2$ ... CH_3

(i) $CH_3C{=}CHCH_2CH_2$ ... Br

(j) $H{-}C{\equiv}CCH_2CHCH_2CH_2CH_3$ ... CH_2

4-14. Draw structures for each of the following compounds:

(a) *m*-Dinitrobenzene
(b) 2-Bromobenzenesulfonic acid
(c) 3,4-Dimethyl-*sec*-butylbenzene
(d) Styrene
(e) 2-Chloro-4-cyanostyrene

(f) Phenylacetylene
(g) 2-Ethyl-5-aminophenol
(h) 3,4-Dichloro-6-phenyl-2-octene
(i) 2-Benzyl-5-hydroxy-3-heptene
(j) 2,7-Dimethylnaphthalene

4-15. Give the structures of the product(s) of the reaction of isopropylbenzene with each of the following groups. Indicate if no reaction occurs.

(a) $K_2Cr_2O_7$, H^+, heat
(b) HNO_3, H_2SO_4
(c) H_2SO_4, SO_3
(d) Cl_2, Fe

(e) Cl_2, ultraviolet light
(f) Br_2, $FeBr_3$
(g) CH_3CH_2Cl, $AlCl_3$

4-16. Predict the products of the Friedel-Crafts type of reaction of benzene with each of the following reagents:

(a) CH_3CHCH_3/H^+ ... OH

(b) $CH_3CH{=}CH_2/H^+$

4-17. Predict the major product of the free-radical chlorination of

CH_3
$C{-}CH_2{-}CH_3$
H

4-18. Predict the point(s) of ring substitution in the bromination of each of the following molecules:

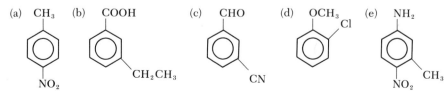

(a) CH_3 ... NO_2

(b) $COOH$ ... CH_2CH_3

(c) CHO ... CN

(d) OCH_3 Cl

(e) NH_2 ... NO_2 CH_3

4-19. Using the data in the following two tables, answer the questions below:

Compound	Type of aromatic substitution	Percent ortho	Percent para	Percent meta
C_6H_5Z	Nitration	41	59	0.2
C_6H_5Z	Chlorination	42	51	7
C_6H_5Z	Bromination	13	85	2

Compound	Type of aromatic substitution	Relative rate
C_6H_5Z	Halogenation	0.08
C_6H_6	Halogenation	1.0

(a) What type of director is the substituent Z? Explain.
(b) Is the substituent Z an activator or deactivator? Explain.
(c) Is it the substituent Z or the electrophilic reagent involved that determines the kind of orientation? Explain.
(d) What substituent do you think Z might be? Explain.

4-20. Draw the structure of the electrophile in each of the following reactions:

(a) Nitration
(b) Sulfonation
(c) Bromination
(d) Friedel-Crafts alkylation with ethyl bromide

4-21. Draw a structure that fits each of the following descriptions:

(a) Intermediate carbonium ion in the chlorination of nitrobenzene
(b) Electrophile in nitration
(c) Most stable cation formed in the nitration of phenol
(d) Least stable cation formed when nitrobenzene undergoes bromination in the *ortho* position.

4-22. Indicate the steps in the synthesis of the following compounds from benzene or toluene:

(a) *p*-Bromotoluene
(b) *o*-Nitrotoluene
(c) *m*-Bromonitrobenzene
(d) Benzoic acid
(e) *m*-Chlorobenzoic acid
(f) Benzyl chloride
(g) *p*-Chlorobenzyl chloride
(h) Ethylbenzene
(i) Styrene
(j) 1,2-Dibromo-1-phenylethane

4-23. Using the following analogies, explain the resonance hybrid concept:

(a) A mule is a hybrid of a horse and a donkey.

Horse   Mule   Donkey

(b) A rhinoceros is a hybrid of a unicorn and a dragon.

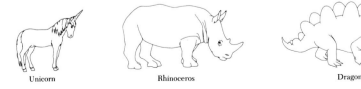

Unicorn   Rhinoceros   Dragon

(c)  is a hybrid of ... .

Self-test

1. Give a correct name for each of the following molecules:

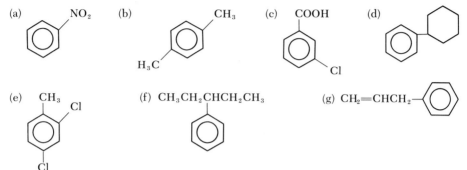

2. Give the structure of the organic product of each of the following reactions:

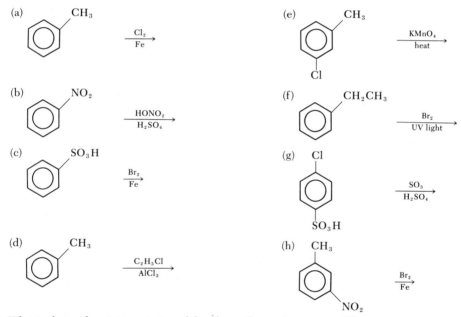

3. What is the modern interpretation of the phrase "aromatic compound"?
4. Draw the resonance structure that explains why —OH is a *para* director and the resonance structure that explains why —NO$_2$ is not a *para* director.
5. Write equations illustrating the synthesis of each of the following compounds from benzene:

   (a) Bromobenzene
   (b) *m*-Chloronitrobenzene
   (c) *o*-Nitrobenzoic acid

6. Give the correct answer for each of the following questions:

   (a) What is the typical reaction of benzene and other aromatic compounds?
   (b) How many electrons does the π cloud of benzene contain?
   (c) How many disubstitution products of benzenes are possible?
   (d) What is the more commonly used name for methylbenzene?
   (e) What type of director is the —CH$_2$CH$_3$ group?

(f) Is —SO_3H a ring activator or deactivator?

(g) What geometry is associated with benzene?

(h) What is the structural formula for the benzyl free radical?

(i) What is the structural formula for the electrophile in nitration?

(j) Which hydrogen in toluene is most easily abstracted by a chlorine atom?

Answers to self-test

1. (a) Nitrobenzene
   (b) *p*-Xylene or *p*-methyltoluene
   (c) 3-Chlorobenzoic acid or
   m-chlorobenzoic acid
   (d) Cyclohexylbenzene or phenylcyclohexane

   (e) 2,4-Dichlorotoluene
   (f) 3-Phenylpentane
   (g) 3-Phenylpropene

2.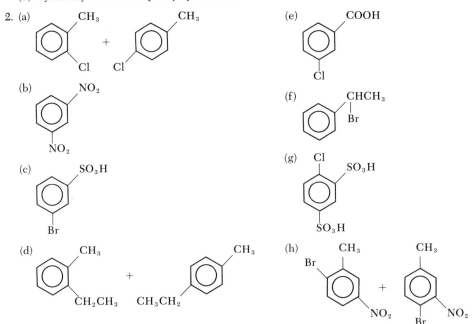

3. An aromatic compound possesses a planar cyclic structure, has unusual stability because of delocalization of π electrons, and undergoes substitution rather than addition reactions.

4.

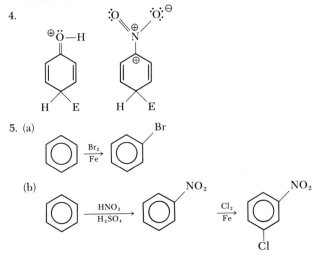

5. (a)

(c)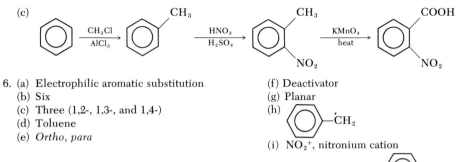

6. (a) Electrophilic aromatic substitution
   (b) Six
   (c) Three (1,2-, 1,3-, and 1,4-)
   (d) Toluene
   (e) *Ortho, para*

   (f) Deactivator
   (g) Planar
   (h) (ring)–$\overset{\cdot}{C}H_2$

   (i) NO_2^+, nitronium cation

   (j) Benzylic hydrogen, (ring)–CH_3

chapter 5

Stereochemistry

She puzzled over this for some time, but at last a bright thought struck her. 'Why, it's a Looking-Glass book, of course! And, if I hold it up to a glass, the words will all go the right way again."

LEWIS CARROLL
Through the Looking-Glass

A molecular formula indicates the number and kinds of atoms in a molecule. The next step after determining the molecular formula of a compound is to determine which atoms are bonded to which other atoms, that is, to determine the chemical constitution. In principle, there are a number of ways in which atoms of a molecule can be bonded to other atoms. This involves the concept of isomerism. An *isomer* is any one of a group of compounds that have the same molecular formula but different structural formulas and that display different physical or chemical properties. There are two types of isomerism: *constitutional isomerism* and *stereoisomerism*. Stereoisomerism can be further subdivided into *configurational* and *conformational isomerism*. *Enantiomers, diastereomers, meso compounds,* and *geometric isomers* are types of configurational isomers. These relationships are illustrated on the opposite page.

Constitutional isomerism

Constitutional isomers have the same molecular formula but differ in the way the atoms in the molecule are bonded together, regardless of direction in space. For example, there are two constitutional isomers corresponding to the molecular formula C_2H_6O: ethyl alcohol (CH_3CH_2OH) and methyl ether (CH_3OCH_3). In ethyl alcohol, an oxygen atom is bonded to a carbon atom and a hydrogen atom; but in methyl ether, an oxygen atom is bonded

114

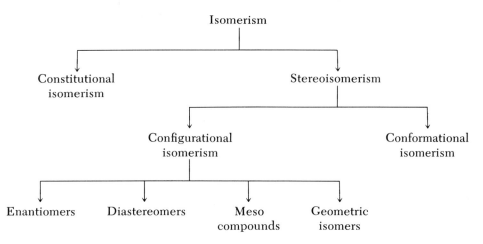

to two carbon atoms. A second difference is the bonding of the two carbons to one another, which occurs in ethyl alcohol but not in methyl ether.

Constitutional isomers also have different physical and chemical properties. Ethyl alcohol belongs to a class of compounds called alcohols. Methyl ether belongs to a class of compounds called ethers. (The organic chemistry of alcohols is discussed in Chapter 6 and that of ethers in Chapter 8.) Table 5-1 gives examples of constitutional isomers.

Table 5-1
Constitutional isomers and corresponding molecular formulas

Molecular formula	Constitutional isomers		
C_4H_{10}	$CH_3CH_2CH_2CH_3$	and	CH_3CHCH_3 $\overset{|}{CH_3}$
C_3H_8O	$CH_3CH_2CH_2OH$	and	CH_3CHCH_3 $\overset{|}{OH}$
C_3H_7Cl	$CH_3CH_2CH_2Cl$	and	CH_3CHCH_3 $\overset{|}{Cl}$
C_5H_{10}	$\underset{CH_2-CH_2}{\overset{CH_2}{\underset{\diagdown}{CH_2}\overset{\diagup}{CH_2}}}$	and	$CH_3CH_2CH_2CH=CH_2$
$C_3H_6O_3$	$CH_3\overset{\overset{O}{\|}}{CH}\overset{|}{C}-OH$ $\overset{|}{OH}$	and	$HOCH_2\overset{\overset{O}{\|}}{CH}C-H$ $\overset{|}{OH}$

5-1. Draw the constitutional isomers of C_6H_{14}.

Answer: Of the five isomers, one possibility is $CH_3(CH_2)_4CH_3$.

5-2. Draw the constitutional isomers of $C_5H_{11}Cl$.

Answer: Of the seven isomers, one possibility is $CH_3CH_2CH_2CH_2CH_2Cl$.

Tetrahedral carbon: projection formulas

A carbon atom that is bonded to four other atoms has sp^3 hybrid orbitals and is tetrahedrally oriented. In Chapter 1, a tetrahedral carbon was depicted

as C where the solid lines indicate bonds in the plane of the page, the

dashed line indicates a bond behind the plane, and the wedged line indicates a bond directed out of the plane toward the reader.

Another common notation for a tetrahedral carbon that is often used in

discussion of stereochemistry is the circle and line picture: . The circle

represents a carbon atom, the lines that touch the circle represent atoms or groups behind the plane of the paper or going away from the reader, and the lines drawn into the circle indicate atoms or groups coming out of the plane toward the reader.

A Fischer projection is a planar projection of the circle and line tetrahedral model. In the Fischer, or planar, projection, the vertical lines represent bonds going away from the reader behind the plane of the paper, the horizontal lines represent bonds coming toward the reader out of the plane of the paper, and the point at which the two lines cross is the carbon atom. The following examples illustrate the relationship between the circle and line tetrahedral model and the planar projection:

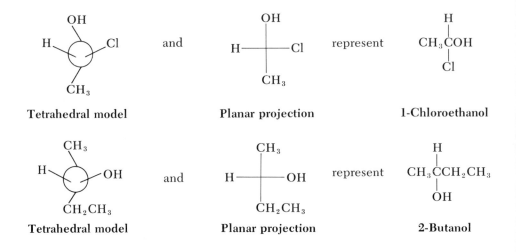

| Tetrahedral model | Planar projection | 1-Chloroethanol |

| Tetrahedral model | Planar projection | 2-Butanol |

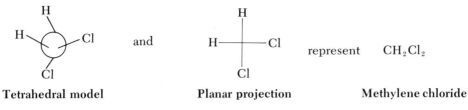

Tetrahedral model	**Planar projection**	**Methylene chloride**

The spatial arrangement of atoms or groups about a carbon atom is an important factor in stereochemistry.

Stereoisomerism

Stereoisomers are molecules that have the same constitution but differ in the spatial arrangements of their atoms. Stereoisomers can be further classified as configurational isomers or conformational isomers. *Configurational isomers,* which may or may not be related as object and mirror image, can be interconverted only by the breaking and making of bonds. *Conformational isomers,* which are the infinite number of arrangements of the atoms in space that can arise from rotation about a single bond, can be interconverted by rotation of one part of the molecule with respect to the rest of the

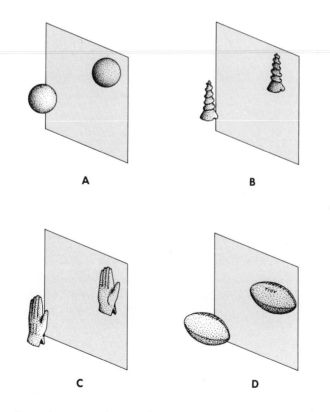

Fig. 5-1
Mirror-image relationships. **A,** Sphere and its mirror image. **B,** Screw and its mirror image. **C,** Left-handed glove and right-handed glove (mirror image). **D,** Football and mirror image.

molecule about a single bond joining these two parts. For example, for the configuration (and conformation) of 2-bromobutane shown as follows, several conformations are shown at the right:

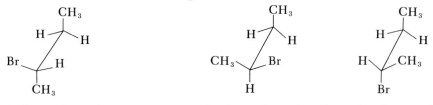

Particular configuration   Conformations of configuration shown at left

Configurational isomerism

Enantiomers. These are stereoisomers that are nonsuperimposable mirror images.

An object is superimposable on its mirror image if, when it is placed on its mirror image, the corresponding parts lie together. A sphere and its mirror image are superimposable, but a right-handed glove and its mirror image (which is a left-handed glove) are not superimposable. Some mirror image relationships are shown in Fig. 5-1.

Similarly, two molecules may have a mirror image relationship and may or may not be superimposable. If a molecule and its mirror image are super-imposable, then they are two molecules of the same compound. For example, the mirror image of the planar projection of methane is superimposable on the planar projection. Hence, the planar projection (sometimes called the object) and its mirror image represent two different molecules of the same compound. This is illustrated as follows where the vertical dashed line represents the mirror:

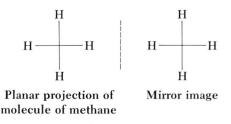

Planar projection of   Mirror image
molecule of methane

If, on the other hand, a molecule and its mirror image are not super-imposable, they are molecules of two different compounds. For example, the planar projection of a molecule of 2-butanol and its mirror image are not superimposable and represent two different molecules:

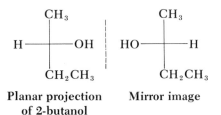

Planar projection   Mirror image
of 2-butanol

The following rules must be followed in testing for superimposability with the planar projections: (1) the planar projections should be mentally moved over one another, or (2) the planar projection should be rotated end for end *but only in the plane of the paper.* If the mirror image of methane is mentally moved over the planar projection, it becomes clear that the object and its mirror image are superimposable. If the mirror image of 2-butanol is mentally moved over the planar projection, a methyl group lies on a methyl group and an ethyl group on an ethyl group, but an HO lies over an H and an H over an HO group. If the planar projection

is rotated end for end, it appears as

$$\overset{\displaystyle CH_2CH_3}{\underset{\displaystyle CH_3}{HO \underset{|}{\overline{}} H}}$$

If the mirror image is mentally moved over this "new" object, an HO group lies on an HO group and an H on an H, but a methyl group lies on an ethyl group and an ethyl group on a methyl group. Clearly, then, the mirror image of 2-butanol is not superimposable on the planar projection. Thus, they represent two different molecules.

Following are several other examples of superimposable and nonsuperimposable molecules:

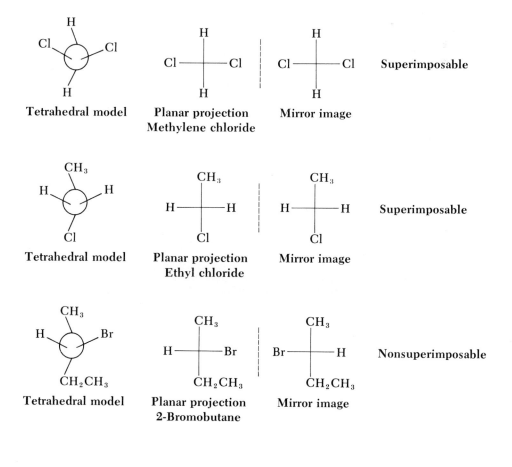

Tetrahedral model	Planar projection Methylene chloride	Mirror image	Superimposable
Tetrahedral model	Planar projection Ethyl chloride	Mirror image	Superimposable
Tetrahedral model	Planar projection 2-Bromobutane	Mirror image	Nonsuperimposable

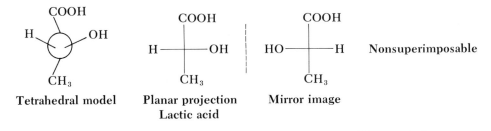

Tetrahedral model **Planar projection** **Mirror image** **Nonsuperimposable**
Lactic acid

Since the two forms of 2-butanol, 2-bromobutane, and lactic acid are non-superimposable mirror images, they are enantiomers. More recently, molecules that are nonsuperimposable on their mirror image are referred to as being chiral.

Chiral, or asymmetric, carbon atom. This type of carbon atom has four different groups bonded to it. The following molecules have no chiral, or asymmetric, carbon atoms:

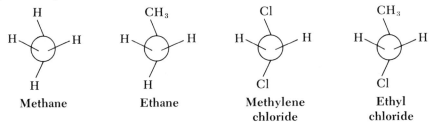

Methane Ethane Methylene chloride Ethyl chloride

On the other hand, the following molecules have one chiral or asymmetric carbon atom, which is underlined in the condensed formula:

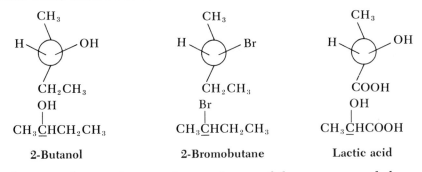

2-Butanol 2-Bromobutane Lactic acid

As shown in the previous section, each one of these compounds has a non-superimposable mirror image and thus exists as a pair of enantiomers. It follows, then, that all compounds with one chiral, or asymmetric, carbon atom can exist as a pair of enantiomers.

5-3. Circle all the chiral or asymmetric carbons, if there are any, in the following molecules:

(a) $CHCl_3$ *Answer:* There are no chiral carbons.

(b) $CH_3CH_2CH{=}CH_2$ *Answer:* There are no chiral carbons.

(c)

H
|
$CH_3CH_2C{-}Cl$ *Answer:* There is one chiral carbon: C2.
|
CH_3

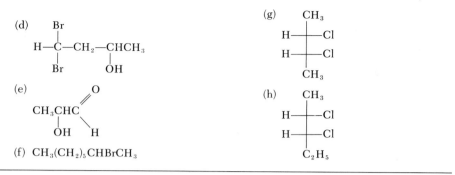

(d)

$$\underset{\underset{\text{Br}}{\displaystyle |}}{\overset{\overset{\text{Br}}{\displaystyle |}}{\text{H}-\text{C}}}-\text{CH}_2-\underset{\underset{\text{OH}}{\displaystyle |}}{\text{CHCH}_3}$$

(e)

$$\underset{\underset{\text{OH}}{\displaystyle |}}{\text{CH}_3\text{CHC}}\overset{\displaystyle \nearrow \text{O}}{\underset{\displaystyle \searrow \text{H}}{}}$$

(f) $\text{CH}_3(\text{CH}_2)_5\text{CHBrCH}_3$

(g)

(h)

Optical activity. Compounds that can rotate the plane of polarized light are said to be optically active. A plane of polarized light is light vibrating in a single plane as opposed to ordinary light, which vibrates in many random vibrations:

Polarized light   **Ordinary light**

The method of determining whether or not a compound is optically active is as follows. Ordinary light is converted to a plane of polarized light by being passed through a polarizer. The plane of polarized light vibrating in a vertical direction is then passed through a tube containing the sample (Fig. 5-2). The light emerging from the tube is passed through an analyzer (which is similar to the polarizer). If the sample is optically inactive, the angle of rotation will be 0°; that is, the polarizer and the analyzer will be parallel for maximum intensity of transmitted light. On the other hand, if the substance is optically active, the plane of polarized light will be rotated by α degrees; that is, the analyzer must be rotated an equal number of degrees so that the maximum intensity of transmitted light can be obtained. The angle of rota-

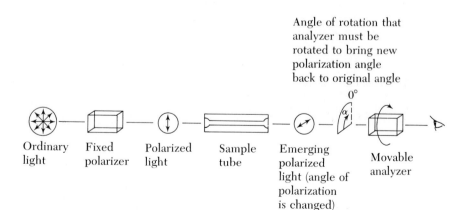

Angle of rotation that analyzer must be rotated to bring new polarization angle back to original angle

0°

Ordinary   Fixed   Polarized   Sample   Emerging
light   polarizer   light   tube   polarized
                                                            light (angle of
                                                            polarization
                                                            is changed)

Movable
analyzer

Fig. 5-2
Schematic drawing of a polarimeter.

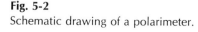

tion, α, through which the analyzer needs to be rotated to obtain maximum intensity of transmitted light is equal to the rotation of the light by the sample. The instrument used for measuring is the polarimeter, a schematic diagram of which is shown in Fig. 5-2.

If the analyzer must be rotated to the right, the substance rotated the beam of polarized light to the right. Such a substance is said to be *dextrorotatory* (Latin: right) and is represented by the symbol d or $(+)$. Similarly, a substance that causes rotation to the left is *levorotatory* (Latin: left) and is represented by the symbol l or $(-)$.

The actual rotation to the right or left is reported as the specific rotation, $[\alpha]$, which is determined by the following formula:

$$[\alpha] = \frac{\alpha}{l \cdot c}$$

where α is the angle measured, l is the length of the tube containing the substance, and c is the concentration of the substance. For example, the specific rotation of a lactic acid solution is $[\alpha]_D^{15} = +3.82$ where 15 is the temperature, D is the wavelength of the light used in the measurement (in this case, the D line of sodium, 5893 Å), and 3.82 is the value of the specific rotation. The sample is d-lactic acid, or $(+)$-lactic acid.

A molecule that has a superimposable mirror image is *optically inactive*, whereas a molecule that has a nonsuperimposable mirror image is *optically active*. Nonsuperimposable mirror image isomers, called entaniomers, are individually optically active. A pair of enantiomers are alike in all respects except the direction of rotation of the plane of polarized light. One enantiomer is the d-isomer and rotates the plane of polarized light to the right. The other is the l-isomer and rotates the plane of polarized light to the left. The only way to distinguish the d-isomer from the l-isomer in a pair of enantiomers is to measure the specific rotation in the polarimeter.

5-4. Which of the following molecules are optically active?

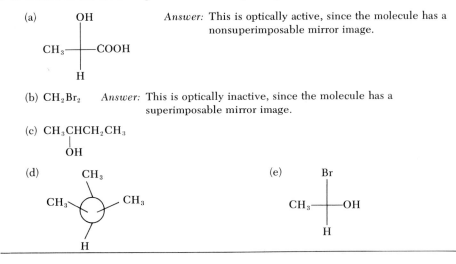

(a)

CH$_3$—C(OH)(H)—COOH

Answer: This is optically active, since the molecule has a nonsuperimposable mirror image.

(b) CH$_2$Br$_2$ *Answer:* This is optically inactive, since the molecule has a superimposable mirror image.

(c) CH$_3$CHCH$_2$CH$_3$
         |
        OH

(d)   CH$_3$

CH$_3$—C(CH$_3$)—CH$_3$

H

(e)   Br
       |
CH$_3$—C—OH
       |
       H

Diastereomers. These are stereoisomers that do not have a mirror image relationship.

In a compound containing two chiral carbon atoms, the chiral carbons can be identical or different. For example, 3-chloro-2-butanol has two different chiral carbons, and 2,3-dichlorobutane has two identical chiral carbons:

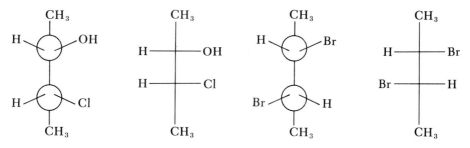

Tetrahedral model   **Planar projection**   **Tetrahedral model**   **Planar projection**

3-Chloro-2-butanol   **2,3-Dibromobutane**

Let us consider first 3-chloro-2-butanol, in which the two chiral carbons are different. The mirror image and the planar projection are represented as follows:

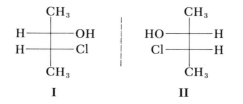

Compounds I and II are nonsuperimposable mirror images and therefore are enantiomers. Each is individually optically active, since it has a nonsuperimposable mirror image. There are two other stereoisomers of 3-chloro-2-butanol, which are represented as follows:

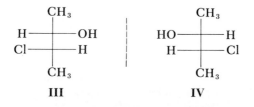

The relationship between compounds III and IV is that of nonsuperimposable mirror images. Thus they are enantiomers. Furthermore, each is individually optically active. A compound containing two different chiral carbons will exist as two pairs of enantiomers.

On the other hand, compounds I and III are stereoisomers but are not mirror image isomers. Thus they are diastereomers. Also, compounds II and III, I and IV, and II and IV are diastereomers.

Next let us consider 2,3-dibromobutane, in which the chiral carbon atoms are identical. The stereoisomers are as follows:

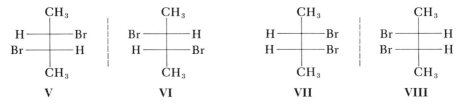

V   VI   VII   VIII

Compounds V and VI are nonsuperimposable mirror images and thus are enantiomers. Both are capable of optical activity. Compounds VII and VIII are mirror image isomers but are superimposable. They are two molecules of the same kind and cannot be enantiomers. Furthermore, they cannot be optically active. A compound that contains an chiral center but that is optically inactive is called a *meso compound*. A characteristic of a *meso* compound is its plane of symmetry, which can be recognized by the fact that one half of the molecule is the mirror image of the other half:

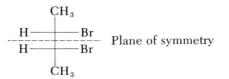

Thus a compound containing two identical chiral carbons will exist as a pair of enantiomers and a *meso* compound.

Also, compounds V and VII and VI and VII are diastereomers.

___

5-5. Answer the following questions in regard to this molecule:

(a) How many chiral carbon atoms does it contain?
   Answer: Two.
(b) Draw the mirror image of the molecule.
(c) Is the molecule optically active? Why or why not?
   Answer: No. The molecule has a superimposable mirror image.
(d) Does the molecule have a plane of symmetry?   *Answer:* Yes
(e) What name is given to this particular kind of stereoisomer?
(f) Draw a diastereomer of the molecule.
(g) Is the diastereomer optically active? Why or why not?

___

Racemic modification and resolution. If equal amounts of the (+)-isomer and (−)-isomer of lactic acid are mixed together, the resulting mixture is optically inactive because the rotation caused by one isomer is exactly canceled by the equal and opposite rotation caused by the other isomer. A mixture of equal amounts of enantiomers is called a *racemic modification.*

There are several ways of forming a racemic modification. One is simply to physically mix equal amounts of enantiomers. Also, there is a chemical method. For example, the reduction of pyruvic acid gives racemic lactic acid. This reaction involves the addition of a molecule of hydrogen to the carbonyl group (C=O) in the presence of a nickel catalyst:

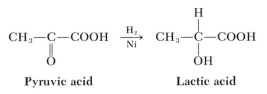

Pyruvic acid   **Lactic acid**

Since the probability that the hydrogen will add on one side or the other of the molecule is the same, then equal amounts of enantiomers should be formed:

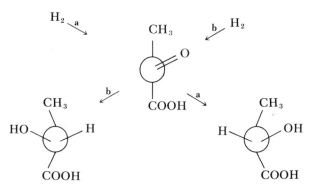

Pyruvic acid contains no chiral carbon atoms, and in this reaction an asymmetric center (that is, a chiral carbon atom) is generated. The conversion of a compound with no chiral center into one containing an chiral center results in the formation of enantiomers in equal amounts. The resulting mixture is therefore optically inactive.

5-6. Using the mechanism you learned in Chapter 3, explain why the free-radical chlorination of butane gives racemic 2-chlorobutane. *Answer:* The mechanism involves the formation of a planar free radical.

The process of separating a racemic modification into its enantiomers is called *resolution.* This cannot be done by ordinary laboratory separation procedures. For example, a racemic modification cannot be separated by distillation because the enantiomers have the same boiling points. Also, because their solubilities in optically inactive solvents (such as benzene, ether, and chloroform) are identical, they cannot be separated by fractional crystallization. However, enantiomers can be separated by biochemical and chemical methods. In biochemical resolution, a microorganism is used to preferentially destroy one of the isomers. In chemical resolution, which is the better of the two methods, a racemic modification is converted into diastereomers, which can be separated on the basis of their different physical properties such as distillation and solubility.

As an example of chemical resolution, racemic lactic acid can be resolved by conversion to a salt by use of an optically active base. Quinine, which is represented as follows, is an optically active amine base commonly used to resolve a racemic acid:

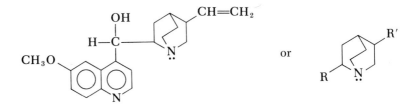

The levorotatory isomer of quinine is used routinely. The reaction between an acid and an amine forms an ammonium salt:

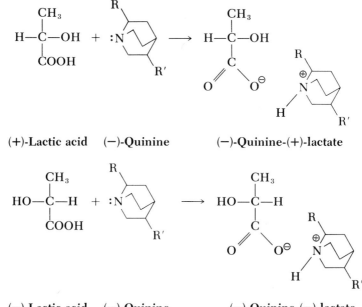

The products formed in these two reactions are not mirror images, since the base portions, (−)-quinine, are not mirror images. Therefore, they are diastereomers, *not* enantiomers. The diastereomeric salts can be separated on the basis of difference in solubility. Once the diastereomers are separated, the enantiomeric pair can be regenerated usually by a simple chemical reaction:

(−)-quinine-(+)-lactic acid + HCl ⟶ (+)-lactic acid + (−)-quinine ·HCl
(−)-quinine-(−)-lactic acid + HCl ⟶ (−)-lactic acid + (−)-quinine ·HCl

Configuration notation. The configuration of a molecule is the arrangements in space of the atoms that can be changed only by the breaking and making of bonds. The notations d- or (+) and l- or (−) do not designate configuration, but rather the direction of rotation of the plane of polarized light. The conventional designations for configuration are the letters D- and L-.

The standard compound chosen to which all other configurations are related is glyceraldehyde, the simplest sugar known, which contains one asymmetric carbon atom. The two configurations of glyceraldehyde are as follows:

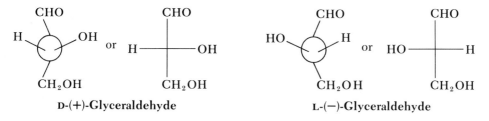

D-(+)-Glyceraldehyde   L-(−)-Glyceraldehyde

D-Glyceraldehyde rotates the plane of polarized light to the right, and L-glyceraldehyde rotates the plane of polarized light to the left. However, not all compounds whose configuration is L rotate the plane of polarized light to the left, and not all whose configuration is D rotate it to the right. For example, glyceric acid and lactic acid have the D configuration but are levorotatory:

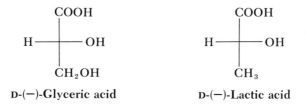

D-(−)-Glyceric acid   D-(−)-Lactic acid

The direction of rotation must always be determined by the use of a polarimeter, whereas the configuration can be obtained from the relation of the compound to D- or L-glyceraldehyde.

By convention, the number one carbon (in the IUPAC nomenclature system) is placed at the top in the planar Fischer projection. If the group on the next to the last carbon (shaded in the following projections) lies to the left, the compound belongs to the L series; but if it lies to the right, it belongs to the D series. Following are some examples:

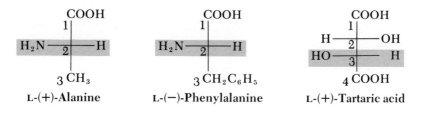

L-(+)-Alanine   L-(−)-Phenylalanine   L-(+)-Tartaric acid

5-7. Which of the compounds belong to the D configuration, and which belong to the L configuration?

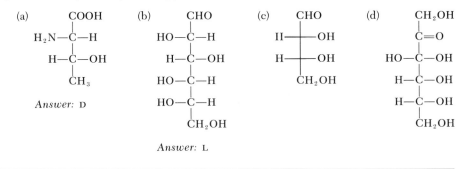

(a)

$$COOH$$
$$H_2N-\underset{|}{\overset{|}{C}}-H$$
$$H-\underset{|}{\overset{|}{C}}-OH$$
$$CH_3$$

Answer: D

(b)

$$CHO$$
$$HO-\underset{|}{\overset{|}{C}}-H$$
$$H-\underset{|}{\overset{|}{C}}-OH$$
$$HO-\underset{|}{\overset{|}{C}}-H$$
$$HO-\underset{|}{\overset{|}{C}}-H$$
$$CH_2OH$$

Answer: L

(c)

$$CHO$$
$$H-\!\!\!-OH$$
$$H-\!\!\!-OH$$
$$CH_2OH$$

(d)

$$CH_2OH$$
$$C=O$$
$$HO-\underset{|}{\overset{|}{C}}-OH$$
$$H-\underset{|}{\overset{|}{C}}-OH$$
$$H-\underset{|}{\overset{|}{C}}-OH$$
$$CH_2OH$$

More recently, the letters R and S were introduced to specify configuration. The use of these letters is a convenient way of designating a configuration without always having to draw its picture. The specification is based on a sequence of priority of the four groups bonded to the chiral carbon. Priority depends on atomic number: the atom of higher atomic number has higher priority. Let us consider the following molecule as an example:

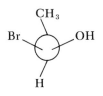

The order of increasing atomic number of the atoms directly attached to the chiral carbon is Br, 35 > O, 16 > C, 6 > H, 1. To specify the configuration as R or S about the chiral carbon, you must orient the tetrahedral model with the group of lowest priority directed away from you. You should mentally take hold of the hydrogen atom (the group of lowest priority) and move it away from yourself so that it projects from the rear of the chiral carbon. This causes the other three groups to move out toward you:

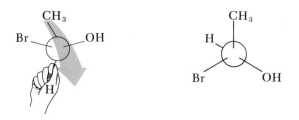

Then the group of highest priority, Br, is located and the group of next highest priority, OH. If you travel in a clockwise direction from the group of highest priority to the group of second priority to the third group, the configuration is specified as R (Latin: *rectus*, right). If you travel in a counterclockwise direction, the configuration is specified as S (Latin: *sinister*, left). In the example, we travel in a counterclockwise direction. Thus the configuration is S. The more complete name for the molecule is S-1-bromoethanol:

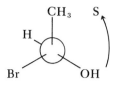

Following are further examples. D-(−)-Glyceric acid has the R configuration:

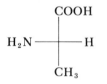

Planar projection

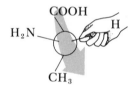

Tetrahedral model

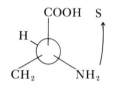

Priority: OH > COOH > CH_2OH
Configuration: R
R-(−)-glyceric acid

L-(+)-alanine has the S configuration:

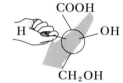

Planar projection

Tetrahedral model

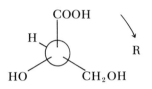

Priority: NH_2 > COOH > CH_3
Configuration: S
S-(+)-alanine

5-8. Label the following compounds as R or S:

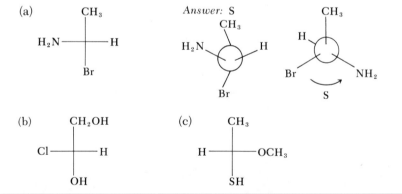

(a)

Answer: S

(b)

(c)

Geometric isomers. These are stereoisomers, since they differ from one another only in the way the atoms are oriented in space. Furthermore, they are configurational isomers, since they can be interconverted only by the breaking and making of bonds.

Geometric isomerism occurs in alkenes of the type WXC=CZY where X ≠ W and Z ≠ Y. Two configurations are possible:

For example, the configurations for 1,2-dibromoethane and 2-butene are as follows:

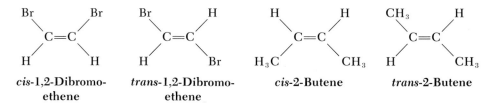

cis-1,2-Dibromo-ethene   *trans*-1,2-Dibromo-ethene   *cis*-2-Butene   *trans*-2-Butene

These configurations are distinguished by the use of the prefixes *cis*- (Latin: on the same side) and *trans* (Latin: across). Thus, in *cis*-1,2-dibromoethene, the two bromine atoms are on the same side (of the planar double bond).

If W = X or Y = Z in alkenes, then geometric isomerism is not possible. For example:

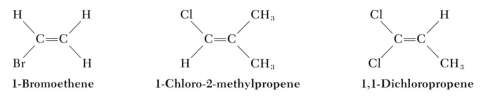

1-Bromoethene   1-Chloro-2-methylpropene   1,1-Dichloropropene

For trisubstituted alkenes, a different method of specification is used for distinguishing the geometric isomers. The groups on *each* carbon of the carbon–carbon double bond are arranged on the basis of priority, which depends on atomic numbers: the group of higher atomic number has higher priority. The priority of the two groups on each carbon of the carbon–carbon double bond is determined independently. If the two groups of higher priority are on the same side of the plane of the carbon–carbon double bond, the isomer is labeled Z (German: *zusammen*, together). If the two groups of higher priority are on opposite sides of the molecule, the isomer is labeled E (German: *entgegen*, opposite). For example, the two isomers of 2-bromo-1-chloropropene are as follows:

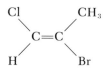

E-2-Bromo-1-chloropropene   Z-2-Bromo-1-chloropropene
(priority: Cl > H, Br > CH$_3$)   (priority: Cl > H, Br > CH$_3$)

Following are further examples:

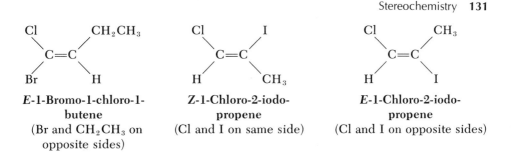

E-1-Bromo-1-chloro-1-butene
(Br and CH$_2$CH$_3$ on opposite sides)

Z-1-Chloro-2-iodo-propene
(Cl and I on same side)

E-1-Chloro-2-iodo-propene
(Cl and I on opposite sides)

The carbon–carbon double bond consists of a σ bond and a π bond, which is formed by overlap of the two p orbitals on adjacent carbons. These bonds are shown as follows for 1,2-dibromoethene:

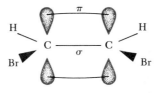

The π bond hinders the rotation about the double bond. If the *cis* isomer is to be converted to the *trans* isomer or vice versa, the π bond must be broken, a 180° rotation about the σ bond must occur, and a π bond must again be formed:

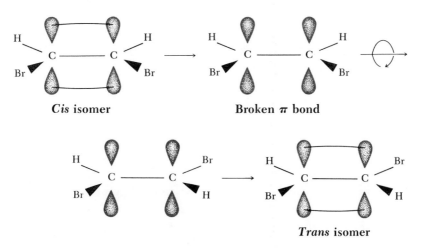

Cis isomer   Broken π bond

Trans isomer

Thus isomerism in alkenes is due to hindered rotation about the carbon–carbon double bond.

Since the *cis* and *trans* isomers are configurational isomers that do not have a mirror image relationship, they are diastereomers, which have different physical properties and similar chemical properties. The chemical properties are similar because the isomers are members of the same family. The chemical properties are not identical because the structures of the isomers are not identical. Usually, the isomers react with the same reagent at different rates. The properties of *cis* and *trans* isomers are illustrated in Table 5-2.

Table 5-2
Properties of geometric isomers

Isomer	cis-1,2-Dibromoethene	trans-1,2-Dibromoethene
Structure	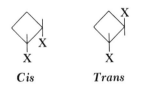	
Boiling point (°C)	110	108
Melting point (°C)	-53	-6
Dipole moment (Debye units)	1.35	0
Product of reaction with bromine	$CHBr_2CHBr_2$	$CHBr_2CHBr_2$
Reaction rate with bromine	Faster than for *trans* isomer	Slower than for *cis* isomer

5-9. Which of the following compounds can exist as geometric isomers?

(a) $CH_3CH_2CH=CH_2$ *Answer:* There are no geometric isomers, since there are like groups on a carbon atom of double bond.

(b) 2-Pentene   *Answer:* Geometric isomers are possible: *cis*-2-pentene and *trans*-2-pentene.

(c) $ClCH_2CH=CHCH_2Cl$

(d) $CH_3CH_2CBr=CHBr$

(e) $C_6H_5CH=CHCH_3$

Geometric isomers are also possible in cyclic structures:

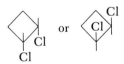

In the discussion of geometric isomerism in cyclic structures, the ring is considered to be planar. If the two substituents lie on the same side of the ring, the configuration is *cis*. If the two substituents lie on opposite sides of the plane of the ring, the configuration is *trans*. Following are some examples:

cis-1,2-Dichlorocyclobutane

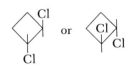

trans-1,2-Dichlorocyclobutane

trans-1,2-Dimethylcyclopropane

cis-1-Bromo-2-chlorocyclohexane

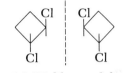

trans-1,3-Dibromocyclopentane

1,1-Dichlorocyclobutane
(no *cis* and *trans* isomers)

Cyclic *cis* and *trans* isomers are configurational isomers, just as are the *cis* and *trans* isomers of alkenes, and can be interconverted only by the breaking and making of bonds.

For 1,2-dichlorocyclobutane, the following configurations are possible:

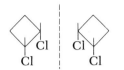

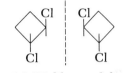

cis-1,2-Dichlorocyclobutane

trans-1,2-Dichlorocyclobutane

The *cis* isomer has a superimposable mirror image. Furthermore, it has a plane of symmetry, with one half of the molecule being the mirror image of the other half:

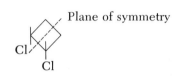

Plane of symmetry

The *trans* compound, on the other hand, exists as enantiomers, since it has a nonsuperimposable mirror image. The *cis* isomer is referred to as *cis-meso*, and the *trans* isomer is referred to as *trans*-racemic. The relationship between either *trans* enantiomer and the *meso* compound is that of diastereomer. The *meso* compound is optically inactive, and either *trans* enantiomer alone is optically active. A mixture of equal amounts of the *trans* enantiomers is optically inactive, whereas any diastereomeric mixture is optically active.

5-10. Which of the following compounds can exist as geometric isomers?

(a) 1,1-Dimethylcyclopropane   *Answer:* There are no geometric isomers.
(b) 1,4-Dibromocyclohexane   *Answer:* Geometric isomers are possible.
(c) 1,2-Dichlorocyclopentane
(d) 1,2-Dichlorocyclopropane
(e) 1-Chloro-2-methylcyclobutane

5-11. Answer the following questions in regard to this compound:

(a) Write a correct name for the compound.   *Answer:* *trans*-1,2-Dichloro-
cyclopropane

(b) Draw the structure of the other geometric isomer.
(c) Is the isomer shown above optically active? Why or why not?
  Answer: Yes. It has a nonsuperimposable mirror image.
(d) Is the isomer in part b optically active? Why or why not?
  Answer: No. It has a plane of symmetry.
(e) Which isomer is the *meso* compound?
(f) Draw the structure of an isomer that cannot be labeled as *cis* or *trans*.
(g) What name, other than geometric isomers, is given to the relationship between *cis* and *trans* isomers.

Stereochemistry of other carbon-containing compounds. The essential requirement for optical activity is not the presence of a chiral carbon atom in a molecule but that the molecule have a nonsuperimposable mirror image. Several examples of other molecules that display optical activity are as follows:

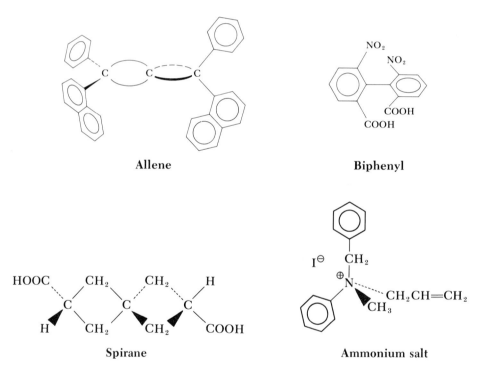

Allene

Biphenyl

Spirane

Ammonium salt

In the case of the allene, the two π bonds to the central carbon atom are perpendicular to each other; an allene of this type is not superimposable on its mirror image. The biphenyl is optically active, since the substituents on carbons 2 and 6 of each ring prevent the two biphenyl rings from lying in the same plane. The two rings are perpendicular to each other; thus there is no plane of symmetry. Similarly, the two rings of the spirane are perpendicular to one another, and a plane of symmetry is absent. The ammonium salt is an example of a nitrogen compound exhibiting optical activity. Nitrogen has sp^3 hybrid orbitals in the ammonium salt.

Conformational isomerism

Stereoisomers that arise by rotation about single bonds are called conformational isomers.

Conformations of ethane. The ethane molecule serves as a good introduction to conformational isomerism. An ethane molecule consists of two methyl groups joined by a single σ bond. Rotation about this bond is not completely free.

In the discussion of conformations, a Newman projection is often used. The planar projection, tetrahedral model, and Newman projection for ethane are as follows:

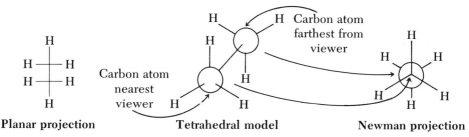

Planar projection   **Tetrahedral model**   **Newman projection**

You can obtain the Newman projection from the tetrahedral model by looking down the carbon–carbon bond. The carbon atom nearest your eye is represented as a point, whereas the carbon atom farthest from your eye is represented as a circle. The actual angle between the hydrogen atoms is 109.5°, but it appears to be 120° when viewed in the Newman projection.

If the position of one carbon atom is maintained while the other carbon atom is rotated with respect to it, an infinite number of possible arrangements (or conformations) are possible. Six of these conformations are as follows. The position of carbon 1 is maintained, and carbon 2 is rotated 60° each time through a total of 360°:

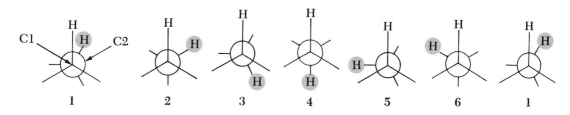

1   2   3   4   5   6   1

Conformations 1, 3, and 5 are called *eclipsed conformations.* Conformations 2, 4, and 6 are called *staggered conformations.* In eclipsed conformations, the hydrogen atoms are in line with one another. (They are drawn slightly to either side of the hydrogen atom on carbon 1 for convenience in seeing them.) At one time it was believed that the rotation about the single bond was essentially free or at least that it was too fast to measure so that the arrangements shown above could not be detected. It was later calculated that the rotation about the single bond was not free and that the energy of rotation reached a maximum when the hydrogen atoms on different carbon atoms were

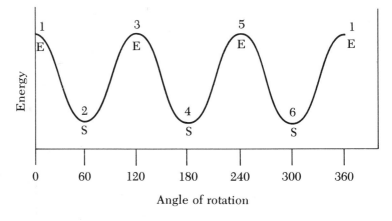

Fig. 5-3
Energy changes during rotation about carbon–carbon bond in ethane. E = eclipsed:
S = staggered.

in line with each other, that is, in an eclipsed conformation. Furthermore, the energy was at a minimum when the hydrogen atoms were as far apart as possible, that is, in a staggered conformation. Since the hydrogen atom is very small, the hindrance to rotation is partially due to steric interactions of the hydrogen atoms in the eclipsed conformation but is mainly due to the repulsion of the electrons in the bonds.

A plot of potential energy versus angle of rotation of one methyl group with respect to the other in ethane is shown in Fig. 5-3. The energy maximums correspond to eclipsed conformations, and the energy minimums correspond to staggered conformations. Between conformation 1 (eclipsed), in which the two hydrogen atoms are closest, and conformation 2 (staggered), in which the two hydrogen atoms are farthest apart, the energy decreases. Then the energy increases as the hydrogen atoms come close together again (between conformations 2 and 3). Between conformations 1 and 2, there are at least 60 other conformations of varying degrees of energy (and stability), but none is as unstable (higher energy) as conformation 1 or as stable (lower energy) as conformation 2. The energy barrier in ethane is about 3 kcal/mole, and the thermal energy of molecules is 0.6 kcal/mole. If the barrier were 0.6 kcal/mole, the rotation would be completely restricted. The rotation in ethane, then, is not free or completely restricted; it is hindered.

Conformations of alkanes. For molecules larger than ethane, a large number of conformations are possible. Following are some examples of these in which only the carbon skeleton formula is used. For convenience all the carbon atoms are kept in the plane of the paper.

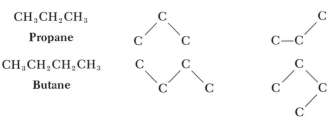

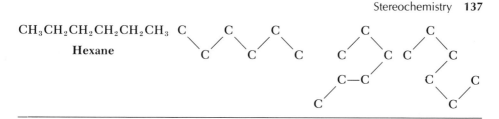

$CH_3CH_2CH_2CH_2CH_2CH_3$
Hexane

5-12. Draw a Newman projection formula for an eclipsed and a staggered form of each of the following compounds:

(a) Ethyl bromide   *Answer:*

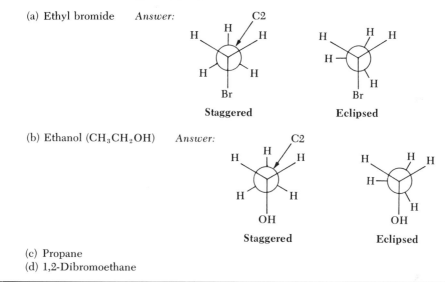

Staggered   Eclipsed

(b) Ethanol (CH_3CH_2OH)   *Answer:*

Staggered   Eclipsed

(c) Propane
(d) 1,2-Dibromoethane

Conformations of cyclohexane. Strictly speaking, the conventional structure used for cyclohexane in previous chapters is not correct. Cyclohexane is not a planar molecule. It is said to be puckered. Two conformations are possible for it:

Boat   Chair

The boat form is the less stable of the two forms and can be physically converted to the chair form by the "flipping" of carbon 6 below carbon 5 and carbon 1. The Newman projections show why the chair conformation is more stable:

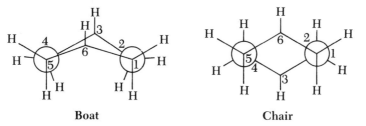

Boat   Chair

In the chair conformation, all the hydrogen atoms are staggered, whereas in the boat conformation, all the hydrogen atoms are eclipsed.

The hydrogen atoms in the chair conformation of cyclohexane are classified as *axial* hydrogens, or as *equatorial* hydrogens, which point sideways around the molecule:

Substituted cyclohexanes. In a monosubstitution product of cyclohexane, a substituent can occupy an axial or an equatorial position:

   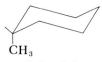

<table>
<tr><td align="center">1-Methylcyclohexane
(equatorial methyl)</td><td align="center">1-Methylcyclohexane
(axial methyl)</td></tr>
</table>

The more stable of these chair conformations is the one in which the substituent occupies an equatorial position.

Disubstitution products of cyclohexanes can exist as *cis* and *trans* configurational isomers and as conformational isomers. The planar cyclohexane ring is used for illustrating the configurational isomers, but the puckered ring is used for illustrating the conformational isomers. In the chair conformation, the isomer is *cis* if the two groups point in the same direction (upward *or* downward), and it is *trans* if they point in opposite directions (upward *and* downward). The following structures illustrate these concepts for 1,2-dimethylcyclohexane:

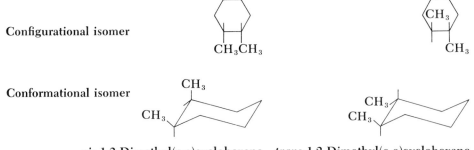

Configurational isomer

Conformational isomer

cis-1,2-Dimethyl(e,a)cyclohexane   *trans*-1,2-Dimethyl(e,e)cyclohexane

The position of the methyl group as equatorial or axial is indicated in the name of the compound as e or a, respectively. The conformation with both substituents in an equatorial position is more stable than that with only one in an equatorial position.

Simple sugars have cyclic structures, and the preferred conformation is the chair form. These concepts are discussed further in Chapter 13.

5-13. Draw conformational structures for each of the following compounds:

(a) *trans*-1,3-Dimethyl (a,a) cyclohexane   *Answer:*

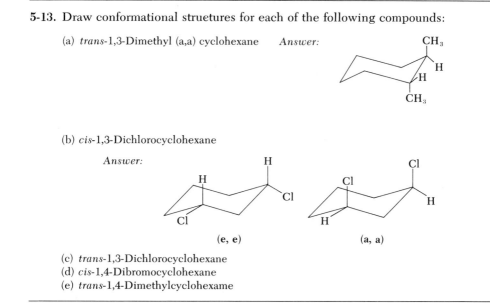

(b) *cis*-1,3-Dichlorocyclohexane

Answer:

(e, e)   (a, a)

(c) *trans*-1,3-Dichlorocyclohexane
(d) *cis*-1,4-Dibromocyclohexane
(e) *trans*-1,4-Dimethylcyclohexame

A stereospecific world

Stereochemistry is important in our world. Following are some examples.
Almost all the amino acids found in the proteins that are so vital to all living systems are L-amino acids, which have the following structure:

$$
\begin{array}{c}
COOH \\
| \\
H_2N\text{---}\overset{|}{C}\text{---}H \\
| \\
R
\end{array}
$$

Recently, a meteorite discovered on earth was analyzed and found to contain amino acids. Scientists were hopeful that these amino acids would give a clue to life in outer space. The results of the analysis showed that nearly all the amino acids were racemic. This could mean only (1) that the amino acids were formed randomly, (2) that the life of outer space is not the same as that found on earth, or (3) that the amino acids racemized by the heat that developed as the meteor passed through the earth's atmosphere.

In the late 1960's, it was discovered that during fossilization the amino acids in the shell proteins of mollusks slowly racemized from the optically active L-isomer to a mixture of D- and L-isomers. L-Isoleucine has two chiral carbons and has four optically active isomers:

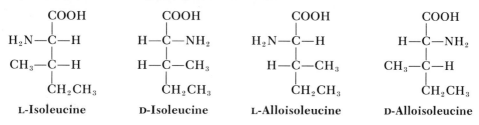

L-Isoleucine   D-Isoleucine   L-Alloisoleucine   D-Alloisoleucine

It was discovered that L-isoleucine in the original shell proteins did not racemize to D-isoleucine but rather to D-alloisoleucine. In the older fossils, there is a mixture of L-isoleucine and D-alloisoleucine. These two isomers can be resolved, and it is possible to estimate the relative ages of fossils from the ratio of alloisoleucine to isoleucine. Thus it may be possible to estimate the age of sediments in deep-sea cores from the residual amino acids in the microscopic shells of foraminifers.

All the naturally occurring sugars, except one, are D-isomers. Furthermore, of the 32 stereoisomers of $C_6H_{12}O_6$, the most abundant one is β-D-glucose, possibly because it is the one in which all the substituents are in equatorial positions:

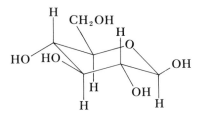

It is interesting that β-L-glucose does not occur naturally although, like β-D-glucose, all the substituents are in equatorial positions. Furthermore, it is interesting that glucose is not racemic. Perhaps at some time during evolution, the living system became stereospecific, and β-D-glucose was the particular conformation selected.

In many instances, one isomer of an insect pheromone (Chapter 2) is physiologically active and the other is inactive. For example, the *cis* isomer shown below, the sex attractant of the female boll weevil, is active; but the *trans* isomer is inactive:

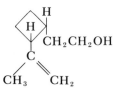

Inositol, one of the nine isomeric hexahydroxycyclohexanes, is thought to stimulate adipose tissue growth:

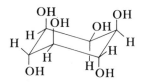

Thyroxine is a naturally occurring amino acid. L-Thyroxine is used as therapy for thyroid deficiency, whereas D-thyroxine is used for reducing the cholesterol level in the blood:

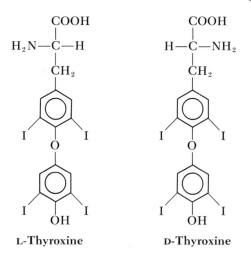

L-Thyroxine   D-Thyroxine

The active ingredient in hemlock, which was the poison Socrates drank, is coniine, which occurs naturally in the dextrorotatory form ($[\alpha]_D^{25°} = +15.7$):

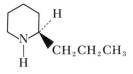

LSD (lysergic acid diethylamide) is a drug that can produce a temporary schizophrenic state and possibly some permanent brain damage. It is currently thought that LSD modifies the balance of serotonin in the brain. Serotonin is a vasoconstrictor and a brain impulse transmitter. Excess amounts of it are thought to bring about stimulation of cerebral activity, whereas a deficiency of it produces a depressant effect. One of the D-isomers is the active isomer of LSD:

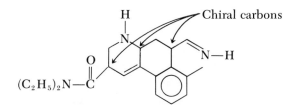

The dextrorotary isomer of amphetamine, more commonly known as an upper, is much more active than the levorotatory isomer:

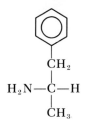

On the other hand, the levorotatory isomer of epinephrine is more active than the dextrorotatory isomer:

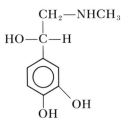

The levorotatory isomer of monosodium glutamate is used as a flavoring additive, whereas the dextrorotatory isomer is not recognized by our taste buds as having any flavoring property:

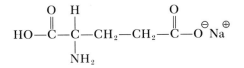

Ascorbic acid is an oxidation product of D-glucose as shown in the following reactions:

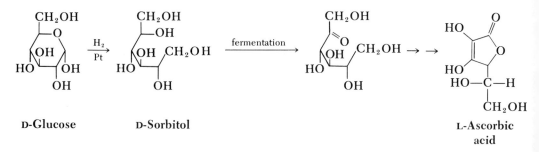

D-Glucose   D-Sorbitol   L-Ascorbic
                                                                               acid

Such a transformation is interesting in that it represents the conversion of a D-compound to an L-compound, that is, a change in configuration. However, the change in configuration is due to a change in the frame of reference in the nomenclature system rather than a change in the configuration about a particular carbon atom. L-Ascorbic acid, better known as vitamin C, and some of its derivatives are used as antioxidants in foodstuffs to prevent rancidity and to prevent the browning of cut apples. A deficiency of vitamin C results in scurvy.

The naturally occurring unsaturated long-chain fatty acids are nearly all of the *cis* configuration. The following example is linolenic acid (*cis,cis,cis-*9,12,15-octadecatrienoic acid):

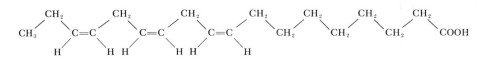

Summary

There are a number of ways in which the atoms of a molecule can be bonded to other atoms. An isomer is any one of a group of compounds that have the same molecular formula but different structural formulas and that display different physical or chemical properties.

Constitutional isomers have the same molecular formula but differ in the way the atoms in the molecule are bonded together, regardless of direction in space.

Stereoisomers have different spatial arrangements of their atoms. They can be classified as configurational isomers or conformational isomers.

Configurational isomers can be classified as enantiomers or diastereomers. Enantiomers are stereoisomers that are nonsuperimposable mirror images. A compound with a nonsuperimposable mirror image is optically active and can rotate the plane of polarized light. A racemic modification is a combination of equal amounts of enantiomers. Resolution is the process of separating a racemic modification into the enantiomers. Stereoisomers that do not have a mirror image relationship are called diastereomers. They can be interconverted only by the breaking or making of bonds. Geometric isomers are diastereomers. They can be interconverted only by the breaking and remaking of σ bonds (in the case of ring compounds) or π bonds (in the case of alkenes). A chiral carbon is a carbon atom bonded to four different

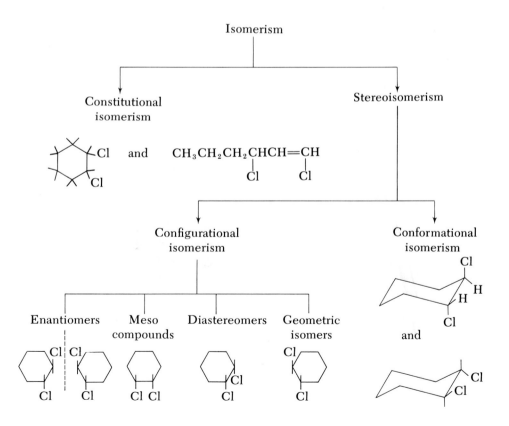

groups. A *meso* compound is superimposable on its mirror image even though it contains asymmetric carbon atoms.

Conformational isomers are the infinite number of arrangements of the atoms in space that can arise from rotation about a single bond. They can be interconverted by rotation of one part of the molecule with respect to the rest of the molecule about a single bond joining these two parts.

Each type of isomerism is illustrated at the bottom of p. 143 with 1,2-dichlorocyclohexane as the example.

Important terms

Axial group	Molecular formula
Boat conformation	Newman projection
Chair conformation	Optical activity
Chiral carbon	Optically active
Cis-trans isomerism	Optically inactive
Configurational isomers	Planar projection
Conformational isomers	Plane of polarized light
Constitutional isomers	Plane of symmetry
D-	Polarimeter
d-	R
Diastereomers	Racemic modification
E	Resolution
Eclipsed conformation	S
Enantiomers	Specific rotation
Equatorial group	Staggered conformation
Geometric isomers	Stereoisomers
Isomer	Superimposability
Isomerism	Z
L-	(+)-
l-	(−)-
Meso compound	

Problems

5-14. Which of the following items have a nonsuperimposable mirror image?

(a) Screw
(b) Shoe
(c) Sock
(d) Hammer
(e) Spool of thread

(f) Rifle
(g) Your nose
(h) Yourself
(i) Knife
(j) Werewolf

5-15. What is required for a molecule to exhibit optical activity?

5-16. Of what limitation must one be aware in using planar projections in determining whether or not a compound is capable of exhibiting optical activity?

5-17. Circle the chiral carbon atom(s), if there are any, in each of the following molecules:

(a) $CH_3CHClCHClCH_3$
(b) $(CH_3)_2CHBr$

(c)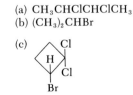

(d) $CH_3CHCH_2CHBrCH=CH_2$
$\quad\quad\quad |$
$\quad\quad\ CH_3$

(e) $CH_3CHOHCH_2CHOHCH_2COOH$

5-18. Which of the following compounds can exist as geometric isomers? Draw their structures.

(a) 1-Heptene
(b) 2-Heptene
(c) 3-Heptene

(d) 1,2-Diethylcyclopropane
(e) 1,1-Diethylcyclopropane
(f) 2,3-Dibromo-2-pentene

5-19. Which of the following molecules have a plane of symmetry?

(a) 1,1-Dichlorocyclobutane
(b) *cis*-1,2-Dichlorocyclobutane
(c) *trans*-1,2-Dichlorocyclobutane
(d)

$$
\begin{array}{c}
COOH \\
| \\
H-C-OH \\
| \\
H-C-OH \\
| \\
COOH
\end{array}
$$

(e)

$$
\begin{array}{c}
CH_2OH \\
| \\
C=O \\
| \\
CH_2 \\
| \\
CH_2OH
\end{array}
$$

(f)

$$
\begin{array}{c}
COOH \\
| \\
HO-C-H \\
| \\
HO-C-H \\
| \\
CH_2OH
\end{array}
$$

5-20. Answer the following questions in regard to these molecules:

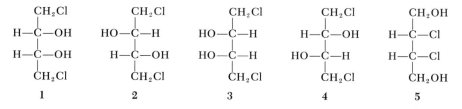

$$
\begin{array}{ccccc}
CH_2Cl & CH_2Cl & CH_2Cl & CH_2Cl & CH_2OH \\
H-C-OH & HO-C-H & HO-C-H & H-C-OH & H-C-Cl \\
H-C-OH & H-C-OH & HO-C-H & HO-C-H & H-C-Cl \\
CH_2Cl & CH_2Cl & CH_2Cl & CH_2Cl & CH_2OH \\
\mathbf{1} & \mathbf{2} & \mathbf{3} & \mathbf{4} & \mathbf{5}
\end{array}
$$

(a) What is the relationship between molecules 1 and 2?
(b) What is the relationship between molecules 2 and 4?
(c) What is the relationship between molecules 2 and 3?
(d) What is the relationship between molecules 1 and 4?
(e) Which is (are) a *meso* compound(s)?
(f) Which is (are) optically active?
(g) Which is (are) optically inactive?
(h) What is a mixture of equal amounts of molecules 2 and 4 called?
(i) Is a mixture of equal amounts of molecules 2 and 4 optically active or inactive? Why?
(j) What is the relationship between molecules 1 and 5, 2 and 5, 3 and 5, and 4 and 5?

5-21. On what principle is the concept of resolution based?

5-22. Answer the following question in regard to this molecule:

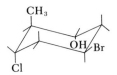

Is the methyl group *cis* or *trans* to the (a) chloro group, (b) bromo group, and (c) hydroxyl group?

5-23. (−)-Lyxoflavine has the structure shown as follows. It is known to be a growth-promoting agent. Of what configuration is lyxoflavine? What is the nature of its specific rotation?

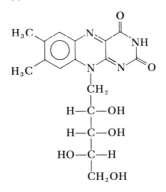

5-24. Indicate whether the molecules in each set are the same compound, enantiomers, diastereomers, *cis* or *trans* isomers, *meso* compounds, or conformational isomers. More than one answer may be possible.

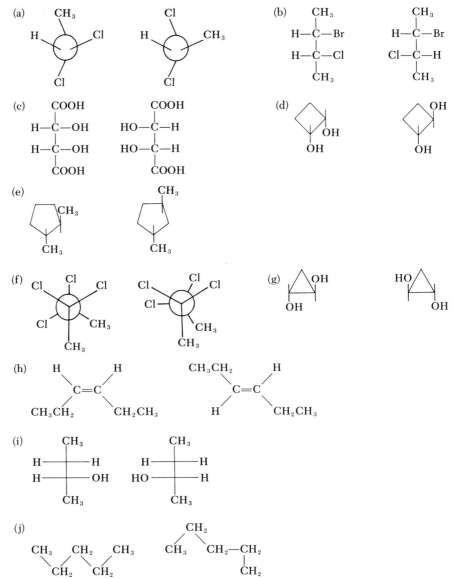

Self-test

1. Draw the structural formulas for all the constitutional isomers of C_5H_{12}. Give each isomer a correct IUPAC name.
2. Answer the following questions concerning the tetrahedral model of 1-bromoethanol:

   (a) Draw the planar projection.
   (b) Draw the mirror image.
   (c) Is the molecule optically active? Why or why not?
   (d) Designate the molecule as R or S.
   (e) How can the optical activity of the molecule be determined?
3. Answer the following questions in regard to the planar projection of D-ribose:

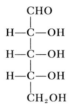

   (a) Draw the mirror image.
   (b) Draw a diastereomer.
   (c) Is D-ribose optically active?
   (d) Is the diastereomer optically active?
   (e) Is D-ribose a *meso* compound?
4. For each structure in column A, draw a structure that fits the description in column B:

A	**B**
(a) 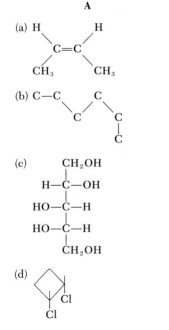	*Trans* isomer
(b)	Another conformation
(c)	*Meso* structure
(d)	Optically active diastereomer

A   **B**

(e) 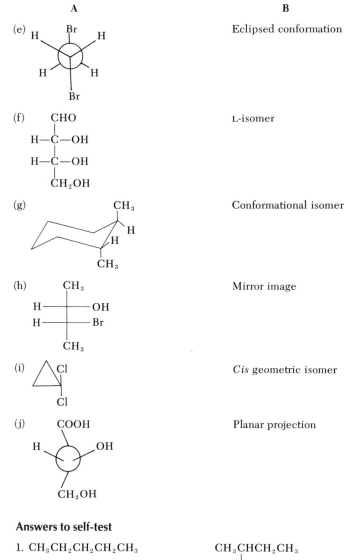 Eclipsed conformation

(f) CHO   L-isomer

H—C—OH

H—C—OH

CH₂OH

(g)   CH₃   Conformational isomer

(h) CH₃   Mirror image

H——————OH

H——————Br

CH₃

(i) Cl   *Cis* geometric isomer

Cl

(j) COOH   Planar projection

H   OH

CH₂OH

Answers to self-test

1. $CH_3CH_2CH_2CH_2CH_3$ $CH_3CHCH_2CH_3$ CH_3
                                          | CH_3CCH_3
                                         CH_3 |
                                                                    CH_3

   Pentane   **2-Methylbutane**   **2,2-Dimethylpropane**

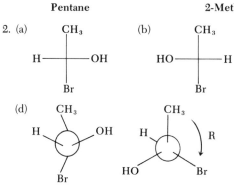

2. (a)   CH₃   (b)   CH₃   (c) Yes. The object and mirror image are
                                                              nonsuperimposable.
   H—————OH   HO—————H

        Br   Br

   (d)   CH₃   CH₃   (e) Through use of a polarimeter

   H   OH   H   R

        Br   HO   Br

3. (a)

```
        CHO
        |
  HO—C—H
        |
  HO—C—H
        |
  HO—C—H
        |
      CH₂OH
```

(b)

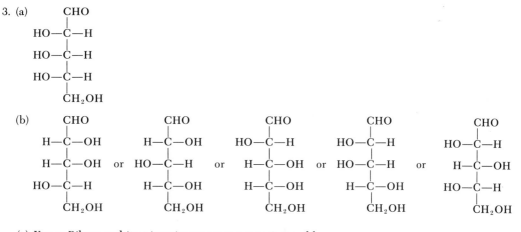

```
    CHO              CHO              CHO              CHO              CHO
    |                |                |                |                |
  H—C—OH           H—C—OH         HO—C—H           HO—C—H          HO—C—H
    |                |                |                |                |
  H—C—OH  or      HO—C—H    or     H—C—OH  or      HO—C—H    or     H—C—OH
    |                |                |                |                |
 HO—C—H            H—C—OH           H—C—OH           H—C—OH          HO—C—H
    |                |                |                |                |
  CH₂OH            CH₂OH            CH₂OH            CH₂OH            CH₂OH
```

(c) Yes. D-Ribose and its mirror image are nonsuperimposable.

(d) All the diastereomers are optically active.

(e) No

4. (a)

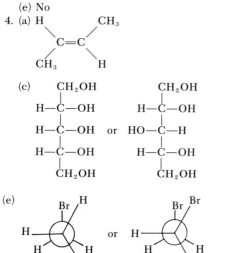

```
  H          CH₃
   \        /
    C=C
   /        \
 CH₃         H
```

(b)

```
      C       C       C
     /       / \     /
    C       C   C   C
```

(c)

```
    CH₂OH            CH₂OH
    |                |
  H—C—OH           H—C—OH
    |                |
  H—C—OH  or       HO—C—H
    |                |
  H—C—OH           H—C—OH
    |                |
  CH₂OH            CH₂OH
```

(d)

```
     ___
    |   \ Cl
    |___/
     |
     Cl
```

(e)

```
     Br  H                  Br  Br
      \ /                     \ /
  H——(   )——H    or    H——(   )——H
     / \                   / \
    H   H                 H   H
      Br                    H
```

(f)

```
        CHO
        |
  HO—C—H
        |
  HO—C—H
        |
      CH₂OH
```

(g)

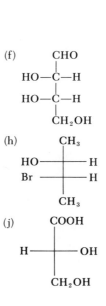

```
         ___
        /   \___
       /        \ CH₃
      |          /
       \        / CH₃
        _____/
           |
```

(h)

```
           CH₃
           |
   HO——————H
   Br ——————H
           |
           CH₃
```

(i)

```
      /\
     /__\
    Cl   Cl
```

(j)

```
        COOH
        |
   H————OH
        |
      CH₂OH
```

chapter 6

Alcohols and phenols

Structure of alcohols and phenols

Structurally, alcohols and phenols can be considered to be derivatives of water:

$$H—\overset{..}{\underset{..}{O}}—H \qquad\qquad R—\overset{..}{\underset{..}{O}}—H \qquad\qquad Ar—\overset{..}{\underset{..}{O}}—H$$

$$\textbf{Water} \qquad\qquad\qquad \textbf{Alcohol} \qquad\qquad\qquad \textbf{Phenol}$$

The functional group common to alcohols and phenols is the hydroxyl group (—OH). They differ in that the —OH in alcohols is bonded to any alkyl group or substituted alkyl group, whereas in phenols the —OH is bonded to an aromatic ring or substituted aromatic ring (Ar—).

Let us consider the kind of orbital oxygen uses when it occurs in alcohols and phenols, that is, when it is bonded to two other atoms. The ground state electron configuration for oxygen is $1s^2 2s^2 2p^2 2p^1 2p^1$. Oxygen forms sp^3 hybrid orbitals similar to those of a carbon atom. The configuration of the carbon atom with sp^3 hybrid orbitals is $1s^2 2[(sp^3)^1(sp^3)^1(sp^3)^1(sp^3)^1]$, and the configuration of the oxygen atom with sp^3 hybrid orbitals is $1s^2 2[(sp^3)^1(sp^3)^1(sp^3)^2(sp^3)^2]$. In the case of carbon, hybridization of the four electrons in the outermost energy level leads to the formation of four equivalent sp^3 hybrid orbitals, each containing one electron. Each electron is available for bond formation. In the case of oxygen, hybridization of the six electrons in the outermost energy level leads to the formation of two filled sp^3 orbitals and two sp^3 orbitals that contain one electron each. Each of the unfilled sp^3 orbitals is available for bond formation. The bond formations of carbon and oxygen atoms with sp^3 orbitals are illustrated as follows:

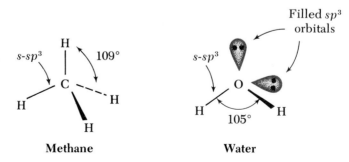

Methane   Water

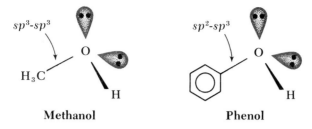

Methanol   **Phenol**

Although bond angles of 109° are generally associated with an atom with sp^3 hybrid orbitals, the two filled sp^3 orbitals of oxygen compress the bond angles in water, alcohols, and phenols to a value less than 109°.

Nomenclature of alcohols and phenols

Alcohols are classified according to the kind of carbon that bears the —OH group. If the —OH group is attached to a primary, secondary, or tertiary carbon atom, then the alcohol is classified as primary, secondary, or tertiary, respectively:

CH_3—CH_2—CH_2—OH

Primary

CH_3—CH—CH—⟨O⟩
         |
         OH

Secondary

$$CH_3\text{—}\overset{\displaystyle CH_3}{\underset{\displaystyle OH}{\overset{|}{\underset{|}{C}}}}\text{—}CH_3$$

Tertiary

Alcohols are named by either the common or the IUPAC system. The common system, which is generally applicable to the simpler alcohols, consists in naming the alkyl group and adding the word *alcohol.* Following are some examples:

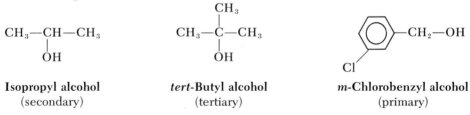

Isopropyl alcohol   ***tert*-Butyl alcohol**   ***m*-Chlorobenzyl alcohol**
(secondary)   (tertiary)   (primary)

In the IUPAC system, the parent structure is named. The name of the parent structure, which is the longest continuous chain that contains the —OH group, is derived by replacing the terminal *-e* of the corresponding alkane with the suffix *-ol.* The position of the —OH group in the parent chain is indicated with the lowest possible number. Then the positions of other groups attached to the parent chain are indicated by numbers. Following are some examples:

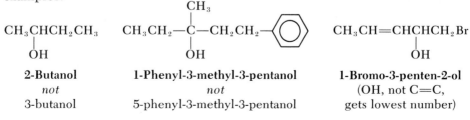

2-Butanol   **1-Phenyl-3-methyl-3-pentanol**   **1-Bromo-3-penten-2-ol**
not   *not*   (OH, not C=C,
3-butanol   5-phenyl-3-methyl-3-pentanol   gets lowest number)

Phenols are generally named as derivatives of the simplest member, phenol:

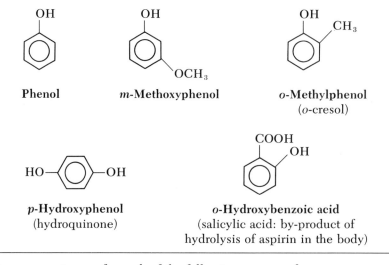

Phenol   *m*-Methoxyphenol   *o*-Methylphenol
(*o*-cresol)

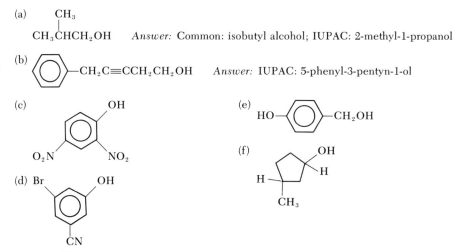

p-Hydroxyphenol   *o*-Hydroxybenzoic acid
(hydroquinone)   (salicylic acid: by-product of
hydrolysis of aspirin in the body)

6-1. Give a correct name for each of the following compounds:

(a)

         CH$_3$
         |
CH$_3$CHCH$_2$OH   *Answer:* Common: isobutyl alcohol; IUPAC: 2-methyl-1-propanol

(b) ⬡—CH$_2$C≡CCH$_2$CH$_2$OH   *Answer:* IUPAC: 5-phenyl-3-pentyn-1-ol

(c)

              OH

O$_2$N   NO$_2$

(d) Br   OH

           CN

(e)

HO—⬡—CH$_2$OH

(f)

             OH

H —   H

      CH$_3$

Physical properties of alcohols and phenols: boiling point and solubility

In contrast to hydrocarbons, which are nonpolar compounds, alcohols and phenols contain a very polar group, —OH. This group is polar because it contains a hydrogen atom bonded to the very electronegative oxygen atom. Such a condition permits *intermolecular hydrogen bonding*, that is, hydrogen bonding between molecules:

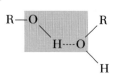

Table 6-1
Structures and boiling points of alcohols

Alcohol	Molecular weight	Boiling point (°C)
Methanol	32	65
Ethanol	46	78.5
1-Propanol	60	97
1-Butanol	74	118.5
2-Methyl-1-propanol	74	108
2-Butanol	74	99.5
2-Methyl-2-propanol	74	83

Table 6-2
Boiling points of compounds of comparable molecular weight

Compound	Structure	Molecular weight	Boiling point (°C)
n-Pentane	$CH_3CH_2CH_2CH_2CH_3$	72	36
Ethyl ether	$CH_3CH_2OCH_2CH_3$	74	35
n-Propyl chloride	$CH_3CH_2CH_2Cl$	79	47
n-Butyraldehyde	$CH_3CH_2CH_2CH_2CHO$	72	76
n-Butyl alcohol	$CH_3CH_2CH_2CH_2OH$	74	118

The effects of this hydrogen bonding are reflected in the physical properties of alcohols and phenols.

For hydrocarbons, it is generally true that boiling points are dependent on molecular weight and shape. A similar trend is noted for alcohols, as shown in Table 6-1. In general, the boiling points of alcohols increase as the molecular weights increase. However, the boiling points decrease as the number of branches from the carbon–oxygen bond increases:

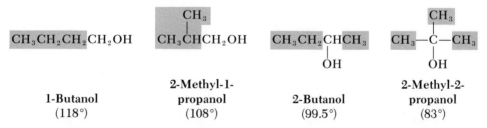

The unusual thing about the boiling points of alcohols is that they are so much higher than those of other molecules of comparable molecular weight. This characteristic, which is illustrated in Table 6-2, is due to the fact that alcohols are associated liquids, whose molecules are held together by hydrogen bonds. Thus the high boiling points of alcohols are due to the greater energy needed for breaking the hydrogen bonds that hold the molecules together as follows:

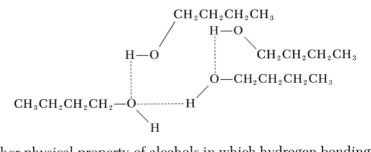

The other physical property of alcohols in which hydrogen bonding plays a role is solubility. In contrast to the hydrocarbons again, the lower alcohols are miscible with water, as shown in Table 6-3. Hydrocarbons are immiscible with water because they are nonpolar. The alcohols of lower molecular weight (methanol, ethanol, and 1-propanol) are miscible because the polar end, —OH, constitutes a large portion of the molecule whereas the hydrocarbon (Hc) portion of the molecule forms a relatively small portion of the molecule:

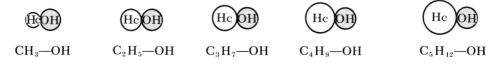

$$CH_3—OH \qquad C_2H_5—OH \qquad C_3H_7—OH \qquad C_4H_9—OH \qquad C_5H_{12}—OH$$

Thus the —OH group of these alcohols is capable of forming a hydrogen bond with water as follows:

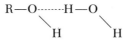

where R is CH_3, C_2H_5, or $n\text{-}C_3H_7$. If a molecule contains more than one hydroxyl group for hydrogen bonding, its solubility in water should reflect this. For example, ethylene glycol, a component of antifreeze, is completely soluble in water, and sugar is more soluble in water than would be expected for a covalent molecule:

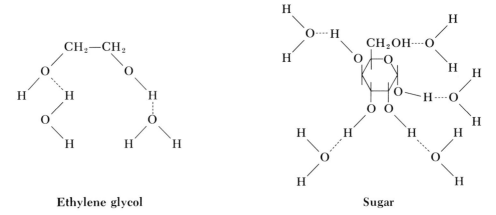

Ethylene glycol   **Sugar**

The boiling points and solubilities of phenols are also affected by hydrogen bonding, as shown in Table 6-4 for nitrophenols.

Table 6-3

Structures and solubilities of alcohols

Compound	Structure	Solubility (g/100 g H$_2$O)
Methanol	CH$_3$OH	Infinite
Ethanol	C$_2$H$_5$OH	Infinite
1-Propanol	CH$_3$CH$_2$CH$_2$OH	Infinite
1-Butanol	CH$_3$CH$_2$CH$_2$CH$_2$OH	7.9
1-Pentanol	CH$_3$CH$_2$CH$_2$CH$_2$CH$_2$OH	2.3
1-Hexanol	CH$_3$CH$_2$CH$_2$CH$_2$CH$_2$CH$_2$OH	0.6

Table 6-4

Properties of nitrophenols

Compound	Boiling point (°C) at 760 mm Hg	Solubility (g/100 g H$_2$O)
o-Nitrophenol	216	0.2
m-Nitrophenol	288	1.4
p-Nitrophenol	279 Decomposes	1.7

The boiling points of *m*- and *p*-nitrophenol are high because of intermolecular hydrogen bonding:

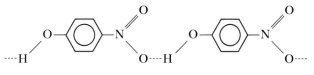

Their solubility is due to hydrogen bonding with water molecules:

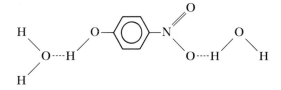

The boiling point and solubility of *o*-nitrophenol differ substantially because the —NO$_2$ and —OH groups can form an *intramolecular hydrogen bond*, a hydrogen bond within the same molecule:

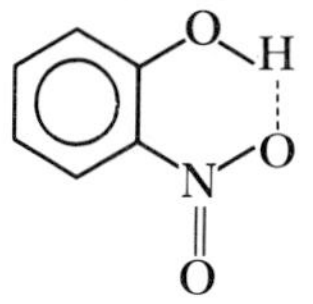

Hence, intermolecular hydrogen bonds with other *o*-nitrophenol molecules or water molecules are not possible.

Recently, hydrogen bonding was proposed as an explanation for the sweet taste of natural sweeteners, such as fructose, and artificial sweeteners, such as saccharin. The sensation of sweetness is generated when hydrogen bonds form between a grouping in the sugar molecule and a complementary grouping at the taste bud receptor site as follows:

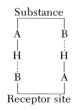

where A and B are electronegative atoms such as oxygen and nitrogen and H is a hydrogen atom attached to A by a covalent bond. The degree of sweetness depends on the spacing and orientation of these groups. The hydrogen bond groupings for fructose and saccharin are illustrated as follows:

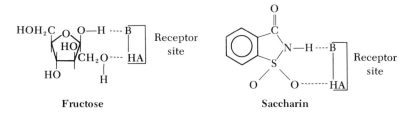

Hydrogen bonding also occurs between the base pairs in the DNA molecule. The hydrogen bonds between the base pairs cytosine and guanine and between the base pairs thymine and adenine are shown in Fig. 14-6.

Uses of alcohols and phenols

Ethyl alcohol, besides being the alcohol of alcoholic beverages (discussed later in the chapter), is widely used as a solvent for paints, perfumes, medicines, and flavorings.

Ethylene glycol (1,2-ethanediol) is a dihydroxy alcohol:

$$\begin{array}{cc} CH_2 & \!\!-CH_2 \\ | & | \\ OH & OH \end{array}$$

It is used as a "permanent" type of antifreeze for automobile radiators. It is useful for this purpose because it is nonvolatile and water soluble.

Glycerol (1,2,3-propanetriol), a trihydroxy alcohol, is a component of lipids (Chapter 13). It is used in liquid medications such as hand lotions because it tends to take up moisture from the air, thereby keeping the skin soft and moist. When glycerol is treated with nitric acid at low temperatures, it forms glyceryl trinitrate, an inorganic ester more commonly known as nitroglycerin:

$$\begin{array}{ccc} CH_2\!-\!CH\!-\!CH_2 & + \; 3HONO_2 & \longrightarrow \quad CH_2\!\!-\!\!-\!CH\!\!-\!\!-\!CH_2 \; + \; 3HOH \\ |\qquad |\qquad | & & |\qquad\quad |\qquad\quad | \\ OH \quad OH \quad OH & & ONO_2 \quad ONO_2 \quad ONO_2 \end{array}$$

Glycerol   **Nitroglycerin**

Nitroglycerin is used as an explosive and is a common constituent of dynamite.

Menthol, a derivative of cyclohexanol, is used in cough drops, shaving lotions, and mentholated cigarettes:

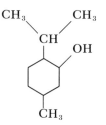

Cholesterol, the chief constituent of gallstones, is an alcohol belonging to a class of compounds called steroids. Some other important steroids are estrogen, the female sex hormone, and testosterone, the male sex hormone:

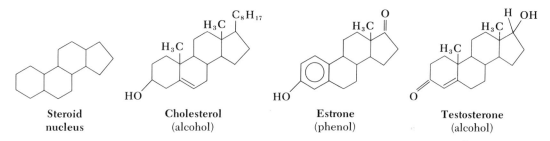

| Steroid nucleus | Cholesterol (alcohol) | Estrone (phenol) | Testosterone (alcohol) |

Phenol was introduced as an antiseptic by Lister in 1867; however, because of its toxic effect on tissue, it is seldom used today. Hexachlorophene is a potent bactericide found in many soaps, shampoos, deodorants, and toothpaste as well as in special preparations used to wash newborn babies (pHisoHex):

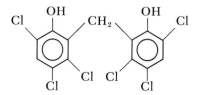

In a recent test, pHisoHex, a 3% solution of hexachlorophene, caused brain lesions in some experimental animals who were bathed with it daily for 90 days.

Uroshiol is an alkylated catechol (1,2-dihydroxybenzene) and may be one of the active constituents of poison ivy:

A mixture of cresols (*o*-, *m*-, and *p*-methylphenol) is used as a wood preservative (creosote).

Some phenols and alcohols are known to be constituents of essential oils of various plants. These oils give the plants their odor or flavor. Following are some examples:

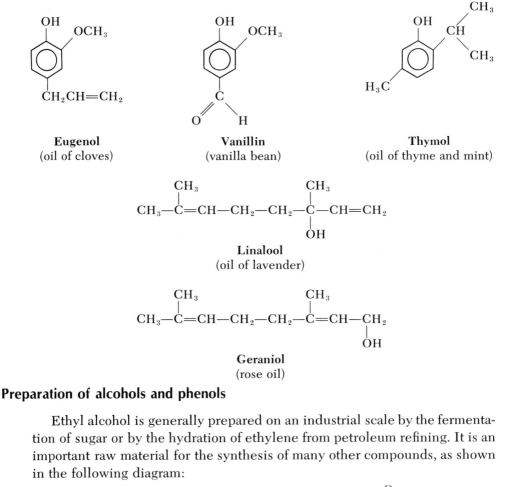

Eugenol
(oil of cloves)

Vanillin
(vanilla bean)

Thymol
(oil of thyme and mint)

Linalool
(oil of lavender)

Geraniol
(rose oil)

Preparation of alcohols and phenols

Ethyl alcohol is generally prepared on an industrial scale by the fermentation of sugar or by the hydration of ethylene from petroleum refining. It is an important raw material for the synthesis of many other compounds, as shown in the following diagram:

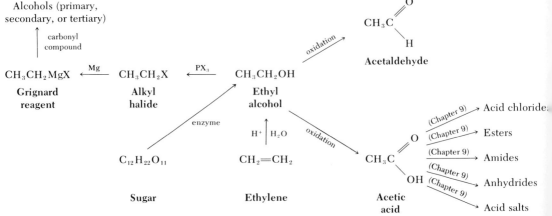

Other alcohols are prepared on an industrial scale by the reduction of fats (Chapter 13) and by the hydration of alkenes. The alcohols obtained from the hydration of alkenes are those consistent with the addition of water according to Markovnikov's rule.

In the laboratory, alcohols and phenols are prepared by hydrolysis of alkyl halides, Grignard synthesis, or hydroboration-oxidation.

Hydrolysis of alkyl halides. The reaction between an alkyl halide and potassium hydroxide in alcohol (C_2H_5OH) forms an alkene. For example:

$$CH_3CH_2CH_2Br + KOH \xrightarrow{\quad C_2H_5OH \quad} CH_3CH{=}CH_2 + KBr + H_2O$$

1-Bromopropane   **1-Propene**

However, if the reaction is carried out in aqueous solution, the resulting product is an alcohol. For example:

$$CH_3CH_2CH_2Br + KOH \xrightarrow{\quad HOH \quad} CH_3CH_2CH_2OH + KBr$$

1-Propanol

These two reactions differ in that the solvent alcohol causes an elimination reaction whereas the solvent water causes a substitution reaction (this is discussed in more detail in Chapter 7). Following are more of the preparation of alcohols and phenols by the hydrolysis of alkyl halides:

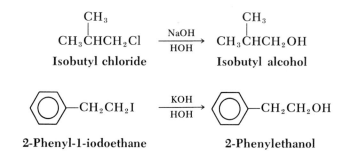

Phenol, however, cannot be prepared by this method. Phenyl halides do not react with aqueous base, as is shown in Chapter 7:

⬡—Br $\xrightarrow[\text{HOH}]{\text{KOH}}$ no reaction

Bromobenzene

Grignard synthesis. This method involves the use of the Grignard reagent, RMgX, which is prepared by the reaction of an organic halide and metallic magnesium as follows:

$$R{-}X + Mg \xrightarrow{\quad \text{dry ether} \quad} R{-}Mg{-}X$$

The halide can be a primary, secondary, or tertiary alkyl halide; a benzyl halide; or a phenyl (aryl) halide. The halogen can be Cl, Br, or I.

Alcohols are prepared by the reaction of Grignard reagents with carbonyl compounds, which contain the carbonyl group (C=O). The types of carbonyl compounds are as follows:

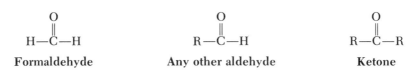

The class of alcohol obtained depends on the type of carbonyl compound used. For example, if formaldehyde is used, a primary alcohol is produced:

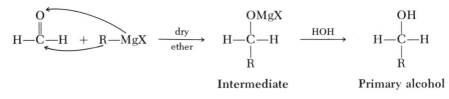

The reaction proceeds very well in anhydrous ether. Then the intermediate addition compound, RCH_2OMgX is converted into the alcohol by the addition of water.

The reaction with any other aldehyde (RCHO) yields a secondary alcohol, and that with a ketone forms a tertiary alcohol:

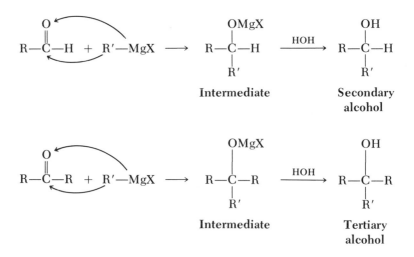

In these general reactions, R can be an alkyl, benzyl, or phenyl group. Following are some examples:

$$CH_3CH_2CH_2MgBr \; + \; \underset{\substack{\\ \text{Formaldehyde}}}{H-\overset{\displaystyle O}{\overset{\|}{C}}-H} \; \longrightarrow \; CH_3CH_2CH_2\overset{\displaystyle H}{\underset{\displaystyle H}{\overset{|}{\underset{|}{C}}}}-OMgBr \; \xrightarrow{\text{HOH}} \; \underset{\text{1-Butanol}}{CH_3CH_2CH_2CH_2OH}$$

**n-Propyl
magnesium
bromide**

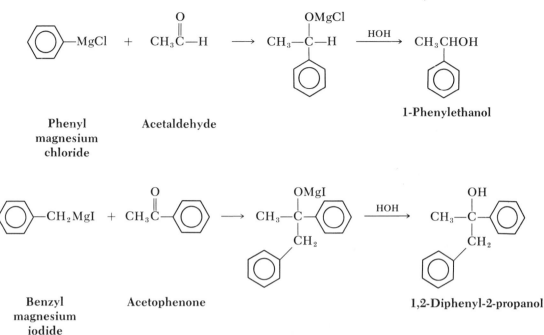

Table 6-5 lists some alcohols and the combinations of aldehydes and Grignard reagents that will yield them.

In a related reaction, a Grignard reagent reacts with ethylene oxide $\left(\begin{array}{c} CH_2\text{——}CH_2 \\ \diagdown \quad \diagup \\ O \end{array} \right)$, a cyclic ether (Chapter 8), instead of with a carbonyl compound. This results in a primary alcohol that contains two more carbon atoms than the original alkyl halide:

$$CH_2\text{——}CH_2 + R\text{—}MgX \longrightarrow R\text{—}CH_2\text{—}CH_2\text{—}OMgX \xrightarrow{\text{HOH}} RCH_2CH_2OH$$

Primary alcohol

For example:

$$CH_2\text{——}CH_2 + CH_3CH_2MgBr \longrightarrow CH_3CH_2CH_2CH_2OMgBr \xrightarrow{\text{HOH}} CH_3(CH_2)_3OH$$

1-Butanol

The electrons in the π bond of the carbonyl group are not equally shared by the carbon and the oxygen atoms because of the different electronegativities of these two atoms. Since oxygen is more electronegative, the electrons of the π bond are pulled more strongly toward this atom:

$$\underset{\diagup}{\overset{\diagdown}{C}} = \overset{\delta^+ \quad \delta^-}{\underset{\cdot\cdot}{O}}:$$

As a result, the carbon atom of the carbonyl group is susceptible to attack by nucleophilic species, and the oxygen atom undergoes attack by electrophilic

Table 6-5
Alcohols prepared by Grignard synthesis*

Alcohol		Carbonyl compound	Grignard reagent
Structure	Class		

CH$_3$—CH$_2$—C(OH)(H)—CH$_2$—CH$_3$	Secondary	CH$_3$CH$_2$CHO	CH$_3$CH$_2$MgBr
		CH$_3$C(=O)—C$_6$H$_5$	CH$_3$MgBr
CH$_3$—C(CH$_3$)(OH)—C$_6$H$_5$	Tertiary	CH$_3$CCH$_3$ (=O)	C$_6$H$_5$—MgBr
CH$_3$—C(CH$_3$)(OH)—C$_6$H$_5$			
CH$_3$CH$_2$CH$_2$CH$_2$CH$_2$OH	Primary	H—C(=O)—H, CH$_2$—CH$_2$(O)	CH$_3$CH$_2$CH$_2$CH$_2$MgBr, CH$_3$CH$_2$CH$_2$MgBr
CH$_3$CH$_2$CH$_2$CH$_2$CH$_2$OH			
cyclopentane(OH)(CH$_3$)	Tertiary	cyclopentanone =O	CH$_3$MgBr

*The boxed-in grouping appears in the Grignard reagent.

species. Such a reaction is called addition to a carbonyl group (Chapter 10). The Grignard reagent contains a polar carbon–magnesium bond:

$$\underset{\delta^-}{R} - \underset{\delta^+}{MgX} \quad \text{or} \quad \underset{\delta^-}{R} \underset{\delta^+}{:MgX}$$

Thus the organic group of the Grignard reagent (R:) becomes bonded to the carbon atom of the carbonyl group, and magnesium (as MgX) becomes bonded to the oxygen atom; the product is the addition compound, which is readily converted to the alcohol in the presence of water:

$$\underset{\delta^+}{\overset{\delta^+}{C}}{=}\underset{\delta^-}{O} + \underset{\delta^-}{R}{:}\underset{\delta^+}{MgX} \longrightarrow R-\overset{|}{\underset{\underset{OMgX}{|}}{C}}- \xrightarrow{HOH} R-\overset{|}{\underset{|}{C}}-OH + Mg(OH)X$$

Phenols cannot be prepared by the Grignard synthesis. No carbonyl compound exists to prepare a phenol. Instead, phenols are prepared by the alkali fusion of sulfonates:

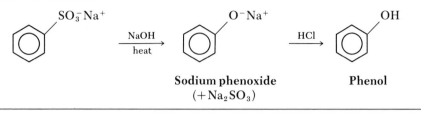

Sodium phenoxide   Phenol
$(+Na_2SO_3)$

6-2. Give the product of the reaction of each of the following compounds with iso-butylmagnesium chloride $[(CH_3)_2CHCH_2MgCl]$:

(a) $CH_3\!\!-\!\!\overset{\displaystyle\|}{\underset{\displaystyle O}{C}}\!\!-\!\!CH_2CH_3$ *Answer:* 3,5-Dimethyl-3-hexanol

(b) $CH_3CH_2\overset{\displaystyle\|}{\underset{\displaystyle O}{C}}\!\!-\!\!H$ *Answer:* 5-Methyl-3-hexanol

(c) $H\!\!-\!\!\overset{\displaystyle\|}{\underset{\displaystyle O}{C}}\!\!-\!\!H$ (d) 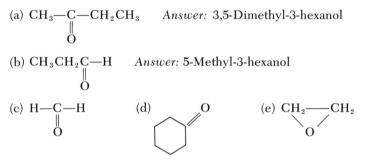   (e) $CH_2\!\!-\!\!\!-\!\!CH_2$
$\overset{\displaystyle \diagdown}{}\overset{O}{}\overset{\displaystyle \diagup}{}$

6-3. What combination of Grignard reagent and carbonyl compound could be used to make each of the following alcohols. Give all possibilities.

(a) 1-Pentanol (two combinations)   *Answer:* (1) *n*-Butylmagnesium bromide and formaldehyde and (2) *n*-propylmagnesium bromide and ethylene oxide

(b) 3-Phenyl-2-butanol
(c) 1-Methylcyclopentanol

Hydroboration-oxidation. Alcohols can be prepared from compounds containing carbon–carbon double bonds by a hydroboration-oxidation process. This process is similar to the Markovnikov addition of water to a carbon–carbon double bond in that the overall reaction is the addition of a molecule of water; however, it is significantly different in that the alcohol obtained is the one corresponding to the anti-Markovnikov addition of water to the carbon–carbon double bond. Furthermore, no rearrangements occur in this

Table 6-6
Markovnikov versus anti-Markovnikov addition of water
to a carbon–carbon double bond

	Markovnikov addition	*Hydroboration-oxidation*
Product	Alcohol	Anti-Markovnikov alcohol
Rearrangements	Yes	No
Conditions	HOH/H^+	(1) $(BH_3)_2$; (2) H_2O_2, OH^-

reaction. These Markovnikov and anti-Markovnikov reactions are compared in Table 6-6.

Hydroboration involves the addition of diborane $[(BH_3)_2]$ to the double bond, which form a trialkylborane; then oxidation gives an alcohol:

$$3R-CH{=}CH_2 + (BH_3)_2 \longrightarrow R-CH_2-CH_2-\underset{\underset{CH_2-CH_2-R}{|}}{B}-CH_2-CH_2-R \xrightarrow{3H_2O_2,\ OH^-} 3RCH_2CH_2OH + 3H_3BO_3$$

Trialkylborane   **Primary alcohol**

The major product is exactly opposite of the one formed in the Markovnikov addition of water in the presence of an acid catalyst. Following are some examples:

$$CH_3CH_2CH{=}CH_2 \xrightarrow[\text{(2) }H_2O_2,\ OH^-]{\text{(1) }(BH_3)_2} CH_3CH_2CH_2CH_2OH$$

1-Butene   **1-Butanol** (primary)

$$\underset{\underset{\substack{\text{2-Methyl-}\\\text{2-butene}}}{}}{\overset{\overset{\displaystyle CH_3}{|}}{CH_3C}{=}CHCH_3} \xrightarrow[\text{(2) }H_2O_2,\ OH^-]{\text{(1) }(BH_3)_2} \underset{\underset{\substack{\text{3-Methyl-}\\\text{2-butanol (secondary)}}}{}}{\overset{\overset{\displaystyle CH_3}{|}}{CH_3\underset{\underset{OH}{|}}{CH}CHCH_3}}$$

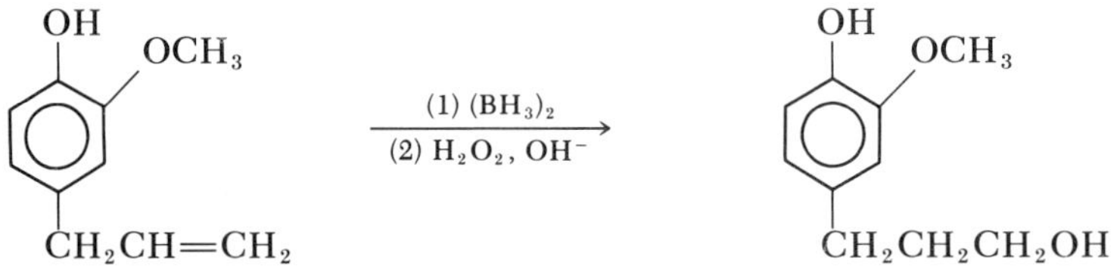

Eugenol   **3-(3-Methoxy-4-hydroxyphenyl)-1-propanol** (primary)

The boron atom in BH_3 with only six electrons is a Lewis acid. Hence, it seeks out the electrons of the π bond in the alkene. In determining to which carbon atom the boron atom will attach, let us consider the following molecule:

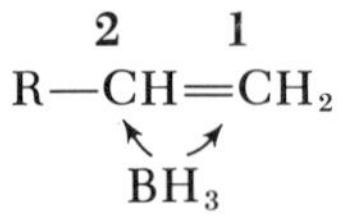

$$\overset{\displaystyle 2\quad\ \ 1}{R-CH{=}CH_2}$$
$$\underset{BH_3}{\nwarrow\ \nearrow}$$

The boron atom probably attaches itself to the least sterically hindered carbon 1. If R on carbon 2 is hydrogen, it makes no difference which carbon of the carbon–carbon double bond becomes bonded to boron, since they are equivalent. However, if R is a methyl or any other alkyl group, its size would encourage attack on the carbon with the fewest substituted groups, that is, carbon 1. The transition state might then have the following structure:

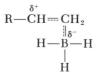

One of the hydrogen atoms is then transferred from boron to the positively charged carbon atom. This is followed by the formation of the dialkylborane and then the trialkylborane by addition of the alkylborane to second and third alkene molecules:

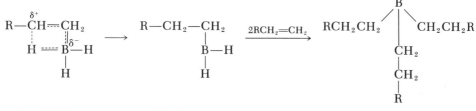

| Transfer of H from B | Alkylborane | Trialkylborane |

Reactions of alcohols and phenols

The reactions of alcohols and phenols involve the breaking of either the carbon–oxygen (C–OH) bond or the oxygen–hydrogen (CO–H) bond.

Breaking of carbon–oxygen bond

The breaking of the carbon–oxygen bond is applicable only to alcohols.
Formation of alkyl halides. Alcohols undergo substitution reaction to form alkyl halides:

$$R—OH \xrightarrow{\text{HX or } PX_3} R—X$$

The various halogen-containing compounds used in preparing these alkyl halides include phosphorus tribromide or triiodide, sodium or potassium bromide with sulfuric acid, and concentrated hydrochloric acid. Following are some examples of the formation of alkyl halides:

$$3CH_3CH_2CH_2OH + PBr_3 \longrightarrow 3CH_3CH_2CH_2Br + H_3PO_3$$

$$2CH_3CH_2OH + 2NaBr + H_2SO_4 \longrightarrow 2CH_3CH_2Br + Na_2SO_4 + 2H_2O$$

$$(CH_3)_3COH + HCl \longrightarrow (CH_3)_3CCl + HOH$$

$$3CH_3\underset{\underset{OH}{|}}{C}HCH_3 + PI_3 \longrightarrow 3CH_3\underset{\underset{I}{|}}{C}HCH_3 + H_3PO_3$$

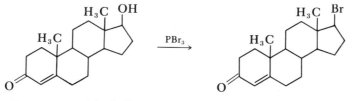

Testosterone (alcohol)

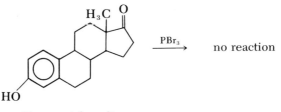

Estrone (phenol)

These reactions occur with each class of alcohol, although the reactivity varies as follows: benzyl > tertiary > secondary > primary > methyl alcohol.

6-4. What is the order of reactivity of the following alcohols with HCl?

    (a) 1-Phenylethanol   *Answer:* Benzylic alcohol: most reactive
    (b) 2-Methyl-3-pentanol
    (c) 1-Pentanol   *Answer:* Primary alcohol: least reactive
    (d) 2-Methyl-2-pentanol

Dehydration. Dehydration of alcohols results in alkene formation (Chapter 3). For example:

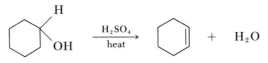

    Cyclohexanol   **Cyclohexene**

A dehydration reaction can also give rise to rearranged products. For example:

$$CH_3-\underset{\underset{OH}{|}}{\overset{\overset{CH_3}{|}}{CH}}-CH-CH_3 \xrightarrow{H_2SO_4} CH_3-\overset{\overset{CH_3}{|}}{CH}-CH=CH_2 + CH_3-\overset{\overset{CH_3}{|}}{C}=CH-CH_3 + CH_2=\overset{\overset{CH_3}{|}}{C}-CH_2-CH_3$$

3-Methyl-2-butanol   **3-Methyl-1-butene**   **2-Methyl-2-butene**   **2-Methyl-1-butene**

The mechanism proceeds through the carbonium ion intermediate (Chapter 3).

6-5. Show the mechanism by which 2-methyl-1-butene is formed in the dehydration of 3-methyl-2-butanol.   *Answer:* 2-Methyl-1-butene arises from a tertiary carbonium ion.

Breaking of oxygen–hydrogen bond

Both alcohols and phenols undergo a number of reactions in which the oxygen–hydrogen bond in broken.

Reactions as acids. Phenols act as acids in aqueous solutions of sodium hydroxide, since salts of phenols are formed. For example:

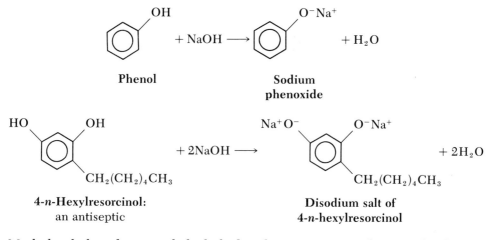

Phenol

Sodium
phenoxide

4-*n*-Hexylresorcinol:
an antiseptic

Disodium salt of
4-*n*-hexylresorcinol

Methyl, ethyl, and *n*-propyl alcohols dissolve in aqueous solutions of sodium hydroxide, since they are water soluble. However, alcohols of five carbons or more do not react with aqueous sodium hydroxide. For example:

$$CH_3(CH_2)_6CH_2OH + NaOH \longrightarrow \text{no reaction}$$
1-Octanol

The strength of an acid depends on its degree of ionization. For example, the ionization reaction of phenol is as follows:

$$C_6H_5OH \longrightarrow C_6H_5O^- + H^+$$

The ionization constant, K_a, for this reaction is calculated as follows:

$$K_a = \frac{[C_6H_5O^-][H^+]}{[C_6H_5OH]}$$

Thus the strength of an acid depends on the ratio of the amount that exists in the ionized form to that which exists in the un-ionized form. The stronger the acid is, the greater the amount of ionized form there is and the larger the value of K_a is. The K_a of phenol is 1×10^{-10} whereas that of the simple alcohol methyl alcohol is 3×10^{-16}.

To determine why the —OH attached to an aromatic ring is more acidic than one attached to a methyl group, let us examine the structures of the reactants and the products of the ionization of methyl alcohol and of phenol:

$$CH_3-\ddot{O}-H \longrightarrow CH_3-\ddot{O}{:}^{\ominus} + H^{\oplus}$$
Methyl alcohol   **Methoxide**
anion

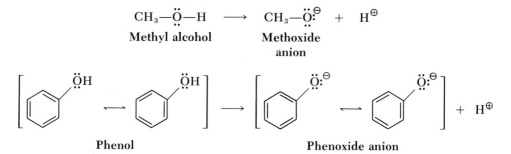

Phenol

Phenoxide anion

Methyl alcohol and the methoxide ion are each satisfactorily represented by single structures. However, phenol and the phenoxide ion are each a resonance hybrid of the structures I to V and Ia to Va, respectively:

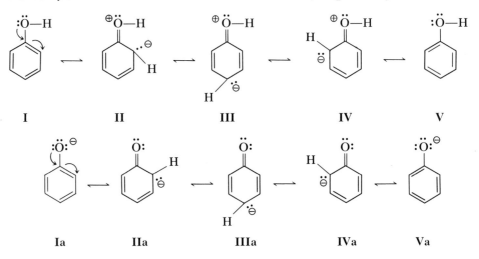

| I | II | III | IV | V |

| Ia | IIa | IIIa | IVa | Va |

In the case of phenol, contribution is made from three polar structures, II to IV, which contain a positively charged oxygen atom and a negatively charged benzene ring. In the case of the phenoxide ion, contribution is made from five polar structures, Ia to Va. The structures of phenol are less stable, since they involve a separation of charge, which requires energy. However, no separation of charges is required in the structures of the phenoxide anion. Thus resonance stabilizes the anion more than phenol. In the phenoxide ion, the contributing structures are actually more stable because of the delocalization of the negative charge. Thus the ionization of phenol requires less energy than the ionization of methyl alcohol. This is illustrated in Fig. 6-1.

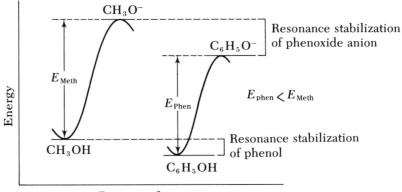

Fig. 6-1
Ionization of phenol versus methyl alcohol.

The acidity of phenol is increased by the presence of an electron-withdrawing group on the ring and decreased by the presence of an electron-donating group. Hence, the order of acidity for phenol and phenols with substituted groups is as follows: phenols with deactivating groups > phenol > phenols with activating groups.

6-6. Using the Kekulé structures shown, predict which phenol with a substituted group is the stronger acid:

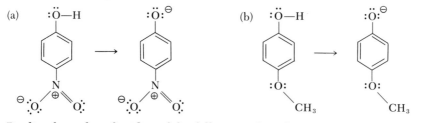

6-7. Predict the order of acidity of the following phenols:

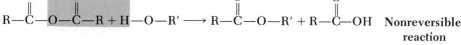

Ester formation. Both alcohols and phenols react with carboxylic acids and carboxylic acid derivatives – acid chlorides and acid anhydrides (Chapter 9) – to form esters:

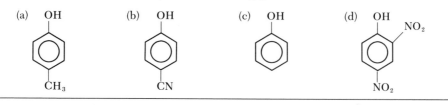

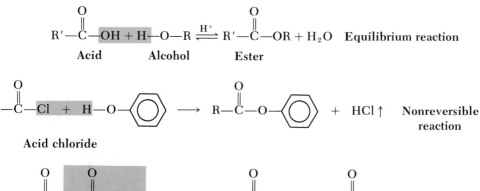

The alkyl groups can be the same (R and R) or different (R and R'). The reaction occurs more easily with acid chlorides and acid anhydrides than with carboxylic acids, since the reaction with carboxylic acids is actually an equilibrium reaction that is usually catalyzed by trace amounts of sulfuric acid. Following are some examples of ester formation:

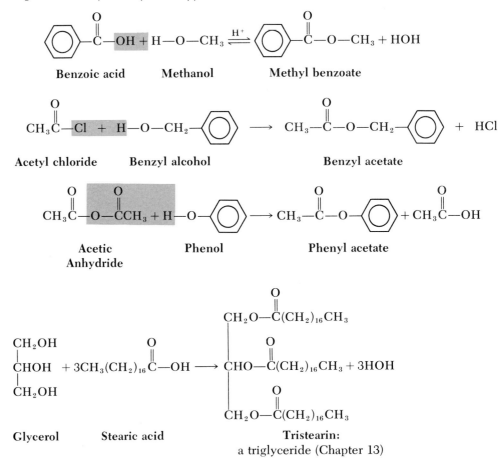

Glycerol   Stearic acid   Tristearin:
                                        a triglyceride (Chapter 13)

Oxidation. The product formed when an alcohol is oxidized depends on the class of alcohol undergoing the reaction and, in one case, on the nature of the oxidizing agent. The schemes of oxidation are as follows:

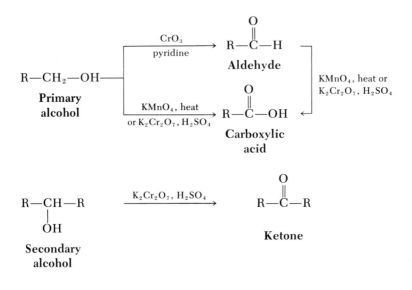

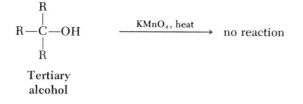

$$R-\underset{\underset{R}{|}}{\overset{\overset{R}{|}}{C}}-OH \xrightarrow{\text{KMnO}_4, \text{ heat}} \text{no reaction}$$

Tertiary
alcohol

Primary alcohols are oxidized to aldehydes when treated with chromium trioxide (CrO_3) in pyridine (C_5H_5N) and yield acids (a more highly oxidized form of alcohol) when a stronger oxidizing agent is used. Secondary alcohols give rise to ketones. Under neutral conditions, tertiary alcohols are not oxidized. Following are some examples:

$$CH_3CH_2CH_2OH \xrightarrow[\text{pyridine}]{\text{CrO}_3} CH_3CH_2\overset{\overset{O}{\|}}{C}-H$$

1-Propanol   Propionaldehyde

$$\underset{\underset{OH}{|}}{CH_3CHCH_3} \xrightarrow{\text{K}_2\text{Cr}_2\text{O}_7, \text{ H}_2\text{SO}_4, \text{ heat}} CH_3\underset{\underset{O}{\|}}{C}CH_3$$

2-Propanol   Acetone

Benzyl alcohol $\xrightarrow{\text{KMnO}_4}$ Benzoic acid

Phenols are also susceptible to attack by oxdizing agents, and a mixture of products is formed.

Roadside breath test. A novel application of the oxidation of ethyl alcohol is found in the roadside breath test, a screening device intended to give a rough measure of the quantity of alcohol in the blood. It is based on the premise that the ethanol absorbed into the blood from the stomach and intestine is continuously transferred into the lungs and exhaled. A volume of exhaled breath is said to contain 1/2100 of the amount of ethanol in an identical volume of blood. It is an offense for a motor vehicle driver to have a blood alcohol level in excess of 150 mg/100 ml of blood (0.15%).

The operation of the Breathalyzer used in this test (Fig. 6-2) is based on the potassium dichromate–sulfuric acid oxidation of ethanol:

$$3C_2H_5OH + 2K_2Cr_2O_7 + 8H_2SO_4 \longrightarrow 3CH_3COOH + 2Cr_2(SO_4)_3 + 2K_2SO_4 + 11H_2O$$
Orange   Green

The particles of silica gel in a sealed gas ampul are impregnated with the reagents. Before the ampul is used, the ends are broken off; then one end is fitted with a plastic mouthpiece, and the other is attached to the neck of a flattened bag. When air containing ethanol is blown through the tube, a

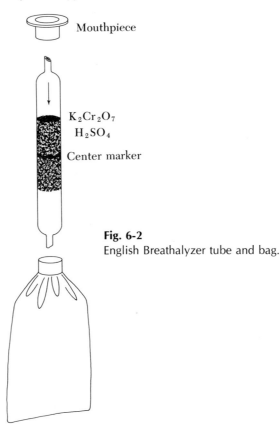

Mouthpiece

$K_2Cr_2O_7$
H_2SO_4

Center marker

Fig. 6-2
English Breathalyzer tube and bag.

chemical reaction takes place (the temperature of the tube increases), and the chromium sulfate produced is shown by the formation of a green color in place of the original orange reagent. If the green color extends beyond the yellow center mark of the tube, the motorist is taken to the police station for further tests (such as the more sensitive gas chromatography described later in the chapter).

The test is rather poor in quantitating ethanol, and it is only capable of picking out nearly all cases of people with more than 150 mg/100 ml of blood (the legal limit). Anomalously high readings for alcohol in the breath occur if the test is carried out less than 20 minutes after the last drink. Smoking prior to the test also leads to incorrect results, since the reagent becomes coated with the brown tars from the tobacco smoke. The main virtues of the Breathalyzer are its cheapness and portability; thus it satisfies the conditions for a screening test.

More recently, law enforcement agencies have adopted a portable dichromate spectrophotometric breath analyzer to detect alcohol intoxication. The alcohol determination is based on the distance light travels in a reaction cell containing dichromate and alcohol exhaled from the lungs compared to a standard cell containing the dichromate solution only. The exhaled breath is blown into a cylinder containing a piston. When the cylinder contains a predetermined amount of breath, the piston allows the air from the lungs to pass through the dichromate solution, where the dichromate changes from orange to green. The distance the light moves through the standard cell compared to the reaction cell is spectrophotometrically measured and is proportional to the amount of alcohol contained in 1 ml of blood.

Electrophilic aromatic substitution in phenols. Phenol undergoes several ring substitution reactions. The hydroxyl group is an *ortho,para* director and, furthermore, is a ring activator, so that all ring substitution reactions go faster with phenol than with benzene. Following are some examples:

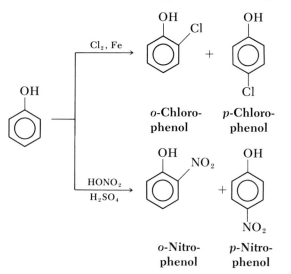

o-Chloro-
phenol   *p*-Chloro-
phenol

o-Nitro-
phenol   *p*-Nitro-
phenol

Gas chromatography

Gas chromatography is an accurate and rapid method for the separation and analysis of the components of a mixture. A basic gas chromatographic apparatus consists of an injector, a separating column, a carrier gas, a detector, and a recorder (Fig. 6-3). The injector is used for introducing the sample. The separating column is a tube packed with a solid material and coated with a stationary adsorbing liquid. The sample is swept through the column by the carrier gas. A constant flow of carrier gas (usually helium) is maintained by

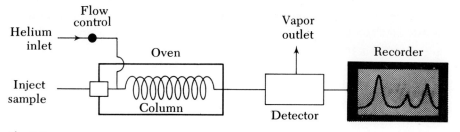

Fig. 6-3
Block diagram for gas chromatography.

a flow control. The detector is a device for detecting and measuring the quantity of the separated components; however, it does not identify the components. The recorder provides a record in the form of a graph or trace in which each component appears as a triangular peak. Such a graph is shown below.

A sample of a mixture is analyzed in the following manner. The sample is introduced into, vaporized in, and mixed with the carrier gas in the injector. The carrier gas pushes the vaporized sample into the column, and the vapor is partitioned between the gas and liquid phases, depending on its solubility in the liquid at the column temperature. This equilibrium is maintained as the carrier gas moves through the column. The rate at which the sample moves through the column is determined by the vapor pressure of the sample, the solubility of the sample in the stationary phase, and the rate at which the carrier gas flows. Since each component of a sample moves at a characteristic rate, the mixture will be completely separated by the time it reaches the detector. As each component reaches the detector, a trace is provided by the recorder. The area under the trace is proportional to the concentration of that component in the sample. A component is identified by its retention time, that is, the length of time it remains in the column. A component in a mixture will have the same retention time as a known sample of that component. Typical records of a mixture and a known sample of a component of the mixture are shown as follows:

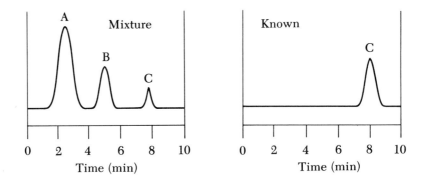

Analysis of alcoholic beverages

Gas chromatography offers a rapid, sensitive, and simple means of separating alcoholic beverages into individual organic components. Many distilleries and breweries find it a practical method for determining the composition of their products.

The four major components of alcoholic beverages are the light ends, ethanol, fusel oils, and materials of higher boiling points. Some beverages such as whiskey, rum, and brandy contain high-molecular-weight tannins, caramel (coloring material), and inorganic materials; these components are not detected and water does not interfere if a special detector is used. The light ends are those components that are eluted before ethanol. Generally,

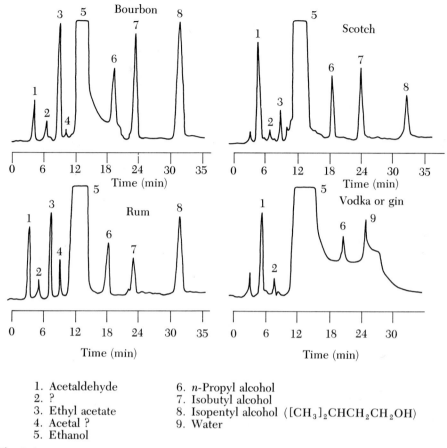

1. Acetaldehyde
2. ?
3. Ethyl acetate
4. Acetal ?
5. Ethanol
6. *n*-Propyl alcohol
7. Isobutyl alcohol
8. Isopentyl alcohol ($[CH_3]_2CHCH_2CH_2OH$)
9. Water

Fig. 6-4

Chromatograms of some common beverages. (From Scott, C., Haddin, N., and Bonelli, E.: Analysis of alcoholic beverages, Walnut Creek, Calif., 1966, Varian Associates. Reprinted with permission of Varian Associates.)

they include acetaldehyde (CH_3CHO), ethyl acetate ($CH_3COOC_2H_5$), acetone (CH_3COCH_3), methanol, and dimethyl acetal ($CH_3CH(OCH_3)_2$). Fusel oil is the mixture of *n*-propyl, isobutyl, and isopentyl alcohols. The materials of higher boiling points include ethyl esters of C_6 to C_{18} acids ($RCOOC_2H_5$ where R is C_5 to C_{17}) and 2-phenylethanol. The organic components in alcoholic beverages range in boiling point (°C) from acetaldehyde (21°), ethyl alcohol (78°), ethyl ocatadecanoate ($C_{17}H_{35}COOC_2H_5$, 215°), to 2-phenylethanol (219°). Fig. 6-4 gives the chromatograms of some common beverages, showing the light ends, ethanol, and fusel oils.

There seems to be some controversy over the exact effect of the fusel oils. Some say that they are responsible for the well-known hangover, whereas others say that they are responsible for the flavors of particular beverages. The student can decide for himself from an examination of the fusel oil concentrations in various beverages listed in Table 6-7.

Alcoholic beverages are prepared by the fermentation of sugars from

Table 6-7

Fusel oil content in various alcoholic beverages, in parts per million

	n-Propyl alcohol	Isobutyl alcohol	Isopentyl alcohol	Total
Georgia Moon	195	386	965	1546
Bourbon	161	270	>1000	1430
Tequila	142	313	873	1330
Brandy	190	310	438	938
Scotch whiskey	228	313	212	753
White lightning	156	440	50	646
Rum (gold)	394	115	50	559
Rum (white)	132	82	125	339
Red wine (homemade)	62	105	140	307
Vermouth	75	79	108	262
White wine	56	66	85	207
Beer	26	30	<50	56
Gin or vodka	<5	<25	<50	

Table 6-8

Products of reactions of alcohols and phenols with various reagents

Reagent	Product of reaction	
	With alcohol (ROH)	With phenol (C_6H_5OH)
Na	RO^-Na^+	$C_6H_5O^-Na^+$
NaOH	No reaction (no alkoxide formation)	$C_6H_5O^-Na^+$
HX(HCl, HBr, HI)	RX (alkyl halide)	No reaction
$PX_3(PCl_3, PBr_3, PI_3)$	RX	No reaction
H_2SO_4, heat	Alkene	No elimination
R'COOH (acid)	R'COOR (ester)	$R'COOC_6H_5$ (ester)
R'COCl	R'COOR	$R'COOC_6H_5$
$(R'CO)_2O$	R'COOR	$R'COOC_6H_5$
CrO_3 in pyridine	RCHO (aldehyde)°	⎧ Complicated
$KMnO_4$, heat or $K_2Cr_2O_7$,	RCOOH° or RCOR (ketone)†	⎨ oxidation product ⎩
Br_2, Fe	—	⎧ Ring substitution
$HONO_2$, H_2SO_4	—	⎨ in *ortho* and
$HOSO_3H$, heat	—	⎩ *para* positions

°If ROH is a primary alcohol.
†If ROH is a secondary alcohol.

various vegetable sources. The reaction is initiated by certain enzymes, such as those found in yeast. The maximum concentration of alcohol resulting from the fermentation process is 12%. The sources of some common alcoholic beverages are as follows: bourbon, corn; brandy, fruits; gin, juniper berries; Irish whiskey, potatoes; rum, molasses; saki, rice; Scotch whiskey, barley; tequila, century plant; vodka, rye; and wine, grapes.

Undenatured alcohol costs about 15 cents per fifth to produce. The tax on it is approximately $4. For tax purposes, the concentration of alcohol is given in terms of proof. Pure alcohol is 200 proof. A solution that is 60% alcohol by volume is 120 proof.

Summary

In the IUPAC system, alcohols are named by replacing the terminal -e of the corresponding alkane with the suffix -ol, and the OH group is given the lowest number and takes precedence in numbering over the carbon–carbon double bond or triple bond. Phenols are named as derivatives of phenols.

Both alcohols and phenols undergo intermolecular hydrogen bonding. In addition, some phenols with substituted groups in the *ortho* position undergo intramolecular hydrogen bonding.

Alcohols, but not phenols, are prepared in the laboratory by hydrolysis of alkyl halides, reactions of Grignard reagents with carbonyl compounds, and hydroboration-oxidation of alkenes. Phenol is synthesized by alkali fusion of sulfonates.

The reactions of alcohols and phenols are summarized in Table 6-8.

Important terms

Acid anhydride	Hydroboration-oxidation
Acid chloride	Intermolecular hydrogen bonding
Alcohol	Intramolecular hydrogen bonding
Aldehyde	Ketone
Carboxylic acid	Phenol
Ester	Primary alcohol
Formaldehyde	Secondary alcohol
Gas chromatography	Tertiary alcohol
Grignard reagent	

Problems

6-8. Draw the structures and give the IUPAC names for the isomeric pentanols ($C_5H_{12}O$).

6-9. Draw the structures and give the IUPAC names for the isomeric phenols of molecular formula $C_8H_{10}O$.

6-10. Give a correct name (common or IUPAC) for each of the following alcohols or phenols:

(a)   OH

(b) $CH_3CH_2CH_2CH_2OH$

(c)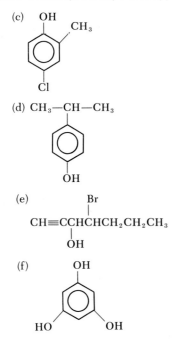

(g)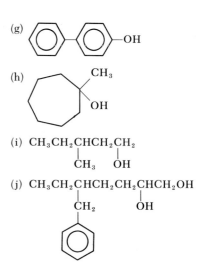

(d) CH₃—CH—CH₃

(e)

Br
|
CH≡CCHCHCH₂CH₂CH₃
|
OH

(f)

OH

HO   OH

(h)

CH₃

OH

(i) CH₃CH₂CHCH₂CH₂
|
CH₃ OH

(j) CH₃CH₂CHCH₂CH₂CHCH₂OH
|
CH₂ OH

6-11. Draw the structure for each of the following compounds:

(a) 3-Cyclopentyl-4-bromo-2-decanol
(b) 2-Isopropyl-5-methylcyclohexanol
(c) 3-Methyl-2-cyclohepten-1-ol
(d) 2-Phenyl-1,3-octanediol
(e) Allyl alcohol (2-propen-1-ol)
(f) Hydroquinone (1,4-dihydroxy-
benzene), photographic developer

(g) *p*-Nitrobenzyl alcohol
(h) *o*-Cresol (2-methylphenol)
(i) 2,4-Dinitrophenol
(j) 2-Penten-1-ol

6-12. Give the product(s) of the reaction of isobutyl alcohol with each of the following reagents. Indicate if no reaction occurs.

(a) K
(b) KOH
(c) HCl
(d) HI
(e) HBr
(f) PBr₃

(g) PI₃
(h)

O
‖
—C—Cl

(i) CH₃CH₂COOH

(j) (CH₃CH₂CO)₂O
(k) Cl₂, Fe
(l) HOSO₃H, heat
(m) HONO₂, H₂SO₄
(n) CrO₃ in pyridine
(o) KMnO₄, heat

6-13. Do problem 6-12 using *p*-cresol.

6-14. For each of the following alcohols, outline a synthesis using the Grignard reagent:

(a) 2-Hexanol (two ways)
(b) 2-Phenyl-1-butanol

(c) 1-Ethylcyclohexanol
(d) 2-Cyclohexyl-1-ethanol

6-15. Give the major organic product of each of the following reactions:

(a) 3-Methyl-2-butanol + H₂SO₄ + heat ⟶
(b) *m*-NO₂C₆H₄OH + NaOH ⟶
(c) 2-Methyl-2-hexene + (1) (BH₃)₂; (2) H₂O₂, OH⁻ ⟶

(d)

—CH₂Br + (1) Mg, dry ether; (2) HCHO ⟶

(e) Cholesterol + CH_3COOH $\xrightarrow{H^+}$

(f) 1,2-Dihydroxybenzene + 2 moles $CH_3CH_2\overset{\displaystyle \|}{\underset{\displaystyle O}{C}}-Cl$ $\longrightarrow$

(g) 2,3-Dimethyl-1-cyclohexene + (1) $(BH_3)_2$; (2) H_2O_2, OH^- $\longrightarrow$

(h) *sec*-Butyl alcohol + PBr_3 $\longrightarrow$

(i) Product of h + Mg, dry ether $\longrightarrow$

(j) Product of i + $C_6H_5COCH_2CH_3$ $\longrightarrow$

(k) Product of j + H_2O $\longrightarrow$

6-16. A student dehydrated 1-phenyl-2-butanol in an attempt to make 1-phenyl-1-butene according to the following reaction:

$$CH_3CH_2\underset{\displaystyle \underset{\displaystyle OH}{|}}{C}HCH_2C_6H_5 \xrightarrow{H^+} CH_3CH_2CH{=}CHC_6H_5 + HOH$$

To test the purity of his product, the student obtained the following gas chromatogram of his product after work-up and drying:

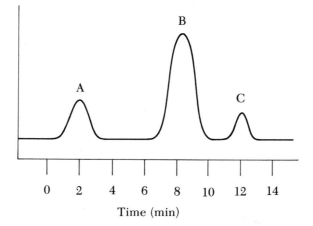

Time (min)

The area under peak A was 25; peak B, 63; and peak C, 12. He tested a sample of the starting compound (the alcohol) by gas chromatography and found it to have a retention time of 12 minutes.

(a) What is peak C?

(b) What percentage of the product mixture is A? What percentage of it is B?

(c) What percentage of the starting compound is unreacted?

(d) What is B?

(e) What is A?   *Hint:* It is a carbon-containing compound. Write the mechanism.

(f) What is the major product?

Self-test

1. Give a correct name for each of the following molecules:

(a) $CH_3CH_2CH_2OH$

(b)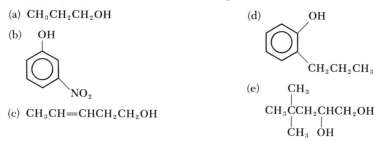

(c) $CH_3CH{=}CHCH_2CH_2OH$

(d)

(e) $CH_3\underset{\displaystyle \underset{\displaystyle CH_3}{|}}{\overset{\displaystyle \overset{\displaystyle CH_3}{|}}{C}}CH_2\underset{\displaystyle \underset{\displaystyle OH}{|}}{C}HCH_2OH$

2. Draw a complete structure corresponding to each of the following compounds:

(a) *tert*-Butyl alcohol

(b) 2,4-Dinitrophenol

(c) 4-Methyl-2-penten-1-ol

(d) Cyclopentyl alcohol

3. Give the structure of the organic product of each of the following reactions:

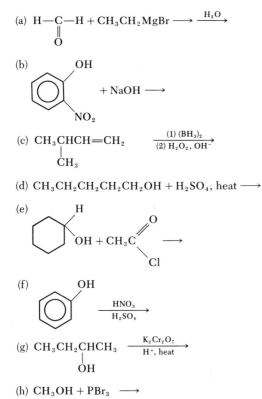

(a) $H-\overset{\overset{\displaystyle \|}{O}}{C}-H + CH_3CH_2MgBr \longrightarrow \overset{H_2O}{\longrightarrow}$

(b) [phenol with OH and NO₂] $+ NaOH \longrightarrow$

(c) $CH_3\underset{\underset{\displaystyle CH_3}{|}}{CH}CH=CH_2 \xrightarrow[\text{(2) H}_2\text{O}_2\text{, OH}^-]{\text{(1) (BH}_3)_2}$

(d) $CH_3CH_2CH_2CH_2CH_2OH + H_2SO_4; \text{ heat} \longrightarrow$

(e) [cyclohexane ring with H and OH] $+ CH_3C\overset{\overset{\displaystyle O}{\diagup\!\!\diagdown}}{\underset{\displaystyle Cl}{}} \longrightarrow$

(f) [phenol with OH] $\xrightarrow[\text{H}_2\text{SO}_4]{\text{HNO}_3}$

(g) $CH_3CH_2\underset{\underset{\displaystyle OH}{|}}{CH}CH_3 \xrightarrow[\text{H}^+\text{, heat}]{\text{K}_2\text{Cr}_2\text{O}_7}$

(h) $CH_3OH + PBr_3 \longrightarrow$

4. Give the correct answer for each of the following questions:

(a) What class of alcohols can be obtained from the reaction of a ketone and a Grignard reagent?

(b) Is the following structure an example of intramolecular or intermolecular hydrogen bonding?

(c) What orbitals make up the C–O bond in CH_3CH_2OH?

(d) Is isobutyl alcohol classified as a primary, secondary, or tertiary alcohol?

(e) Which is more acidic: ethanol or phenol?

(f) What class of organic compounds is formed when an alcohol reacts with an acid anhydride?

(g) What is the common name of the alcohol in the ester $CH_3-\overset{\overset{\displaystyle O}{\|}}{C}-O-CH_2C_6H_5$?

(h) What name is given to the technique of separation and analysis of the components of a mixture?

(i) What is the name of the process of preparing alcoholic beverages from sugars?

(j) What class of organic compounds is formed in the oxidation of a secondary alcohol?

Answers to self-test

1. (a) 1-Propanol or *n*-propyl alcohol
   (b) *m*-Nitrophenol or 3-nitrophenol
   (c) 3-Penten-1-ol

   (d) *o-n*-Propylphenol or 2-*n*-propylphenol
   (e) 4,4-Dimethyl-1,2-pentanediol

2. (a)

   (b)
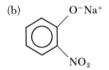

   (c) $HOCH_2CH\!=\!CHCHCH_3$ (with CH₃)

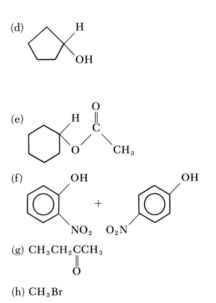

   (c) $HOCH_2CH\!=\!CHCHCH_3$

   (d) [cyclopentane ring with H and OH]

3. (a) $CH_3CH_2CH_2OMgBr \xrightarrow{\ H_2O\ } CH_3CH_2CH_2OH$

   (e) [cyclohexane ring with H and O–C(=O)–CH₃]

   (b) [benzene ring with O⁻Na⁺ and NO₂]

   (f) [benzene ring with OH and NO₂] + [benzene ring with OH and O₂N]

   (c) $CH_3CHCH_2CH_2OH$ (with CH₃)

   (g) $CH_3CH_2CCH_3$ (with =O)

   (d) $CH_3CH_2CH_2CH\!=\!CH_2 +$
   $CH_3CH_2CH\!=\!CHCH_2$

   (h) CH_3Br

4. (a) Tertiary alcohol
   (b) Intramolecular hydrogen bond
   (c) sp^3-sp^3
   (d) CH_3CHCH_2OH: primary alcohol (with CH₃)
   (e) Phenol

   (f) Ester (by-product: carboxylic acid)
   (g) Benzyl alcohol ($C_6H_5CH_2OH$)
   (h) Gas chromatography
   (i) Fermentation
   (j) Ketone

chapter 7

Organic halides

Classes of organic halides

Organic halogen compounds are classified as alkyl (aliphatic), aryl, allyl, or vinyl halides. Alkyl halides are classified as primary, secondary, or tertiary, depending on the type of carbon to which the halogen is bonded; for example:

$$CH_3—CH_2—CH_2—Cl \qquad CH_3—\underset{\underset{I}{|}}{CH}—CH_3 \qquad CH_3—\underset{\underset{CH_3}{|}}{\overset{\overset{CH_3}{|}}{C}}—Br$$

n-Propyl chloride   **Isopropyl iodide**   **tert-Butyl bromide**

(1-chloropropane)   (2-iodopropane)   (2-methyl-2-bromopropane)

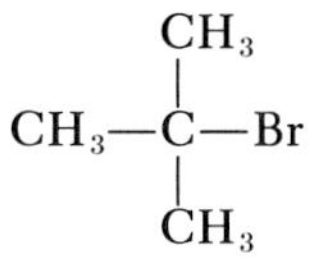

(primary)   (secondary)   (tertiary)

Aryl halides are compounds in which the halogen is attached directly to an aromatic ring. For example:

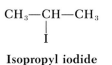

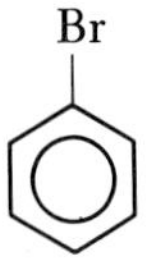

Bromobenzene 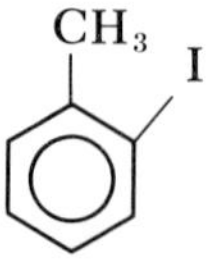   **m-Iodotoluene**

However, not all organic halides containing an aromatic ring are aryl halides. Benzyl bromide, in which the substituent Br is separated from the aromatic ring by a CH_2 group, is an aliphatic halide:

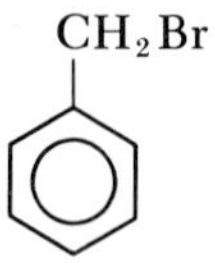

Benzyl bromide
(bromophenylmethane)

In vinyl halides, the halogen is attached directly to a doubly bonded carbon;

whereas in allyl halides, a $-\overset{\displaystyle |}{\underset{\displaystyle |}{C}}-X$ group is attached to a doubly bonded carbon. For example:

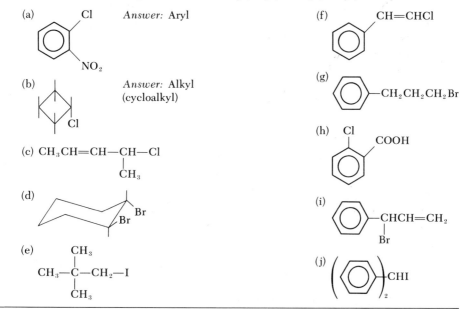

$$CH_2\!\!=\!\!C-\boxed{Cl}$$
$$\overset{\displaystyle |}{H}$$

Vinyl chloride
(1-chloroethene)

$$CH_2\!\!=\!\!C-\boxed{CH_2-Br}$$
$$\overset{\displaystyle |}{H}$$

Allyl bromide
(3-bromo-1-propene)

Of these classes of organic halides, the most important are the alkyl and aryl halides. These classes are discussed in more detail in this chapter.

7-1. Classify the following molecules as alkyl, aryl, benzyl, vinyl, or allyl halides:

(a)   *Answer:* Aryl

(b)   *Answer:* Alkyl
      (cycloalkyl)

(c) $CH_3CH\!\!=\!\!CH-CH-Cl$
$$\qquad\qquad\qquad \overset{\displaystyle |}{CH_3}$$

(d)

(e) $$\qquad\quad CH_3$$
$$\qquad\quad \overset{\displaystyle |}{} $$
$$CH_3-C-CH_2-I$$
$$\qquad\quad \overset{\displaystyle |}{CH_3}$$

(f) $CH\!\!=\!\!CHCl$

(g) $-CH_2CH_2CH_2Br$

(h)   Cl   COOH

(i) $-CHCH\!\!=\!\!CH_2$
$$\qquad \overset{\displaystyle |}{Br}$$

(j) $\left(\!\!\left\langle\overline{\bigcirc}\right\rangle\!\!\right)_{\!2}CHI$

Nomenclature of organic halides

Alkyl and aryl halides are given both common and IUPAC names. The IUPAC nomenclature of these compounds is an extension of the nomenclature of alkanes. The halide substituents are named chloro, bromo, and iodo. Examples of the nomenclature are given on p. 37.

7-2. Give a correct name for each of the compounds in problem 7-1.
   Answer: (a) 2-Nitrochlorobenzene or *o*-chloronitrobenzene
   (b) Cyclobutylchloride or chlorocyclobutane

Preparations of organic halides

The preparation of simple halogenated compounds is discussed in other chapters. One method of preparation is the electrophilic aromatic substitution reaction of benzene and substitution products of benzene with chlorine or bromine (Chapter 4). Following are some examples:

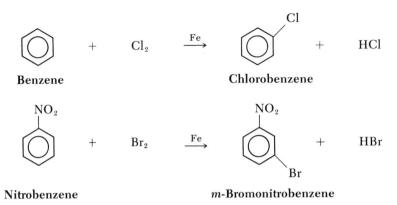

Iodine, being much less reactive, does not react under these conditions. Another method is the free-radical substitution reaction of alkylbenzenes with chlorine or bromine (Chapter 4). Following is an example:

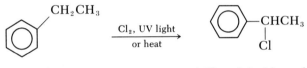

In addition, there are the reactions of alcohols with various halogen-containing compounds to form alkyl halides (Chapter 6). Following are some examples:

$$3CH_3CH_2OH + PBr_3 \longrightarrow 3CH_3CH_2Br + H_3PO_3$$

Ethyl alcohol   **Ethyl bromide**

$$2CH_3CH_2CH_2OH + 2NaBr \xrightarrow{H_2SO_4} 2CH_3CH_2CH_2Br + Na_2SO_4 + 2H_2O$$

n-**Propyl alcohol**   *n*-**Propyl bromide**

$$(CH_3)_2CHCH_2OH + HBr \longrightarrow (CH_3)_2CHCH_2Br + H_2O$$

Isobutyl alcohol   **Isobutyl bromide**

Another method is the electrophilic addition reaction of halogens and hydrogen halides (HX) to alkenes and alkynes (Chapter 3). Following is an example:

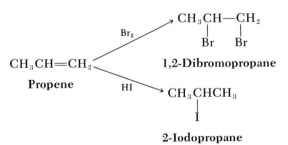

7-3. For each of the following preparations of organic halides, state whether the mechanism is electrophilic addition, electrophilic substitution, nucleophilic substitution, or free radical substitution:

(a) $CH_3CH_2C{\equiv}CH + HCl \longrightarrow CH_3CH_2C{=}CH_2$ *Answer:* Electrophilic addition

$\qquad\qquad\qquad\qquad\qquad\qquad\qquad\qquad\quad$ |
$\qquad\qquad\qquad\qquad\qquad\qquad\qquad\qquad\quad$ Cl

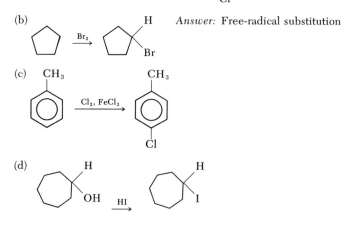

(b) $\xrightarrow{Br_2}$ H   *Answer:* Free-radical substitution

$\qquad\qquad\qquad$ Br

(c) CH_3 $\xrightarrow{Cl_2,\ FeCl_3}$ CH_3

$\qquad\qquad\qquad\qquad\qquad\qquad\qquad$ Cl

(d) H $\xrightarrow[HI]{}$ H

$\qquad$ OH $\qquad\qquad\qquad$ I

Organic fluorine compounds are not prepared with molecular fluorine or hydrogen fluoride since the latter compounds are too reactive and thus it is hazardous to handle them. Therefore, fluorides are prepared by halogen exchange with mercury(I) fluoride (Hg_2F_2) or antimony(III) fluoride (SbF_3). Following are some examples:

$$2CH_3CH_2Cl \xrightarrow{Hg_2F_2} 2CH_3CH_2F + Hg_2Cl_2$$
$\qquad$ **Ethyl chloride** $\qquad\qquad$ **Ethyl fluoride**

$$3CCl_4 \quad + \quad 2SbF_3 \quad \longrightarrow \quad 3CCl_2F_2 \quad + \quad 2SbCl_3$$
Carbon $\qquad\qquad\qquad\qquad$ **Dichlorodifluoromethane**
tetrachloride

Since molecular iodine (I_2) is extremely unreactive, the synthesis of aliphatic organic iodides usually involves an iodide exchange in acetone. The formation of the iodide is thermodynamically favored, since unlike sodium iodide, both sodium chloride and sodium bromide are insoluble in acetone and precipitate. For example:

$$CH_3Br \quad + \quad NaI \xrightarrow{acetone} CH_3I \quad + \quad NaBr$$
$\quad$ **Methyl** $\qquad\qquad\qquad\qquad\quad$ **Methyl iodide** $\quad$ **Insoluble**
$\quad$ **bromide** $\qquad\qquad\qquad\qquad\qquad\qquad\qquad\qquad$ **salt in**
$\qquad\qquad\qquad\qquad\qquad\qquad\qquad\qquad\qquad\qquad\qquad$ **acetone**

The synthesis of aryl iodides is discussed in Chapter 11.

Polyhalogenated compounds are generally prepared by direct halogenation or by exchange reactions. For example, the successive halogenation

Table 7-1
Methods of synthesizing organic halides

Organic starting material	Reagent/condition	Product
C_6H_6	Br_2/Fe	C_6H_5Br
	Cl_2/Fe	C_6H_5Cl
$C_6H_5CH_2R$	Cl_2/UV light	$C_6H_5CH(Cl)R$
	Br_2/UV light	$C_6H_5CH(Br)R$
RCH_2OH	PCl_3	RCH_2Cl
	$P + I_2$	RCH_2I
	$NaBr/H^+$	RCH_2Br
	HBr	RCH_2Br
$RCH{=}CH_2$	Br_2/CCl_4	$RCHBrCH_2Br$
	Cl_2/CCl_4	$RCHClCH_2Cl$
	HCl	$RCHClCH_3$
	HBr	$RCHBrCH_3$
	HI	$RCHICH_3$
RCH_2Br	NaI/acetone	RCH_2I
RCH_2Cl	NaI/acetone	RCH_2I
RCH_2Cl	Hg_2F_2	RCH_2F
RCH_2Br	Hg_2F_2	RCH_2F

of methane in the presence of ultraviolet (UV) light forms the following products:

$$CH_4 \xrightarrow[\text{UV light}]{Cl_2} CH_3Cl \xrightarrow[\text{UV light}]{Cl_2} CH_2Cl_2 \xrightarrow[\text{UV light}]{Cl_2} CHCl_3 \xrightarrow[\text{UV light}]{Cl_2} CCl_4$$

| Methyl chloride + HCl | Methylene chloride + HCl | Chloro- form + HCl | Carbon tetra- chloride + HCl |

The various methods of synthesizing organic halides are summarized in Table 7-1.

Uses of organic halogen compounds

Many organic halogen compounds play an important role in the chemical industry. Organofluorine compounds are used extensively as refrigerants (Freon); as propellants (Freon) for aerosol sprays in paints, hair sprays, deodorants, and whipped creams; and as plastics (Teflon). Following are some examples:

CCl_2F_2

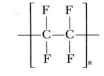

Dichlorodifluoromethane (Freon-12: refrigerant)   **Octafluorocyclobutane** (Freon-C318)   **Tetrafluoroethylene** (polymer, Teflon)

The Freon compounds are nontoxic and are very stable to hydrolysis. Teflon does not conduct electricity, is resistant to high temperatures, and is noted for its nonsticking properties in utensils. Recently scientists have expressed concern over the environmental effects of Freon 11 ($CFCl_3$) and 12 on the earth's protective ozone layer. In the stratosphere these Freons absorb ultraviolet light and liberate chlorine atoms. The latter remove ozone from the stratosphere. Destruction of the ozone is predicted to lead to additional cases of skin cancer and severe climatic changes. One important organobromine compound is merbromin (Mercurochrome). Some commercially important organochlorine compounds include chloroform (an anesthetic of limited use), carbon tetrachloride (a dry-cleaning solvent), and p-dichlorobenzene (a constituent of mothballs). All these compounds are highly toxic. Another useful organic halogen polymer is that formed from 1,1-dichloroethene; it has widespread use as the plastic Saran Wrap.

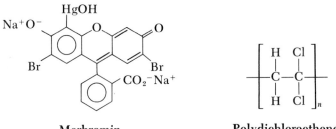

Merbromin   **Polydichloroethene**

Organochlorine insecticides

The most controversial organochlorine compounds are those used as insecticides. Man has long tried to thwart the multitude of pests that strive to deprive him of food, fiber, shelter, and health. One way of controlling insects is to use DDT [1,1,1-trichloro-2,2-bis(p-chlorophenyl)ethane], a chemical found to be capable of acting as an insecticide. DDT was used to combat typhus, yellow fever, malaria, and dysentery during World War II by killing the flies, mosquitoes, lice, and ticks that spread these diseases. As a result of his work in discovering the insecticidal properties of DDT, Paul Müller was awarded the Nobel Prize in 1948.

However, some insects, such as mosquitoes, have developed a resistance to DDT. Scientists eventually succeeded beyond their wildest dreams in developing new insecticides to which insects are not resistant. Following are some examples of organochlorine insecticides. In general, these compounds have been found to be as potent as or more potent than DDT:

Structure   *Name*

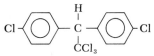   DDT

Structure	*Name*
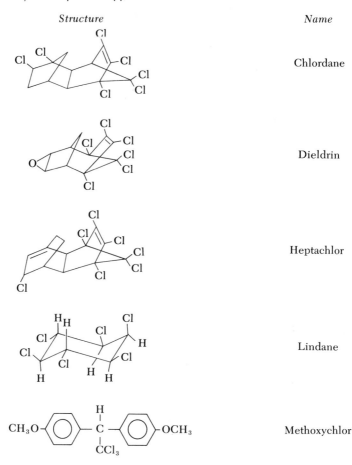	Chlordane
	Dieldrin
	Heptachlor
	Lindane
	Methoxychlor

Since DDT is a covalent organic molecule (see structure on p. 187), it is unaffected by and insoluble in rainwater. When an insect eats the leaves of plants or birds eat the berries of plants that have been sprayed with DDT, the DDT becomes soluble in the nonpolar fats present in their systems and harmful effects are thereby produced in the insects and birds. DDT is capable of changing some of the enzymes found in birds, and the "new" enzymes attack the sex hormones and upset the birds' hormonal balance. Some birds, for example, produce eggs whose shells contain insufficient calcium, and the soft-shelled eggs do not permit survival of the offspring.

In 1966 Rachel Carson through *The Silent Spring* made DDT an infamous chemical. She espoused the contention that the continued use of DDT and other insecticides would lead to disasters ranging from the death of birds (there would eventually come a spring in which there would be no chirping of birds signaling a new spring, that is, a silent spring) to the total upsetting of the ecology of nature. Many authorities, including the World Health Organization, suspect that she overstated the dangers of insecticides and warn that the banning of DDT before cheap and effective substitutes are found would be a "disaster to world health." Norman E. Borlaug, the 1970 Nobel

Peace Prize winner, assailed environmentalists for trying to block the use of such chemicals as DDT, which, he said, are vital to adequate food production: "If agriculture is denied their use because of unwise legislation . . . then the world will be doomed not by chemical poisoning, but from starvation. . . . Recalling that 50% of the present world population is undernourished and that . . . 65% is malnourished, no room is left for complacency. . . . No chemical has done as much as DDT to improve the health, economic and social benefits of the people of developing nations." Nevertheless, indiscriminate spraying with poisonous chemicals has declined, and scientists are researching less dangerous and safer means of dealing with insects. Two such methods currently receiving much attention are the use of insect pheromones (Chapter 2) and the use of organophosphate esters (Chapter 10).

Reactions of alkyl halides

Alkyl halides undergo two types of reactions: nucleophilic substitution and elimination. We will discuss each of these types of reactions and show how they are related to one another.

Nucleophilic substitution

In nucleophilic substitution reactions of alkyl halides, the halogen is substituted by a nucleophile. *Nucleophiles* are basic, electron-rich reagents. They can be charged or uncharged; for example: $:\overset{..}{O}H^-$ and $\overset{..}{N}H_3$. These reactions are of two types: S_N1 and S_N2. The S_N1 *mechanism (substitution, nucleophilic, unimolecular)* is most common for tertiary alkyl halides, whereas the S_N2 *mechanism (substitution, nucleophilic, bimolecular)* is most common for primary alkyl halides. Secondary halides react by either mechanism.

S_N1 reaction. Let us consider the *mechanism* of this reaction, that is, the pathway by which a tertiary alkyl halide reacts. Since the S_N1 mechanism is unimolecular, the rate of the reaction depends on the concentration of only one of the reactant species; that is:

$$\text{rate of } S_N1 \text{ reaction} \quad \alpha \quad [\text{one reacting species}]$$

As an example, consider the hydrolysis of *tert*-butyl bromide:

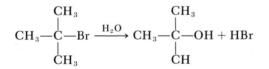

In this reaction, the water molecule, the nucleophilic attacking group, displaces (is substituted for) the bromide ion, the leaving group. It has been found that the concentration of the water does not affect the rate at which the hydrolysis of *tert*-butyl bromide proceeds; however, if the concentration of *tert*-butyl bromide is varied, the rate of the reaction changes. Hence, the rate of the reaction is dependent on the concentration of *tert*-butyl bromide.

A mechanism that satisfactorily explains this is as follows:

(1)

$$CH_3—\underset{\underset{CH_3}{|}}{\overset{\overset{CH_3}{|}}{C}}—Br \longrightarrow \underset{\underset{CH_3}{|}}{\overset{\overset{CH_3 \quad CH_3}{\diagdown\diagup}}{C\oplus}} + Br^-$$

tert-Butyl   Carbonium

bromide   ion

Slow

(2)

$$CH_3—\underset{\underset{CH_3}{|}}{\overset{\overset{CH_3}{|}}{C}}\oplus \ + :\ddot{O}H_2 \longrightarrow CH_3—\underset{\underset{CH_3}{|}}{\overset{\overset{CH_3}{|}}{C}}—\overset{\oplus}{\underset{..}{O}}H_2$$

Protonated

alcohol

Fast

(3)

$$CH_3—\underset{\underset{CH_3}{|}}{\overset{\overset{CH_3}{|}}{C}}————\overset{\oplus\,..}{O}—H \longrightarrow CH_3—\underset{\underset{CH_3}{|}}{\overset{\overset{CH_3}{|}}{C}}—OH + H^{\oplus}$$

tert-Butyl

alcohol

Fast

The rate of any reaction is never greater than the slowest step. In other words, the slow step determines the rate of the reaction and thus is called the *rate-determining step*. Thus step 1, which involves the ionization of *tert*-butyl bromide, is the rate-determining step. Step 1 can be described as formation of a planar carbonium ion. Step 2, a faster step, involves the attack of the nucleophilic water molecule on the carbonium ion. Step 3, another fast step, involves the loss of a proton by the protonated alcohol.

The energy diagram for the S_N1 reaction of *tert*-butyl bromide is shown in Fig. 7-1. In order to react, the molecules of *tert*-butyl bromide and water must acquire a certain amount of energy, that is, the energy of activation (E_A). The first maximum represents the bond-breaking process, which eventually gives rise to the stable intermediate, the carbonium ion. The water molecule then approaches the carbonium ion, and the bond-making process is complete when the product state is reached. The reaction is exothermic, since the product is at a lower energy level than the reactants.

Now let us consider what the consequences of the S_N1 reaction are if an optically active starting material is used. As an example, consider the hydrolysis of (+)-*sec*-butyl bromide:

$$(+)—CH_3CH_2\underset{\underset{Br}{|}}{C}HCH_3 + H_2O \longrightarrow (?)—CH_3CH_2\underset{\underset{OH}{|}}{C}HCH_3 + HBr$$

It would be predicted that the product mixture would be a racemic modification, that is, half (+)-*sec*-butyl alcohol and half (−)-*sec*-butyl alcohol, since

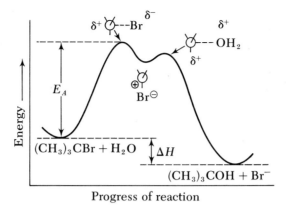

Fig. 7-1
S_N1 mechanism profile. There are two transition states: in the first, the carbon–bromine bond is breaking; in the second, the carbon–oxygen bond is forming. The intermediate carbonium ion is of lower energy than either transition state.

the carbonium ion is planar (flat) and the attack by the water molecule from either side would be expected to be purely random. This is explained by the following mechanism:

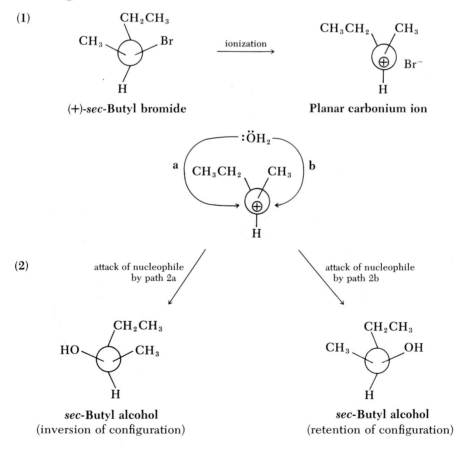

(1)   ionization

(+)-sec-Butyl bromide   **Planar carbonium ion**

(2)   attack of nucleophile   attack of nucleophile
          by path 2a   by path 2b

sec-Butyl alcohol   **sec-Butyl alcohol**
(inversion of configuration)   (retention of configuration)

In step 2a, the water molecule attacks the back side and gives rise to inversion of the configuration. In step 2b, the water molecule attacks the front side and gives rise to retention of the configuration. It is found that, when this reaction is actually carried out in the laboratory, the product is a mixture of a greater amount of the inverted compound and a lesser amount of its enantiomer. Thus the reaction proceeds with partial rather than complete racemization. Obviously, then, the attack by the water molecule on the carbonium ion in step 2 is not random. The explanation for this is that the water molecule begins to attack the tertiary carbon before the bromine has completely left the vicinity of the carbonium ion. The bromine, which leaves the front side, therefore minimizes front-side attack by the water molecule, which then moves to the rear to attack the carbonium ion.

Finally, tertiary halides react faster than primary halides in the S_N1 reaction, since they form more stable carbonium ions. The formation of a carbonium ion is the rate-determining step; thus the halide that can form the most stable carbonium ion reacts fastest by the S_N1 mechanism. The order of ease of reaction is the same as the order of stability of carbonium ions: benzylic > tertiary > secondary > primary > methyl cation.

In summary, the S_N1 reaction has a rate dependent only on the concentration of the reacting halide, proceeds with partial racemization, involves the formation of a carbonium ion, and proceeds most easily with benzylic and tertiary halides.

S_N2 reaction. Let us consider the S_N2 mechanism by which a primary halide reacts. As an example, consider the basic hydrolysis of methyl bromide:

$$CH_3Br + OH^- \longrightarrow CH_3OH + Br^-$$

In this S_N2 reaction, the hydroxide ion (OH^-) is the attacking nucleophile, and the bromine is the leaving group. It has been found experimentally that, if the concentration of either the methyl bromide or the OH^- is varied, the rate varies; that is, the rate is dependent on the concentrations of both the reacting halide and the OH^-:

$$\text{rate of } S_N2 \text{ reaction} \quad \alpha \quad [\text{methyl bromide}]\,[OH^-]$$

Thus the mechanism is bimolecular. The following mechanism satisfactorily explains the rate variations:

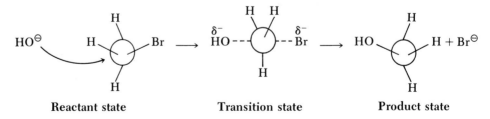

Reactant state   Transition state   Product state

Since the hydrogen atom is much smaller than the bromine or any other atom, the attacking OH^- group will show a greater tendency to attack the back side of the primary carbon than to attack the front side, where the bromine is located. As the mechanism shows, the OH^- group does not become completely bonded to the carbon atom and the carbon–bromine bond is not completely broken at once. As the OH^- group becomes bonded to the carbon atom, it loses part of the negative charge it had in the reactant state and is left with a partial negative charge (δ^-). On the other hand, the bromine in the reactant state possesses no charge; however, as the OH^- group attacks the carbon atom, the carbon–bromine bond weakens, and the bromine begins to acquire a partial negative charge because it begins to acquire electrons from the carbon. This state in which the OH^- and the Br^- groups are partially bonded to carbon is called a *transition state*. The carbon atom is described as being pentavalent. In this state, the $-OH$ and $-Br$ are as far apart as possible. The three hydrogen atoms and the carbon atom lie in a single plane. If we consider the carbon–hydroxide and carbon–bromine bonds to lie along the axle of a wheel, then the three hydrogen atoms form the spokes of the wheel. In the product state, the bromine has acquired all of the negative charge; it is a bromide ion.

The energy diagram for the S_N2 reaction of methyl bromide is shown in Fig. 7-2. Before the methyl bromide and the OH^- group can react, they must acquire a certain amount of energy, the energy of activation. The transition state, in which some bonds are breaking and some are forming, corresponds to the top of the energy barrier. This particular reaction is an exothermic reaction, since the products lie at a lower energy level than the reactants.

Now let us consider what happens in the S_N2 reaction if an optically active reactant is used. As an example, consider the hydrolysis of (+)-*sec*-butyl bromide:

$$(+)—CH_3CH_2\underset{\underset{\displaystyle Br}{|}}{CH}CH_3 + OH^- \longrightarrow (?)—CH_3CH_2\underset{\underset{\displaystyle OH}{|}}{CH}CH_3 + Br^-$$

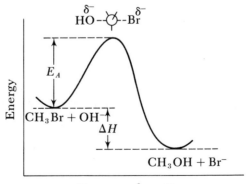

Progress of reaction

Fig. 7-2
S_N2 mechanism energy profile. In the transition state, bonds are broken and formed.

The mechanism for this reaction is as follows:

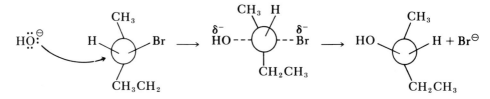

In this mechanism, the OH⁻ group does not take the position occupied by the bromine; rather, it attacks the opposite side. Thus the product alcohol has a configuration opposite to that of the bromide. Such a reaction is said to proceed with inversion of configuration. An S_N2 reaction proceeds with complete inversion; that is, every molecule is inverted.

Finally, let us consider why primary halides react faster than tertiary halides in an S_N2 reaction. Consider the reactant state in both cases:

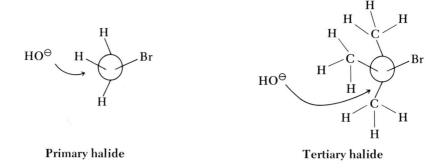

Primary halide   Tertiary halide

The methyl group in the tertiary halide hinders the back-side attack of the OH⁻ group more than the hydrogen atom in the primary halide does. Thus the attack of the OH⁻ group on a primary halide will be more likely to occur than the attack of the OH⁻ group on a tertiary halide. Hence, the order of reactivity of organic halides in an S_N2 reaction is methyl > primary > secondary > tertiary.

In summary, the S_N2 reaction depends on the concentration of both the halide and the nucleophilic reagent, is a one-step reaction, is more common for primary halides, involves a pentavalent carbon in the transition state, and proceeds with complete inversion.

•   •   •

In the discussions of the S_N1 and S_N2 mechanisms, either hydroxide or water was used as the nucleophilic reagent. Table 7-2 lists other nucleophilic reagents that can be used in the general reaction with alkyl halides and the types of compounds that are formed.

7-4. Predict the product of each of the following nucleophilic substitution reactions. Use Table 7-2 as a reference. Identify the class of organic compound formed in each case:

Table 7-2
Nucleophilic substitution reactions

$R:\ddot{X}:$ Alkyl halide	+	$Nu:^-$ Nucleophile	$\longrightarrow$	$R:Nu$ Product	+	$:\ddot{X}:^-$ Leaving group
		$H_2\ddot{O}$, water		ROH, alcohol		
		$:\ddot{O}R^-$, alkoxide		ROR, ether		
		$:CN^-$, cyanide		RCN, nitrile		
		$:\ddot{S}H^-$, sulfide		RSH, thiol		
		$:\ddot{S}R^-$, thioalkoxide		RSR, thioether		
		$:C{\equiv}CH$, acetylide		$R{-}C{\equiv}CH$, alkyne		
		$:\ddot{I}:^-$, iodide, in acetone		RI, alkyl iodide		
		$\ddot{N}H_2^-$, amide		RNH_2, amine		
		$\ddot{N}H_3$, ammonia		$RNH_3^{\oplus}$, amine salt		

(a) $CH_3CH_2CH_2Cl + NaSH \longrightarrow$ *Answer:* $CH_3CH_2CH_2SH$, thiol

(b) Isobutyl iodide + $KOCH_3 \longrightarrow$ *Answer:* $(CH_3)_2CHCH_2OCH_3$, ether

(c) $CH_3Br + CH_3C{\equiv}C:^{\ominus}\ \overset{\oplus}{Na} \longrightarrow$

(d) $C_6H_5CH_2Br \xrightarrow{NH_3}$

(e) Ethyl chloride + sodium iodide in acetone $\longrightarrow$

(f) $CH_3Cl + \ddot{N}H_2^{\ominus} \longrightarrow$

(g) $CH_3(CH_2)_2CH_2Br + CH_3CH_2SNa \longrightarrow$

Nature of nucleophile and solvent polarity. The nature of the nucleophile in nucleophilic substitution is quite varied, as show in Table 7-2. The nature of the nucleophile has very little effect on the rate of an S_N1 reaction, since the nucleophile is not involved in the rate-determining step. On the other hand, the rate of an S_N2 reaction depends very much on the nature of the nucleophile. The degree to which a reagent is nucleophilic depends on its basicity, polarizability, size, solvation energy, and charge. Since these effects cannot be generalized by simple rules, we will not be concerned with them in this text.

The effects of solvents on S_N1 and S_N2 reactions are a little easier to predict. In an S_N1 mechanism, the rate-determining step is ionization of the alkyl halide to form a carbonium ion intermediate. The polar transition state formed is favored by polar solvents. The reactant and product states of the S_N2 mechanism involve a nucleophilic species that is highly solvated in a polar solvent. For the reaction to reach the transition state, this solvation must be destroyed. Since the charge in the transition state is dispersed over several atoms and since solvation destruction is an energy-requiring process, the rate of an S_N2 reaction will decrease with increasing solvent polarity.

Elimination

There are two types of elimination mechanisms: *E1 (elimination, unimolecular)* and *E2 (elimination, bimolecular)*. Generally, the E1 reaction accompanies the S_N1 reaction, and the E2 reaction accompanies the S_N2 reaction.

Following are some examples of elimination reactions:

$$CH_3CH_2CHCH_3 \xrightarrow{\text{NaOCH}_3} CH_3CH=CHCH_3 + CH_3CH_2CH=CH_3 + NaBr + CH_3OH$$
$$\underset{\displaystyle Br}{|}$$

2-Bromobutane   **2-Butene**   **1-Butene**

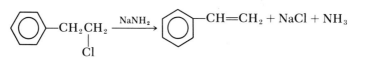

2-Phenyl-   **Styrene**
1-chloroethane   **(phenylethene)**

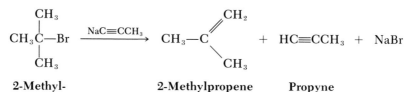

2-Methyl-   **2-Methylpropene**   **Propyne**
2-bromopropane

Both E1 and E2 reactions belong to the class of elimination reaction known as *1,2-elimination* or *β-elimination*. In a 1,2-elimination, the groups are lost from adjacent carbons. Following are some examples:

$$CH_3CH_2CH-CH_2 \xrightarrow{\text{OH}^-} CH_3CH_2CH=CH_2$$

$$\underset{2 \qquad 1}{\overset{}{\text{H} \qquad \text{Br}}}$$

$$CH_3CH-CHCH_3 \xrightarrow{\text{OH}^-} CH_3CH=CHCH_3$$

$$\underset{2 \qquad 1}{\overset{}{\text{H} \qquad \text{Br}}}$$

E1 reaction. The rate of this reaction, like that of S_N1 reaction, depends on the concentration of only one reacting species, the alkyl halide. Hence:

$$\text{rate of E1 reaction} \quad \alpha \quad [\text{alkyl halide}]$$

An E1 reaction therefore can complete with an S_N1 reaction. Tertiary halides undergo E1 reactions more easily than primary halides. The mechanism that best accounts for the rate data is as follows:

(1)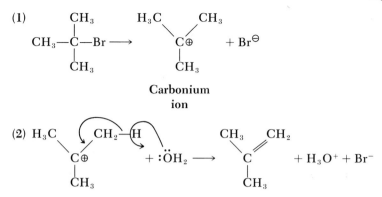

Carbonium
ion

(2)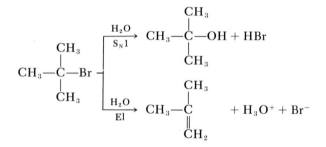

Step 1 is the ionization of the halide to form the planar carbonium ion. This is exactly the same as in an S_N1 reaction. In Step 2, the water molecule (nucleophile) abstracts a proton from the carbonium ion, and the electrons of the carbon–hydrogen bond move to form the carbon–carbon double bond.

Both the S_N1 and the E1 reaction mechanisms involve the formation of a carbonium ion in the rate-determining step. Thus the two reactions can compete with one another. The product of the S_N1 reaction is an alcohol; the product of the E1 reaction is an alkene. As an example, consider the hydrolysis and dehydrohalogenation of *tert*-butyl bromide:

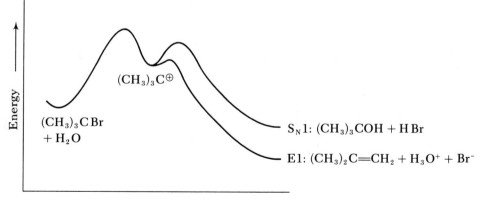

Fig. 7-3
Energy profile for S_N1 reaction versus E1 reaction. For a tertiary alkyl halide, the predominant reaction is elimination.

The energy diagram in Fig. 7-3 shows that the paths for the S_N1 and E1 reactions leading to the intermediate carbonium ion are identical. The E1 and S_N1 transition states leading to the products are lower in energy than the transition state leading to the carbonium ion, but they are of different energies. The carbonium ion can eliminate a proton (E1 reaction), or it can undergo attack by a nucleophile (S_N1 reaction). In either case, the rate of the reaction is determined by the concentration of the alkyl halide. The E1 product is the predominant product in the case of a tertiary halide.

The product of an E1 reaction of an alkyl halide is an alkene. If more than one alkene can be formed, then the major product is the olefin with more substituted groups (Chapter 2). Hence, 2-methyl-2-butene is the major product in the dehydrohalogenation of 3-methyl-2-chlorobutane:

$$CH_3-\overset{\overset{\displaystyle CH_3}{|}}{\underset{\overset{|}{H}\ \ \overset{|}{Cl}}{C}}-CH-CH_3 + H_2O \longrightarrow CH_3-\overset{\overset{\displaystyle CH_3}{|}}{C}\text{=}CH-CH_3 + CH_3-\overset{\overset{\displaystyle CH_3}{|}}{CH}-CH\text{=}CH_2 + Cl^- + H_3O^+$$

3-Methyl-2-chlorobutane	2-Methyl-2-butene (major product)	3-Methyl-1-butene

E2 reaction. The rate of this reaction is dependent on both species, the reacting halide and the nucleophile. Thus:

rate of E2 reaction ∞ [reacting halide] [nucleophile]

An E2 reaction therefore accompanies, or competes with, an S_N2 mechanism. The mechanism most consistent with this data is as follows for the dehydrohalogenation of *n*-propyl bromide:

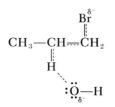

$$CH_3-CH-CH_2 \longrightarrow CH_3CH\text{=}CH_2 + Br^- + H_2O$$

The mechanism is concerted; that is, it has one step. The hydroxide group (the nucleophile) abstracts a proton at the same time that the carbon–carbon double bond forms and the bromine leaves with the electrons that formed the carbon–bromine bond. The transition state for the E2 reaction is probably as follows:

$$\overset{\overset{\displaystyle \overset{\delta^-}{Br}}{\|}}{CH_3-CH\text{---}CH_2}$$

An energy profile for an E2 reaction is shown on page 39.

An E2 reaction competes with an S_N2 reaction. The product of the S_N2

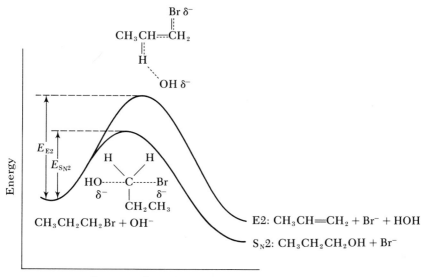

Fig. 7-4

Energy profile for E2 reaction versus S_N2 reaction. For a primary alkyl halide, the predominant reaction is substitution.

reaction is an alcohol; the product of the E2 reaction is an alkene. As an example, consider the hydrolysis and dehydrohalogenation of *n*-propyl bromide:

$$CH_3CH_2CH_2Br + OH^- \underset{E2}{\overset{S_N2}{\longrightarrow}} \begin{array}{l} CH_3CH_2CH_2OH + Br^- \\ CH_3CH=CH_2 + Br^- + HOH \end{array}$$

The energy profiles for these two competing reactions are shown in Fig. 7-4. The energy profiles for the two paths leading to the products are different. The energy of activation for the substitution process is less than that for the elimination process. Thus, under ordinary conditions, primary alkyl halides undergo substitution more readily than elimination. In either case, however, the rate of the reaction is dependent on the concentrations of both the alkyl halide and the nucleophilic reagent.

Just as in an E1 reaction, when the product of an E2 reaction is a mixture of alkenes, the more stable alkene is the major product. In an E1 reaction the more stable alkene results from the more stable carbonium ion. The order of reactivity of alkyl halides in E2 reactions is the same as that of alkyl halides in E1 reactions in forming the more stable carbonium ion: tertiary > secondary > primary. In an E2 reaction, the most stable alkene (the one with the most substituted groups) is formed from the tertiary alkyl halide.

7-5. Draw the energy profile for the E2 reaction of 2-bromobutane + OH^-.

Table 7-3
Comparison of S_N1, S_N2, E1, and E2 reactions

	S_N1	E1
Reactivity of RX	$3° > 2° > 1°$	$3° > 2° > 1°$
Intermediate	Planar carbonium ion	Planar carbonium ion
Kinetics	Unimolecular	Unimolecular
Stereochemistry	Partial racemization	—
Reactivity of halogen	$RI > RBr > RCl$	$RI > RBr > RCl$

	S_N2	E2
Reactivity of RX	$CH_3 > 1° > 2° > 3°$	$3° > 2° > 1°$
Transition state		
Kinetics	Bimolecular	Bimolecular
Stereochemistry	Complete inversion	—
Reactivity of halogen	$RI > RBr > RCl$	$RI > RBr > RCl$

In summary, in an E2 reaction, the mechanism is concerted, the order of reactivity of alkyl halides is tertiary > secondary > primary, and the major product is the more stable alkene.

•   •   •

In Table 7-3, the S_N1, S_N2, E1 and E2 reactions are compared and summarized. In all four mechanisms, alkyl iodides react more rapidly than alkyl bromides, which react faster than alkyl chlorides.

Carbene formation

This is another type of elimination reaction. The two groups are lost from the same carbon. Such an elimination reaction is called *1,1-elimination* or *α-elimination*. For example, when chloroform is treated with a mixture of potassium *tert*-butoxide in *tert*-butyl alcohol, a 1,1-elimination occurs, and a new carbon-containing intermediate is formed:

$$CHCl_3 \xrightarrow[\text{(CH}_3)_3\text{COH}]{\text{(CH}_3)_3\text{CO}^-\text{K}^+} HCl \; + \; Cl-\ddot{C}-Cl$$

Chloroform   **Dichloromethylene**

A species in which carbon is divalent and has two unshared electrons is called a *carbene*. Although carbenes are electrically neutral, they contain a carbon atom with a sextet of electrons and hence are electrophilic. They are named as derivatives of methylene, the parent member of the series. Following are some examples of naming:

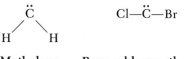

Methylene   Bromochloromethylene

They are very reactive and, like free radicals and carbonium ions, have short lifetimes. They can undergo dimerization, insertion between a carbon–hydrogen or carbon–carbon bond, and addition to an unsaturated bond. Following are some examples:

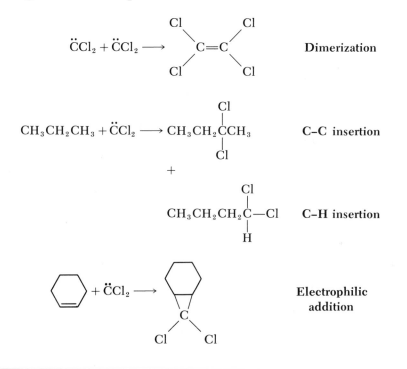

7-6. Predict the product of each of following reactions if $\overset{..}{C}HCl$ is the reactive carbene:

(a) $CH_3CH{=}NCH_3 \xrightarrow{CH_2Cl_2,\ (CH_3)_3CO^-K^+}$ *Answer:* Addition

(b) $(CH_3)_4C \xrightarrow{CH_2Cl_2,\ (CH_3)_3CO^-K^+}$ *Answer:* Insertion

Reactions of aryl halides

Aryl halides undergo two important types of substitution reactions: electrophilic and nucleophilic.

Electrophilic aromatic substitution

This reaction is discussed in Chapter 4. The halogens, F, Cl, Br, and I, are *ortho, para* directors but are deactivators. Hence, a reaction with benzene proceeds faster than the comparable reaction with the halobenzene. Following are examples of electrophilic aromatic substitutions of chlorobenzenes:

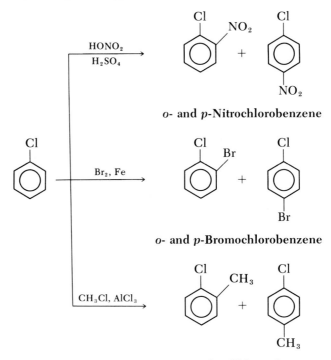

o- and p-Nitrochlorobenzene

o- and p-Bromochlorobenzene

o- and p-Chlorotoluene

Nucleophilic aromatic substitution

Aromatic halides are much less reactive than alkyl halides in nucleophilic substitution. Table 7-4 gives a comparison of the conditions for displacement of halogen in some representative aromatic and aliphatic compounds. Several conclusions can be drawn from this comparison. Alkyl chlorides undergo substitution under less severe conditions than do aryl chlorides. Substitution

Table 7-4
Comparison of reaction conditions for aliphatic and aromatic substitutions

Halide	Reaction conditions	Product
$CH_3(CH_2)_2CH_2Cl$	NaOH, H_2O, reflux	$CH_3(CH_2)_2CH_2OH$
⟨C₆H₅⟩—Cl	10% NaOH, $> 300°$ C, very high pressure	⟨C₆H₅⟩—OH
O_2N—⟨C₆H₄⟩—Cl	NaOH, H_2O; $160°$ C	O_2N—⟨C₆H₄⟩—OH
⟨C₆H₄⟩—Cl with NO_2	NaOH, H_2O; $> 300°$ C, very high pressure	⟨C₆H₄⟩—OH with NO_2

occurs faster in chlorobenzene with a nitro group present in the *para* position than in chlorobenzene or chlorobenzene with a nitro group substituted in the *meta* position.

The reasons for this unusual lack of reactivity of aryl chlorides can be summarized as follows:

1. The carbon–chlorine bond in chlorobenzene has some double-bond character as a result of resonance; consequently, it is stronger than in alkyl chlorides. Two of the resonance structures are as follows:

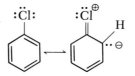

2. The aromatic ring is a center of high electron density because of the π electrons. A nucleophile is discouraged from attacking a center of high electron density.

3. The orbitals of the carbon atom in the carbon–chlorine bond in an aryl chloride are sp² hybridized. Thus the carbon–chlorine bond of an aryl chloride is stronger than the carbon–chlorine bond of an alkyl chloride with sp^3 hybrid orbitals.

4. Aryl halides cannot form the transition state required for the S_N2 reaction of an alkyl halide (p. 192), since back-side attack is impossible; and they do not ionize into aryl cations required by the S_N1 mechanism (p. 190).

Nucleophilic aromatic substitution is a bimolecular displacement mechanism. For chlorobenzene, the mechanism by which an OH group substitutes for a Cl group is as follows:

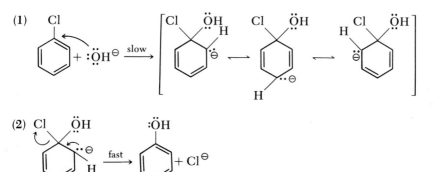

Step 1 is attack by the hydroxide ion on the carbon atom bearing the halogen. It is the slow, or rate-determining, step. In the resonance hybrid structures formed, the carbon atom bears a negative charge. A species in which a carbon atom bears a negative charge is called a *carbanion*. Hence the rate-determining step is sometimes referred to as carbanion formation. Step 2, which is rapid, is the loss of chloride ion to form the product.

In nucleophilic aromatic substitution, a substituent that stabilizes the negative charge (through delocalization of electrons) will facilitate carbanion formation in the ring relative to the compound with no substituted groups or to the compound with a substituent that cannot stabilize the negative charge. The negative charge of the carbanion is formed on the *ortho* and *para* carbon atoms. The substituent that stabilizes the negative charge, then, should occupy one, two, or all three of these positions. Substituents that can stabilize the carbanion are the *meta* directors, such as —NO₂, —CN, and —SO₃H. The effect of such a substituent can be illustrated by the reaction of *p*-nitrochlorobenzene with hydroxide ion. The structure of the carbanion is described as follows in terms of resonance structures I to IV:

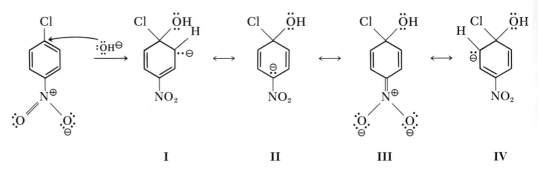

Structure III is particularly important, since the negative charge is carried by oxygen, which readily accommodates the negative charge because of its high electronegativity.

A structure such as III cannot be drawn for the case in which the nitro group is in the *meta* position. Thus the presence of the substituent in the *meta* position has very little effect on the reactivity of the chlorobenzene. This is illustrated as follows:

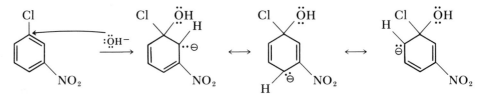

Therefore, it can be said that the presence of a deactivator in the *ortho* or *para* position activates the ring in aromatic nucleophilic substitution.

Following are some examples of nucleophilic substitution reactions of aryl halides:

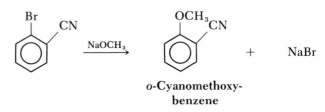

o-Cyanomethoxy-
benzene

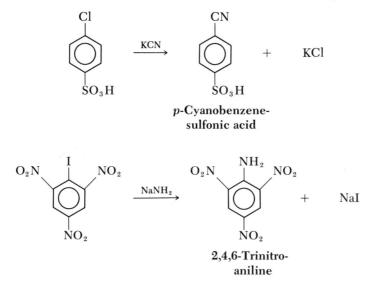

p-Cyanobenzene-
sulfonic acid

2,4,6-Trinitro-
aniline

7-7. Predict whether product A or B would be the major product and why if the following reaction were carried out in the laboratory:

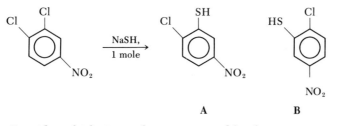

A   B

Answer: Consider which ring carbon is activated by the nitro group.

Substitution through benzyne intermediate

If there is no deactivator in the ring or if an activator, such as —OCH_3 or —CH_3, is present, nucleophilic substitution occurs very slowly, if it occurs at all. However, substitution can be made to take place by the use of a very strong base, such as sodium amide ($NaNH_2$) in ammonia (NH_3) or phenyl lithium (C_6H_5Li). The substitution occurs by a different mechanism than described for aromatic nucleophilic substitution. It involves the formation of an intermediate, the *benzyne* molecule. Following is an example of this type of substitution:

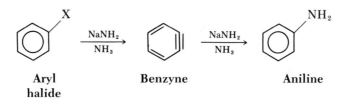

Aryl   Benzyne   Aniline
halide

The new π bond formed does not overlap with the π bond of the ring; in fact, it is perpendicular to the π system of the ring:

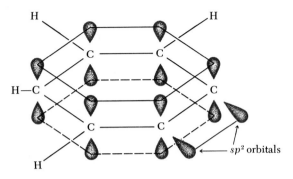

Therefore, it does not interact with the π system of the ring. Furthermore, it is not like the π bond in ethylene, which is as follows:

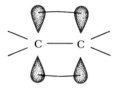

Instead, the electron density is concentrated on one side of the σ bond between the two ring carbons involved, and the overlap of the two sp^2 orbitals of the carbon–carbon bond is poorer than that of the two p orbitals of ethylene. Hence, this bond is very weak, and the intermediate is highly reactive.

Following are some reactions involving the benzyne intermediate:

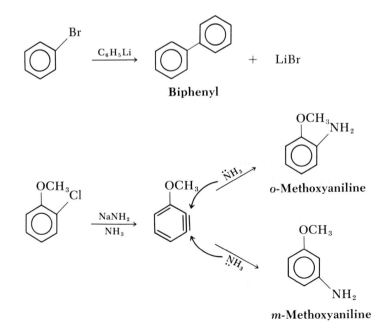

Summary

Aliphatic halides are prepared by a variety of reactions, which are summarized in Table 7-1. Aryl chlorides and bromides are prepared through electrophilic substitution reactions.

Alkyl halides undergo two types of reactions: nucleophilic substitution and elimination. Primary alkyl halides undergo mainly substitution reactions. Tertiary halides undergo both substitution and elimination reactions. Weak nucleophilic reagents and highly polar solvents encourage substitution. One type of elimination reaction, the 1,1-elimination, results in the formation of carbenes, which are highly reactive divalent carbon species that undergo dimerization reactions, insertions between saturated bonds, and electrophilic addition to unsaturated bonds.

Aryl halides undergo electrophilic substitution at the *ortho* and *para* positions slower than benzene undergoes the comparable substitution reaction. In regard to nucleophilic substitution, aryl halides are somewhat unreactive compared to alkyl halides. However, nucleophilic substitution in aryl halides occurs if there is a deactivating substituent *ortho* or *para* to the halogen. There is evidence that nucleophilic substitution is possible in aryl halides in the presence of a strong base through the benzyne intermediate.

Important terms

Alkyl halide	Nucleophilic aromatic substitution
Allyl halide	Primary halide
Aryl halide	Racemization
Benzyne	Secondary halide
Carbene	Substitution, nucleophilic, bimolecular
Elimination, bimolecular	Substitution, nucleophilic, unimolecular
Elimination, unimolecular	Tertiary halide
Inversion of configuration	Vinyl halide
Nucleophile	

Problems

7-8. Give a correct name for each of the following molecules:

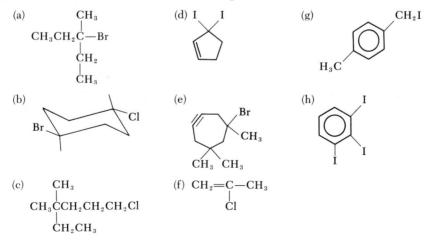

7-9. Which compounds from problem 7-8 fit the following descriptions?

    (a) Allyl halide
    (b) Saturated cyclic halide
    (c) Alkyl halide
    (d) Vinyl halide
    (e) Benzyl halide
    (f) Aryl halide

7-10. Draw the structure for each of the following compounds:

    (a) *sec*-Butyl chloride
    (b) *p*-Bromotoluene
    (c) 3-Iodo-1-hexyne
    (d) *m*-Nitrochlorobenzene
    (e) *m*-Bromophenol

    (f) 3-Chloro-2-methyl-1-pentene
    (g) Cycloheptyl fluoride
    (h) 2-Iodo-1-phenylpropane
    (i) 2,3,6-Tribromonitrobenzene
    (j) 1,2-Diiodo-3-ethyldecane

7-11. Give the organic product(s) of each of the following reactions:

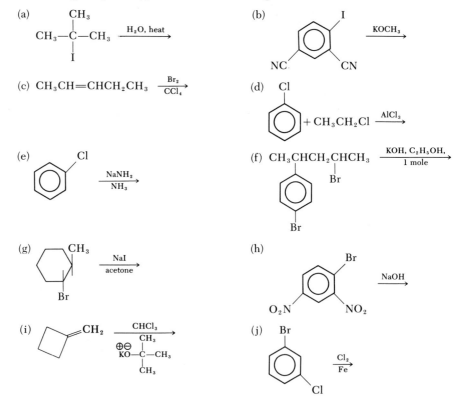

7-12. Which reactions in problem 7-11 fit the following descriptions:

    (a) Carbene addition
    (b) E2
    (c) Electrophilic substitution
    (d) Electrophilic addition

    (e) S_N2
    (f) Benzyne mechanism
    (g) Aromatic nucleophilic substitution
    (h) S_N1

7-13. You performed the following reaction in the laboratory:

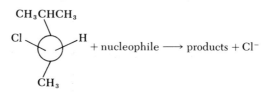

The following statements are your observations. Analyze each statement and indicate *all* the types of reactions (S_N1, S_N2, E1, and E2) to which it applies or indicate that none applies:

(a) Two-step reaction
(b) Reaction rate slower when reactant is 2-chloro-2-methylbutane
(c) Reaction rate slower for 2-bromo-3-methylbutane
(d) Decreasing concentration of alkyl halide by one half decreases rate by one half
(e) Gives an optically active product
(f) Involves bond formation in rate-determining step
(g) Gives some 2-methyl-1-butene

7-14. Predict the *major* product of each of the following reactions:

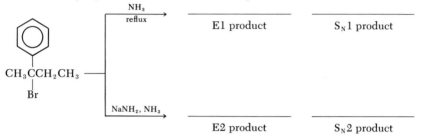

7-15. Each of the following structures represents the transition state or intermediate in an S_N1, S_N2, E1, or E2 reaction. Identify the type of reaction and the product(s) resulting from it for each structure:

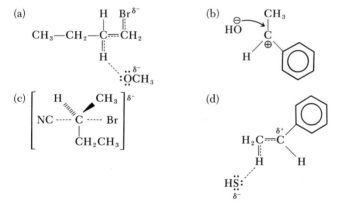

7-16. A student carried out the following synthesis in the laboratory:

$$CH_3CH_2CHCH_3 \xrightarrow[C_2H_5OH]{KOH} CH_3CH_2CH{=}CH_2 + CH_3CH{=}CHCH_3$$
with Br on the third carbon.

The student decided to obtain a gas chromatogram of the product to determine the ratios of the two alkenes. The student observed that the chromatogram had four peaks instead of two. One of the extra peaks was the starting material, as determined from retention time. The component that caused the other extra peak was isolated; it did not decolorize bromine in carbon tetrachloride. Identify the component that did not decolorize bromine in carbon tetrachloride. *Hint:* consider the competing reaction.

Self-test

1. Give a correct name for each of the following molecules:

(c) $CH_3CHCH_2CH_2CH_2Br$
    |
    CH_3

(d) $CH_3CH=CHCH_2Cl$

2. Draw a complete structure corresponding to each of the following compounds:

   (a) Isobutyl chloride
   (b) 2,2-Dimethyl-1-iodohexane
   (c) Benzyl bromide
   (d) 2,4-Dinitrofluorobenzene

3. Give the structure of the organic product of each of the following nucleophilic substitution reactions:

   (a) $CH_3CH_2CH_2CH_2Br + NaOH \xrightarrow{H_2O}$

   (b)
    $O_2N-\!\!\bigcirc\!\!-Cl + NaOCH_3 \longrightarrow$

   (c)
    $\bigcirc\!\!-CH_2Br + NaI,$ acetone $\longrightarrow$

   (d) $CH_3CH_2CH_2CH_2Cl + NaC\equiv CH \longrightarrow$

4. Give the reagent and/or condition that will cause each step to occur as written in the following transformation:

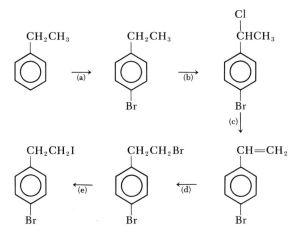

5. Give the correct answer for each of the following questions:

   (a) What is the general name given to the intermediate in an S_N1 reaction?
   (b) What is the name of the class of organic compounds formed when an alkyl halide undergoes an S_N2 reaction with $NaNH_2$?
   (c) What is the name of the class of organic compounds formed when an alkyl halide undergoes an E2 reaction with $NaNH_2$?
   (d) Through which substitution reaction would an optically active alkyl halide form a partially racemized product?
   (e) Which halogen atom in 2,5-dichloronitrobenzene would be more easily substituted?
   (f) Which bromine in 1,3-dibromopropene is considered to be an allylic bromine?
   (g) What type of alkyl halide is most likely to undergo an S_N2 reaction?
   (h) The major product in competing S_N2 and E2 reactions is the S_N2 product. Which reaction, the S_N2 or E2, possesses the lowest energy of activation?
   (i) Give the name and formula for any nucleophile.
   (j) What is the common name for 2-bromobutane?

Answers to self-test

1. (a) *o*-Dichlorobenzene or 1,2-dichlorobenzene
   (b) *tert*-Butyl iodide or 2-methyl-2-iodopropane
   (c) 4-Methyl-1-bromopentane
   (d) 1-Chloro-2-butene

2. (a) 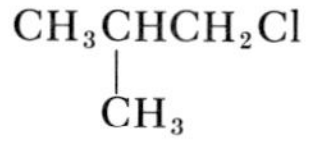

   (b) 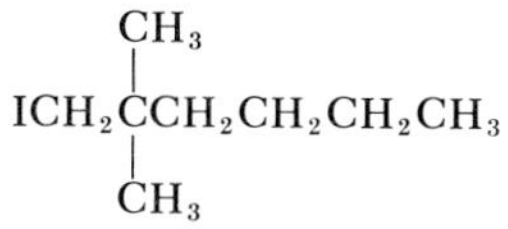

   (c) 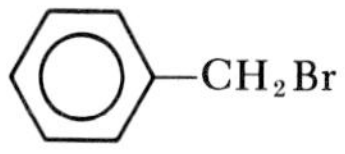

   (d) 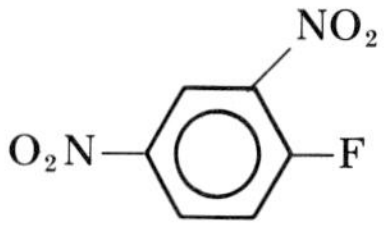

3. (a) $CH_3CH_2CH_2CH_2OH$

   (b) 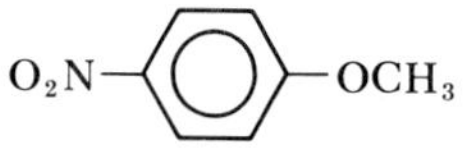

   (c) 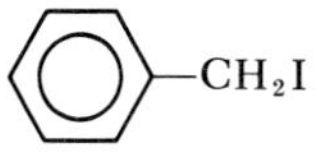

   (d) $CH_3CH_2CH_2CH_2C{\equiv}CH$

4. (a) Br_2, Fe
   (b) Cl_2, light
   (c) KOH, CH_3CH_2OH
   (d) HBr, peroxides
   (e) NaI, acetone

5. (a) Carbonium ion
   (b) Amine
   (c) Alkene
   (d) S_N1
   (e) 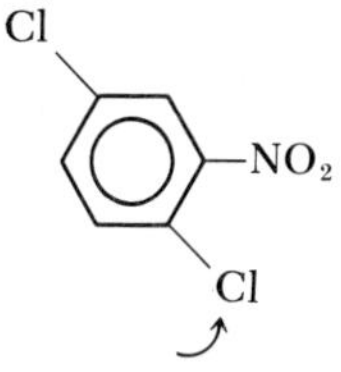

   (f) $BrCH{=}CHCH_2Br$
   Vinylic   Allylic
   (g) Methyl or primary
   (h) S_N2
   (i) Any nucleophile from Table 7-2
   (j) *sec*-Butylbromide

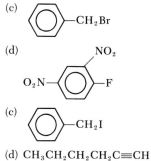

chapter 8

Ethers and epoxides

Structure and nomenclature of ethers

Ethers, like alcohols and phenols, are oxygen-containing compounds. Ethers differ from alcohols and phenols in that the oxygen atom is singly bonded to two other carbon atoms. Each of these types of compounds is represented as follows where R is an alkyl group:

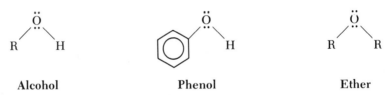

| Alcohol | Phenol | Ether |

Ethers are similar to alcohols and phenols in that the orbitals of the oxygen atom are sp^3 hybridized, with the two unshared pairs of electrons on oxygen occupying sp^3 hybrid orbitals as shown in Fig. 8-1.

8-1. What is the orbital makeup of the indicated σ bond in each of the following ethers:

(a) $H_3C{-}O{-}CH_3$ *Answer: $sp^3{-}sp^3$*   (b)

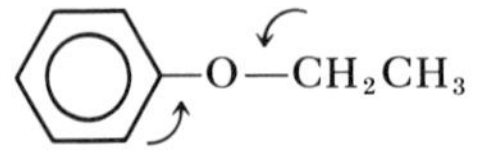

The oxygen atom of an ether can be bonded to two alkyl groups, an alkyl group and a phenyl group, or two phenyl groups. A phenyl group can have substituted groups. If the two groups bonded to the oxygen are identical, the ether is classified as *symmetrical;* otherwise, it is *unsymmetrical.* Ethers containing a phenyl group or phenyl group with substituted groups bonded to oxygen are *aromatic ethers.* Ethers containing only alkyl groups are called *aliphatic ethers.* Following are some examples:

R—O—R   ⬡—O—R   ⬡—O—⬡

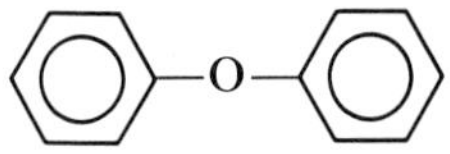

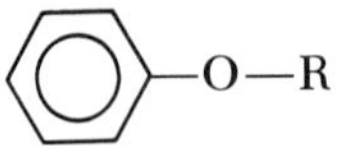

Symmetrical   Unsymmetrical   Symmetrical
aliphatic   aromatic   aromatic

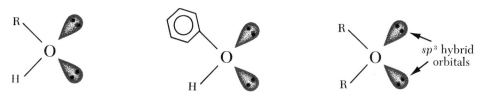

Fig. 8-1
Unshared electrons of oxygen occupy an *sp³* hybrid orbital in alcohols, phenols, and ethers.

In the naming of ethers, the names of the two groups that are attached to oxygen are followed by the word *ether:*

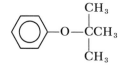

CH₃OCHCH₃
      |
      CH₃

CH₃CH₂OCH₂CH₃

Methyl isopropyl ether
(unsymmetrical
aliphatic ether)

Phenyl *tert*-butyl ether
(unsymmetrical
aromatic ether)

(Di)ethyl ether°
(symmetrical
aliphatic ether)

Ethers can also be named as *alkoxy* derivatives. Some common alkoxy groups (—OR) are listed in Table 8-1. Some examples in which the alkoxy names are used are as follows. The simplest aromatic ether has the common name of *anisole*.

CH₃
|
CH₃CH₂CH₂CHOCH₃

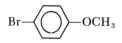

2-Methoxypentane

**Methoxybenzene
methyl phenyl ether
anisole**

***p*-Bromomethoxybenzene
p-bromophenyl methyl ether
p-bromoanisole**

°It is more correct to say diethyl ether to indicate both groups; but if they are identical, it suffices to say ethyl ether.

Table 8-1
Alkyl and alkoxy groups

Alkyl group	Structure	Alkoxy group	Structure
Methyl	CH₃—	Methoxy	CH₃O—
Ethyl	C₂H₅—	Ethoxy	C₂H₅O—
n-Propyl	CH₃CH₂CH₂—	*n*-Propoxy	CH₃CH₂CH₂O—
Isopropyl	(CH₃)₂CH—	Isopropoxy	(CH₃)₂CHO—
Phenyl		Phenoxy	O—

8-2. Give the names and structures of the ethers that fit the following descriptions:

(a) $C_{12}H_{10}O$: symmetrical, aromatic   *Answer:* $C_6H_5OC_6H_5$
(b) $C_4H_{10}O$: isomeric with methyl isopropyl ether, unsymmetrical
   Answer: $CH_3OCH_2CH_2CH_3$
(c) C_3H_6O: unsymmetrical, unsaturated
(d) $C_9H_{12}O$: unsymmetrical, contains a benzylic carbon
(e) $C_{12}H_{12}O$: symmetrical, contains a ring, nonaromatic

Comparison of physical properties of ethers and alcohols

The boiling points of ethers are much lower than those of the corresponding alcohols. However, their boiling points are about the same as those of alkanes of comparable molecular weight. Furthermore, ethers are less soluble in water than alcohols. A comparison of boiling points and solubilities is given in Table 8-2. The boiling points of ethers are low because the hydrogen bonding between ether molecules is very weak to nonexistent:

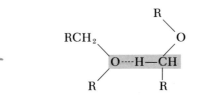

The solubilities of ethers in water are slight because the hydrogen bonding between ether and water molecules is very weak (as shown on the left below) to moderately effective (as shown on the right below):

The insolubility of ethers, particularly ethyl ether, makes them useful as solvents for extracting organic materials from aqueous solutions.

Table 8-2
Physical properties of some alkanes, alcohols and ethers

Compound	Molecular weight	Boiling point (°C)	Solubility
Pentane	72	32	Slight
Ethyl ether	74	35	Slight
1-Butanol	74	118	Soluble
Butane	58	−0.5	Slight
Methyl ethyl ether	60	11	Slight
1-Propanol	60	97	Infinite

Uses of ethers

Vanillin, the principal flavoring ingredient of the vanilla bean; anethole, the chief ingredient of oil of aniseed; and eugenol, the chief ingredient in oil of cloves, have ether linkages:

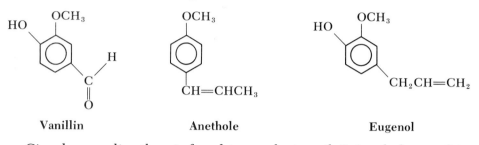

Vanillin	**Anethole**	**Eugenol**

Cineole, a cyclic ether, is found in eucalyptus oil. It is colorless, and its pungent liquid is used as an expectorant in cough medicine. Guaiacol carbonate, also used in cough remedies, contains a methoxy group. The structures of these compounds are as follows:

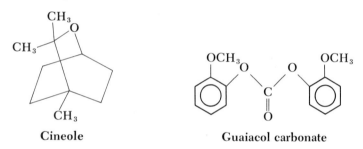

Cineole	**Guaiacol carbonate**

Safrole and piperonal are diethers. Safrole occurs in the oil of sassafras, and piperonal is used in the perfume industry:

Safrole	**Piperonal**

Thioethers are ethers in which the oxygen atom has been replaced by a sulfur atom. They have the general formula RSR. Allyl sulfide $[(CH_2{=}CHCH_2)_2S]$ is the chief constituent of oil of garlic and is present in onions. During World War I, mustard gas $(ClCH_2CH_2SCH_2CH_2Cl)$, a dichlorinated thioether, was used as a poisonous war gas. Methyl sulfide (CH_3SCH_3) is used as an odorant in natural gas. The aromatic ether and thioether shown as follows is a constituent of fresh coffee aroma:

Marijuana and hashish both contain as their physiologically active constituent the isomeric tetrahydrocannabinols, which have ether, phenolic hydroxyl, and alkene functional groups. The total tetrahydrocannabinol content in marijuana is 1.2%. This is a relatively high value for a physiologically active compound. The structures of the isomeric tetrahydrocannabinols are as follows:

Lignin, which is the major noncarbohydrate constituent of wood and woody plants, functions as a natural plastic binder for the cellulose fibers.

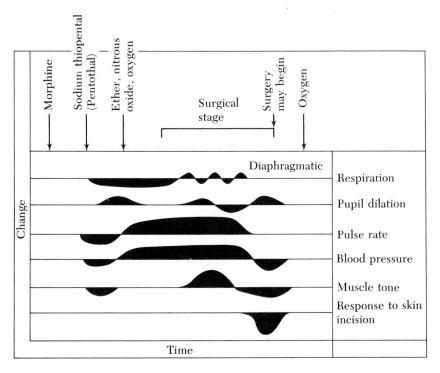

Fig. 8-2
Response of patient during journey into painless sleep. Horizontal line represents normal response of patient. Shaded areas represent deviations from normal after injections of morphine and sodium thiopental (Pentothal) and the administration of a mixture of ether, nitrous oxide, and oxygen. An injection of morphine is used as an analgesic. Then an injection of sodium thiopental puts the patient in a twilight sleep. Shortly thereafter, the patient's respiration decreases. After the patient is given the mixture of anesthetic gases, respiration becomes very irregular, and eventually it must be aided with a diaphragm. After surgery, the lungs are flooded with oxygen to remove the anesthesia from the body, and respiration returns to normal. The other five responses are interpreted in a similar manner.

Although its complete structure is complicated, it is known that it contains a number of methoxyl groups.

Ethyl ether: classic anesthetic

Ethyl ether was first used as an anesthetic in 1842. It is one of several effective drugs capable of inducing the deep, relaxed sleep necessary for major surgery. Ether has a good margin of safety. An inhaled concentration of 6% to 8% in air maintains surgical anesthesia, whereas a concentration of 12% to 15% is required to produce respiratory arrest. Furthermore, it undergoes no chemical changes while in the body. Ether is seldom used alone but is often mixed with another anesthetic, since alone it can irritate the respiratory system.

The journey into the deep, relaxed sleep necessary for major surgery is generally accomplished as follows. A pleasant, calm relaxation is induced in the patient by an injection of morphine. This is followed by an intravenous injection of sodium thiopental (Pentothal), which brings on the first stages of anesthesia, a light doze commonly called a twilight sleep. Complete unconsciousness generally requires the use of some drug as strong as ether. The anesthesiologist inserts a plastic tube into the trachea and feeds to the lungs a mixture of ether, nitrous oxide, and oxygen. As soon as the patient's central nervous system is completely insensitive to pain, the anesthesiologist indicates to the surgeon that the operation may begin.

During surgery, the anesthesiologist watches the patient very closely for any complication that may arise. After surgery, the anesthesiologist restores his patient as smoothly and painlessly as possible. Oxygen is flooded into the lungs, forcing the anesthetic gases out of the patient. Oxygen also eases the work of the heart and the respiratory system as the patient awakens.

Some of the important changes occurring in the patient before and during the administering of anesthesia are summarized in Fig. 8-2.

Preparation of ethers

Aliphatic ethers are prepared by the *Williamson reaction*. It can be used to make symmetrical as well as unsymmetrical ethers. This method involves the reaction of a sodium (or potassium) alkoxide with a primary alkyl halide:

$$Na^{+-}O{-}R \;+\; R{-}X \;\longrightarrow\; R{-}O{-}R \;+\; Na^+X^-$$

$$\text{Alkoxide} \qquad \text{Primary} \qquad\quad \text{Ether}$$
$$\text{halide}$$

For example, methyl ethyl ether can be prepared by either of the following methods:

$$CH_3O^-Na^+ \;+\; C_2H_5I$$
$$\text{Sodium methoxide} \quad \text{Ethyl iodide}$$

$$\Big\} \longrightarrow \quad CH_3OC_2H_5 \;+\; Na^+I^-$$
$$\text{Methyl ethyl ether}$$

$$C_2H_5O^-Na^+ \;+\; CH_3I$$
$$\text{Sodium ethoxide} \quad \text{Methyl iodide}$$

Because of their unreactivity (p. 202), aryl halides do not undergo the reaction:

$$\langle\bigcirc\rangle{-}Br + CH_3O^-Na^+ \xrightarrow{\;\times\;} \langle\bigcirc\rangle{-}OCH_3 + NaBr$$

The Williamson reaction is a nucleophilic substitution of alkoxide ion for halide ion. It follows the S_N2 mechanism:

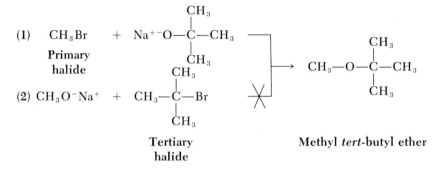

This is accompanied by the competing elimination reaction. Tertiary halides undergo elimination in the presence of alkoxides (strong bases). Thus, even though there are two possible paths for synthesizing methyl *tert*-butyl ether, only one is feasible, since the second method forms 2-methylpropene [$(CH_3)_2C=CH_2$] through elimination rather than the expected ether:

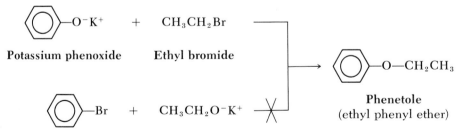

8-3. Write the mechanism for the formation of 2-methylpropene from sodium methoxide and *tert*-butyl bromide.

Answer: The concepts presented in Chapter 7 on elimination reactions should be helpful.

Aromatic ethers also can be formed by the Williamson synthesis. However, there are two limitations: (1) the phenoxide ion should not be reacted with a tertiary alkyl halide, since an elimination reaction will occur; and (2) only one combination of alkoxide ion and halide is useful, namely, phenoxide ion and primary alkyl halide (not alkoxide ion and aryl halide, since aryl halides do not react). Following is an example:

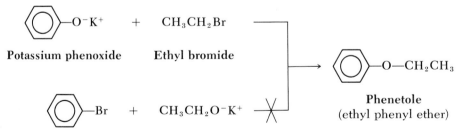

Methyl aryl ethers are commonly prepared with dimethyl sulfate, which is considerably less expensive than the methyl halides. For example:

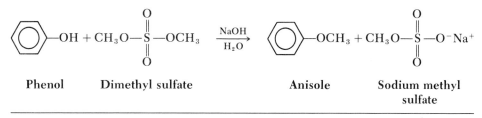

| Phenol | Dimethyl sulfate | | Anisole | Sodium methyl sulfate |

8-4. Why is the preceding reaction carried out in the presence of NaOH?

8-5. Using alcohols and phenols, outline a synthesis for each of the following ethers:

(a) *n*-Propyl phenyl ether   *Answer:* *n*-Propyl bromide and sodium phenoxide
(b) *n*-Butyl cyclohexyl ether   *Answer:* *n*-butyl bromide and sodium cyclohexyl oxide

(c) Ethyl isobutyl ether
(d) Benzyl *p*-nitrophenyl ether

Reactions of ethers

There are only a few reactions of ethers, since ethers are stable toward bases, most oxidizing agents, and reducing agents.

Oxidation of ethers to peroxides

Although ethers are stable toward most oxidizing agents, many aliphatic ethers are converted slowly into peroxides when they stand in contact with air. Peroxides, which have the general formula ROOR, are unstable compounds and are capable of causing violent explosions during the distillations after extraction with ether.

Cleavage by hydroiodic acid and hydrobromic acid

Ethers can be cleaved by acids. Vigorous conditions are required to cause cleavage: concentrated acids, hydroiodic acid (HI) and hydrobromic acid (HBr), and high temperatures. With HI or HBr, the product is an alkyl halide. Following are some examples:

$$CH_3CH_2-O-CH_2-CH_3 \xrightarrow[\text{heat}]{\text{HI}} 2CH_3CH_2I$$

Ethyl ether   Ethyl iodide

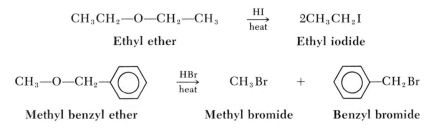

Methyl benzyl ether   Methyl bromide   Benzyl bromide

Actually, the cleavage of an aliphatic ether forms an alcohol and an alkyl halide, but the alcohol can react further to form another mole of alkyl halide:

$$CH_3CH_2-O-CH_2CH_3 \xrightarrow{\text{HI}} CH_3CH_2OH + CH_3CH_2I$$
$$\downarrow \text{HI}$$
$$CH_3CH_2I$$

An aryl alkyl ether undergoes cleavage to form a phenol and an alkyl halide. The bond between the oxygen atom and the carbon atom of the ring is very strong (unreactive). Thus phenols do not react with concentrated HI. Following are some examples:

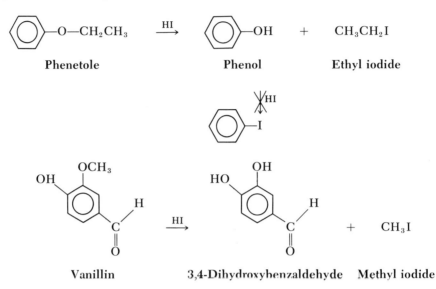

Electrophilic substitution

Aromatic ethers can undergo electrophilic aromatic substitution reactions, such as chlorination, bromination, nitration, and sulfonation. The alkoxyl group is a strongly activating *ortho,para-* director. Thus substitution occurs faster in anisole than in benzene. Following is an example:

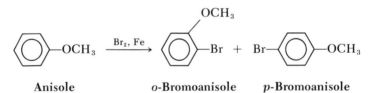

The carbonium ions resulting from *ortho* and *para* attack are stabilized by contributions from the following resonance structures, respectively:

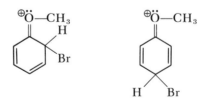

These structures are especially stable, since in them every atom, except hydrogen, has a completed octet and the oxygen atom is able to accommodate the positive charge.

8-6. What is the major organic product formed when *p*-cyanoanisole is nitrated? Explain your answer.

> *Answer:* It is necessary to consider the type of director each substituent is, that is, each substituent's orientation effect. (See Chapter 4.)

Epoxides: cyclic ethers

Epoxides are a class of cyclic ethers with a three-membered ring, one atom of which is an oxygen atom:

They are unusually reactive because of the ease of opening of the ring. The ring opens easily because of the strain inherent in a three-membered ring, with average bond angles of 60°, which is considerably less than the normal tetrahedral carbon angle of 109.5° or the oxygen angle of 110° for open-chain ethers. These bond angles are illustrated as follows:

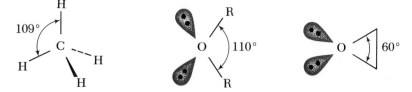

Epoxides undergo ring opening in reactions with a variety of nucleophilic substances. Some of these reactions are illustrated as follows for the simplest epoxide, ethylene oxide:

$$H\ddot{O}H \quad + \quad \underset{\underset{O}{\diagdown\diagup}}{CH_2-CH_2} \quad \xrightarrow{H^+} \quad \underset{\underset{OH \quad OH}{|\quad\quad|}}{CH_2-CH_2}$$

$$\text{Ethylene oxide} \qquad\qquad \text{Ethylene glycol:}$$
$$\text{component of antifreeze}$$

$$CH_3\ddot{O}H \quad + \quad \underset{\underset{O}{\diagdown\diagup}}{CH_2-CH_2} \quad \longrightarrow \quad \underset{\underset{CH_3O \quad OH}{|\quad\quad|}}{CH_2-CH_2}$$

$$\text{2-Methoxyethanol}$$

$$\ddot{N}H_3 \quad + \quad \underset{\underset{O}{\diagdown\diagup}}{CH_2-CH_2} \quad \longrightarrow \quad \underset{\underset{NH_2 \quad OH}{|\quad\quad|}}{CH_2-CH_2}$$

$$\text{2-Aminoethanol}$$

Primary alcohols can be prepared from the reaction of a Grignard reagent with ethylene oxide. The unique feature of this reaction is that the product contains two more carbon atoms than the alkyl or aryl group of the Grignard reagent:

$$R-Mg-X + CH_2-CH_2 \longrightarrow R-CH_2-CH_2-OMgX \xrightarrow{H_2O,\ H^+} RCH_2CH_2OH$$
$$\underset{O}{\diagdown\diagup}$$

Primary
alcohol

Following is an example:

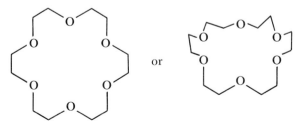

⬡—MgBr + CH$_2$—CH$_2$ $\longrightarrow$ ⬡—CH$_2$—CH$_2$—OMgX $\xrightarrow{H_2O,\ H^+}$ ⬡—CH$_2$CH$_2$OH

2-Phenylethanol

Phenylmagnesium
bromide

Commerical application of ethylene oxide

The products of the reactions of ethylene oxide with various nucleophiles generally have some commercial application. Following are some examples. Ethylene glycol, the product of the reaction between ethylene oxide and water, is used as antifreeze. Carbitol, formed from ethylene oxide and 2-ethoxyethanol, is a solvent for quick-drying varnishes, enamels, and wood stains:

$$CH_2-CH_2 + CH_3CH_2OCH_2CH_2OH \longrightarrow CH_3CH_2OCH_2CH_2OCH_2CH_2OH$$
$$\underset{O}{\diagdown\diagup}$$

2-Ethoxyethanol   Carbitol

8-7. How would you make 2-ethoxyethanol from ethylene oxide?

Diethylene glycol, the product of the reaction between ethylene oxide and ethylene glycol, is used as an antifreeze in sprinkler systems and as a lubricating and finishing agent for wood:

$$CH_2-CH_2 + CH_2-CH_2 \longrightarrow HO-CH_2-CH_2-O-CH_2-CH_2-OH$$
$$\underset{O}{\diagdown\diagup} \quad \underset{OH\ \ OH}{|\ \ \ |}$$

Ethylene   Diethylene glycol
glycol

Crown ethers

Crown ethers, so named because their structure resembles a crown, are recently synthesized large cyclic polyethers. For example, 1,4,7,10,13,16-hexaoxacyclooctadecane (18-crown-6) is an 18-member ring with six oxygen atoms in the ring:

These compounds are of interest to chemists and biochemists because they form stable complexes with salts of alkali metals, which are used in organic synthesis. It has been shown that crown ethers can help to overcome the limited solubility of these alkali metal cations in organic solvents by forming a strong complex with the alkali metal cation, which is enclosed in the cavity of the crown. This is illustrated as follows:

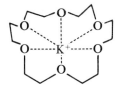

The metal cation–ether complex is very soluble in nonpolar organic solvents because the exterior of the crown possesses hydrocarbon character. Hence, it is now possible for the organic chemist to carry out reactions with ionic materials in nonpolar solvents. One example is the oxidation of toluene by potassium permanganate ($KMnO_4$) in benzene rather than in an aqueous solution.

Biochemists have shown an interest in crown ethers because certain antibiotics have structural features that can influence the transport of Na^+ and K^+ across cell membranes, which is a fundamental process of the living system. It has been proposed that these antibiotics possess complexing properties similar to those of crown ethers.

Summary

Ethers are a class of compounds in which an oxygen atom is bonded to two carbon atoms. They have the general formula R—O—R or C_6H_5OR. Ethers are classified as symmetrical or unsymmetrical. The orbitals of the oxygen atom in an ether are sp^3 hybridized.

Unlike alcohols, ethers do not undergo hydrogen bonding with other ether molecules. The insolubility in water of ethers, particularly ethyl ether, makes them useful solvents for extracting organic compounds from aqueous solutions.

Ethers are prepared by the Williamson reaction of a sodium alkoxide or phenoxide and a primary alkyl halide. They cannot be prepared from aryl halides or tertiary halides by the Williamson reaction.

Ethers are fairly inert organic compounds but can be cleaved by HI or HBr to alkyl iodides or alkyl bromides, respectively. Aromatic ethers undergo electrophilic substitution reactions.

The epoxides are cyclic ethers that undergo a facile ring-opening reaction with various nucleophilic reagents. Many of the products of these reactions have commercial uses.

Important terms

Anisole	Methyl sulfate
Aromatic ether	Phenetole
Epoxides	Symmetrical ether
Ether	Thioether
Ethylene oxide	Williamson reaction

Problems

8-8. For each of the following compounds, give as many names as possible:

(a) CH_3OCH_3

(b) 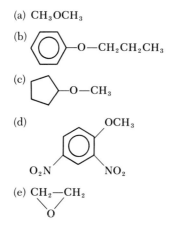 —O—$CH_2CH_2CH_3$

(c) ⬠—O—CH_3

(d)

O_2N ⬡ NO_2 with OCH_3

(e) CH_2—CH_2 with O (epoxide)

(f)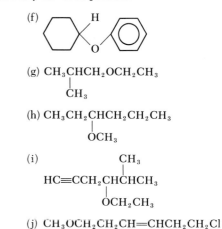

(g) $CH_3CHCH_2OCH_2CH_3$ with CH_3

(h) $CH_3CH_2CHCH_2CH_2CH_3$ with OCH_3

(i) $HC{\equiv}CCH_2CHCHCH_3$ with CH_3 and OCH_2CH_3

(j) $CH_3OCH_2CH_2CH{=}CHCH_2CH_2Cl$

8-9. Draw the structure corresponding to each of the following compounds:

(a) *tert*-Butyl ether
(b) Anisole
(c) 3-Ethoxyoctane
(d) Ethylene oxide
(e) Isobutyl benzyl ether
(f) 2,4-Dichlorophenyl ethyl ether
(g) Ethyl *n*-propyl ether
(h) 3,4-Dibromophenetole
(i) 4-Methoxy-2-pentene
(j) 3-Isopropoxy-4-chloronitrobenzene

8-10. Which ether(s) in problems 8-8 and 8-9 is (are):

(a) Aliphatic
(b) Aromatic
(c) Symmetrical
(d) Unsymmetrical
(e) An epoxide

8-11. Write equations to illustrate how each of the following ethers can be prepared through the Williamson reaction:

(a) Ethyl ether
(b) Phenyl *n*-butyl ether
(c) Phenyl benzyl ether
(d) Methyl cyclopentyl ether
(e) 4-Nitroanisole

8-12. Give the structures of the products of each of the following reactions:

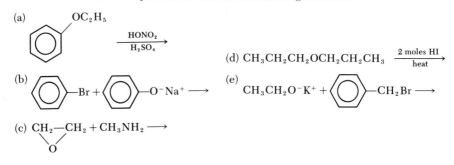

(a) ⬡—OC_2H_5 $\xrightarrow[\text{H}_2\text{SO}_4]{\text{HONO}_2}$

(b) ⬡—Br + ⬡—O^-Na^+ ⟶

(c) CH_2—CH_2 (epoxide) + CH_3NH_2 ⟶

(d) $CH_3CH_2CH_2OCH_2CH_2CH_3$ $\xrightarrow[\text{heat}]{\text{2 moles HI}}$

(e) $CH_3CH_2O^-K^+$ + ⬡—CH_2Br ⟶

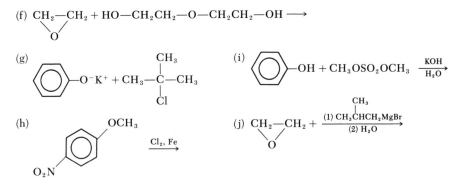

(f) $CH_2\!-\!CH_2 + HO\!-\!CH_2CH_2\!-\!O\!-\!CH_2CH_2\!-\!OH \longrightarrow$

(g) ... $-O^-K^+ + CH_3\!-\!\overset{\overset{\displaystyle CH_3}{|}}{\underset{\underset{\displaystyle Cl}{|}}{C}}\!-\!CH_3$

(i) ... $-OH + CH_3OSO_2OCH_3 \xrightarrow[H_2O]{KOH}$

(h) ... OCH_3 $\xrightarrow{Cl_2,\ Fe}$ O_2N

(j) $CH_2\!-\!CH_2 + \xrightarrow[(2)\ H_2O]{(1)\ CH_3\overset{\overset{\displaystyle CH_3}{|}}{CH}CH_2MgBr}$

8-13. An ether of the formula $C_4H_{10}O$, when treated with a hot, concentrated solution of hydro-iodic acid, gives two structurally different alkyl iodides. What is the structure of the ether?

8-14. An aromatic ether of the formula C_7H_8O gives a methyl iodide as the only alkyl iodide when treated with a hot, concentrated solution of hydroiodic acid. What is the structure of the ether?

8-15. An ether of the formula C_4H_8O gives only 1,4-diiodobutane when treated with a hot, con-centrated solution of hydroiodic acid. What is the structure of the ether?

8-16. A compound of formula $C_4H_{10}O_2$ gives 1 mole of 1,3-dibromopropane and 1 mole of methyl bromide when treated with hot, concentrated hydrobromic acid. What is the structure of the compound?

8-17. Write the structure of the product of the reaction of each of the following compounds with ethylene oxide:
(a) Phenol
(b) Methylmagnesium bromide and then water
(c) Isobutyl alcohol
(d) Aniline ($C_6H_5NH_2$)
(e) Carbitol

Self-test

1. Give a correct name for each of the following molecules:

(a) 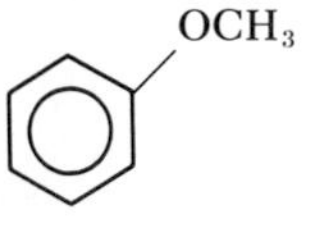 ... OCH_3

(d) 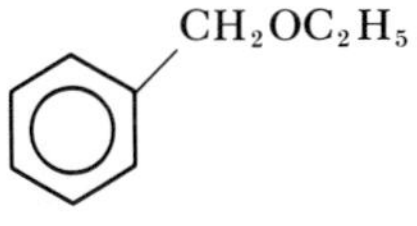 ... $CH_2OC_2H_5$

(b) $CH_3OCH(CH_3)_2$

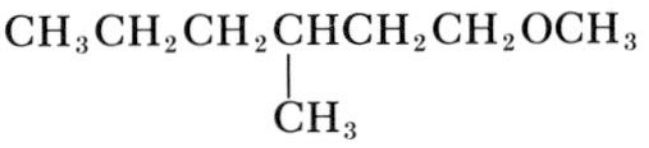

(c) 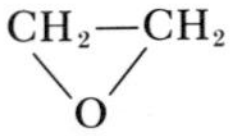 $CH_2\!-\!CH_2$

(e) $CH_3CH_2CH_2\overset{\overset{\displaystyle CH_3}{|}}{CH}CH_2CH_2OCH_3$

2. Draw a complete structure corresponding to each of the following compounds:

(a) Ethyl isobutyl ether   (b) Benzyl ether   (c) Anisole
(d) p-Bromoethoxybenzene   (e) 3-Methoxy-1-hexene

3. Give the structure of the organic product of each of the following reactions:

(a) $CH_3Br + CH_3CH_2O^-K^+ \longrightarrow$
(b) $(CH_3)_2CHCl + CH_3O^-Na^+ \longrightarrow$
(c) 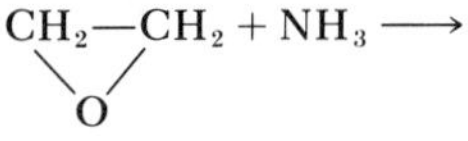 $CH_2\!-\!CH_2 + NH_3 \longrightarrow$

(d) 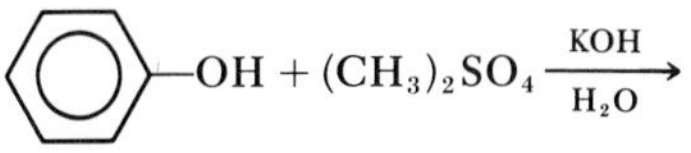 ... $-OH + (CH_3)_2SO_4 \xrightarrow[H_2O]{KOH}$

(e) $CH_3CH_2OCH_2CH_3 + HI \longrightarrow$

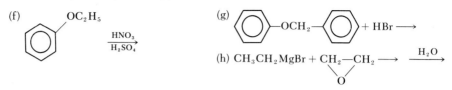

(f) [benzene ring with OC_2H_5] $\xrightarrow{\text{HNO}_3}_{\text{H}_2\text{SO}_4}$

(g) [benzene ring]$-OCH_2-$[benzene ring] $+ HBr \longrightarrow$

(h) $CH_3CH_2MgBr + CH_2{-}CH_2$ [O bridge] $\longrightarrow \xrightarrow{\text{H}_2\text{O}}$

4. Give a correct answer for each of the following questions:

(a) What is the hybridization of the orbitals of oxygen in methyl ether?
(b) Draw the structural formula of an aromatic ether.
(c) Draw the structural formula of a symmetrical ether.
(d) What is a common use of ethyl ether?
(e) What organic product is formed in the reaction between a primary alkyl halide and a sodium alkoxide?
(f) To what positions on the ring does a methoxy group direct an incoming group in electrophilic substitution?
(g) Which would undergo bromination fastest: benzene, anisole, or nitrobenzene?
(h) What is the name given to a class of compounds known as cyclic ethers?
(i) What class of compounds is formed when an alkyl ether reacts with excess HI?
(j) What type of alcohol is formed in the reaction between a Grignard reagent and ethylene oxide?

Answers to self-test

1. (a) Methoxybenzene or anisole
   (b) Methyl isopropyl ether or 2-methoxypropane
   (c) Ethylene oxide

   (d) Benzyl ethyl ether
   (e) 1-Methoxy-3-methylhexane

2. (a) $CH_3CH_2OCH_2CHCH_3$
            $\overset{|}{C}H_3$

   (b) 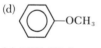 [benzene ring]$-CH_2OCH_2-$[benzene ring]

   (c) [benzene ring]$-OCH_3$

   (d) $Br-$[benzene ring]$-OC_2H_5$

   (e) $CH_2{=}CHCHCH_2CH_2CH_3$
                  $\overset{|}{O}CH_3$

3. (a) $CH_3OCH_2CH_3$
   (b) $CH_3CH{=}CH_2 + CH_3OH$
   (c) $HOCH_2CH_2NH_2$

   (d) [benzene ring]$-OCH_3$

   (e) $2CH_3CH_2I$

   (f)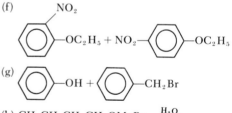
       [benzene ring with NO_2]$-OC_2H_5 + NO_2-$[benzene ring]$-OC_2H_5$

   (g) [benzene ring]$-OH +$ [benzene ring]$-CH_2Br$

   (h) $CH_3CH_2CH_2CH_2OMgBr \xrightarrow{\text{H}_2\text{O}}$
                  $CH_3CH_2CH_2CH_2OH$

4. (a) sp^3

   (b) [benzene ring]$-O-R$

   (c) $R-O-R$
   (d) Anesthetic

   (e) Ether
   (f) *Ortho* and *para*
   (g) Anisole
   (h) Epoxides
   (i) Alkyl iodide
   (j) Primary alcohol

chapter 9

Carboxylic acids, carboxylic acid derivatives, and dicarboxylic acids

Structure and nomenclature of carboxylic acids

Carboxylic acids contain the carboxyl group (COOH), and the general formula is RCO_2H, RCOOH, or

$$R-C\overset{\displaystyle O}{\underset{\displaystyle OH}{\diagup}}$$

. If the carboxyl group is attached to an alkyl group, the acid is an *aliphatic acid;* if it is attached to an aromatic ring, it is an *aromatic acid:*

$$R-CO_2H$$

 $-CO_2H$

Aliphatic   Aromatic

Carboxylic acids have common and IUPAC names. The common names refer to sources or associated odors rather than to chemical structures. For example, formic acid is found in ants, and acetic acid is a component of vinegar. Butyric and caproic acid odors are associated with rancid butter and goats, respectively. Palmitic acid is found in palm oil, and myristic acid is found in nutmeg. The common names of some carboxylic acids are given in Table 9-1.

In the common naming system, branched-chain carboxylic acids are named as derivatives of the straight-chain acids. Greek letters are used to indicate the position of substitution. The α-carbon is the carbon atom attached to the carboxyl carbon. Following are some examples:

$$\overset{\beta CH_3}{\underset{\alpha}{\beta CH_3-CH}}-COOH$$

$$CH_3-\overset{CH_3}{\underset{\gamma}{CH}}-\overset{}{\underset{|\beta}{CH}}-COOH \atop \underset{CH_3}{}$$

Isobutyric acid
α-methylpropionic acid

α,β-Dimethylbutyric acid

Table 9-1
Names of some carboxylic acids

Formula	Common name	IUPAC name
HCOOH	Formic acid	Methanoic acid
CH_3COOH	Acetic acid	Ethanoic acid
CH_3CH_2COOH	Propionic acid	Propanoic acid
$CH_3(CH_2)_2COOH$	Butyric acid	Butanoic acid
$CH_3(CH_2)_3COOH$	Valeric acid	Pentanoic acid
$CH_3(CH_2)_4COOH$	Caproic acid	Hexanoic acid
$(CH_3)_2CHCOOH$	Isobutyric acid	2-Methylpropanoic acid
$CH_3(CH_2)_{10}COOH$	Lauric acid	Dodecanoic acid
$CH_3(CH_2)_{12}COOH$	Myristic acid	Tetradecanoic acid
$CH_3(CH_2)_{14}COOH$	Palmitic acid	Hexadecanoic acid
$CH_3(CH_2)_{16}COOH$	Stearic acid	Octadecanoic acid

α-Methyl-γ-bromo-δ-phenylvaleric acid

The IUPAC system is based on the naming of the parent structure, which is the longest chain carrying the carboxyl group. The parent structure is named by replacing the final -*e* of the corresponding alkane with the suffix -*oic acid*. The position of a substituent is indicated by a number, but unlike in the common system, the carboxyl carbon is number 1:

$$C-C-C-\overset{\frown}{C}OOH \qquad \text{Carboxyl}$$
$$4 \quad 3 \quad 2 \quad 1 \qquad\qquad \text{carbon}$$

Following are some examples. For comparison, the same three compounds given as examples of the common naming system are used:

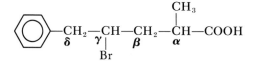

2-Methylpropanoic acid   2,3-Dimethylbutanoic acid

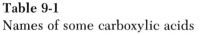

2-Methyl-4-bromo-5-phenylpentanoic acid

The aromatic acids are named as derivatives of benzoic acid. Following are some examples:

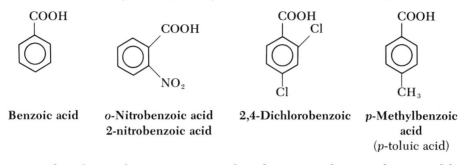

Benzoic acid   ***o*-Nitrobenzoic acid**   **2,4-Dichlorobenzoic**   ***p*-Methylbenzoic**
                        2-nitrobenzoic acid   **acid**
                                                                                      (*p*-toluic acid)

Dicarboxylic acids contain two carboxyl groups. They can be named by the common and IUPAC systems. Following are some examples:

HOOCCOOH   HOOCCH$_2$COOH

Common: **Oxalic acid**   **Malonic acid**
IUPAC: **Ethanedioic acid**   **Propanedioic acid**

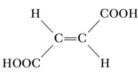

Maleic acid   **Fumaric acid**
cis-**Butenedioic acid**   *trans*-**Butenedioic acid**

Dicarboxylic acids with substituted groups are named by the common and IUPAC systems as the carboxylic acids with substituted groups are. Following are some examples:

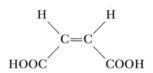

HOOCCH$_2$CHCOOH   HOOCCH$_2$CHCH$_2$CH$_2$COOH

Common: **α-Methylsuccinic acid**   **β-Chloroadipic acid**
IUPAC: **2-Methylbutanedioic acid**   **3-Chlorohexanedioic acid**

9-1. Draw the structure corresponding to each of the following compounds, and give another name for each acid:

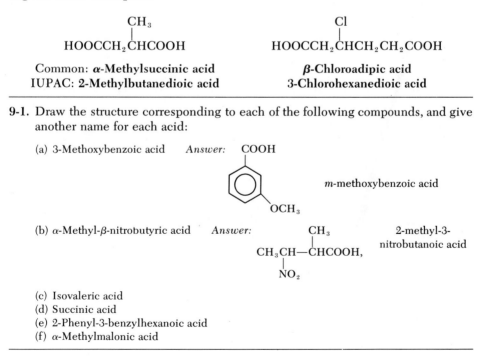

(a) 3-Methoxybenzoic acid   *Answer:*   COOH

                                                          m-methoxybenzoic acid

                                                          OCH$_3$

(b) α-Methyl-β-nitrobutyric acid   *Answer:*   CH$_3$ 2-methyl-3-
                                            CH$_3$CH—CHCOOH,   nitrobutanoic acid
                                                          NO$_2$

(c) Isovaleric acid
(d) Succinic acid
(e) 2-Phenyl-3-benzylhexanoic acid
(f) α-Methylmalonic acid

Structures and nomenclature of acid derivatives

The derivatives of acids that are of interest to organic chemists are amides, esters, anhydrides, and acid chlorides. The structural formulas of these classes of compounds are as follows:

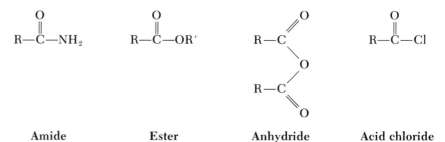

| Amide | Ester | Anhydride | Acid chloride |

These acid derivatives are named as indicated in Table 9-2. Even though

Table 9-2
Naming of acid derivatives

Acid derivative	Rule for naming from common name of acid[°]	Example
Amide	Change -ic acid to amide	O $\parallel$ $CH_3CH_2C—NH_2$ Propionic acid to propionamide
Ester	Change -ic acid to -ate, preceded by name of alcohol	O $\parallel$ $CH_3C—OCH_2—\bigcirc$ Acetic acid to benzyl acetate
Anhydride	Change acid to anhydride	CH_3C ... O, O CH_3C ... O Acetic acid to acetic anhydride
Acid chloride	Change -ic acid to -yl chloride	$\bigcirc—\overset{O}{\overset{\parallel}{C}}—Cl$ Benzoic acid to benzoyl chloride

[°]IUPAC names are similarly obtained except that for an amide the -oic acid of the acid's IUPAC name is changed to -amide.

these derivatives are named differently, the following basic steps can be used in naming them:

1. Give the common name of the acid from which the compound is derived.
2. Name the type of acid derivative.
3. Follow the rule in Table 9-2 for naming the particular derivative.

Following are several examples:

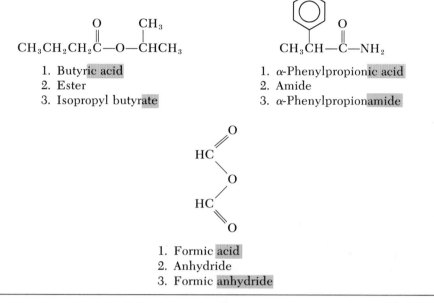

$$CH_3CH_2CH_2\overset{O}{\underset{\|}{C}}\!-\!O\!-\!\overset{CH_3}{\underset{|}{C}}HCH_3$$

1. Butyric acid
2. Ester
3. Isopropyl butyrate

$$CH_3\overset{}{\underset{|}{C}}H\!-\!\overset{O}{\underset{\|}{C}}\!-\!NH_2$$

1. α-Phenylpropionic acid
2. Amide
3. α-Phenylpropionamide

1. Formic acid
2. Anhydride
3. Formic anhydride

9-2. Classify each of the following molecules as acid, ester, amide, acid halide, or anhydride, and name each compound:

(a) $(CH_3)_2CHCH_2\overset{}{\underset{\|}{C}}\!-\!NH_2$ *Answer:* Amide, β-methylbutyramide or
 3-methylbutanamide (IUPAC)
O

(b) $CH_3CH_2CHBr\overset{}{\underset{\|}{C}}\!-\!OH$ *Answer:* Acid, α-bromobutyric acid or
 2-bromobutanoic acid (IUPAC)
O

(c) $H\!-\!\overset{}{\underset{\|}{C}}\!-\!OCH_3$
O

(d) $\langle\bigcirc\rangle\!-\!CH_2\!-\!\overset{}{\underset{\|}{C}}\!-\!Br$
O

(e) $CH_3\!-\!\overset{}{\underset{\|}{C}}\!-\!O\!-\!\overset{}{\underset{\|}{C}}\!-\!CH_3$
OO

9-3. Classify and draw the structure for each of the following compounds:

(a) β-Chlorovaleric acid   *Answer:* $CH_3CH_2CH(Cl)CH_2COOH$

(b) *sec*-Butyl acetate   *Answer:*
$$CH_3\overset{O}{\underset{\|}{C}}\!-\!OCH(CH_3)CH_2CH_3$$

     (c) Propionic anhydride
     (d) Butyryl chloride
     (e) *n*-Propyl propionate
     (f) *m*-Chlorobenzamide

Preparation of carboxylic acids

The methods of preparing carboxylic acids are oxidation of primary alcohols, oxidation of alkylbenzenes, carbonation of Grignard reagents, and hydrolysis of nitriles.

Oxidation

Following are an example of oxidation of a primary alcohol:

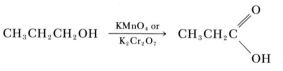

       1-Propanol   **Propanoic acid**

and an example of oxidation of an alkylbenzene:

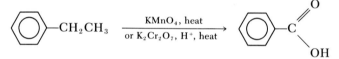

       Ethylbenzene   **Benzoic acid**

Carbonation of Grignard reagents

In addition to reacting with compounds containing the carbonyl group to form various classes of alcohols, Grignard reagents react with carbon dioxide to form carboxylic acids:

$$\text{RMgX} + \text{C}{=}\text{O} \longrightarrow \underset{\overset{\|}{\text{O}}}{\text{R}-\text{C}}-\text{OMgX} \xrightarrow{\text{H}_2\text{O, H}^+} \underset{\overset{\|}{\text{O}}}{\text{R}-\text{C}}-\text{OH} + \text{MgX(OH)}$$

For example:

$$\text{CH}_3\text{CH}_2\text{MgCl} + \text{C}{=}\text{O} \longrightarrow \underset{\overset{\|}{\text{O}}}{\text{CH}_3\text{CH}_2\text{C}}-\text{OMgCl} \xrightarrow{\text{H}_2\text{O, H}^+} \underset{\overset{\|}{\text{O}}}{\text{CH}_3\text{CH}_2\text{C}}-\text{OH} + \text{MgCl(OH)}$$

Ethylmagnesium   **Propionic acid**
   chloride

The carbon dioxide used in this synthesis is solid carbon dioxide (dry ice), although the gas works much better since dry ice most often is very wet.

Hydrolysis of nitriles

Hydrolysis of a nitrile, or alkyl cyanide, (R—C≡N) in the presence of boiling aqueous acid or base results in the formation of an acid:

$$RCN \xrightarrow{H_2O, H^+} RCOOH$$

For example:

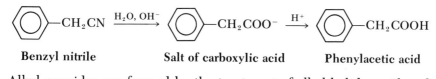

Benzyl nitrile   **Salt of carboxylic acid**   **Phenylacetic acid**

Alkyl cyanides are formed by the treatment of alkyl halides with sodium or potassium cyanide. Cyanide ion, being the conjugate base of the weak acid HCN, is a very strong base. It is so strong that it can cause elimination rather than substitution to occur in tertiary alkyl halides:

$$CH_3CH_2CH_2Cl + NaCN \longrightarrow CH_3CH_2CH_2CN \qquad \text{Substitution in primary halide}$$

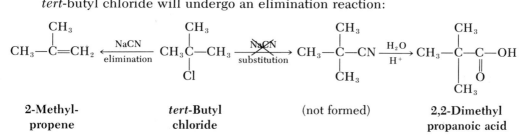

Elimination in tertiary halide

Thus 2,2-dimethylpropanoic acid cannot be made by the treating of *tert*-butyl chloride, a tertiary halide, with NaCN followed by hydrolysis, since *tert*-butyl chloride will undergo an elimination reaction:

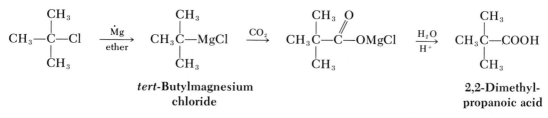

2-Methyl-   **tert-Butyl**   (not formed)   **2,2-Dimethyl**
propene   **chloride**   **propanoic acid**

On the other hand, 2,2-dimethylpropanoic acid can be made through the reaction of *tert*-butylmagnesium chloride with carbon dioxide:

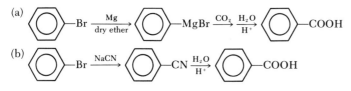

tert-Butylmagnesium   **2,2-Dimethyl-**
chloride   **propanoic acid**

9-4. Which synthetic method gives benzoic acid in good yield. Explain why. (*Hint:* see Chapter 7.)

(a)

(b)

9-5. Predict the carboxylic acid formed in each of the following reactions:

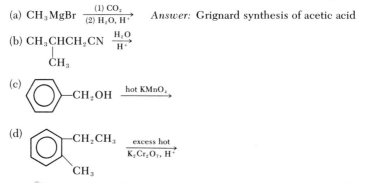

(a) $CH_3MgBr \xrightarrow[\text{(2) } H_2O, H^+]{\text{(1) } CO_2}$ *Answer:* Grignard synthesis of acetic acid

(b) $CH_3CHCH_2CN \xrightarrow[H^+]{H_2O}$
$\quad\quad\; |$
$\quad\quad CH_3$

(c) $\text{(ring)} - CH_2OH \xrightarrow{\text{hot KMnO}_4}$

(d) $\text{(ring)} - CH_2CH_3 \xrightarrow[K_2Cr_2O_7,\ H^+]{\text{excess hot}}$
$\quad\quad\quad\; |$
$\quad\quad\quad CH_3$

Reactions of carboxylic acids

Aliphatic and aromatic carboxylic acids undergo reactions at the carboxyl group. In addition, aromatic carboxylic acids undergo ring substitution.

Reduction

When a carboxylic acid is treated with lithium aluminum hydride ($LiAlH_4$), the carboxyl group is reduced to a $—CH_2OH$ group. This reaction thus is the reduction of a carboxylic acid to a primary alcohol.
Following is an example:

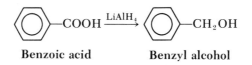

$$\text{(ring)} - COOH \xrightarrow{LiAlH_4} \text{(ring)} - CH_2OH$$

Benzoic acid   **Benzyl alcohol**

However, this reduction is a difficult process. Generally, carboxylic acids are converted to acid derivatives such as acid chlorides or anhydrides, and then the derivatives are reduced. The reduction of several acid derivatives is discussed later in the chapter.

Salt formation

Salt formation is the most characteristic reaction of an acid, for it is the property that gives acids their name. Acids react with bases to form salts by neutralization:

$$\text{Acid} + \text{Base} \xrightarrow{\text{neutralization}} \text{Salt} + \text{Water}$$

Following are some examples:

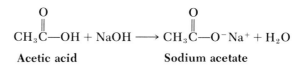

$$\overset{\displaystyle O}{\underset{\displaystyle \|}{CH_3C}}—OH + NaOH \longrightarrow \overset{\displaystyle O}{\underset{\displaystyle \|}{CH_3C}}—O^-Na^+ + H_2O$$

Acetic acid   **Sodium acetate**

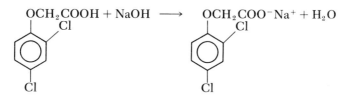

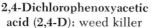

2,4-Dichlorophenoxyacetic   **Sodium salt of 2,4-D**
acid (2,4-D): weed killer

Carboxylic acids are stronger acids than phenols and alcohols, and they react with both $NaOH$ and $NaHCO_3$. Following are some examples:

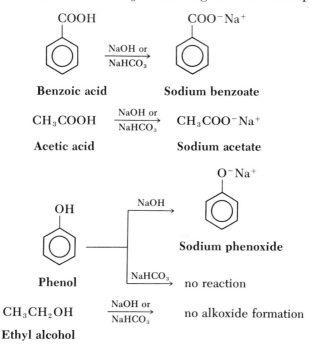

Acetic acid   Sodium acetate

Phenol

Ethyl alcohol

The acidity of a compound is a measure of its ability to donate a proton to water. For a carboxylic acid, the equilibrium established in aqueous solution is as follows:

$$RCOOH \rightleftharpoons RCOO^- + H^+$$

Thus the ionization (or equilibrium) constant, K, is given as follows:

$$K = \frac{[RCOO^-][H^+]}{[RCOOH]}$$

Mathematically, the greater the concentration of ions is, the larger the value of K and hence the greater the ionization of the acid is. Thus the stronger the acid is, the larger the value of K is. Table 9-3 lists the acidities of some common acids.

Compared to sulfuric acid, benzoic acid, phenol, and ethanol are only slightly ionized in solution. However, benzoic acid is more highly ionized

Table 9-3
Acidities of some acids

Acid	K
H_2SO_4	K_1: very large K_2: 2×10^{-2}
⬡—COOH	6.3×10^{-5}
⬡—OH	1.1×10^{-10}
CH_3CH_2OH	$\sim 10^{-16}$

than phenol or ethanol. The explanation for this is as follows. The ionization of ethanol forms the ethoxide ion, in which the negative charge resides on a single oxygen atom:

$$CH_3CH_2{-}\ddot{\underset{..}{O}}{-}H \longrightarrow H^+ + CH_3CH_2{-}\ddot{\underset{..}{O}}{:}^-$$

The ionization of phenol forms the phenoxide ion, in which the charge resides on the oxygen atom as well as on the benzene ring:

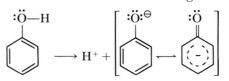

On the other hand, in the carboxylate ion, which is formed when the carboxyl group ionizes, the negative charge is delocalized over two oxygen atoms:

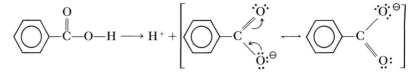

The carboxylate ion is stablized by resonance and shows even less tendency to recombine with the proton than the phenoxide ion or the ethoxide ion. That the negative charge in the carboxylate ion is equally delocalized over the two oxygen atoms is indicated by the fact that the carbon–oxygen bonds are 1.3 Å, whereas the carbon–oxygen single bond length is 1.42 Å and the carbon–oxygen double bond length is 1.2 Å:

Carboxyl group   **Carboxylate ion**

Table 9-4

Acidities of some carboxylic acids

Acid	K
CH_3CH_2COOH	1.3×10^{-5}
$ClCH_2COOH$	136×10^{-5}
$ClCH_2CH_2COOH$	9×10^{-5}
Cl_3CCOOH	23200×10^{-5}
C_6H_5COOH	6.0×10^{-5}
$p\text{-}NO_2C_6H_4COOH$	36×10^{-5}
$p\text{-}CH_3C_6H_4COOH$	4.0×10^{-5}
$o\text{-}ClC_6H_4COOH$	130×10^{-5}
$p\text{-}ClC_6H_4COOH$	10×10^{-5}

Table 9-4 lists the ionization constants of some aliphatic and aromatic carboxylic acids.

The strength of aliphatic carboxylic acids is determined by the inductive effect of the substituent, which depends on the kind of substituent, the position of the substituent, and the number of substituents. Thus chloroacetic acid (Cl—CH_2—COOH) is a stronger acid than propionic acid (CH_3—CH_2—COOH), since the chlorine atom has more electron-withdrawing inductive effect than the methyl group. Chloroacetic acid is a stronger acid than β-chloropropionoic acid (Cl—CH_2—CH_2—COOH), since the electron-withdrawing inductive effect of the halogen is more strongly felt when it is closer to the carboxyl group. Also, trichloroacetic acid

$$\left(\begin{array}{c} Cl \\ | \\ Cl-C-COOH \\ | \\ Cl \end{array} \right)$$

is a stronger acid than chloroacetic acid, since the electron-withdrawing inductive effect of three chloro groups is greater than that of one chloro group. Other groups that increase the acidity of aliphatic carboxylic acids because of their electron-withdrawing inductive effect are —OCH_3, —NO_2, and —CN.

The acidity of benzoic acids is increased by electron-withdrawing groups (such as halogen, —NO_2, and —CN) and decreased by electron-donating groups (such as —OCH_3, —CH_3, —NH_2, and —OH). The electron-withdrawing groups increase the acidity of benzoic acids just as they increase that of phenols, by dispersing the negative charge on the carboxylate oxygen throughout the ring and onto the substituent. Also, the electron-withdrawing inductive effect of the substituent is more strongly felt when it is closer to the carboxyl group. Thus o-chlorobenzoic acid (on the left as follows) is a stronger acid than p-chlorobenzoic acid (on the right as follows):

However, both of these are stronger than benzoic acid.

9-6. Which acid in each pair is stronger?

(a) H_2SO_4 or CH_3COOH *Answer:* H_2SO_4: K is very large; CH_3COOH: $K = 1.8 \times 10^{-5}$

(b) CH_3CH_2COOH or $CH_3CHBrCOOH$ *Answer:* 2-Bromopropanoic acid because of inductive effect of Br— group

(c)
O_2N—⬡—COOH or H_3C—⬡—COOH   *Answer:* p-Nitrobenzoic acid

(d)
⬡—COOH or ⬡—OH

(e)
$CH_3CH_2CH_2OH$ or ⬡—COOH

(f)
Cl—⬡—COOH or ⬡—COOH

The reaction between a carboxylic acid and a base such as $NaHCO_3$ to give a carboxylic acid salt can be used in the extraction of a carboxylic acid from neutral, basic substances (such as amines) or weakly acidic substances (such as phenols). The carboxylic acid is converted to a water-soluble salt, and the organic and aqueous layers are separated. Then the carboxylic acid is recovered from the aqueous layer after it is made acidic, usually with hydrochloric acid:

$$RCOO^-Na^+ + H^+ \longrightarrow RCOOH + Na^+$$

Formation of functional derivatives

The preparation of esters, amides, anhydrides, and acid chlorides involves the replacement of the —OH group by another group (G):

$$R-\overset{\overset{\displaystyle O}{\|}}{C}-OH \xrightarrow{G} R-\overset{\overset{\displaystyle O}{\|}}{C}-G$$

Acid chlorides. These compounds are formed by the reaction of an acid with phosphorus trichloride (PCl_3), phosphorus pentachloride (PCl_5), or thionyl chloride ($SOCl_2$). The general reaction is as follows:

$$R-\overset{\overset{\displaystyle O}{\|}}{C}-OH \xrightarrow{PCl_3,\ PCl_5,\ or\ SOCl_2} R-\overset{\overset{\displaystyle O}{\|}}{C}-Cl$$

Following are some examples:

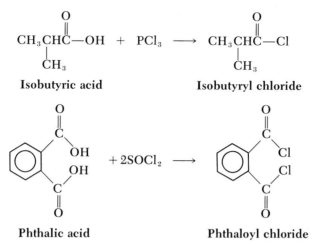

Isobutyric acid   Isobutyryl chloride

Phthalic acid   Phthaloyl chloride

Esters. These compounds are formed by the reaction of an acid with an alcohol. However, the reaction is reversible. It is generally catalyzed by traces of sulfuric acid:

$$RC\text{—}OH + R'OH \xrightarrow{H^+} RC\text{—}OR' + H_2O$$
$$\quad\quad \overset{\|}{O} \quad\quad\quad\quad\quad \overset{\|}{O}$$

Following are some examples:

$$CH_3\overset{O}{\overset{\|}{C}}\text{—}OH + CH_3OH \rightleftharpoons CH_3\overset{O}{\overset{\|}{C}}\text{—}OCH_3 + H_2O$$

Acetic acid   Methyl alcohol   Methyl acetate

$$HOOCCH_2COOH + 2C_2H_5OH \xrightarrow{H^+} C_2H_5OOCCH_2COOC_2H_5 + 2H_2O$$

Malonic acid   Diethyl malonate

The esterification reaction occurs according to the following mechanism:

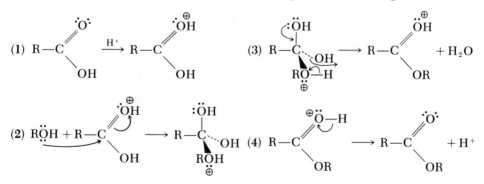

In step 1, the ester undergoes protonation at the carboxyl oxygen; this renders the carboxyl carbon more susceptible to nucleophilic attack by the alcohol in step 2, through which a tetrahedral intermediate is formed (the sp^2 hybrid orbitals of the carboxyl carbon become sp^3 hybridized). In step 3, a proton is transferred from the alcohol portion to the hydroxyl portion, and a molecule of water is lost; thus the orbitals of the carboxyl carbon return to their original sp^2 hybridized state. Loss of a proton in step 4 regenerates the catalyst and forms the ester.

9-7. The bond angles in the ester (see step 4) are probably very close to 120°, whereas those in the tetrahedral intermediate (see step 2) are 109°. Thus there is some compression of the bond angles in ester formation. With this in mind, decide which of the following two methyl esters is formed more easily:

$$
\underset{\text{}}{CH_3CH_2\overset{\displaystyle O}{\overset{\|}{C}}{-}OCH_3} \qquad \text{or} \qquad CH_3{-}\overset{\displaystyle CH_3}{\underset{\displaystyle CH_3}{\overset{|}{\underset{|}{C}}}}{-}\overset{\displaystyle O}{\overset{\diagup\!\!\diagup}{C}}{-}OCH_3
$$

Unfortunately, the substance that catalyzes the forward reaction also catalyzes the reverse reaction, the hydrolysis of the ester. Hydrolysis can be minimized by the use of an apparatus in the synthesis that removes the water as it is formed. A better method is to prepare the ester through the reaction of an acid chloride (via the acid) and an alcohol. The by-product, HCl, which is a gas, escapes from the reaction mixture; the reaction is therefore driven to the right and is nonreversible:

$$
\underset{\displaystyle O}{R{-}\overset{\|}{C}{-}Cl} + R'OH \longrightarrow \underset{\displaystyle O}{R{-}\overset{\|}{C}{-}OR'} + HCl\uparrow
$$

For example:

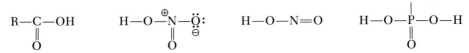

Benzoyl chloride   Ethyl alcohol   **Ethyl benzoate**

Esters can also be formed by the reaction of an alcohol with an inorganic acid, such as nitric acid, nitrous acid, or phosphoric acid. These acids resemble carboxylic acids in that they all possess hydroxyl groups:

$$
\underset{\displaystyle O}{R{-}\overset{\|}{C}{-}OH} \qquad \underset{\displaystyle O}{H{-}O{-}\overset{\oplus}{\underset{\ominus}{N}}{-}\overset{\displaystyle ..}{\underset{..}{O}}{:}} \qquad H{-}O{-}N{=}O \qquad \underset{\displaystyle O}{H{-}O{-}\overset{\displaystyle O{-}H}{\overset{|}{P}}{-}O{-}H}
$$

Carboxylic acid   **Nitric acid**   **Nitrous acid**   **Phosphoric acid**

Following are some examples of the reaction between alcohols and these inorganic acids:

$$
\begin{array}{l}
CH_2{-}O{-}H \\
|\\
CH{-}O{-}H \\
|\\
CH_2{-}O{-}H
\end{array}
+
\begin{array}{l}
H{-}O{-}NO_2 \\
H{-}O{-}NO_2 \\
H{-}O{-}NO_2
\end{array}
\longrightarrow
\begin{array}{l}
CH_2{-}O{-}NO_2 \\
|\\
CH{-}O{-}NO_2 \\
|\\
CH_2{-}O{-}NO_2
\end{array}
+ 3H_2O
$$

     Glycerol   **Nitric acid**   **Glyceryl trinitrate**
                                      (nitroglycerin):
                                        explosive

$$\underset{\text{Isoamyl alcohol}}{CH_3\overset{\overset{\displaystyle CH_3}{|}}{C}HCH_2CH_2OH} + \underset{\text{Nitrous acid}}{H\text{—}ONO} \longrightarrow \underset{\substack{\text{Isoamyl nitrite: for} \\ \text{relief of pain due to} \\ \text{angina pectoris}}}{CH_3\overset{\overset{\displaystyle CH_3}{|}}{C}HCH_2CH_2ONO} + H_2O$$

The esters of phosphoric acid are probably the most important compounds in this group. Monoalkyl, dialkyl, and trialkyl esters of phosphoric acid have the following general structure:

$$\underset{\underset{\displaystyle OH}{|}}{RO\text{—}\overset{\overset{\displaystyle O}{\|}}{P}\text{—}OH} \qquad \underset{\underset{\displaystyle OH}{|}}{RO\text{—}\overset{\overset{\displaystyle O}{\|}}{P}\text{—}OR} \qquad \underset{\underset{\displaystyle OR}{\|}}{RO\text{—}\overset{\overset{\displaystyle O}{\|}}{P}\text{—}OR}$$

These esters are important in a number of reactions in the body, especially in providing useful chemical energy. (See Chapters 13 and 14.) Commercially, they are useful as insecticides.

Organophosphate insecticides. One possible method of minimizing the hazards and problems involved with the organochlorine insecticides (Chapter 7) is to replace the organochlorine compounds with organophosphates. The organophosphates are still highly toxic compounds; but the advantage in using them is that they are less persistent in the environment than the organochlorine compounds, since they can be hydrolyzed to less toxic substances. This hydrolysis is a reaction characteristic of esters.

The organophosphate group of insecticides also includes esters in which the doubly bonded oxygen has been replaced by sulfur. These compounds are called thioesters or thiophosphates.

Following are some common organophosphate and thiophosphate insecticides:

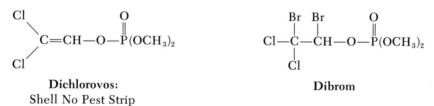

Dichlorovos:
Shell No Pest Strip

Dibrom

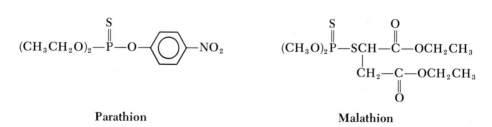

Parathion

Malathion

Dibrom is about six times more toxic than dichlorovos, and Malathion is 100 times less toxic than parathion.

Amides. These compounds are acid derivatives in which the —OH group has been replaced by the —NH₂ group. They are usually prepared in one of the two following ways. One method is to neutralize the acid with ammonia (NH_3) and then to heat the resulting ammonium salt strongly to drive off water:

$$R-\underset{\underset{O}{\|}}{C}-OH \xrightarrow{NH_3} R-\underset{\underset{O}{\|}}{C}-O^{\ominus}NH_4^{\oplus} \xrightarrow{\text{heat}} R-\underset{\underset{O}{\|}}{C}-NH_2 + H_2O$$

For example:

$$CH_3-\underset{\underset{O}{\|}}{C}-OH \xrightarrow{NH_3} CH_3-\underset{\underset{O}{\|}}{C}-O^{\ominus}NH_4^{\oplus} \xrightarrow{\text{heat}} CH_3-\underset{\underset{O}{\|}}{C}-NH_2 + H_2O$$

Acetic acid   **Ammonium acetate**   **Acetamide**

The other method is to prepare amides from the reaction of an acid chloride (via the acid) and ammonia:

$$R-\underset{\underset{O}{\|}}{C}-Cl + NH_3 \longrightarrow R-\underset{\underset{O}{\|}}{C}-NH_2 + HCl$$

For example:

$$CH_3CH_2\underset{\underset{O}{\|}}{C}-Cl + NH_3 \longrightarrow CH_3CH_2\underset{\underset{O}{\|}}{C}-NH_2 + HCl$$

Propionyl chloride   **Propionamide**

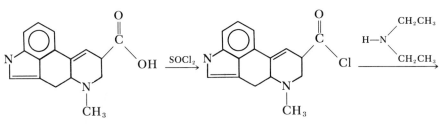

Lysergic acid

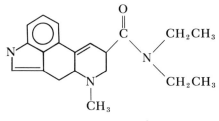

**Lysergic acid
diethylamide (LSD)**

The amide bond ($\overset{\overset{O}{\|}}{C}$—N) occurs in a large number of biomolecules. These biomolecules are referred to as peptides and proteins. The following example is a dipeptide:

$$\underset{\text{R O}}{\overset{\text{R O}}{H_2NCHC-NHCHCOH}}$$

Peptides and proteins are discussed in more detail in Chapters 13 and 14.

Acetic anhydride. This is probably the most often used anhydride. It is prepared by the reaction of ketene with acetic acid. Ketene is synthesized by the dehydration of acetic acid at high temperature, which is catalyzed by aluminum phosphate ($AlPO_4$):

$$CH_3\overset{O}{\overset{\|}{C}}-OH \xrightarrow[\text{heat}]{AlPO_4} CH_2=C=O + HOH$$

Acetic acid   **Ketene**

Following is an example of the preparation of acetic anhydride:

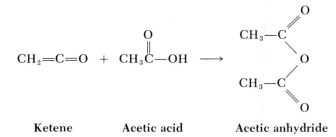

Ketene   **Acetic acid**   **Acetic anhydride**

The net effect of this preparation is the removal of a molecule of water from two molecules of acetic acid:

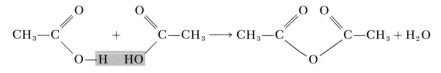

A similar reaction can occur between acetic acid and an inorganic acid, such as phosphoric acid:

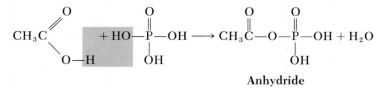

Anhydride

The living system makes use of this type of reaction in the conversion of 3-phosphoglyceric acid to 1,3-diphosphoglyceric acid:

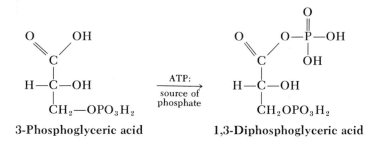

3-Phosphoglyceric acid   **1,3-Diphosphoglyceric acid**

This reaction is accomplished in the living system when 3-phosphoglyceric acid is reduced to 3-phosphoglyceraldehyde. The usefulness of the reaction lies in the ease of reduction of the anhydride, which is discussed later in the chapter.

Noncarboxyl group substitution

There are two types of substitution reactions of carboxylic acids in which the —OH group is not substituted by another group. One reaction is the *Hell-Volhard-Zelinsky reaction;* it occurs in aliphatic carboxylic acids that have at least one α-hydrogen. The reagent is chlorine or bromine and the reaction is catalyzed by a trace of phosphorus. It is a specific reaction in that only the α-hydrogen in the carboxylic acid is substituted:

$$
\begin{array}{c}
\qquad\qquad\qquad O \qquad\qquad\qquad\qquad\qquad\qquad\qquad O \\
\qquad\qquad\qquad \| \qquad\qquad\qquad\qquad\qquad\qquad\qquad\quad \| \\
R-CH_2-C-OH \;+\; Cl_2 \;\xrightarrow{\;P\;}\; R-CH-C-OH \\
\qquad\qquad\qquad\qquad\qquad\qquad\qquad\qquad\qquad\qquad\quad | \\
\qquad\qquad\qquad\qquad\qquad\qquad\qquad\qquad\qquad\qquad\;\; Cl
\end{array}
$$

For example:

$$
\begin{array}{c}
\qquad O \qquad\qquad\qquad\qquad\qquad\qquad\qquad\qquad\qquad\qquad\quad O \\
\qquad \| \qquad\qquad\qquad\qquad\qquad\qquad\qquad\qquad\qquad\qquad\quad \| \\
CH_3CH_2C-OH \;+\; Br_2 \;\xrightarrow{\;P\;}\; CH_3CHC-OH \\
\qquad\qquad\qquad\qquad\qquad\qquad\qquad\qquad\qquad\qquad\qquad\quad | \\
\qquad\qquad\qquad\qquad\qquad\qquad\qquad\qquad\qquad\qquad\qquad\;\; Br
\end{array}
$$

Propanoic acid   **2-Bromopropanoic acid**
                                       α-bromopropionic acid

The other type of substitution reaction occurs in aromatic acids and is therefore called ring substitution. The —COOH group is a deactivator and a *meta* director in electrophilic substitution. For example, nitration of benzoic acid yields *m*-nitrobenzoic acid, and bromination gives *m*-bromobenzoic acid.

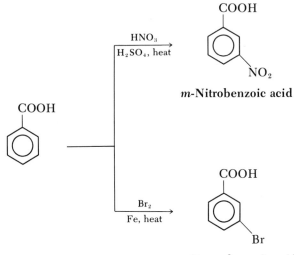

m-Nitrobenzoic acid

m-Bromobenzoic acid

Reactions of carboxylic acid derivatives

Hydrolysis

The typical reaction of a carboxylic acid derivative is *nucleophilic acyl substitution*. The mechanism of this substitution, illustrated as follows, involves the substitution of the group or atom attached to the acyl carbon (L) by some other group (Nu). Typical nucleophiles (Nu) include $H_2\ddot{O}:$, $:\ddot{O}H^-$, $R\ddot{O}H$, and $\ddot{N}H_3$.

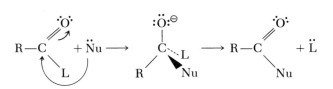

The nucleophile, $\ddot{N}u$, attacks the acyl carbon with sp^2 hybrid orbitals, thus forming an intermediate in which the previously trigonal carbon becomes tetrahedral and oxygen bears a negative charge. Since L, the leaving group, is a weaker base than Nu, it leaves more easily from the intermediate. Thus a product is formed in which the carbon again becomes trigonal. The energy changes that occur are shown in Fig. 9-1.

The acyl derivatives are more reactive in nucleophilic substitution than their aliphatic analogs. For example, acetyl chloride hydrolyzes rapidly in water, whereas ethyl chloride is, for all practical purposes, inert. The explanation for this can be seen from a comparison of the mechanisms of these two reactions on the next page.

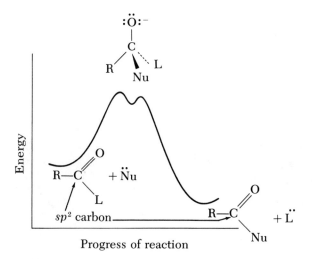

Fig. 9-1
Energy profile of nucleophilic acyl substitution. If L were a stronger base than Nu, the substitution would not occur.

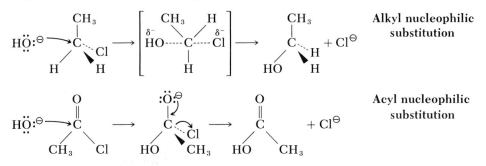

Alkyl nucleophilic substitution

Acyl nucleophilic substitution

Nucleophilic attack in ethyl chloride involves attack on a tetrahedral carbon to form a crowded transition state in which carbon is pentavalent, whereas nucleophilic attack in acetyl chloride involves attack on a trigonal carbon to form a less crowded intermediate in which carbon is tetrahedral. It takes less energy to form an intermediate in which carbon is tetrahedral than a transition state in which carbon is pentavalent. These energy requirements are shown in Fig. 9-2.

For similar reasons, amides are more reactive than amines (RNH_2), and esters are more reactive than ethers (ROR).

Esters undergo hydrolysis to yield the corresponding carboxylic acid and an alcohol:

$$R-\underset{\underset{O}{\parallel}}{C}-OR' \xrightleftharpoons{H_2O, H^+} R-\underset{\underset{O}{\parallel}}{C}-OH + R'OH$$

For example:

$$CH_3-\underset{\underset{O}{\parallel}}{C}-^{18}OCH_2CH_3 \rightleftharpoons^{H_2O, H^+} CH_3-\underset{\underset{O}{\parallel}}{C}-OH + CH_3CH_2{}^{18}OH$$

Ethyl acetate   **Acetic acid**   **Ethyl alcohol**

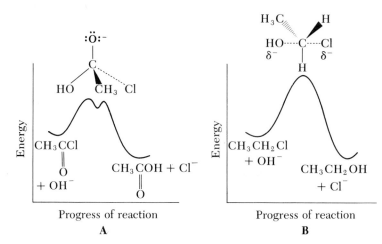

Fig. 9-2
A, Acyl nucleophilic substitution for $CH_3COCl + OH^-$. **B,** Alkyl nucleophilic substitution for $CH_3CH_2Cl + OH^-$.

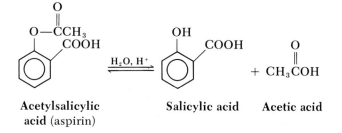

Acetylsalicylic   **Salicylic acid**   **Acetic acid**
acid (aspirin)

In this reaction, bond cleavage occurs between the acyl carbon–oxygen bond rather than the alkyl carbon–oxygen bond:

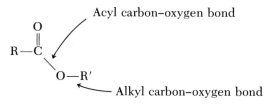

This is evident from the hydrolysis of ethyl acetate just illustrated, in which ^{18}O, an isotope of oxygen, is used for one of the oxygens to label it. The labeled oxygen is found in the ethyl alcohol ($CH_3CH_2{}^{18}OH$), indicating that acyl carbon–oxygen cleavage occurred. If alkyl carbon–oxygen cleavage had occurred, the labeled oxygen atom would have been found in acetic acid ($CH_3C^{18}OH$).

The hydrolysis of acid chlorides occurs as follows:

$$R-C-Cl + H_2O \longrightarrow R-C-OH + HCl$$
$$\quad\ \ \| \qquad\qquad\qquad\qquad \|$$
$$\quad\ \ O \qquad\qquad\qquad\qquad O$$

For example:

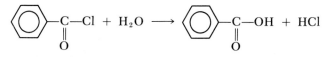

Benzoyl chloride   **Benzoic acid**

The hydrolysis of amides occurs as follows:

$$R-C-NH_2 + H_2O \xrightarrow{H^+} R-C-OH + NH_4{}^+$$
$$\quad\ \ \| \qquad\qquad\qquad\qquad\quad \|$$
$$\quad\ \ O \qquad\qquad\qquad\qquad\quad O$$

For example:

$$CH_3CH_2CH_2C-NH_2 + H_2O \xrightarrow{H^+} CH_3CH_2CH_2C-OH + NH_4{}^+$$
$$\qquad\qquad\qquad \| \qquad\qquad\qquad\qquad\qquad\quad \|$$
$$\qquad\qquad\qquad O \qquad\qquad\qquad\qquad\qquad\quad O$$

Butyramide   **Butyric acid**

The hydrolysis of anhydrides occurs as follows:

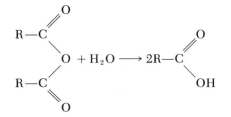

For example:

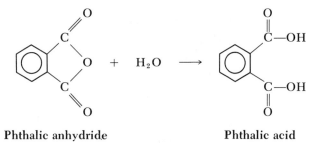

Phthalic anhydride   **Phthalic acid**

The order of ease of hydrolysis of acid derivatives is acid chlorides > anhydrides > esters > amides.

Alcoholysis

Esters react with alcohols to yield other esters. This reaction, which is called *transesterification*, is essentially the displacement of the alcohol group in the ester by another alcohol:

$$\underset{\underset{O}{\|}}{R-C}-OR' + R''OH \overset{H^+}{\rightleftharpoons} \underset{\underset{O}{\|}}{R-C}-OR'' + R'OH$$

For example:

$$\underset{\underset{O}{\|}}{CH_3C}-OCH_3 + CH_3CH_2CH_2OH \overset{H^+}{\rightleftharpoons} \underset{\underset{O}{\|}}{CH_3C}-OCH_2CH_2CH_3 + CH_3OH$$

Methyl acetate   *n*-Propyl alcohol   *n*-Propyl acetate   Methyl alcohol

Anhydrides react with alcohols to form esters and carboxylic acids:

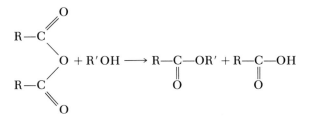

For example:

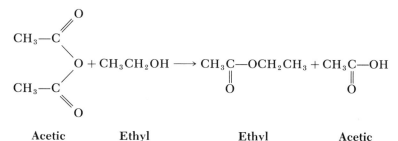

| Acetic anhydride | Ethyl alcohol | Ethyl acetate | Acetic acid |

Acid chlorides react with alcohols to form esters:

$$R-\underset{\underset{O}{\|}}{C}-Cl + R'OH \longrightarrow R-\underset{\underset{O}{\|}}{C}-OR' + HCl$$

For example:

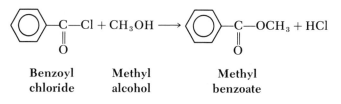

| Benzoyl chloride | Methyl alcohol | Methyl benzoate |

Reduction

Acid derivatives can be reduced. Alcohols are the reduction products in the case of acid chlorides, anhydrides, and esters. Amines are the reduction products of amides. Lithium aluminum hydride is the reducing agent commonly used. The order of reactivity of acid derivatives in reduction is the same as that in hydrolysis: acid chlorides > anhydrides > esters > amides. Some typical reduction reactions of acid derivatives are as follows:

$$\underset{\text{Acetic anhydride}}{CH_3\underset{\underset{O}{\|}}{C}-O-\underset{\underset{O}{\|}}{C}CH_3} \xrightarrow{\text{LiAlH}_4} \underset{\text{Ethanol}}{2CH_3CH_2OH}$$

$$\underset{\text{Methyl acetate}}{CH_3\underset{\underset{O}{\|}}{C}-O-CH_3} \xrightarrow{\text{LiAlH}_4} \underset{\text{Ethanol}}{CH_3CH_2OH} + \underset{\text{Methanol}}{CH_3OH}$$

The living system takes advantage of the ease of reduction of acid derivatives over carboxylic acids. In biochemical reactions, the carboxyl group can be reduced to an aldehyde group rather than to an alcohol group. This reduction occurs in a stepwise manner. The acid is converted to an anhydride, which is then transformed through hydrolysis to the aldehyde. These reactions are illustrated as follows for the conversion of 3-phosphoglyceric acid

to 3-phosphoglyceraldehyde. The overall reaction is the reduction of a carboxyl group to a carbonyl group:

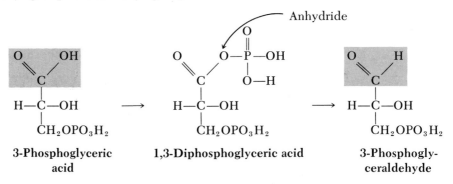

| 3-Phosphoglyceric acid | 1,3-Diphosphoglyceric acid | 3-Phosphogly- ceraldehyde |

If the reduction of a carboxyl group to an aldehyde group is carried out in
the laboratory, the energy formed is released as heat. In the living system,
however, the energy formed is stored by the living system to be used to drive
some other energy-requiring reaction or even to do mechanical work, such as
muscle contraction. These processes are discussed in more detail in Chapter 14.

Reformatsky reaction

The Reformatsky reaction is a useful synthetic reaction between an ester
and a carbonyl compound. The product is a β-hydroxy ester. The reaction
involves the formation of an organozinc intermediate from an α-bromoester
and zinc:

$$BrCH_2C{-}OC_2H_5 \quad + \quad Zn \quad \xrightarrow{\text{ether}} \quad BrZnCH_2C{-}OC_2H_5$$
$$\underset{O}{\|} \qquad\qquad\qquad\qquad\qquad\qquad\qquad \underset{O}{\|}$$

Ethyl bromoacetate   **Organozinc intermediate**

Then, in the presence of an aldehyde or ketone, the intermediate forms a
β-hydroxy ester. The overall reaction is illustrated as follows:

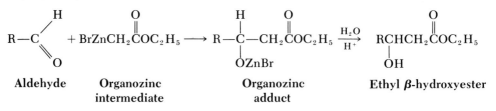

Aldehyde   **Organozinc intermediate**   **Organozinc adduct**   **Ethyl β-hydroxyester**

The formation of the organozinc adduct and its subsequent hydrolysis to
the ester are similar to the formation of the organomagnesium adduct from
a Grignard reagent and an aldehyde or ketone and the subsequent hydrolysis
of the organomagnesium adduct to an alcohol.

A wide variety of β-hydroxy esters can be obtained if the proper esters and
carbonyl compounds are used. The following reactions are examples of the
versatility of the Reformatsky reaction:

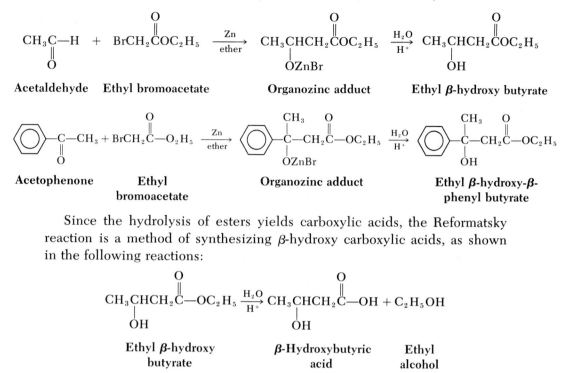

Since the hydrolysis of esters yields carboxylic acids, the Reformatsky reaction is a method of synthesizing β-hydroxy carboxylic acids, as shown in the following reactions:

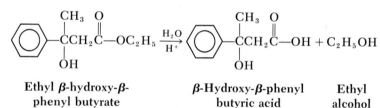

| Ethyl β-hydroxy butyrate | β-Hydroxybutyric acid | Ethyl alcohol |

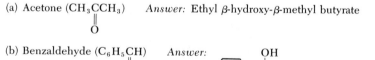

| Ethyl β-hydroxy-β-phenyl butyrate | β-Hydroxy-β-phenyl butyric acid | Ethyl alcohol |

9-8. Draw the structures of the β-hydroxy esters that are obtained when ethyl bromoacetate reacts with the following carbonyl compounds:

(a) Acetone (CH_3CCH_3)   *Answer:* Ethyl β-hydroxy-β-methyl butyrate
           ‖
           O

(b) Benzaldehyde (C_6H_5CH)   *Answer:*
           ‖
           O

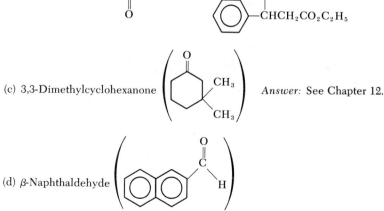

(c) 3,3-Dimethylcyclohexanone   *Answer:* See Chapter 12.

(d) β-Naphthaldehyde

Dicarboxylic acids

Malonic acid

From a synthetic point of view, malonic acid is perhaps the most important dicarboxylic acid. The *malonic ester synthesis* is a good method of preparing carboxylic acids via diethyl malonate ($CH_3CH_2O\overset{O}{\overset{\|}{C}}CH_2\overset{O}{\overset{\|}{C}}OCH_2CH_3$). This method of synthesis of carboxylic acids depends on the following factors: (1) the increased acidity of the α-hydrogens of diethyl malonate and (2) the ease with which the resulting malonic acid decarboxylates, that is, loses a molecule of carbon dioxide. These two factors are illustrated by the following reaction:

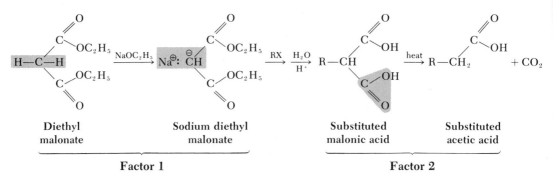

| Diethyl malonate | Sodium diethyl malonate | Substituted malonic acid | Substituted acetic acid |

Factor 1   Factor 2

The synthesis proceeds as follows. When treated with sodium ethoxide ($Na^+{}^-OC_2H_5$) in dry ethanol, diethyl malonate (I) is converted into its sodium salt (II); then the reaction between the salt and an alkyl halide (RX), which is a nucleophilic substitution reaction, forms a monosubstituted malonic ester (III):

$$
\begin{array}{ccc}
\begin{matrix} CO_2C_2H_5 \\ | \\ CH_2 \\ | \\ CO_2C_2H_5 \end{matrix}
& \xrightarrow[100\% \ C_2H_5OH]{NaOC_2H_5}
& \overset{\oplus}{Na} \begin{matrix} CO_2C_2H_5 \\ | \\ :\overset{\ominus}{C}H \\ | \\ CO_2C_2H_5 \end{matrix}
\end{array}
\xrightarrow{RX}
\begin{matrix} CO_2C_2H_5 \\ | \\ R-CH \\ | \\ CO_2C_2H_5 \end{matrix}
$$

Diethyl malonate   Sodium diethyl malonate   Monosubstituted diethyl malonate

I   II   III

Hydrolysis of the malonic ester (III) yields a monosubstituted malonic acid (IV). Then heating to induce decarboxylation forms a monosubstituted acetic acid (V). Alternatively, since the monosubstituted malonic ester (III) still contains an acidic hydrogen, when it is treated with sodium ethoxide, it is converted to a salt (VI). The reaction of this salt with an alkyl halide – either the same as the first (RX) or different (R′X) – produces a disubstituted malonic ester (VII). Hydrolysis then converts the disubstituted malonic ester to the disubstituted malonic acid (VIII), which when heated loses carbon dioxide and forms a disubstituted malonic acid (IX). Since the original diethyl malonate contains only two ionizable hydrogens, a malonic ester synthesis yields an acetic acid in which either one or two hydrogens have been replaced by alkyl groups:

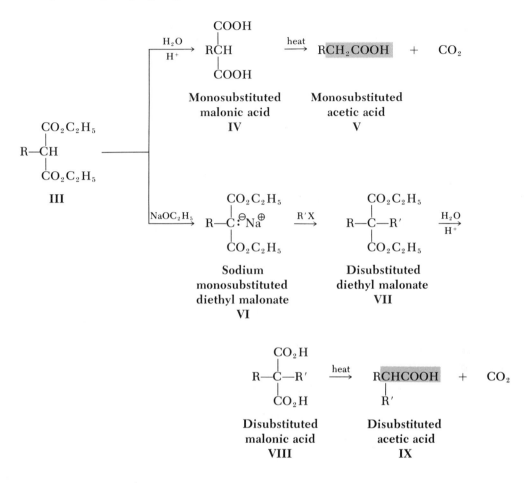

To plan a malonic ester synthesis of a carboxylic acid, you need only to choose the proper alkyl halide or halides to substitute. The main consideration in the selection is that the carboxylic acid formed is a substituted acetic acid:

For example, butyric acid can be considered to be an acetic acid in which one hydrogen has been replaced by an ethyl group:

$$CH_3CH_2CH_2COOH$$

The sequence of steps, then, in the malonic ester synthesis of butyric acid is as follows:

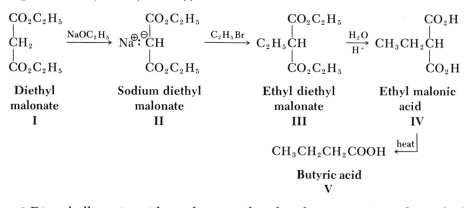

α,β-Dimethylbutyric acid can be considered to be an acetic acid in which one hydrogen has been replaced by a methyl group and the second hydrogen has been replaced by an isopropyl group:

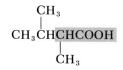

The sequence of steps is as follows:

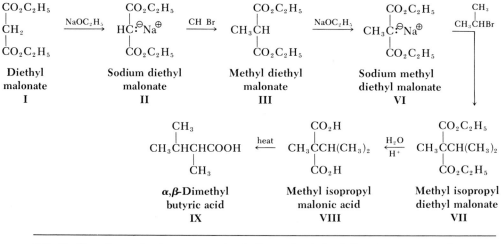

9-9. Outline the malonic ester synthesis for each of the following compounds:

(a) α-Methyl butyric acid   Answer: I $\longrightarrow$ II $\xrightarrow{CH_3Br}$ III $\longrightarrow$ VI $\xrightarrow{C_2H_5Br}$ VII $\longrightarrow$ VIII $\longrightarrow$ IX

(b) β-Methyl butyric acid
(c) Dibenzylacetic acid

Diethyl malonate also undergoes an important reaction with urea. Urea is a unique amide; it is a diamide:

$$NH_2-\underset{\underset{O}{\|}}{C}-NH_2$$

The reaction between diethyl malonate and urea, in which two molecules of ethyl alcohol are released, forms barbituric acid:

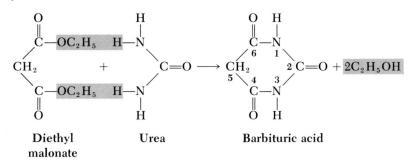

| Diethyl malonate | Urea | Barbituric acid |

Barbituric acids in which there are two substituents on the number five carbon are useful as hypnotics and sedatives. These 5,5-disubstituted barbituric acids are called barbiturates:

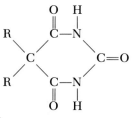

Following are some examples:

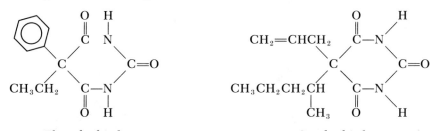

Phenobarbital: long-acting sedative

Secobarbital: short-acting sedative

Other important dicarboxylic acids

Another important dicarboxylic acid is adipic acid. Adipic acid is used in the industrial synthesis of Nylon 66, a polyamide. In this polymerization, adipic acid is reacted with the amine hexamethylenediamine:

$$HO-\overset{O}{\overset{\|}{C}}-(CH_2)_4-\overset{O}{\overset{\|}{C}}-OH + H_2N(CH_2)_6NH_2 \xrightarrow{heat}$$

Adipic acid   **Hexamethylene-diamine**

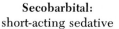

$$\left(-\overset{O}{\overset{\|}{C}}(CH_2)_4\overset{O}{\overset{\|}{C}}-\underset{H}{N}(CH_2)_6\underset{H}{N}-\overset{O}{\overset{\|}{C}}(CH_2)_4\overset{O}{\overset{\|}{C}}-\underset{H}{N}(CH_2)_6\underset{H}{N}-\right)_n$$

Nylon 66: a synthetic fiber

Terephthalic acid is one of the three benzene dicarboxylic acids:

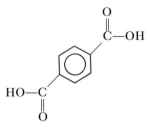

When the dimethyl ester of terephthalic acid is treated with ethylene glycol in acid solution, a polyester known as Dacron is formed:

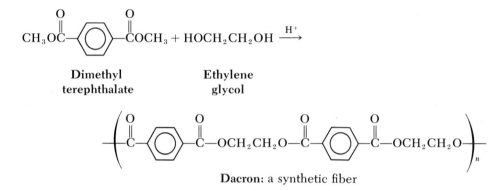

Dimethyl terephthalate + **Ethylene glycol**

Dacron: a synthetic fiber

Chemicals in the foods we eat: food additives

Food additives have been around for a long time. Salt (NaCl), sugar ($C_6H_{12}O_2$), vinegar (CH_3COOH), pepper (shown as follows), and other spices have been used for centuries.

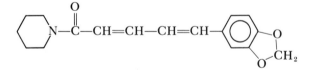

Piperine: pepper

It is not safe for everyone to consume such substances. For example, many people have to curtail sugar and salt in their diets. Most people consume a great many additives in their food without observable harm. It is sometimes argued that the well-known additives are all right because they are natural whereas chemical or artificial additives are the villains. This argument makes little scientific sense, since the flavoring agents in such common additives as vanilla (Chapter 6), pepper, and cloves (Chapter 8) are complex organic molecules as sophisticated as any synthetic molecule produced in a test tube. Furthermore, some of the most potent poisons known to man, such as strychnine and curare, are natural compounds; and several natural compounds, such as the flavoring agents safrole (Chapter 6) and coumarin, have been banned as additives since laboratory tests proved them to be toxic.

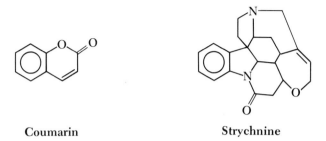

Coumarin   Strychnine

The controlled use of some additives can be beneficial without introducing a significant risk. In many cases, food additives are used because they provide clear-cut economic benefits. For example, calcium propionate $[(CH_3CH_2COO^-)_2Ca^{2+}]$ prevents the growth of mold in bread and cheese. Besides being safe, the calcium in the calcium propionate can aid in the growth of teeth and bones.

Other additives have positive benefit to health. Rickets, a bone disease caused by a deficiency of vitamin D (Chapter 13), has been virtually eliminated since dairies began adding amounts of vitamin D to milk. The addition of niacin to cornmeal and bread was instrumental in eliminating pellagra, the vitamin-deficiency disease that used to afflict people dependent on a cereal diet.

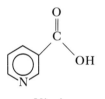

Niacin

Furthermore, many products would not exist or would be priced beyond reach if artificial coloring and flavoring were not used in place of natural food. Expensive brands of ice cream, candy, and yogurt may contain real fruit or fruit extract, but cheaper brands have to rely on color and flavor additives to provide the desired taste and appearance. These additives are beneficial as long as they are known to be safe. For example, the use of imitation fruit flavoring permits the marketing of useful, tasty products at prices the majority of people can afford. Esters blended with other chemicals and essential oils produce a fruity flavor and are used commonly in fruit flavorings; some examples are given in Table 9-5. Such additives are referred to as *flavor enhancers*.

Another type of food additive, the *acidity-controlling agent*, is found in foods from butter to canned fruits and vegetables. Citric acid is the most popular acid additive, and tartaric acid is ideal for augmenting fruit flavors. In general, these additives impart a tartness to foods.

Table 9-5
Ester additives

Name	Structure	Flavor
Allyl caproate	$CH_3(CH_2)_4COOCH_2CH{=}CH_2$	Pineapple
Amyl propionate	$CH_3CH_2COO(CH_2)_4CH_3$	
Ethyl butyrate	$CH_3(CH_2)_2COOCH_2CH_3$	
Ethyl caproate	$CH_3(CH_2)_4COOCH_2CH_3$	
Amyl acetate	$CH_3COOCH_2(CH_2)_3CH_3$	Banana
Methyl o-aminobenzoate	$o\text{-}H_2NC_6H_4COOCH_3$	Grape
Methyl butyrate	$CH_3(CH_2)_2COOCH_3$	Apple
Octyl acetate	$CH_3COO(CH_2)_7CH_3$	Orange
3-Methyl-2-butenyl ethanoate	$CH_3COOCH_2CH{=}C(CH_3)_2$	Juicy Fruit gum

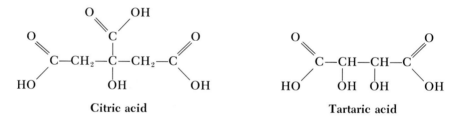

Citric acid   **Tartaric acid**

Antioxidants are used to keep fatty products from smelling and tasting bad. Rancidity is a common type of food spoilage and is recognized by an off odor and on off taste. Ascorbic acid (vitamin C) and BHA (butylated hydroxyanisole) are examples of a natural and a synthetic antioxidant, respectively:

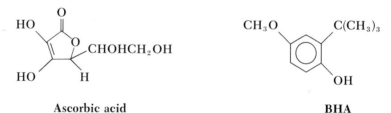

Ascorbic acid   **BHA**

Preservatives include sugars, which keep jams and jellies from spoiling, and calcium propionate, which retards mold formation in bread and cheese.

Color and *flavor stabilizers* include EDTA (ethylenediaminetetraacetic acid), which prevents trace metallic compounds from spoiling the flavor and color of soft drinks. EDTA has solved two problems for beer drinkers: chill haze and gushing. By stabilizing the traces of iron usually contributed by the malt, EDTA prevents gushing, or the sudden violent release of carbon dioxide when a beer bottle is opened. EDTA can also prevent chill haze, which appears to be the result of the reaction of trace amounts of copper with proteins in the malt, without adding any taste of its own to the beer.

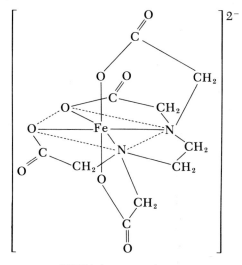

EDTA-iron complex

Emulsifiers keep substances in food from separating; and *thickeners* and *stabilizers,* most of which are plant extracts such as gum arabic, add thickness to the consistency of ice cream, cake mixes, and cooking oils. One example is propylene glycol monostearate, an emulsifier found in Dream Whip:

$$CH_3(CH_2)_{16}\overset{\overset{O}{\|}}{C}\!-\!OCH_2\!-\!\overset{\overset{OH}{|}}{C}HCH_3$$

Propylene glycol monostearate

Food additives can be placed in one of the following risk categories:

1. Chemicals that occur in nature and are easily metabolized by the body. These are the safest. They include proteins, which tenderize meat and speed the brewing of beer; vitamins and minerals, which increase the nutritional value of foods; starches such as cornstarch, which is a wholesome thickening agent; and sugars. If taken in excess, however, any of these chemicals can be dangerous to life. A high sugar intake, for example, can add fattening calories, cause dental caries, and lead to diabetes.

2. Chemicals that are not absorbed by the body. These are complex chemicals that cannot be digested by enzymes and therefore do not enter the bloodstream, where they can damage the vital organs. Cellulose, a major component of broccoli, lettuce, and other vegetables, is typical of these chemicals, which are generally safe and act as natural laxatives.

3. Chemicals that occur naturally and are known to affect the body's operation. This category includes caffeine, which occurs naturally in coffee, tea, and chocolate and is added to cola soft drinks. It is a powerful stimulant. Quinine (Chapter 5), a soft drink flavoring, can in large doses control malaria and also bring on blindness and severe heart

damage. Scientists are unsure of the long-term effect of these chemicals on the body.

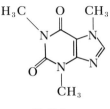

Caffeine

4. The synthetics, that is, chemicals that do not occur in nature and are absorbed into the body. These chemicals involve high risk. They include many artificial colors and some preservatives. The chemicals are transported to the liver, where they are detoxified before being excreted. However, this process does not always work satisfactorily. Examples of compounds in this category are cyclamate sodium and methyl tetra-O-methyl carminate, the yellow dye found in oleo:

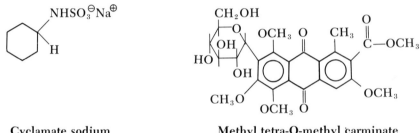

Cyclamate sodium   **Methyl tetra-O-methyl carminate**

Ideally, every additive should be tested so that it can be determined what its exact chemical makeup is, how the body metabolizes it, and whether or not long-term ingestion of it can have harmful effects. However, this is a time-consuming and expensive task. It does not follow that if a chemical harms an animal, it will harm a human being. The animal could differ from human beings in some crucial metabolic reaction, so that the animal could be more (or less) sensitive to an additive; thus a false result could be obtained. This hazard is minimized if the chemical is fed to more than one species, but still the question of similarity to human beings remains.

9-10. Find and identify in each of the compounds referred to as a food additive a functional group discussed in this chapter, such as carboxylic acid, ester, amide, acid salt, and anhydride. For example:

Carboxylic acid group

$$HO_2C—CH—CH—CO_2H$$
$$\qquad | \qquad |$$
$$\qquad OH \quad OH$$

Tartaric acid

Summary

The carboxyl group is the functional group of a carboxylic acid. The general formulas of carboxylic acids and their derivatives are carboxylic acid, RCOOH; ester, RCOOR; anhydride, RCOOCOR; acid chloride, RCOCl; and amide, $RCONH_2$.

Carboxylic acids can be named by a common name and an IUPAC name. In the common system, carbons are given Greek letter designations, the carbon bonded to the carboxyl group being the α-carbon. In the IUPAC system, carbons are indicated by numbers, carbon 1 being the carboxyl carbon.

Carboxylic acids can be prepared by oxidation of primary alcohols, carbonation of a Grignard reagent, hydrolysis of nitriles, oxidation of aromatic hydrocarbons containing an alkyl group, and the malonic ester synthesis.

Carboxylic acids are reduced to primary alcohols; react with alcohols to form esters (esterification); form acid chlorides with PCl_5, PCl_3, or $SOCl_2$; give amides when treated with ammonia; and if they have an α-hydrogen, undergo the Hell-Volhard-Zelinsky reaction.

Carboxylic acids are stronger acids than phenols and alcohols. The acid derivatives undergo a number of reactions with various nucleophilic reagents via an acyl nucleophilic substitution mechanism.

The reactions of carboxylic acid and some carboxylic acid derivatives are summarized in Table 9-6.

Table 9-6
Reactions of carboxylic acids and derivatives

Carboxylic compound	Reagent	Product
Acid°		
RCOOH	Na, NaOH, or $NaHCO_3$	$RCOO^-Na^+$
	PCl_3, PCl_5, or $SOCl_2$	RCOCl
	$R'OH/H^+$	RCOOR'
	NH_3	$RCOO^-NH_4^+$ ($\underline{\text{heat}}$, $RCONH_2$)
	$LiAlH_4$	RCH_2OH
RCH_2COOH	Br_2, P	RCH(Br)COOH
Acid derivative		
RCOCl	R'OH	RCOR'
	NH_3	$RCONH_2$
	H_2O	RCOOH
	$LiAlH_4$	RCH_2OH
RCOOR'	H_2O/H^+	RCOOH + R'OH
	R"OH	RCOOR" + R'OH
$RCONH_2$	H_2O/H^+	$RCOOH + NH_4^+$
$(RCO)_2O$	$LiAlH_4$	RCH_2OH

°Dicarboxylic acids undergo the same types of reactions; but since they have two carboxyl groups, they require twice as much reagent as carboxylic acids.

Important terms

α-Carbon	Hell-Volhard-Zelinsky reaction
Acid chloride	Inductive effect
Amide	Malonic ester synthesis
Anhydride	Monosubstituted acetic acid
Benzoic acid	Nitrile
Carboxyl group	Nucleophilic acyl substitution
Carboxylic acid	Nucleophilic alkyl substitution
Decarboxylate	Phosphate ester
Dicarboxylic acid	Reformatsky reaction
Disubstituted acetic acid	Salt of carboxylic acid
Ester	Transesterification
Esterification	Urea

Problems

9-11. Give a correct name for each of the following molecules:

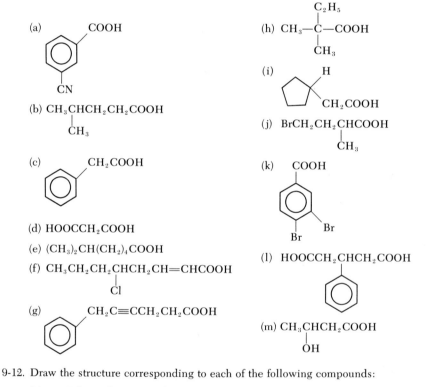

(b) $CH_3CHCH_2CH_2COOH$
        $|$
        CH_3

(d) $HOOCCH_2COOH$

(e) $(CH_3)_2CH(CH_2)_4COOH$

(f) $CH_3CH_2CH_2CHCH_2CH$=$CHCOOH$
                    $|$
                    Cl

(j) $BrCH_2CH_2CHCOOH$
                $|$
                CH_3

(l) $HOOCCH_2CHCH_2COOH$

(m) CH_3CHCH_2COOH
            $|$
            OH

9-12. Draw the structure corresponding to each of the following compounds:

(a) β,β-Dibromobutyric acid
(b) 2-Ethyl-4-heptenoic acid
(c) 3-Methoxy-4-fluorobenzoic acid
(d) Phenylacetic acid
(e) α-o-Nitrophenylcaproic acid
(f) 3,4-Dichlorohexanedioic acid

(g) 2-Dimethyl-5-bromo-7-octynoic acid
(h) β-Benzyl-α-bromovaleric acid
(i) 2,4-Hexenedioic acid
(j) m-Phenylbenzoic acid
(k) Cyclohexylacetic acid
(l) Butanedioic acid

9-13. Using equations, show how each of the following molecules can be converted to isobutyric acid:

(a) $(CH_3)_2CHCN$ (b) $(CH_3)_2CHBr$ (two ways)   (c) $(CH_3)_2CHCH_2OH$

9-14. Using equations, show the products of the reaction of benzoic acid with each of the following reagents:

(a) NaOH
(b) NH_3
(c) Product of b + heat
(d) PCl_3
(e) PCl_5
(f) $LiAlH_4$

(g) $SOCl_2$
(h) Isopropyl alcohol + H^+
(i) HNO_3, H_2SO_4
(j) SO_3, H_2SO_4

9-15. Which acid in each of the following pairs is stronger?

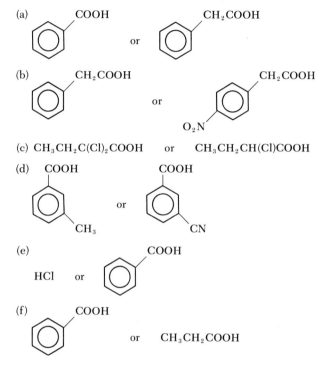

(a)   COOH   or CH_2COOH

(b) CH_2COOH or CH_2COOH (O_2N)

(c) $CH_3CH_2C(Cl)_2COOH$ or $CH_3CH_2CH(Cl)COOH$

(d)   COOH (CH_3)   or   COOH (CN)

(e)   HCl   or   COOH

(f)   COOH   or CH_3CH_2COOH

9-16. Draw the structure for each of the following acid derivatives:

(a) Sodium benzoate
(b) Isopropyl acetate
(c) Oxaloyl chloride
(d) Propionic anhydride
(e) Phenylacetamide
(f) Diethyl malonate

(g) Acetyl chloride
(h) Calcium propionate
(i) Succinic anhydride
(j) 5-Methyl-5-isopropyl barbituric acid

9-17. Give the products of each of the following reactions:

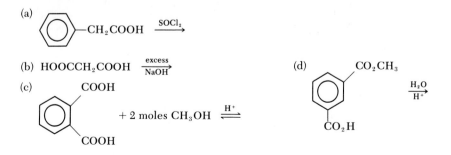

(a) $-CH_2COOH$ $\xrightarrow{SOCl_2}$

(b) $HOOCCH_2COOH$ $\xrightarrow[\text{NaOH}]{\text{excess}}$

(c)   COOH ($COOH$)   + 2 moles CH_3OH $\rightleftharpoons^{H^+}$

(d) CO_2CH_3 / CO_2H $\xrightarrow[H^+]{H_2O}$

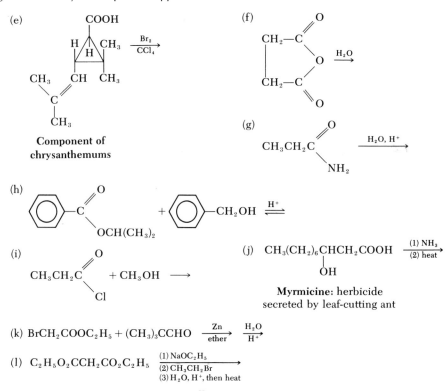

(e) Component of chrysanthemums

(j) $CH_3(CH_2)_6CHCH_2COOH$ $\xrightarrow[\text{(2) heat}]{\text{(1) NH}_3}$
    with OH on the CH

Myrmicine: herbicide
secreted by leaf-cutting ant

(k) $BrCH_2COOC_2H_5 + (CH_3)_3CCHO$ $\xrightarrow[\text{ether}]{\text{Zn}}$ $\xrightarrow[\text{H}^+]{\text{H}_2O}$

(l) $C_2H_5O_2CCH_2CO_2C_2H_5$ $\xrightarrow[\substack{\text{(2) CH}_3\text{CH}_2\text{Br} \\ \text{(3) H}_2\text{O, H}^+\text{, then heat}}]{\text{(1) NaOC}_2\text{H}_5}$

(m) $(CH_2{=}CH(CH_2)_8COO^-)_2\ Zn^{2+}$ $\xrightarrow{\text{H}^+}$

   Zinc undecylenate: fungicide for athlete's foot

9-18. Give the synthesis of each of the following molecules, using benzene, toluene, and alcohols of four carbons or less and any needed inorganic reagents:

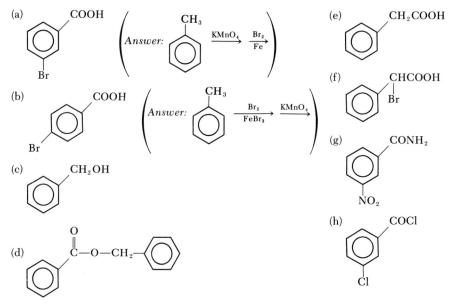

9-19. An alcohol of the formula C_7H_7ClO, when oxidized by permangate solution, produces *m*-chlorobenzoic acid. What is the structure of the alcohol?

9-20. An acid of the formula $C_7H_6O_3$ consumes 2 moles of NaOH but only 1 mole of $NaHCO_3$. What is the structure of the acid?

9-21. An amide of the formula C_3H_7NO forms propionic acid and ammonia in hydrolysis. What is the structure of the amide?

9-22. An ester undergoes hydrolysis to form 2,5-dichlorobenzoic acid and isopropyl alcohol. What is its structure?

9-23. Write the structural formulas and give correct names for the anhydride, methyl ester, amide, chloride, and sodium salt of butyric acid.

9-24. Some suntan lotions contain β-ethoxyethyl p-methoxycinnamate as the absorber of ultraviolet light. It is said to absorb the peak tanning and burning wavelength of light:

$$CH_3O-\underset{}{\bigcirc}-CH=CH\overset{\overset{\displaystyle O}{\parallel}}{C}OCH_2CH_2OC_2H_5$$

Which portion of this molecule will react with each of the following reagents? Give the structure of each product.

(a) Bromine   (b) HBr   (c) H_2O/H^+

Self-test

1. Give a correct name for each of the following compounds:

(a) $CH_3CH_2CH_2COOH$

(b) COOH

    (benzene ring with NO_2)

(c) $HOOCCH_2COOH$

(d) $CH_3CH=CHCH_2COOH$

(e) $CH_3\overset{\overset{\displaystyle O}{\parallel}}{C}OCH_2CH_3$

(f) $CH_3\overset{\overset{\displaystyle }{}}{\underset{\underset{\displaystyle O}{\parallel}}{C}}-Cl$

2. Draw the complete structure corresponding to each of the following compounds:

(a) Propionic acid
(b) α-Chlorovaleric acid
(c) Formic anhydride

(d) Butyramide
(e) 3,5-Dimethylbenzoic acid
(f) Butenedioic acid

3. Give the structure of the organic product of each of the following reactions:

(a) $CH_3CH_2MgBr + CO_2 \xrightarrow{H_2O}$

(b)

(cyclobutane) H, CH_2CH_2OH $\xrightarrow[H_2SO_4]{K_2Cr_2O_7}$

(c) (benzene ring)$-C\equiv N \xrightarrow[H_2SO_4]{H_2O}$

(d) $CH_3CH_2COOH + NaOH \longrightarrow$

(e) $CH_3\overset{\overset{\displaystyle O}{\parallel}}{C}OCH(CH_3)_2 \xrightarrow{H^+, H_2O}$

(i) $C_2H_5OOCCH_2COOC_2H_5 \xrightarrow{NaOC_2H_5} \xrightarrow{CH_3CH_2Cl}$

(f) (benzene ring)$-COOH \xrightarrow{PCl_5}$

(g) $H-\overset{\overset{\displaystyle O}{\parallel}}{C}-Cl + NH_3 \longrightarrow$

(h) $CH_3\overset{\overset{\displaystyle O}{\parallel}}{C}-O-\overset{\overset{\displaystyle O}{\parallel}}{C}CH_3 + C_6H_5CH_2OH \longrightarrow$

4. Indicate the reagent and/or condition that will cause each step to occur as written in the following transformation:

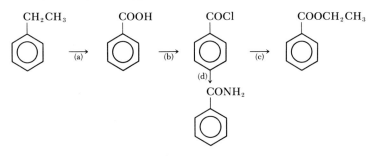

5. Give the correct answer for each of the following questions:

   (a) How many carbon atoms are there in caproic acid?
   (b) An oxygen-containing compound is soluble in $NaHCO_3$. Is the compound phenol or benzoic acid?
   (c) Which is the stronger acid: α-chloropropionic acid or β-chloropropionic acid?
   (d) What is the hybridization of the orbitals of the carbon atom in the intermediate in acyl nucleophilic substitution?
   (e) What is (are) the position(s) on the ring taken by an incoming group in electrophilic substitution in benzoic acid?
   (f) What is the name of the process by which an ester is converted to another ester by an alcohol?
   (g) What is the name given to the general class of compounds formed through the Reformatsky reaction?
   (h) What is the type of compound formed from diethyl malonate, sodium ethoxide, and a primary alkyl halide?
   (i) Barbituric acids are the class of compounds formed when diethyl malonate reacts with what compound?
   (j) Which is more reactive in nucleophilic substitution toward water: ethyl chloride or acetyl chloride?
   (k) What is the reduction product of benzoyl chloride by $LiAlH_4$?

Answers to self-test

1. (a) Butyric acid or butanoic acid
   (b) *m*-Nitrobenzoic acid or 3-nitrobenzoic acid
   (c) Malonic acid or propanedioic acid
   (d) 3-Pentenoic acid
   (e) Ethyl acetate or ethyl ethanoate
   (f) Acetyl chloride or ethanoyl chloride

2. (a) CH_3CH_2COOH

   (b) $CH_3CH_2CH_2CHCOOH$
       with Cl substituent

   (c)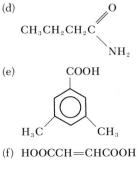

   (d) $CH_3CH_2CH_2C$ double bond O, bonded to NH_2

   (e)
       COOH ring with H_3C and CH_3

   (f) $HOOCCH{=}CHCOOH$

3. (a) $CH_3CH_2COOMgBr \longrightarrow CH_3CH_2COOH$

   (b)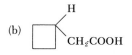
       H ... CH_2COOH

   (c)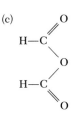
       —COOH

   (d) $CH_3CH_2COO^-Na^+$

(e) $CH_3COOH + (CH_3)_2CHOH$

(f)

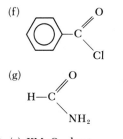

(g)

$$H-C\overset{O}{\underset{NH_2}{\diagdown}}$$

(h)

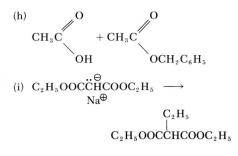

(i) $C_2H_5OOC\overset{..\ominus}{C}HCOOC_2H_5 \longrightarrow$
$\qquad\qquad\quad Na^{\oplus}$

$$\qquad\qquad\qquad\qquad\qquad\qquad \overset{C_2H_5}{\underset{|}{}}$$
$$C_2H_5OOC\overset{|}{C}HCOOC_2H_5$$

4. (a) $KMnO_4$, heat
   (b) PCl_5 or $SOCl_2$
5. (a) Six
   (b) Benzoic acid
   (c) α-Chloropropionic acid
   (d) sp^3
   (e) *Meta*
   (f) Transesterification

   (c) CH_3CH_2OH
   (d) NH_3
   (g) β-Hydroxy ester
   (h) Monosubstituted or disubstituted acetic acid
   (i) Urea
   (j) Acetyl chloride
   (k) Benzyl alcohol

chapter 10

Aldehydes and ketones

Structure and bonding

Aldehydes and ketones are often called carbonyl compounds, since they contain the carbonyl group (C=O). Their general structures are as follows:

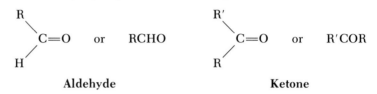

Aldehydes differ from ketones in that the carbonyl group has a hydrogen atom attached to it.

The carbonyl group determines the chemistry of aldehydes and ketones. The difference between aldehydes and ketones causes a difference in their chemical properties. Aldehydes are more easily oxidized and are more re-active in nucleophilic addition, which is the characteristic reaction of alde-hydes and ketones.

The carbonyl group, like the carbon–carbon double bond of an alkene, is composed of one σ bond and one π bond. The orbitals of the carbon atom of the carbonyl group are sp^2 hybridized. The orbitals of the carbonyl oxygen atom, on the other hand, are unhybridized: $1s^2 2s^2 2p^2 2p^1 2p^1$. The σ bond is formed by the overlap of an sp^2 orbital of carbon and a p orbital of oxygen. The π bond is formed by the side-by-side overlap of the p orbital of carbon with the other p orbital of oxygen. One of the unshared electron pairs of oxygen occupies an s orbital of oxygen, and the other unshared electron pair occupies a p orbital of oxygen. The bond formations of the carbonyl group are illustrated as follows:

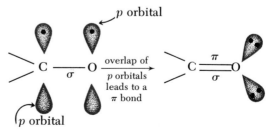

Since the orbitals of the carbonyl carbon are sp^2 hybridized, the three atoms attached to the carbonyl carbon all lie in a plane 120° apart as follows:

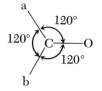

10-1. Identify the orbital makeup of all the bonds in the following molecule:

$$CH_3CH_2\overset{\overset{\displaystyle O}{\|}}{C}{-}H$$

Which atoms lie in the same plane?

Answer: There is one sp^3-sp^2, one sp^3-sp^3, five sp^3-s, one sp^2-s, and one sp^2-p bond.

The electrons in the π bond of the carbonyl group are not equally shared. In fact, they are pulled more toward the more electronegative oxygen atom; as a result the bond is somewhat polarized, with the oxygen atom being slightly negative (δ^-) and the carbon atom being slightly positive (δ^+):

$$\underset{}{\overset{\delta^+ \quad \delta^-}{C{=}O}}$$

Nomenclature of aldehydes and ketones
Aldehydes

Aldehydes can be named by the common and the IUPAC systems. In the common system, the *-ic acid* ending of the corresponding carboxylic acid is replaced by the suffix *-aldehyde*. For example:

Acid	*Common name*	*Aldehyde*	*Common name*
HCOOH	Form*ic acid*	HCHO	Form*aldehyde*
CH_3COOH	Acet*ic acid*	CH_3CHO	Acet*aldehyde*
CH_3CH_2COOH	Propion*ic acid*	CH_3CH_2CHO	Propion*aldehyde*
$CH_3(CH_2)_2COOH$	Butyr*ic acid*	$CH_3(CH_2)_2CHO$	Butyr*aldehyde*
$CH_3(CH_2)_3COOH$	Valer*ic acid*	$CH_3(CH_2)_3CHO$	Valer*aldehyde*
$(CH_3)_2CHCOOH$	Isobutyr*ic acid*	$(CH_3)_2CHCHO$	Isobutyr*aldehyde*
C_6H_5COOH	Benzo*ic acid*	C_6H_5CHO	Benz*aldehyde*

Benzaldehyde and benzaldehyde with substituted groups are aromatic aldehydes. Following are examples of aromatic aldehydes:

m-**Nitrobenzaldehyde**

o-**Hydroxybenzaldehyde**
(salicyaldehyde)

Phenylacetaldehyde

Occasionally, Greek letters are used to indicate the positions of substituted groups. The α-carbon is the carbon atom bonded to the carbonyl carbon. Following are some examples:

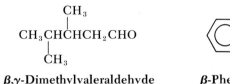   

β,γ-Dimethylvaleraldehyde   **β-Phenylpropionaldehyde**

In the IUPAC system, the longest chain carrying the —CHO group is the parent compound. The -e ending of the corresponding alkane is replaced by the suffix -al. The positions of substituents are indicated by numbers, the carbonyl carbon always being carbon number 1. Following are some examples:

Aldehyde	IUPAC name
HCHO	Methanal
CH_3CHO	Ethanal
CH_3CH_2CHO	Propanal
$CH_3(CH_2)_2CHO$	Butanal
$CH_3(CH_2)_3CHO$	Pentanal

$$CH_3CHCH_2CH_2CHCHO$$

with Cl and CH_3 substituents

2-Methyl-5-chlorohexanal

$$CH_3CH{=}CHCHCHO$$
with cyclopropyl—H

2-Cyclopropyl-3-pentenal

$$C_6H_5{-}C{\equiv}C{-}CHO$$

3-Phenyl-2-propynal

benzene ring with CHO, Cl, Cl substituents

3,4-Dichlorobenzaldehyde

Ketones

Ketones are also named by the IUPAC and the common systems. In the common system, the names of the two groups attached to the carbonyl carbon are followed by the word *ketone*. The simplest ketone, containing three carbons, is most often called acetone. If one of the groups is aromatic, the suffix -*phenone* is used. Following are some examples:

Ketone	Common name
CH_3COCH_3	Acetone [(di)methyl ketone*]
$CH_3COCH_2CH_3$	Methyl ethyl ketone
$CH_3CH_2COCH(CH_3)_2$	Ethyl isopropyl ketone
$C_6H_5COCH_3$	Acetophenone (phenyl methyl ketone)
$C_6H_5CH_2COCH_3$	Benzyl methyl ketone
$C_6H_5COC_6H_5$	Benzophenone [(di)phenyl ketone]

*It is not necessary to use the prefix di- if the two groups are identical.

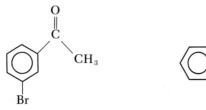

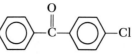

m-**Bromoacetophenone**   *p*-**Chlorobenzophenone**

In the IUPAC system, the longest chain carrying the C=O group is the parent compound. The *-e* ending of the corresponding alkane is replaced by the suffix *-one*. The carbonyl carbon is given the lowest number, and the positions of substituents are indicated by numbers. Following are some examples:

Ketone	*IUPAC name*
CH_3COCH_3	Propanone°
$CH_3COCH_2CH_3$	Butanone°
$CH_3CH_2COCH_2CH_3$	3-Pentanone

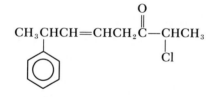

$$CH_3CH_2\overset{O}{\overset{\|}{C}}CH_2CH_2CH_3$$
$$\underset{CH_3}{|}$$

2-Methyl-3-hexanone

$$CH_3CH_2C\equiv C\overset{O}{\overset{\|}{C}}CH_3$$

3-Hexyn-2-one

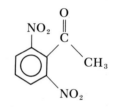

2,6-Dinitroacetophenone

$$CH_3CHCH=CHCH_2\overset{O}{\overset{\|}{C}}-CHCH_3$$

2-Chloro-7-phenyl-5-octen-3-one

10-2. Draw the structure and give another name for each of the following compounds.

(a) 3-Pentanone   *Answer:* $CH_3CH_2\underset{\underset{O}{\|}}{C}CH_2CH_3$, ethyl ketone

(b) α,α-Dimethylpropionaldehyde   *Answer:* $CH_3C(CH_3)_2CHO$, 2,2-dimethylpropanal

(c) Methyl benzyl ketone
(d) 3-Nitrobenzaldehyde
(e) *sec*-Butyl *tert*-butyl ketone
(f) 2,3-Dichloropentanal

°The number 2 to locate the position of the carbonyl carbon is not necessary; if the carbonyl carbon were the terminal carbon, the compound would be an aldehyde.

Occurrence and uses of carbonyl compounds

Aldehydes and ketones are widely distributed in nature and are used in a number of ways.

Benzaldehyde has a pleasant odor and is the chief constituent of oil of bitter almond. Vanillin is the essence of vanilla flavor, which occurs naturally in the vanilla bean; it is easily synthesized and is used in artificial vanilla extracts. Cinnamaldehyde is the chief constituent of the oil of cinnamon. Citral, the oil of citrus fruits, is found in the rinds of lemons, limes, and oranges. The structures of these compounds are as follows:

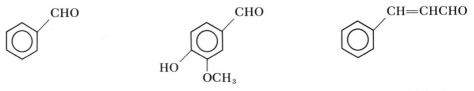

Benzaldehyde   Vanillin   Cinnamaldehyde

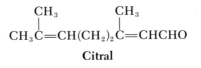

Citral

The sex attractants of several insects are aldehydes, such as undecanal, found in the wax moth, and the *cis* and *trans* isomers found in the boll weevil:

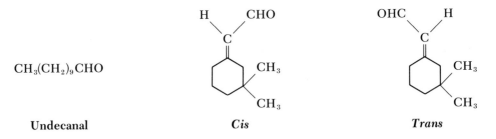

$CH_3(CH_2)_9CHO$

Undecanal   *Cis*   *Trans*

A solution of formaldehyde in water has potent antibacterial properties. It is also used as a preservative of anatomical specimens, since it has a hardening effect on the tissue protein.

Chloral hydrate, the stable hydrate of trichloroacetaldehyde, is used in medicine as a hypnotic and narcotic. It is a quick-acting soporific commonly known as knock-out drops. A proper dose of the drug causes sedation in about 10 minutes and sleep usually within an hour.

$$Cl_3CCHO \; + \; H_2O \; \longrightarrow \; Cl_3CCH(OH)_2$$

Trichloro-   Chloral hydrate
acetaldehyde

Methyl ethyl ketone is an excellent solvent for fingernail polish and thus is used as a polish remover.

Diacetyl is found in butter in minute quantities and gives butter its characteristic odor:

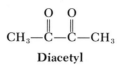

Diacetyl

Ninhydrin is used in the identification of amino acids and proteins. A color ranging from clear deep blue to a violet-pink indicates the presence of at least one free amino group and one free carboxyl group.

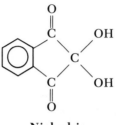

Ninhydrin

Menthone has a strong peppermint odor and is used as a flavoring. Camphor is used as a medicinal, incense, and plasticizer (that is, softener) and in embalming fluid.

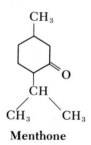

Menthone

Camphor

The male and female sex hormones are ketones. Testosterone, the male sex hormone, causes development of secondary male characteristics such as facial hair and deep voice. Progesterone, the female sex hormone, is important in regulating the course of pregnancy and is known to inhibit the release of ova from the ovaries (Chapter 2). The structures of these compounds are shown in Chapter 6.

The basic components of MACE are chloroacetophenone (tear gas) (0.9%), kerosene (4%), and some Freon propellants. It is commercially sold in pocket- and standard-size spray cans. Generally, a compound with a halogen atom next to a carbonyl group has a tendency to hydrolyze and thus can be used as a lacrimator. The moisture present in the eyeball or in the lungs is capable of hydrolyzing a lacrimator. The hydrolysis of chloroacetophenone liberates concentrated hydrochloric acid, which causes the glands of the eye to secrete tears to wash away the irritating acid. In some instances, law enforcement groups have used chloroacetophenone as a tear gas.

$$\underset{\text{Chloroacetophenone}}{\langle\!\!\bigcirc\!\!\rangle\!\!-\!\!\overset{\overset{\text{O}}{\|}}{\text{C}}\!\!-\!\!\text{CH}_2\text{Cl}}$$

Chloroacetophenone

Ketones: sources of perfumes

The male musk deer, a small animal about 20 inches high that inhabits China, Siberia, Tibet, and India, produces musk from a small gland located near the stomach. The function of the gland secretion is to attract the female deer. About 2 ounces (57 g) of musk is obtained from a single male musk deer. Depending on the quality, prices of musk, a substance highly prized as a perfume, range from $100 to $175 per ounce. The active ingredient of musk is muscone, a 15-member cyclic saturated ketone. Musk has the ability to enhance, enrich, and blend the odors of other perfume components, but it has been largely supplanted by synthetic musks like 4-*tert*-butyl-2,6-dimethyl-3,5-dinitroacetophenone (musk ketone).

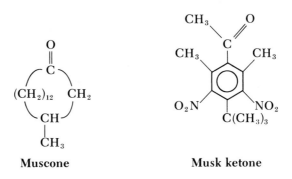

Muscone   Musk ketone

Civetone is the ingredient in the gland secretion of the African civet cat, the equivalent of the American skunk. The pure molecule is extremely obnoxious and nauseating in odor; however, civetone is amazingly useful when present in minute amounts in perfumes, since it gives body and lasting quality to the perfume. Muscone and civetone are two of the largest naturally occurring cyclic compounds known.

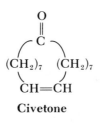

Civetone

β-Ionone is the fragrant substance responsible for the odor of violets; it is used in artificial violet perfumes. α-Ionone, on the other hand, has the aroma of cedar in high concentrations and that of violets in low concentrations.

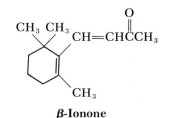

β-Ionone

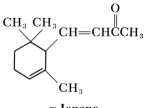

α-Ionone

Jasmone is a ketone that gives jasmine its characteristic odor.

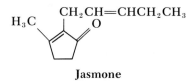

Jasmone

Preparation of aldehydes and ketones

Aldehydes

Aldehydes can be prepared by oxidation of primary alcohols, reduction of acid chlorides, or hydrolysis of geminal dihalides.

The selective oxidation of a primary alcohol to an aldehyde is usually carried out under anhydrous conditions with chromium trioxide (CrO_3) in dry pyridine (C_5H_5N):

$$RCH_2OH \xrightarrow[\text{pyridine}]{CrO_3} RCHO$$

For example:

$$CH_3CH_2CH_2OH \xrightarrow[\text{pyridine}]{CrO_3} CH_3CH_2CHO$$

n-Propyl alcohol   Propionaldehyde

Other oxidizing agents, such as $K_2Cr_2O_7$ and $KMnO_4$, are too strong and oxidize the aldehyde to the corresponding acid. For example:

$$CH_3CH_2CH_2OH \xrightarrow{KMnO_4} CH_3CH_2CHO \longrightarrow CH_3CH_2COOH$$

n-Propyl alcohol   Propionaldehyde   Propionic acid
                              (some)   (major product)

Both aliphatic and aromatic aldehydes can be prepared from the corresponding acid via reduction of the acid chloride. The reducing agent is generally lithium aluminum tri-*tert*-butoxide [$LiAlH(O\text{-}tert\text{-}C_4H_9)_3$]:

$$RCOCl \xrightarrow{LiAlH(O\text{-}tert\text{-}C_4H_9)_3} RCHO$$

Following is an example of the preparation of an aldehyde from an acid:

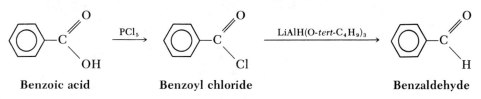

Benzoic acid   Benzoyl chloride   Benzaldehyde

Aromatic aldehydes can be prepared by the hydrolysis of geminal dihalides. Geminal substituents are two identical substituents located on the same carbon atom:

Geminal dihalides are prepared by the free radical halogenation of alkylated benzenes such as toluene:

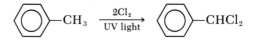

The hydrolysis of geminal dihalides in basic solution leading to aldehydes occurs as follows:

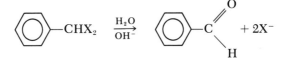

For example:

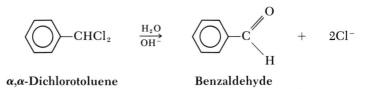

    α,α-Dichlorotoluene   **Benzaldehyde**

Ketones

Two of the most common methods of preparing ketones are the oxidation of secondary alcohols and Friedel-Crafts acylation.

Secondary alcohols can be oxidized by $KMnO_4$ or $K_2Cr_2O_7$ to ketones and, unlike aldehydes, are not easily oxidized further:

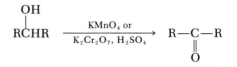

For example:

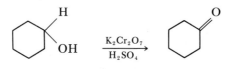

    Cyclohexanol   **Cyclohexanone**

The Friedel-Crafts acylation is a method of introducing an acyl group
$\left(\begin{smallmatrix} O \\ \| \\ R-C- \end{smallmatrix} \right)$ into an aromatic ring. The product formed is a ketone. The mechanism is that of electrophilic aromatic substitution. The reaction is catalyzed by a Lewis acid such as $AlCl_3$ and involves the use of an acid chloride:

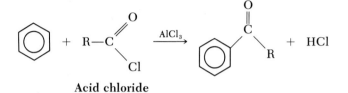

Acid chloride

For example:

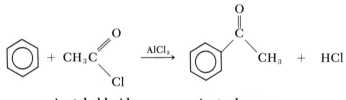

Benzene   Acetyl chloride   Acetophenone

This reaction is very useful in synthetic organic chemistry. The acyl group can be converted into an alkyl group by amalgamated zinc and concentrated HCl. This reaction is known as the *Clemmensen reduction*. Following is an example:

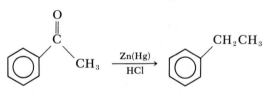

Acetophenone   Ethylbenzene

A straight-chain alkyl group longer than an ethyl group cannot be substituted onto an aromatic ring by the Friedel-Crafts alkylation via an alkyl halide (Chapter 4) because of rearrangement of the initially formed carbonium ion. For example:

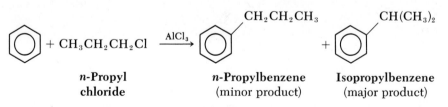

| *n*-Propyl chloride | *n*-Propylbenzene (minor product) | Isopropylbenzene (major product) |

A group longer than an ethyl group can be introduced onto a benzene ring by Friedel-Crafts acylation via an acid chloride followed by the Clemmensen reduction. For example:

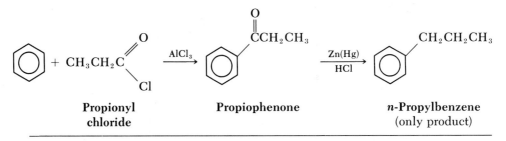

| Propionyl chloride | Propiophenone | *n*-Propylbenzene (only product) |

10-3. Indicate how each of the following syntheses of aldehydes and ketones is carried out:

(a) Hexanal via oxidation of 1-hexanol   *Answer:* $CH_3(CH_2)_4CH_2OH$

$$\xrightarrow[\text{pyridine}]{CrO_3} CH_3(CH_2)_4CHO$$

(b) Menthone via oxidation of corresponding secondary alcohol, menthol
(c) Ethyl phenyl ketone via Friedel-Crafts acylation
(d) *p*-Nitrobenzaldehyde via geminal dichloride
(e) Butyraldehyde via reduction of corresponding acid chloride

Reactions of aldehydes and ketones

The carbon–carbon double bond undergoes electrophilic addition. On the other hand, the characteristic reaction of the carbon–oxygen double bond is *nucleophilic addition*. The electrons in the carbonyl group are not equally shared: the carbonyl carbon is electron deficient (δ^+), and the carbonyl oxygen is electron rich (δ^-) (p. 269). The carbonyl carbon is therefore susceptible to attack by nucleophilic (electron-rich) reagents.

The addition of a nucleophile to a carbonyl group can occur under either acidic or basic conditions. In a basic medium, a nucleophile attacks the electron-deficient carbon atom of the carbonyl group:

$$\overset{\diagdown}{\underset{\diagup}{C}}=\ddot{O}: +\,^{\ominus}Nu \longrightarrow -\overset{\overset{\textstyle Nu}{|}}{\underset{|}{C}}-\ddot{O}:^{\ominus}$$

Such an intermediate is, in general, unstable, and the solvent added during work-up donates a proton to form a neutral product:

$$-\overset{\overset{\textstyle Nu}{|}}{\underset{|}{C}}-\ddot{O}:^{\ominus}+\,H-Nu \longrightarrow -\overset{\overset{\textstyle Nu}{|}}{\underset{|}{C}}-\ddot{O}H +\,^{\ominus}Nu$$

In an acid solution, the carbonyl oxygen undergoes protonation:

$$\overset{\diagdown}{\underset{\diagup}{C}}=\ddot{O}: +\,H^+ \longrightarrow \overset{\diagdown}{\underset{\diagup}{C}}=\overset{\oplus}{\ddot{O}}-H$$

This is followed by attack of a nucleophile to form a neutral product:

$$\overset{\diagdown}{\underset{\diagup}{C}}=\overset{\oplus}{\ddot{O}}-H +\,^{\ominus}Nu \longrightarrow -\overset{|}{\underset{\underset{\textstyle Nu}{|}}{C}}-\ddot{O}-H$$

Examples of nucleophilic reagents that add to the carbonyl group are $RMgX$, HCN, H_2O, and NH_3. The types of nucleophilic addition reactions that the carbonyl group undergoes are discussed as follows.

Addition of Grignard reagents

The addition of Grignard reagents with carbonyl compounds as a method of obtaining various classes of alcohols is discussed in Chapter 6. The class of alcohol obtained depends on the type of carbonyl compound used. Table 10-1 summarizes the reactions between Grignard reagents ($RMgX$) and various carbonyl compounds. Following are examples of the versatility of the reaction of Grignard reagents with carbonyl compounds:

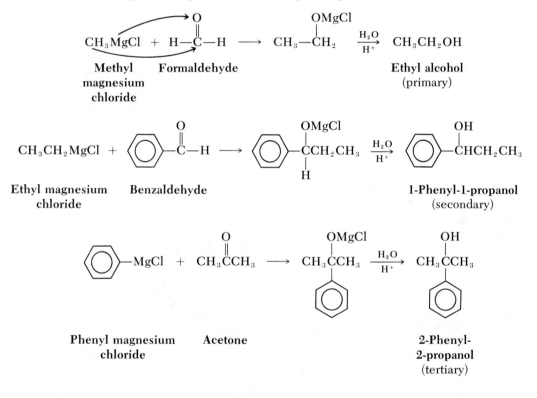

Table 10-1
Alcohol products formed by reaction between Grignard reagents (RMgX)
and carbonyl compounds

		Alcohol
Carbonyl compound	*Structure*	*Class*
HCHO	RCH_2OH	Primary
R'CHO	$RCH(OH)R'$	Secondary
R'COR'	$(R')_2RCOH$	Tertiary

10-4. Give the structure of the carbonyl compound(s) and Grignard reagent(s) that produce each of the following alcohols:

(a)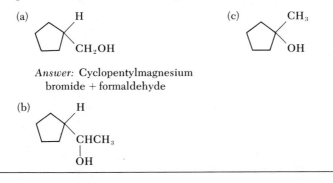

Answer: Cyclopentylmagnesium
bromide + formaldehyde

(b)

(c)

Addition of cyanide

When a carbonyl compound is treated with an aqueous solution of sodium cyanide in the presence of mineral acid (H^+), a type of compound called *cyanohydrin* is formed:

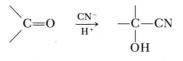

Cyanohydrin

The mechanism most likely involves two steps: protonation of oxygen, followed by nucleophilic attack by cyanide on carbon:

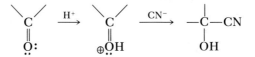

Following are some examples:

$$CH_3CHO \xrightarrow[H^+]{CN^-} CH_3-\overset{\overset{\displaystyle H}{|}}{\underset{\underset{\displaystyle OH}{|}}{C}}-CN$$

Acetaldehyde   **Acetaldehyde
cyanohydrin**

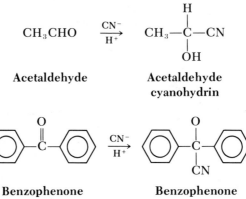

Benzophenone   **Benzophenone
cyanohydrin**

The real utility of this reaction lies in the fact that the cyano group (CN) can be converted into the carboxyl group (COOH) by hydrolysis. Thus the product of hydrolysis of a cyanohydrin is an α-hydroxy carboxylic acid. Following are examples of the formation of carboxylic acids by hydrolysis of the corresponding cyanohydrins:

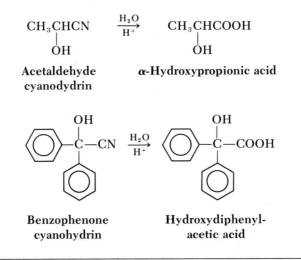

$$\underset{\substack{\text{Acetaldehyde}\\\text{cyanodydrin}}}{\text{CH}_3\text{CHCN}} \xrightarrow[\text{H}^+]{\text{H}_2\text{O}} \underset{\alpha\text{-Hydroxypropionic acid}}{\text{CH}_3\text{CHCOOH}}$$

Benzophenone cyanohydrin

Hydroxydiphenyl-acetic acid

10-5. Write the equations for the formation of benzaldehyde cyanohydrin and for the hydrolysis of the cyanohydrin.

Answer: Final product is α-hydroxyphenylacetic acid.

Addition of alcohols

Alcohols in the amount of 1 mole react with aldehydes or ketones in the presence of anhydrous acids to form a class of compounds called *hemiacetals* (from aldehydes) or *hemiketals* (from ketones); alcohols in the amount of 2 moles react to form *acetals* (from aldehydes) or *ketals* (from ketones):

$$\underset{}{\text{R}-\overset{\text{O}}{\overset{\|}{\text{C}}}-\text{H}} \underset{}{\overset{\text{R}'\text{OH, dry HCl}}{\rightleftharpoons}} \underset{\text{Hemiacetal}}{\text{R}-\overset{\text{OH}}{\underset{\text{H}}{\overset{|}{\underset{|}{\text{C}}}}}-\text{OR}'} \overset{\text{R}''\text{OH}}{\rightleftharpoons} \underset{\text{Acetal}}{\text{R}-\overset{\text{OR}''}{\underset{\text{H}}{\overset{|}{\underset{|}{\text{C}}}}}-\text{OR}'} + \text{HOH}$$

Hemiacetals are usually too unstable to isolate. Acetal and ketal formations are reversible reactions. The forward reaction is catalyzed by dry hydrogen chloride, and the reverse reaction is catalyzed by hydrochloric acid. Following are some examples:

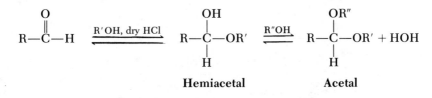

$$\underset{\text{Propionaldehyde}}{\text{CH}_3\text{CH}_2\text{C}\overset{\text{O}}{\diagup}_{\text{H}}} + 2\text{CH}_3\text{OH} \overset{\text{dry HCl}}{\rightleftharpoons} \underset{\substack{\text{Methyl acetal of}\\\text{propionaldehyde}}}{\text{CH}_3\text{CH}_2\overset{\text{OCH}_3}{\underset{\text{H}}{\overset{|}{\underset{|}{\text{C}}}}\text{OCH}_3}} + \text{H}_2\text{O}$$

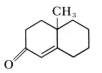

$$\underset{\text{2-Butanone}}{CH_3CH_2\overset{\overset{\displaystyle O}{\|}}{C}CH_3} + 2CH_3OH \;\underset{}{\overset{\text{dry HCl}}{\rightleftharpoons}}\; \underset{\substack{\text{Methyl ketal of}\\\text{2-butanone}}}{CH_3CH_2\overset{\overset{\displaystyle OCH_3}{|}}{\underset{\underset{\displaystyle OCH_3}{|}}{C}}CH_3} + H_2O$$

Acetal and ketal groups are useful as blocking or protecting groups in many synthetic reactions. For example, a chemist wishes to reduce the double bond in the following molecule without reducing the carbonyl group:

This can be accomplished as follows: a cyclic ketal is formed by ethylene glycol, hydrogenation reduces the carbon–carbon double bond, and then hydrolysis (the reverse of ketal formation) removes the ketal group:

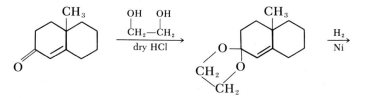

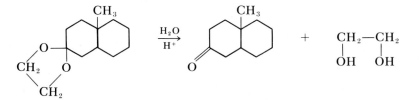

If the carbonyl group were not protected, the final product would be the saturated alcohol:

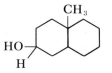

The chemistry of hemiacetals and acetals is discussed in the study of carbohydrates in Chapter 13.

Addition of compounds derived from ammonia

Compounds derived from ammonia can be added to carbonyl compounds. The initial addition product (adduct) is unstable and eliminates a molecule of water to give a product containing a carbon–nitrogen double bond. The reaction between a carbonyl compound and the simplest ammonia derivative, hydroxylamine (NH_2OH), is as follows:

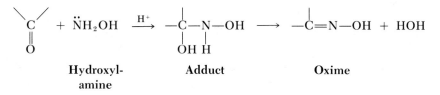

Hydroxyl-amine   Adduct   Oxime

For example:

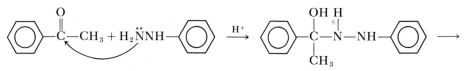

Benzaldehyde   Benzaldehyde oxime

Some other derivatives of ammonia and the products of reactions with carbonyl compounds are listed in Table 10-2. Following are some examples of these reactions:

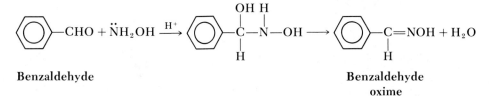

Acetophenone   Phenylhydrazine

Acetophenone phenylhydrazone

Table 10-2
Products of reactions between carbonyl groups and ammonia derivatives

Ammonia derivative		Product with C=O	
Structure	Name	Structure	Class of compound
$H_2\ddot{N}NH_2$	Hydrazine	$C{=}NNH_2$	Hydrazone
$H_2\ddot{N}NHC_6H_5$	Phenylhydrazine	$C{=}NNHC_6H_5$	Phenylhydrazone
$H_2\ddot{N}NHCNH_2$ with O below	Semicarbazide	$C{=}NNHCNH_2$ with O below	Semicarbazone

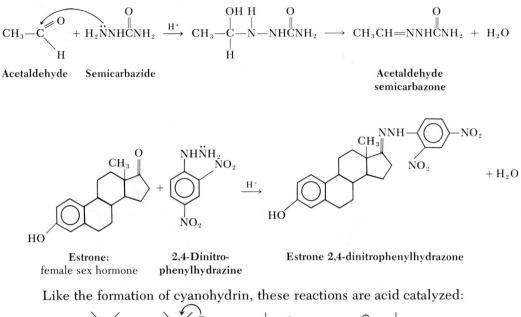

Acetaldehyde   Semicarbazide

Acetaldehyde
semicarbazone

Estrone: 2,4-Dinitro-   Estrone 2,4-dinitrophenylhydrazone
female sex hormone   phenylhydrazine

Like the formation of cyanohydrin, these reactions are acid catalyzed:

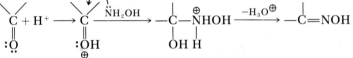

The 2,4-dinitrophenylhydrazones are the most colorful derivatives of aldehydes and ketones. Their colors are shades of red, yellow, and orange.

Hydrazones, phenylhydrazones, and semicarbazones are used for identifying aldehydes and ketones. There is a characteristic melting point for the derivative of each particular aldehyde and ketone. An unknown compound is converted into a derivative by a known chemical reaction. The derivative can be identified by the characteristic melting point, and thus the unknown compound can be determined. Table 10-3 gives the melting points of some useful derivatives of aldehydes and ketones.

10-6. Give the structure of each of the following compounds:

(a) Oxime of 2-pentanone   *Answer:* $CH_3CH_2CH_2\overset{\overset{\displaystyle CH_3}{|}}{C}=NOH$

(b) 2,4-Dinitrophenylhydrazone of cyclohexanone

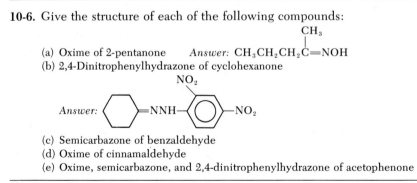

(c) Semicarbazone of benzaldehyde
(d) Oxime of cinnamaldehyde
(e) Oxime, semicarbazone, and 2,4-dinitrophenylhydrazone of acetophenone

---

Addition of carbanions (aldol condensation)

The α-hydrogens of aldehydes and ketones are fairly acidic. The α-hydrogens of a carbonyl compound are the hydrogens on a carbon atom bonded to the carbonyl carbon:

Table 10-3
Melting points of derivatives of some carbonyl compounds

| Name | Carbonyl compound | | Melting point (°C) of derivative | | |
	Melting point (°C)	Boiling point (°C)	Oxime	Semicarbazone	2,4-Dinitrophenyl-hydrazone
Ketones					
Acetone		56	59	190	126
3-Pentanone		102	69	139	156
2-Pentanone		102	58	112	144
Cyclopentanone		131	56	203	146
Cyclohexanone		156	90	167	162
2-Octanone		173		123	58
Acetophenone		203	59	199	240
Dibenzyl ketone	34	330	125	146	100
Benzophenone	48		141	167	239
p-Bromoacetophenone	51		128	208	235
Aldehydes					
Propionaldehyde		49	40	154	156
Butyraldehyde		75		106	123
2-Methylbutanal		93		103	120
Hexanal		131	51	106	104
Benzaldehyde		179	35	222	237
o-Methylbenzaldehyde		200	49	212	195
p-Methylbenzaldehyde		204	79	215	234
Cinnamaldehyde		252	138	215	255
Vanillin	81		117	230	271

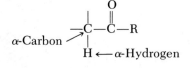

When a carbonyl compound containing an α-hydrogen is treated with dilute sodium hydroxide, a proton is abstracted by the base, and a species with the α-carbon bearing a negative charge is formed. Such a species is called a *carbanion* (also formed in the malonic ester synthesis of carboxylic acids):

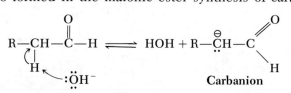

The carbanion formed in such a reaction can condense with another carbonyl compound via base-catalyzed nucleophilic addition to form either a β-hydroxyaldehyde or a β-hydroxyketone:

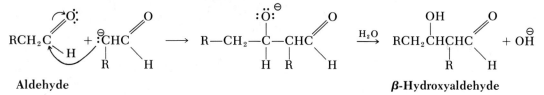

Aldehyde   **β-Hydroxyaldehyde**

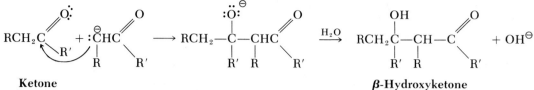

Ketone   **β-Hydroxyketone**

This reaction is referred to as the *aldol condensation. Aldols* are a class of compounds containing a carbonyl group and a β-hydroxy substituent. Following are some examples of the aldol condensation. The shaded grouping represents that portion of the molecule that is the attacking carbanion:

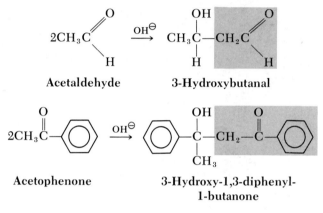

Acetaldehyde   3-Hydroxybutanal

Acetophenone   3-Hydroxy-1,3-diphenyl-
1-butanone

10-7. What carbanion is formed in each of the preceding reactions?

    Answer: For 3-hydroxybutanal, the carbanion is $^{\ominus}$:CH_2CHO.

The dehydration of a β-hydroxyaldehyde or β-hydroxyketone leads to the formation of an α,β-unsaturated aldehyde or ketone. Following are some examples:

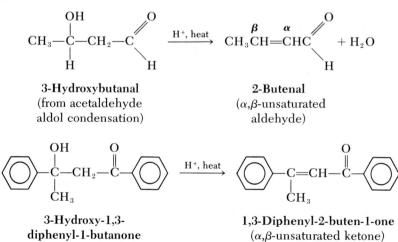

3-Hydroxybutanal
(from acetaldehyde
aldol condensation)

                      2-Butenal
                     (α,β-unsaturated
                      aldehyde)

**3-Hydroxy-1,3-
diphenyl-1-butanone**
(from acetophenone
aldol condensation)

               1,3-Diphenyl-2-buten-1-one
               (α,β-unsaturated ketone)

Oxidation

Aldehydes can be distinguished from ketones by oxidation with *Tollens'* *reagent*, a solution of silver nitrate in ammonium hydroxide [actually $Ag(NH_3)_2OH$]. Tollens' reagent is capable of oxidizing an aldehyde but not a ketone to the corresponding acid salt. The silver ion is also reduced to free silver. If the test is done in a clean test tube, the deposited silver resembles a mirror; thus the test is sometimes called the silver mirror test. The reaction occurs as follows:

RCHO	+	$Ag(NH_3)_2OH$	$\longrightarrow$	$RCOO^-$	+	$Ag\downarrow$
Aldehyde		**Tollens' reagent**		**Acid salt**		**Silver mirror**

For example:

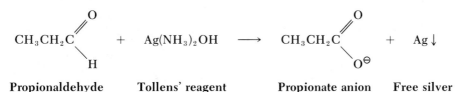

Propionaldehyde	Tollens' reagent	Propionate anion	Free silver

The acid can be regenerated from its salt by treatment with mineral acid:

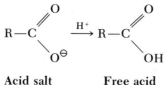

Acid salt   Free acid

For example:

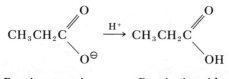

Propionate anion   Propionic acid

Aliphatic aldehydes but not aromatic aldehydes or ketones react with *Fehling's solution*, an alkaline solution of cupric ion complexed with sodium potassium tartrate:

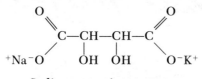

Sodium potassium tartrate

The aliphatic aldehyde undergoes oxidation to an acid salt, and the complexed, deep blue cupric ion is reduced to red cuprous oxide:

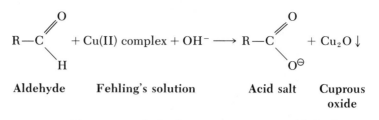

Aldehyde   Fehling's solution   Acid salt   Cuprous
                                                   oxide

Reducing sugars (that is, carbohydrates containing aldehyde groups) are oxidized by Fehling's solution. In fact, this reaction is sometimes used as a quantitative test for sugar in the urine.

In addition to sugars many other reducing substances are found in urine, but they occur seldom and their concentration remains fairly constant. Glucose, a reducing sugar, is the primary sugar of pathological significance found in urine. The healthy, fasting individual has no glucose in the urine; its presence in urine is an indication of abnormal glucose metabolism such as that in diabetes. Its detection is based on the fact that glucose (an aldehyde) reduces Fehling's solution to red cuprous oxide.

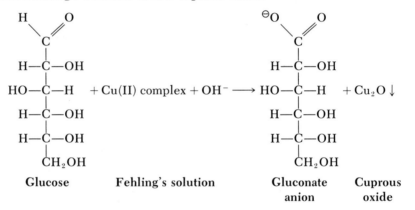

Glucose   Fehling's solution   Gluconate   Cuprous
                                             anion   oxide

Clinically this test is done by adding to five drops of urine in ten drops of water a tablet containing anhydrous cupric sulfate, sodium hydroxide, citric acid, and sodium bicarbonate (the reagents of Fehling's solution). After standing undisturbed for fifteen seconds, the mixture is shaken and then observed for color. A manufacturer's chart similar to that shown below is used to interpret the result.

Appearance	*Approximate glucose concentration* (gm/100 ml)	
Blue to green, no precipitate	0 to 0.1	Normal
Green with yellow precipitate	0.3	
Olive green	1.0	Abnormal
Brownish orange	1.5	
Brick red	2.0	

Even though Tollens' reagent and Fehling's solution are oxidizing agents, they are mild oxidizing agents. These mild oxidizing agents are useful in reactions in which oxidation of the carbon- carbon double bond is not desired, since both are capable of oxidizing an aldehyde group but not a carbon–carbon double bond. For example:

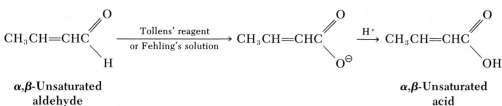

α,β-Unsaturated   **α,β-Unsaturated**
aldehyde   **acid**

Ketones are more stable toward the common oxidizing agents ($KMnO_4$, $K_2Cr_2O_7$, $Ag(NH_3)_2OH$, and Fehling's solution) than are aldehydes. Methyl ketones are oxidized by a solution of iodine in sodium hydroxide. Indication of a reaction is the precipitation of yellow iodoform (CHI_3). The test is sometimes called the *iodoform test*:

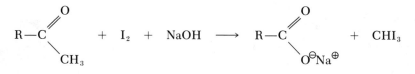

Methyl ketone   **Acid salt**   **Iodoform**

For example:

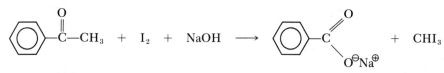

Acetophenone   **Sodium benzoate**   **Iodoform**

10-8. Can you use the iodoform test to distinguish between $CH_3CH_2CCH_2CH_3$ and $CH_3CH_2CH_2CCH_3$? Why or why not?

Answer: Yes. The methyl ketone will react.

10-9. There is evidence that secondary alcohols of the type $\underset{\underset{OH}{|}}{R}CHCH_3$ give a positive

iodoform test but that secondary alcohols of the type $\underset{\underset{OH}{|}}{R}CH_2CHCH_2R$ do not!

Explain.

Answer: Consider the oxidation product of each secondary alcohol.

Reduction

Carbonyl compounds are reduced to alcohols by hydrogen with a nickel catalyst, lithium aluminum hydride ($LiAlH_4$), or sodium borohydride ($NaBH_4$). Aldehydes are reduced to primary alcohols, and ketones are reduced to secondary alcohols:

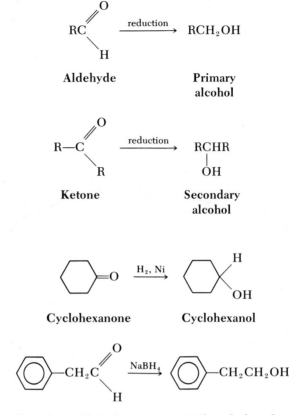

LiAlH$_4$ and NaBH$_4$ will not reduce a carbon–carbon double bond. The NaBH$_4$ reduction of an unsaturated carbonyl compound gives an unsaturated alcohol. For example:

$$CH_3CH{=}CHCH_2\overset{\overset{\displaystyle O}{\|}}{C}H \xrightarrow{\ NaBH_4\ } CH_3CH{=}CHCH_2CH_2OH$$

3-Pentenal   **3-Penten-1-ol**
(unsaturated aldehyde)   (unsaturated alcohol)

Electrophilic aromatic substitution

The carbonyl group is a *meta* director. Hence electrophilic substitution in aromatic aldehydes and ketones gives a product with the substituent in the *meta* position. For example:

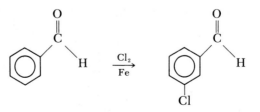

Benzaldehyde   *m*-Chlorobenzaldehyde

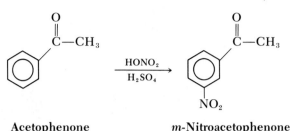

Acetophenone   ***m*-Nitroacetophenone**

These substitutions occur more slowly than the same substitution reactions in benzene, since the carbonyl group is a deactivating group.

Synthetic usefulness of aldol condensation

The aldol condensation of aldehydes or ketones with α-hydrogens gives rise to β-hydroxyaldehydes or β-hydroxyketones, and the dehydration of the β-hydroxyaldehydes or β-hydroxyketones forms α,β-unsaturated aldehydes or ketones. Tollens' reagent or Fehling's solution is capable of oxidizing an aldehyde group in the presence of a carbon–carbon double bond without affecting the carbon–carbon double bond. The reaction of a carbonyl group with either hydrogen and nickel catalyst or the hydride $LiAlH_4$ or $NaBH_4$ causes the carbonyl group to be reduced to the hydroxyl (alcohol) group, and $LiAlH_4$ and $NaBH_4$ do not reduce a carbon–carbon double bond. These reactions allow organic chemists to synthesize a variety of aldehydes, ketones, and alcohols. Following is an example:

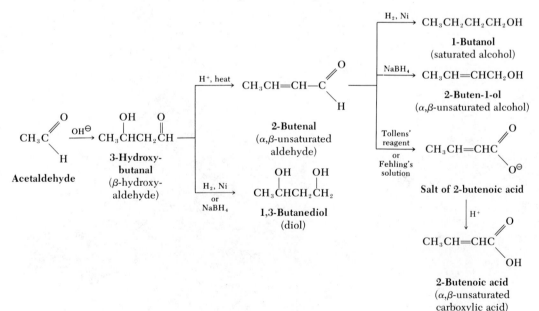

In this scheme, a number of varied compounds can be synthesized from the product obtained from the aldol condensation of acetaldehyde. Acetaldehyde is treated with a dilute solution of sodium hydroxide. The aldol condensation product is 3-hydroxybutanal. The reduction of 3-hydroxybutanal with either hydrogen and nickel or the hydride $NaBH_4$ leads to the formation of the diol 1,3-butanediol. Dehydration, on the other hand, leads to the formation of the α,β-unsaturated aldehyde 2-butenal. 2-Butenal can then undergo either reduction or oxidation. Oxidation with Tollens' reagent or Fehling's solution gives an α,β-unsaturated acid salt, which on acidification gives the free α,β-unsaturated carboxylic acid. Reduction can be accomplished in two ways. If the reducing agent is $NaBH_4$ or $LiAlH_4$, the product is the α,β-unsaturated alcohol 2-buten-1-ol. Reduction of both the carbon–carbon double bond and the carbonyl group gives the saturated primary alcohol 1-butanol.

Summary

The functional group in aldehydes and ketones is the carbonyl group $\left(\diagdown C=O \diagup \right)$. Aldehydes and ketones have the following general structures:

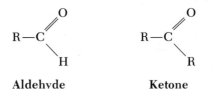

Aldehyde   Ketone

The carbonyl group consists of a σ bond and a π bond. The carbonyl carbon and the three atoms bonded to it lie in a plane 120° apart.

In the IUPAC system of naming carbonyl compounds, the characteristic

Table 10-4
Reactions of carbonyl compounds

Reagent	With aldehyde (RCHO)	With ketone (RCOR)
$R'MgX$	$RCH(OH)R'$	$R_2R'COH$
$Ag(NH_3)_2{}^+$	$RCOO^-$	No reaction
$NaOH + I_2$	No reaction	$RCOO^{-\,\circ}$
$KMnO_4$	$RCOOH$	No reaction
$K_2Cr_2O_7, H_2SO_4$	$RCOOH$	No reaction
H_2, Ni	RCH_2OH	$RCH(OH)R$
$LiAlH_4$	RCH_2OH	$RCH(OH)R$
$NaBH_4$	RCH_2OH	$RCH(OH)R$
CN^-, H^+	$RCH(OH)CN$	$R_2C(OH)CN$
$R'OH(2 \text{ moles})$	$RCH(OR')_2$	$R_2C(OR')_2$
H_2NOH	$RCH=NOH$	$R_2C=NOH$
H_2NNH_2	$RCH=NNH_2$	$R_2C=NNH_2$
$H_2NNHC_6H_5$	$RCH=NNHC_6H_5$	$R_2C=NNHC_6H_5$
$H_2NNHCNH_2$ $\overset{\|}{\underset{O}{}}$	$RCH=NNHCNH_2$ $\overset{\|}{\underset{O}{}}$	$R_2C=NNHCNH_2$ $\overset{\|}{\underset{O}{}}$
$R'CH_2CHO/OH^-$	$RCH(OH)CH(R')CHO$	$R_2C(OH)CH(R')CHO$
$R'CH_2COR'/OH^-$	$RCH(OH)CH(R')COR'$	$R_2C(OH)CH(R')COR'$

°One R of ketone is CH_3.

ending for an aldehyde is *-al*, and that for a ketone is *-one*. Some aldehydes and ketones also have common names.

Aldehydes can be prepared by the oxidation of primary alcohols in CrO_3/pyridine, reduction of acid chlorides, or hydrolysis of geminal dihalides. The last method is particularly suited to the preparation of aromatic aldehydes. Ketones are prepared by the oxidation of secondary alcohols with $KMnO_4$ or $K_2Cr_2O_7$ in H_2SO_4 and by the Friedel-Crafts acylation.

The characteristic reaction of the carbonyl group is nucleophilic addition. The kinds of reagents in nucleophilic addition reactions include Grignard reagents, forming alcohols; cyanide, forming cyanohydrins; alcohols, forming acetals or ketals; compounds derived from ammonia, forming hydrazones, semicarbazones, and phenylhydrazones; and carbanions, forming β-hydroxy carbonyl compounds. In addition, aldehydes and methyl ketones can be oxidized to carboxylic acids. Ketones do not oxidize easily. Also, the carbonyl group can be reduced to the alcohol group. In electrophilic aromatic substitution, the carbonyl group is a *meta* director. A summary of the reactions of carbonyl compounds with various reagents is given in Table 10-4.

Important terms

Acetal	Hydrazone
Aldehyde	Hydroxylamine
Carbanion	Iodoform test
Carbonyl group	Ketal
Clemmensen reduction	Ketone
Cyanohydrin	Nucleophilic addition
Fehling's solution	Oxime
Friedel-Crafts acylation	Phenylhydrazine
Geminal dihalide	Phenylhydrazone
Hemiacetal	Semicarbazide
Hemiketal	Semicarbazone
	Tollens' reagent

Problems

10-10. Give a correct name for each of the following molecules:

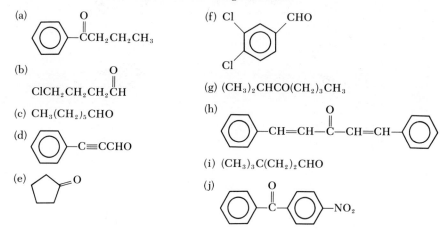

(a)

(b)

(c) $CH_3(CH_2)_5CHO$

(d)

(e)

(f)

(g) $(CH_3)_2CHCO(CH_2)_3CH_3$

(h)

(i) $(CH_3)_3C(CH_2)_2CHO$

(j)

10-11. Draw the structure for each of the following compounds:

(a) α,γ-Dimethylcaproaldehyde
(b) 2,4-Dimethoxyacetophenone
(c) 4-Ethyl-2-octenal
(d) Isopropyl *sec*-butyl ketone
(e) Butyraldehyde
(f) 4-Hexyn-2-one
(g) *p*-Tolualdehyde (*p*-methylbenzaldehyde)
(h) Phenyl benzyl ketone
(i) 3-Chloro-2-butanone
(j) Trimethylacetaldehyde

10-12. Using equations, show the product of the reaction of benzaldehyde with each of the following reagents:

(a) $(CH_3)_2CHMgBr$; then H_2O
(b) NaCN (aqueous), H^+
(c) Product of b + H_2O/H^+
(d) H_2NOH
(e) H_2NNH_2
(f) Phenylhydrazine
(g) Semicarbazide
(h) Ethyl alcohol (2 moles), dry HCl
(i) $K_2Cr_2O_7$, H_2SO_4
(j) Tollens' reagent
(k) Fehling's solution
(l) $LiAlH_4$
(m) Br_2, Fe
(n) HNO_3, H_2SO_4
(o) I_2 + NaOH

10-13. Answer problem 10-12 for acetophenone.

10-14. Draw the structure for each of the following compounds:

(a) Aromatic aldehyde of molecular formula C_7H_6O
(b) Cyclic saturated ketone of molecular formula C_5H_8O
(c) Aromatic ketone of molecular formula $C_9H_{10}O$ that gives a positive iodoform test
(d) Aldehyde formed from oxidation of 2-phenyl-1-ethanol
(e) Ketone that when reacted with methyl magnesium bromide forms *tert*-butyl alcohol
(f) Butyraldehyde cyanohydrin
(g) Ethyl ketal of acetophenone
(h) Benzaldehyde semicarbazone
(i) Aldol condensation product of acetaldehyde
(j) Oxidation product of hexanal by Tollens' reagent
(k) $LiAlH_4$ reduction product of 3-hexen-2-one

10-15. Give the structure of the product of each of the following reactions:

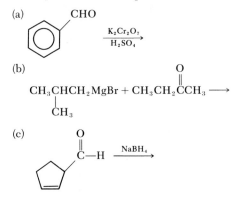

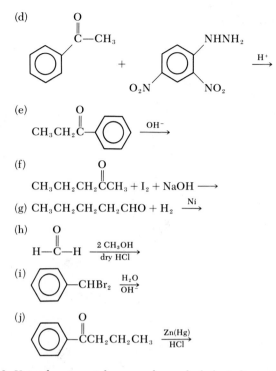

(d)

(e)

(f)

(g) $CH_3CH_2CH_2CH_2CHO + H_2 \xrightarrow{Ni}$

(h)

(i)

(j)

10-16. Using benzene, toluene, and any alcohols, indicate the method of synthesizing each of the following compounds:

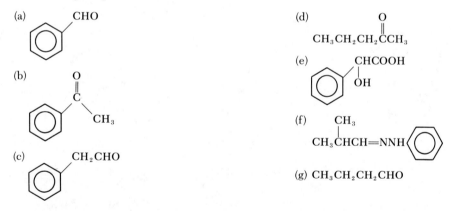

(a) CHO

(b)

(c) CH$_2$CHO

(d) $CH_3CH_2CH_2CCH_3$ (O)

(e) CHCOOH / OH

(f) CH$_3$ / CH$_3$CHCH=NNH

(g) $CH_3CH_2CH_2CHO$

10-17. Using the aldol condensation, indicate the method of synthesizing each of the following compounds:

(a) 2-Methyl-3-hydroxypentanal
(b) 4,4-Diphenyl-4-hydroxy-2-butanone
(c) 2-Ethyl-2-hexenal

10-18. Using the various methods of distinguishing between aldehydes and ketones (with Tollens' reagent, Fehling's solution, iodine and sodium hydroxide, and 2,4-dinitrophenylhydrazine), how would you distinguish between the compounds of each of the following pairs; that is, which compound would give a positive test with one of the aforementioned reagents and which would not? What would you do and see?

(a) Acetaldehyde and ethyl ketone
(b) 2-Pentanone and 3-pentanone

(c) Benzaldehyde and propionaldehyde

(d) Acetone and benzophenone

(e) Acetophenone and benzophenone

10-19. Using Table 10-3, identify each of the following compounds:

(a) A carbonyl compound gives a positive iodoform test, and its 2,4-dinitrophenyl-hydrazone melts at 239° to 240° C.

(b) A carbonyl compound melts at 49° to 50° C, and the presence of bromine is indicated.

(c) The 2,4-dinitrophenylhydrazone derivative of a carbonyl compound melts at 155° to 156° C, and the original carbonyl compound gives a positive Tollens' test.

(d) The semicarbazone of a carbonyl compound melts at 215° C. The carbonyl compound gives a negative Fehling's test.

(e) The melting point of the semicarbazone of a carbonyl compound is 105° to 107° C, and that of the 2,4-dinitrophenylhydrazone of the carbonyl compound is 103° to 105° C.

(f) The oxime of a carbonyl compound melts at 56° to 58° C. The carbonyl compound also gives a positive iodoform test. The semicarbazone melts at 189° to 190° C.

Self-test

1. Give a correct name for each of the following molecules:

(a) CH_3CH_2CHO

(c)

$$CH_3CH_2\overset{\overset{\displaystyle O}{\|}}{C}CH(CH_3)_2$$

(e) $CH_2{=}CHCH_2CH_2\overset{\overset{\displaystyle O}{\|}}{C}CH_3$

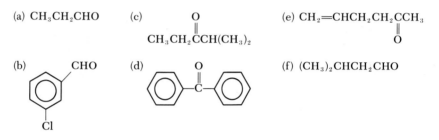

(b)

(d)

(f) $(CH_3)_2CHCH_2CHO$

2. Draw the complete structure corresponding to each of the following compounds:

(a) β-Chlorobutyraldehyde   (c) 4-Methyl-2-pentenal   (e) 5-Methyl-3-chloro-2-hexanone

(b) Acetophenone   (d) Benzyl isopropyl ketone

3. Give the structure of the organic product of each of the following reactions:

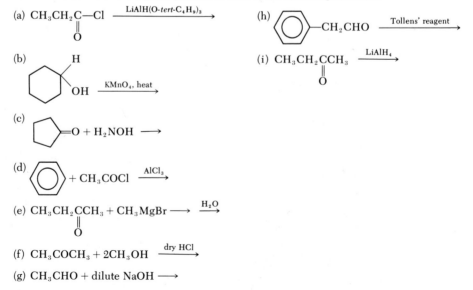

(a) $CH_3CH_2\overset{\overset{\displaystyle O}{\|}}{C}{-}Cl \xrightarrow{\text{LiAlH(O-}tert\text{-C}_4\text{H}_9)_3}$

(b)

(c) $={O} + H_2NOH \longrightarrow$

(d) $+ CH_3COCl \xrightarrow{\text{AlCl}_3}$

(e) $CH_3CH_2\overset{\overset{\displaystyle O}{\|}}{C}CH_3 + CH_3MgBr \longrightarrow \xrightarrow{\text{H}_2\text{O}}$

(f) $CH_3COCH_3 + 2CH_3OH \xrightarrow{\text{dry HCl}}$

(g) $CH_3CHO + \text{dilute NaOH} \longrightarrow$

(h) ${-}CH_2CHO \xrightarrow{\text{Tollens' reagent}}$

(i) $CH_3CH_2\overset{\overset{\displaystyle O}{\|}}{C}CH_3 \xrightarrow{\text{LiAlH}_4}$

4. Write the mechanism for the addition of NaCN to acetone in the presence of dilute acid.
5. Give the reagent and/or condition that will cause each step to occur as written in the following transformation:

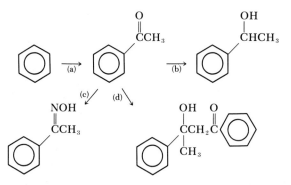

6. Give the correct answer for each of the following questions:

(a) What is the name of the functional group characteristic of aldehydes and ketones?
(b) What is the IUPAC name of ethyl isobutyl ketone?
(c) What is the typical reaction of a carbonyl compound?
(d) What type of alcohol is formed in the reaction of a Grignard reagent with formaldehyde?
(e) What is the general name of the product formed from an aldehyde and 1 mole of alcohol?
(f) What is the name of the intermediate in the aldol condensation?
(g) What is the observable indication of a positive Fehling's test?
(h) What oxidizing agent can be used to distinguish between an aldehyde and a ketone?
(i) What reducing reagent would you use to reduce $CH_3CH=CHCHO$ to $CH_3CH=CHCH_2OH$.
(j) Are acetophenone and methyl phenyl ketone isomers?

Answers to self-test

1. (a) Propionaldehyde or propanal
   (b) *m*-Chlorobenzaldehyde or
       3-chlorobenzaldehyde
   (c) Ethyl isopropyl ketone or
       2-methyl-3-pentanone

   (d) Benzophenone or phenyl ketone
   (e) 5-Hexen-2-one
   (f) Isovaleraldehyde,
       3-methylbutanal, or β-methylbutyraldehyde

2. (a)

$$CH_3CHCH_2C \overset{O}{\underset{\underset{Cl}{|}}{\diagup}} \diagdown H$$

   (b)

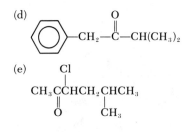

   (c)

$$CH_3CHCH=CHC \overset{O}{\diagup} \\ \underset{CH_3}{|} \qquad \diagdown H$$

   (d)

$$\bigcirc -CH_2-\overset{O}{\overset{||}{C}}-CH(CH_3)_2$$

   (e)   Cl

$$CH_3\overset{}{C}CHCH_2CHCH_3 \\ \underset{O}{||} \qquad \underset{CH_3}{|}$$

3. (a) $CH_3CH_2\overset{}{C}-H$
          $\underset{O}{||}$

   (b)

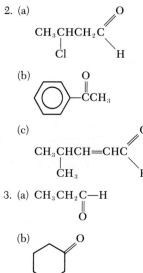

   (c)

   $=NOH$

   (d)

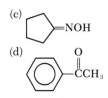

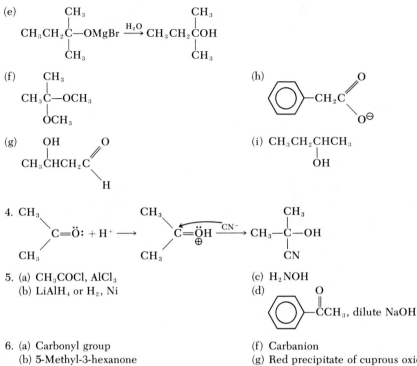

(e)

$$CH_3CH_2\underset{\underset{CH_3}{|}}{\overset{\overset{CH_3}{|}}{C}}-OMgBr \xrightarrow{H_2O} CH_3CH_2\underset{\underset{CH_3}{|}}{\overset{\overset{CH_3}{|}}{C}}OH$$

(f)

$$CH_3\underset{\underset{OCH_3}{|}}{\overset{\overset{CH_3}{|}}{C}}-OCH_3$$

(g)

$$CH_3\underset{\underset{OH}{|}}{\overset{}{C}}HCH_2\overset{\overset{O}{\parallel}}{C}{\diagdown}_H$$

(h)

benzene ring$-CH_2C{\overset{\overset{O}{\parallel}}{\diagdown}}_{O^{\ominus}}$

(i) $CH_3CH_2\underset{\underset{OH}{|}}{C}HCH_3$

4.

$$CH_3{\diagdown}_{CH_3}{\diagup}C{=}\ddot{O}: + H^+ \longrightarrow CH_3{\diagdown}_{CH_3}{\diagup}\underset{\oplus}{C}{=}\overset{..}{O}H \xrightarrow{CN^-} CH_3-\underset{\underset{CN}{|}}{\overset{\overset{CH_3}{|}}{C}}-OH$$

5. (a) CH_3COCl, $AlCl_3$
   (b) $LiAlH_4$ or H_2, Ni

(c) H_2NOH

(d) benzene ring$-\overset{\overset{O}{\parallel}}{C}CH_3$, dilute NaOH

6. (a) Carbonyl group
   (b) 5-Methyl-3-hexanone
   (c) Nucleophilic addition
   (d) Primary alcohol
   (e) Hemiacetal

   (f) Carbanion
   (g) Red precipitate of cuprous oxide
   (h) Tollens' reagent
   (i) $LiAlH_4$ or $NaBH_4$
   (j) No. They are identical.

chapter 11

Organic nitrogen compounds

The organic nitrogen-containing compounds include nitro compounds, nitriles, amides, amines, diazonium salts, and heterocyclic amines. Following are some examples:

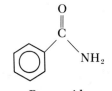

NO₂ structure — Nitrobenzene (nitro compound)

CH₃CH₂C≡N — Propionitrile (nitrile)

Benzamide (amide)

Nitrobenzene
(nitro compound)

Propionitrile
(nitrile)

Benzamide
(amide)

CH₃NH₂

⊕N≡NCl⊖ — Benzenediazonium chloride (diazonium salt)

Pyridine (heterocyclic amine)

Methylamine
(amine)

Benzenediazonium chloride
(diazonium salt)

Pyridine
(heterocyclic amine)

The synthesis of nitro compounds from benzene and of nitriles from primary alkyl halides is discussed in other chapters. The reduction of nitro compounds and nitriles is a convenient method of preparing primary amines. Amides can be prepared from carboxylic acids and can be converted to primary aliphatic and aromatic amines. Amines, especially the aliphatic amines, are the most important bases of organic chemistry and are fairly good nucleophilic reagents. The diazonium salts, which are available from primary aromatic amines, are useful to the organic chemist in syntheses. Heterocyclic amines are those amines in which the nitrogen atom is part of the ring system. Heterocyclic amines and some other amines are important from the standpoint of biological activity. In this chapter we will deal primarily with the chemistry of aliphatic and aromatic amines, although we will discuss the other types of nitrogen compounds.

Structure and nomenclature of amines

Amines are classified as primary, secondary, or tertiary according to the number of groups attached to the nitrogen atom:

| Primary | Secondary | Tertiary |

R is any alkyl or aryl group.

Nitrogen forms hybrid orbitals and covalent bonds similar to those of carbon. When nitrogen is bonded to three other atoms, as in amines, it uses sp^3 hybrid orbitals. The five electrons in the outermost energy level of nitrogen are distributed in the four equivalent sp^3 hybrid orbitals as follows: $(sp^3)^2(sp^3)^1(sp^3)^1(sp^3)^1$. Thus nitrogen has three sp^3 hybrid orbitals that each contain one electron that are capable of forming σ bonds with other atoms. The nitrogen–hydrogen σ bond is formed by the overlap of an s orbital of hydrogen and an sp^3 orbital of nitrogen, and the carbon–nitrogen σ bond is formed by the overlap of two sp^3 orbitals. The fourth sp^3 hybrid orbital contains an unshared pair of electrons. The four sp^3 orbitals of a nitrogen atom point toward the corners of a tetrahedron just as do the orbitals of a carbon atom:

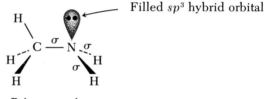

Filled sp^3 hybrid orbital

Primary amine

11-1. Identify the hybridized and unhybridized orbitals that make up the covalent bonds in methyl ethyl amine $\left(\mathrm{CH_3CH_2\underset{\underset{H}{|}}{N}CH_3}\right)$.

Answer: There are three sp^3—sp^3 bonds and nine s—sp^3 bonds.

The names of simple amines are formed by the names of alkyl groups attached to the nitrogen atom followed by the word *-amine*. In the names of more complicated amines, the prefix *amino-* is used as shown in the following examples:

Structure	*Type*	*Example*	*Name*
—NHR	Alkylamino-	—NHCH$_3$	N-Methylamino-
—NR$_2$	Dialkylamino-	—N(CH$_3$)$_2$	N,N-Dimethylamino-
—NR$_3$	Trialkylamino-	—N(CH$_3$)$_2$C$_2$H$_5$	N,N-Dimethyl-N-ethylamino-

N is used to indicate bonding to nitrogen, whereas *n* is used to indicate a straight-chain alkyl group. The simplest aromatic amine is called aniline, and most aromatic amines are named as derivatives of it. Following are some examples of names and classifications of amines:

$CH_3CH_2NH_2$

Ethylamine
(primary)

$CH_3CH_2CH_2NCH(CH_3)_2$
$\qquad\qquad\quad |$
$\qquad\qquad\quad H$

***n*-Propylisopropylamine**
(secondary)

$\qquad\quad CH_3$
$\qquad\quad |$
$CH_3—N—CH_3$

Trimethylamine
(tertiary)

$\qquad\quad CH_3$
$\qquad\quad |$
$CH_3—C—NH_2$
$\qquad\quad |$
$\qquad\quad CH_3$

***tert*-Butylamine**
(primary)

CH_3CH_2N $\boxed{CH_2CH_2CHCH_2CH_3}$
$\qquad\quad |\qquad\qquad\qquad |$
$\qquad\quad H\qquad\qquad\qquad CH_3$

1-(N-Ethylamino)-3-methylpentane
(secondary)

Benzylamine
(primary)

Aniline
(primary aromatic)

p-Aminoaniline
(***p*-phenylenediamine**)
(primary aromatic):
black hair dye

o-Methylaniline
(***o*-toluidine**)
(primary aromatic)

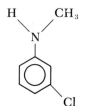

**3-Chloro-N-methyl-
aniline**
(secondary aromatic)

p-Methoxyaniline
(***p*-anisidine**)
(primary aromatic)

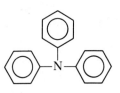

Triphenylamine
(tertiary aromatic)

11-2. Draw the structures and give the names of all the isomeric primary aliphatic amines with the molecular formula $C_5H_{13}N$.

Answer: One of the ten possible isomers is 1-aminopentane or *n*-pentylamine ($CH_3CH_2CH_2CH_2CH_2NH_2$).

11-3. Draw the structures of two secondary aliphatic and two tertiary aliphatic amines with the molecular formula $C_5H_{13}N$.

11-4. Using the molecular formula C_7H_9N, draw and name the structure of each of the following compounds:

(a) Benzylamine

(b) Primary aromatic amine
(c) Secondary aromatic amine
(d) *m*-Toluidine

Preparation of amines

Amines are generally prepared by reduction of nitro compounds, nitriles, or amides and by reductive amination of carbonyl compounds.

Reduction of nitro compounds

Reduction of aromatic nitro compounds is a convenient method of preparing primary aromatic amines. Aromatic nitro compounds are readily available by nitration of aromatic compounds. Following is an example of the preparation of a primary aromatic amine:

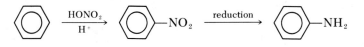

The reduction of aromatic nitro compounds can be accomplished by hydrogen in the presence of a nickel, palladium, or platinum catalyst, (2) by tin or iron and dilute hydrochloric acid, and (3) by lithium aluminum hydride. Following are some typical examples:

Nitrobenzene **Aniline**

***m*-Nitrotoluene** ***m*-Toluidine**

***o*-Nitroanisole** ***o*-Anisidine**

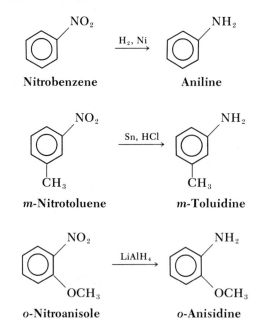

Aliphatic nitro compounds are not readily available; thus this reduction process is generally limited to aromatic nitro compounds.

Reductive amination

Reductive amination is a process by which aldehydes and ketones are converted into amines by treatment with ammonia and hydrogen and a catalyst:

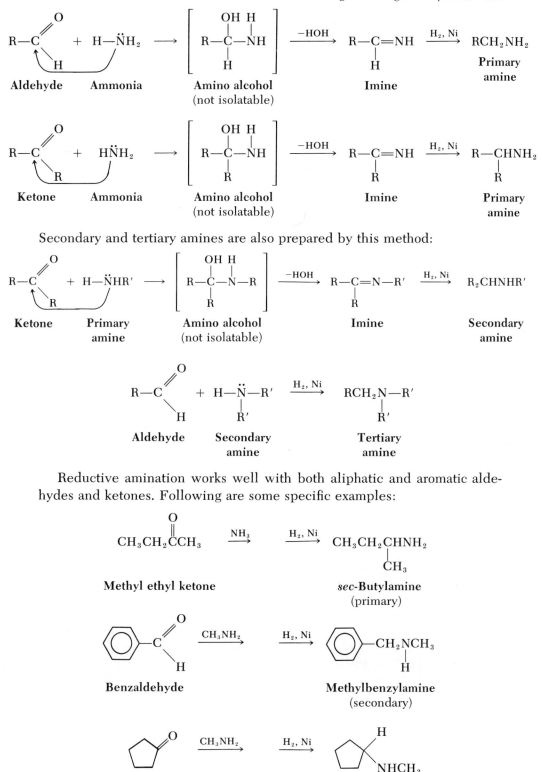

Secondary and tertiary amines are also prepared by this method:

Reductive amination works well with both aliphatic and aromatic aldehydes and ketones. Following are some specific examples:

Methyl ethyl ketone

sec-**Butylamine**
(primary)

Benzaldehyde

Methylbenzylamine
(secondary)

Cyclopentanone

**N-Methylamino-
cyclopentane**
(secondary)

The class of amine obtained depends not on the carbonyl compound used but rather on the amine used:

Amine used	*Product amine*
NH_3	Primary
RNH_2	Secondary
R_2NH	Tertiary
R_3N	No reaction

The mechanism of the reaction is as follows. The nucleophilic nitrogen atom of the amine attacks the electrophilic carbon of the carbonyl group. In other words, the carbonyl group undergoes its typical reaction: nucleophilic addition. The resulting product is an amino alcohol. However, the amino alcohol is unstable and easily loses a molecule of water. The final step in the reaction is the hydrogenation of the carbon–nitrogen double bond of the imine.

The addition of ammonia to aldehydes has theoretical implications for chemical evolution. Scientists have attempted to recreate chemical events in the primitive atmosphere by radiating a mixture of methane, ammonia, hydrogen, and water with ultraviolet radiation from an electric arc. Analyses of the product mixtures have indicated the presence of many common amino acids, which are the monomeric units of proteins. (Amino acids and proteins are discussed in more detail in Chapters 13 and 14.) During the course of the experiments, samples were obtained and analyzed. It has been observed that the appearance of amino acids is associated with the disappearance of ammonia, hydrogen cyanide, and aldehydes. These observations have led investigators to the conclusion that the four original gases were partially transformed to hydrogen cyanide and aldehydes and that the subsequent reaction of the intermediate products formed the amino acids.

The proposed reactions, which are shown as follows, are illustrations of the addition of ammonia to an aldehyde, the addition of hydrogen cyanide to an intermediate imine, and the hydrolysis of a nitrile:

$$CH_4, \; NH_3, \; H_2, \; \text{and} \; H_2O \xrightarrow{\text{energy}} HCN \; + \; RCHO$$

Aldehyde

$$RCHO + NH_3 \longrightarrow \left[\begin{array}{c} OH \\ | \\ R-C-NH_2 \\ | \\ H \end{array} \right] \xrightarrow{-H_2O} RCH{=}NH \qquad \text{Nucleophilic addition}$$

Imine

$$RCH{=}NH + HCN \longrightarrow \begin{array}{c} CN \\ | \\ R-CHNH_2 \end{array} \qquad \text{Electrophilic addition}$$

$$\begin{array}{c} NH_2 \\ | \\ RCHCN \end{array} + 2H_2O \longrightarrow \begin{array}{c} NH_2 \\ | \\ RCHCOOH \end{array} \qquad \text{Hydrolysis of nitrile}$$

α-Amino acid

This proposed synthetic route is not necessarily the manner in which life began. The exact nature of the atmosphere of gases is not crucial in terms of the synthesis of amino acids. Many of the amino acids in proteins can be obtained with other mixtures of gases (such as a mixture of carbon monoxide, ammonia, and water) or when mixtures of gases are exposed to ultraviolet radiation (a better model of primeval energy than an electric arc). In conclusion, it appears that the efforts of scientists to reproduce primitive conditions have led to amino acids, which are so vital to life.

Reduction of nitriles

Nitriles ($RC \equiv N$) are readily prepared from primary alkyl halides and aqueous sodium cyanide. This reaction is considered to follow a nucleophilic substitution mechanism:

$$R{-}X \xrightarrow[\text{aqueous}]{\text{NaCN}} R{-}C \equiv N$$

Primary   **Nitrile**
alkyl halide

The reduction of the nitrile then yields a primary amine:

$$R{-}C \equiv N \xrightarrow{\text{reduction}} R{-}CH_2NH_2$$

Nitrile   **Primary**
                            amine

The reduction can be accomplished either with hydrogen in the presence of a catalyst or with lithium aluminum hydride. Following are some examples:

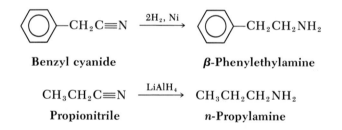

Benzyl cyanide   **β-Phenylethylamine**

$$CH_3CH_2C \equiv N \xrightarrow{\text{LiAlH}_4} CH_3CH_2CH_2NH_2$$

Propionitrile   **n-Propylamine**

The reduction of a nitrile gives a primary amine containing the same number of carbons as the nitrile.

From amides

An amide can be converted to a primary amine by degradation or reduction.

The *Hofmann degradation* is the conversion of an amide to a primary amine containing one less carbon than the original amide. The conversion is accomplished with sodium hypobromite (NaOBr), which is prepared from sodium hydroxide and bromine ($NaOH + Br_2 \longrightarrow NaOBr + HBr$). The Hofmann degradation reaction is, in essence, loss of the carbonyl group and union of the alkyl group with the nitrogen:

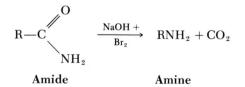

Amide   **Amine**

Following are some examples:

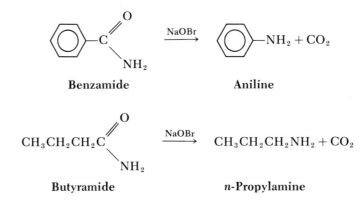

Benzamide   Aniline

$$CH_3CH_2CH_2C \xrightarrow{\text{NaOBr}} CH_3CH_2CH_2NH_2 + CO_2$$

Butyramide   *n*-Propylamine

The mechanism of the Hofmann degradation involves a rearrangement step in which the R group migrates to the nitrogen atom to form an isocyanate and the subsequent loss of the carbonyl group as carbon dioxide:

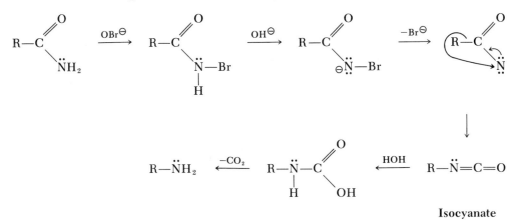

Isocyanate

More recently, it has been shown that amides can be reduced to amines by lithium aluminum hydride. The amine product contains the same number of carbons as the original amide. Following are some examples:

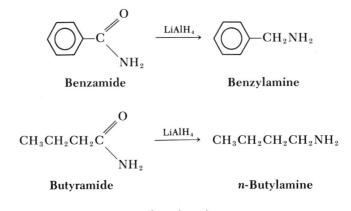

Benzamide   Benzylamine

$$CH_3CH_2CH_2C \xrightarrow{\text{LiAlH}_4} CH_3CH_2CH_2CH_2NH_2$$

Butyramide   *n*-Butylamine

• • •

Some of the reactions discussed in other chapters, especially those of alcohols, can be related to the synthesis of amines. The methods of synthesis of amines from alcohols are summarized as follows:

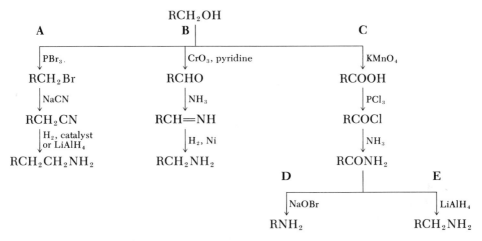

Synthesis by route A allows the preparation of an amine containing one more carbon atom than the original alcohol. The extra carbon atom is introduced in the reaction of the alkyl bromide with sodium cyanide. Synthesis by route B leads to an amine containing the same number of carbon atoms as the original alcohol. Actually, this synthesis can be modified to give secondary and tertiary amines. This can be accomplished by substitution of a primary amine and a secondary amine, respectively, for ammonia. Synthesis by routes C and D permits the preparation of an amine containing one less carbon atom than the original alcohol. The carbon atom is removed in the last step in the Hofmann degradation of the amide. Synthesis by routes C and E includes reduction of an amide to give a primary amine. The primary amine contains the same number of carbons as the amide.

11-5. Write reactions showing how *n*-butylamine can be synthesized by each of these three routes, using the appropriate alcohol as the original material.

Answer: Route A:

$$CH_3CH_2CH_2OH \xrightarrow{PBr_3} CH_3CH_2CH_2Br \xrightarrow{NaCN} CH_3CH_2CH_2CN \xrightarrow{H_2, Ni}$$
$$CH_3CH_2CH_2CH_2NH_2$$

Reactions of amines
Salt formation and basicity

Aliphatic and aromatic amines do not have comparable basicities. Aliphatic amines have about the same basicity as ammonia, but aromatic amines are considerably less basic than ammonia. All aliphatic and aromatic amines and ammonia are weaker bases than sodium hydroxide. On the other hand, they are stronger bases than alcohols (R$\ddot{\text{O}}$H), ethers (R$\ddot{\text{O}}$R), esters (RC$\ddot{\text{O}}$R), alkenes

$$\overset{\|}{\text{O}}$$

(RCH=CH$_2$), and water (H$\ddot{\text{O}}$H); that is, the unshared electron pair of nitrogen in amines is more readily available for bond formation than that in these other compounds. The order of basicity then is NaOH, KOH > aliphatic amines > ammonia > aromatic amines > esters, ethers, alcohols, alkenes water.

The basicity of a compound is a measure of the compound's ability to accept a proton. An amine accepts a proton as follows:

$$R\ddot{N}H_2 + H^{\oplus} \rightleftharpoons R\overset{\oplus}{N}H_3$$

The basicity constant, K_b, for this reaction is calculated as follows:

$$K_b = \frac{[R\overset{\oplus}{N}H_3]}{[R\ddot{N}H_2][H^+]}$$

The larger the value of K_b is, the stronger the base is. If K_b is large, the amine has a high tendency to accept a proton, and most of the amine exists in solution as ions. A comparison of the basicity constants of various amines indicates that aliphatic amines are slightly stronger than ammonia and that aromatic amines are the weakest amines:

Type of amine	K_b
Aliphatic	10^{-3} to 10^{-4}
Ammonia	1.8×10^{-4}
Aromatic	10^{-9} to 10^{-11}

The structures of amines can explain the differences in basicity. Amines are stronger bases than alcohols and ethers because nitrogen is less electronegative than oxygen and can better accommodate the positive charge of the ion:

$$R{-}\ddot{N}H_2 + H^+ \longrightarrow R{-}\overset{\oplus}{N}H_3$$

$$R{-}\ddot{O}H + H^+ \longrightarrow R{-}\underset{\oplus}{\ddot{O}}H_2$$

Aliphatic amines are slightly stronger bases than ammonia. The reactions between ammonia and a proton and between a primary amine and a proton are as follows:

$$\ddot{N}H_3 + H^+ \rightleftharpoons \overset{\oplus}{N}H_4$$

$$R{-}\ddot{N}H_2 + H^+ \rightleftharpoons R{-}\overset{\overset{\displaystyle H}{|}}{\underset{\underset{\displaystyle H}{|}}{N}}\!{}^{\oplus}{-}H$$

The aliphatic amine is stronger because an alkyl group is an electron donor compared to H and therefore makes the electron pair on N in the amine more available for bonding and disperses, although very slightly, the positive charge on N in the ammonium ion. This dispersal of charge and stabilization of the ion are not possible in the ammonium ion with no substituted groups:

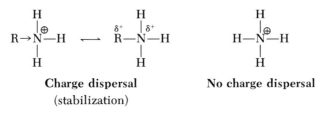

Charge dispersal   **No charge dispersal**
(stabilization)

Aromatic amines are weaker bases than ammonia. The reaction of an aromatic amine with a proton is as follows:

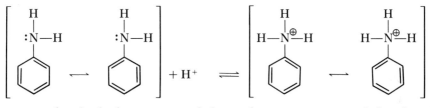

It appears that both the amine and the anilinium ions are stabilized to the same extent by resonance. However, the amine actually has the following resonance structures:

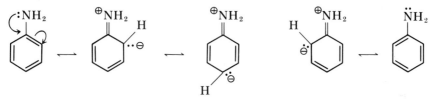

Comparable structures with charge separation cannot be drawn for the anilinium ion. The three additional structures for aniline stabilize the amine, but this same stabilization is not available to the anilinium ion. Hence, the energy content of aniline is lower than that of the anilinium ion. The energy required to overcome the energy barrier is greater for aniline than for ammonia, as shown in the energy profile in Fig. 11-1. Therefore, the ability of aniline to accept a proton is less than that of ammonia.

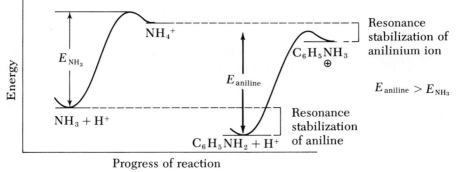

Fig. 11-1
Basicity of ammonia versus aniline.

The basicity of an aromatic amine is increased by the presence of an electron-donating group ($—NH_2$, $—CH_3$, or $—OCH_3$) in the ring, represented by X, because nitrogen is more basic; and it is lowered by the presence of an electron-withdrawing group (NO_2, $—CN$, $—COOH$, or $—Cl$) in the ring, represented by Y, because nitrogen is less basic:

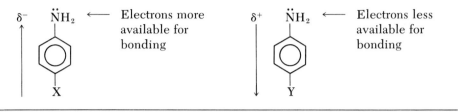

δ^- $\ddot{N}H_2$ ⟵ Electrons more available for bonding

δ^+ $\ddot{N}H_2$ ⟵ Electrons less available for bonding

11-6. Which compound in each of the following pairs is the stronger base in regard to accepting a proton from hydrochloric acid?

(a) HOH or NH_3 *Answer:* Ammonia
(b) $C_2H_5NH_2$ or C_2H_5OH *Answer:* Ethylamine
(c) Cyclohexylamine or aniline
(d)

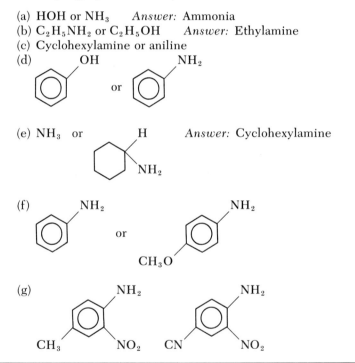

(e) NH_3 or   *Answer:* Cyclohexylamine

(f)   or

(g)

Dilute aqueous solutions of mineral acids (such as HCl and H_2SO_4) and carboxylic acids (RCOOH) convert amines into ammonium salts. This reaction is called neutralization. The types of salts obtained are as follows:

Amine	$\overset{HX}{\rightleftharpoons}$	Salt
RNH_2		$R\overset{\oplus}{N}H_3X^{\ominus}$
R_2NH		$R_2\overset{\oplus}{N}H_2X^{\ominus}$
R_3N		$R_3\overset{\oplus}{N}HX^{\ominus}$

Following are some typical examples of neutralization:

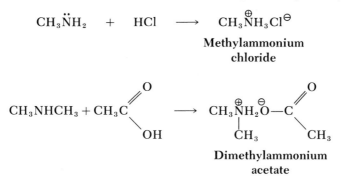

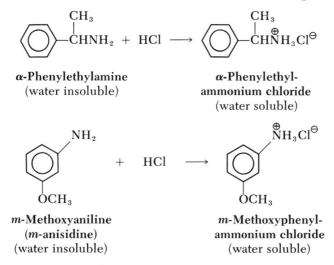

α-Phenylethylamine
(water insoluble)

**α-Phenylethyl-
ammonium chloride**
(water soluble)

m-Methoxyaniline
(**m-anisidine**)
(water insoluble)

**m-Methoxyphenyl-
ammonium chloride**
(water soluble)

Thus basic amines can be extracted from neutral or acidic substances (such as phenols and carboxylic acids) by being converted to water-soluble salts by the neutralization reaction. After the organic and aqueous layers are separated, the amine can be recovered from the aqueous layer, which is made alkaline, usually with sodium hydroxide:

$$R\overset{\oplus}{N}H_3X^{\ominus} + OH^{\ominus} \longrightarrow RNH_2 + H_2O + X^{\ominus}$$

Conversion into quaternary ammonium salts

Alkylation of primary amines results in *quaternary ammonium salts*, with the general formula $R_4N^+X^-$. In these salts, which are the products of the complete alkylation of nitrogen, four alkyl groups are covalently bonded to nitrogen, and the positive charge on nitrogen is balanced by some negative ion like Cl^- or OH^-:

$$RNH_2 \xrightarrow{R'X} RNHR' \xrightarrow{R'X} \underset{\underset{R'}{|}}{RNR'} \xrightarrow{R'X} R-\overset{\overset{R'}{|}}{\underset{\underset{R'}{|}}{\overset{\oplus}{N}}}-R'X^{\ominus}$$

Primary   **Secondary**   **Tertiary**   **Quaternary**

For example:

$$CH_3NH_2 \xrightarrow{CH_3I} CH_3NCH_3 \xrightarrow{CH_3I} CH_3-N-CH_3 \xrightarrow{CH_3I} CH_3-\overset{\underset{|}{CH_3}}{\overset{\oplus}{N}}-CH_3I^{\ominus}$$

| Methyl-amine | Dimethyl-amine | Trimethyl-amine | Tetramethyl-ammonium iodide |

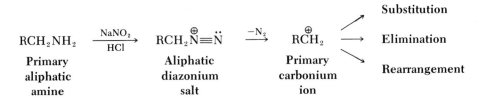

Aniline $\xrightarrow[\text{excess}]{CH_3I}$ Trimethylphenyl-ammonium iodide

Coniine
(component of hemlock)

2-*n*-Propyl-N,N-dimethyl-piperidinium iodide

Treatment of the ammonium halides with moist silver oxide (Ag_2O) forms the corresponding ammonium hydroxides; these hydroxides have basicities comparable to those of NaOH and KOH, since an aqueous solution of any of them contains the hydroxide ion:

$$(CH_3)_4N^{\oplus}I^{\ominus} \xrightarrow[\text{moist}]{Ag_2O} (CH_3)_4N^{\oplus}OH^{\ominus}$$

Tetramethyl-
ammonium
hydroxide

Reaction with nitrous acid: diazonium salts

All classes of aliphatic and aromatic amines react with nitrous acid (HONO), which is unstable and must be prepared in situ from sodium nitrite and hydrochloric acid.

Primary aliphatic amines react with nitrous acid to form unstable aliphatic diazonium salts; these salts easily lose molecular nitrogen to form primary carbonium ions, which can undergo the characteristic reactions of substitution, elimination, and rearrangement to more stable carbonium ions:

$$RCH_2NH_2 \xrightarrow[\text{HCl}]{NaNO_2} RCH_2\overset{\oplus}{N}\!\!\equiv\!\!\overset{..}{N} \xrightarrow{-N_2} RCH_2^{\oplus} \longrightarrow$$

    Substitution
    Elimination
    Rearrangement

| Primary aliphatic amine | Aliphatic diazonium salt | Primary carbonium ion |

Thus the reaction of primary aliphatic amines with nitrous acid gives a mixture of products and is of little synthetic usefulness.

Secondary aliphatic and aromatic amines yield nitroso compounds when treated with nitrous acid. Following are some examples:

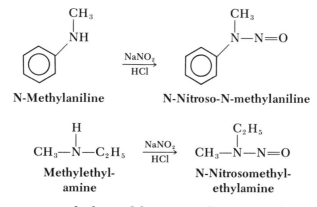

N-Methylaniline	N-Nitroso-N-methylaniline

Methylethyl- amine	N-Nitrosomethyl- ethylamine

These nitroso compounds derived from secondary amines have characteristic yellow colors and are generally used as a qualitative test for secondary amines.

Tertiary aromatic amines undergo electrophilic substitution to form nitroso compounds; for example:

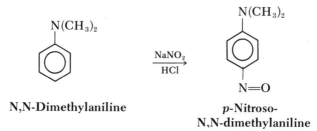

N,N-Dimethylaniline	p-Nitroso- N,N-dimethylaniline

Of the reactions with nitrous acid, the most interesting are the reactions of the primary aromatic amines, which lead to a synthetically useful series of compounds called *diazonium salts*. These salts are formed by treatment of the amines with an aqueous solution of sodium nitrite and HCl below 5° C. Following is an example:

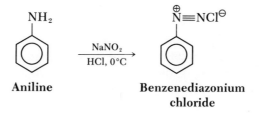

Aniline	Benzenediazonium chloride

The aromatic amines from which the diazonium salts are obtained are easily prepared from the corresponding nitro compounds, which are obtained by direct nitration. There are very few groups present in a molecule that will interfere with diazotization. Following is an example of the preparation of a diazonium salt:

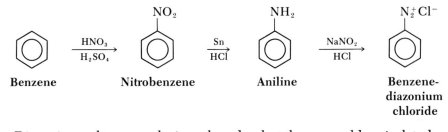

Benzene   Nitrobenzene   Aniline   Benzene-diazonium chloride

Diazonium salts are explosive when dry, but they are seldom isolated and are used in solution immediately after being prepared. They undergo two types of reactions: substitution, in which nitrogen is lost as N_2, and coupling, in which nitrogen is retained in the product.

Substitution of the diazonium group is the best general method of introducing a substituent such as —F, —Cl, —Br, —I, —CN, —OH, or —H into an aromatic ring. Examples of the preparation of compounds by this method are shown on the opposite page.

The replacement of nitrogen with —I can be effected by mixing of the diazonium salt solution with a potassium iodide (KI) solution. The replacement of nitrogen with —F is carried out by the addition of fluoroboric acid (HBF_4) to the diazonium salt solution. The precipitate that forms, $C_6H_5N_2^+BF_4^-$, is isolated and dried. This particular diazonium salt is stable when dry. Decomposition of it by heating yields the fluorobenzene. The use of diazonium salts is the best known way of introducing —I and —F into the aromatic ring. Direct halogenation with iodine and a Lewis acid does not always occur, and electrophilic substitution with fluorine is much too reactive:

$$\overset{I_2}{\underset{AlCl_3}{\longrightarrow}} \quad \text{no reaction}$$

$$\overset{F_2}{\underset{AlCl_3}{\longrightarrow}} \quad \text{mixture of fluorinated products}$$

The reaction between cuprous bromide (CuBr) or cuprous chloride (CuCl) and a diazonium salt is carried out by mixing of the two freshly prepared solutions. If the solution is then allowed to slowly warm to room temperature, there is a gradual evolution of nitrogen. Replacement of nitrogen by these cuprous salts is referred to as the Sandmeyer reaction. This reaction is a simple method of preparation of benzenes with —Br and —Cl substituted groups. Direct halogenation to form these substituted benzenes does occur, but the boiling points of the *ortho* and *para* products are too close to allow them to be separated by distillation; for example:

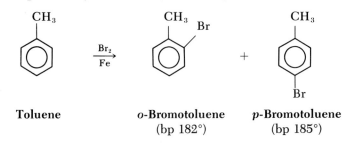

Toluene   *o*-Bromotoluene (bp 182°)   *p*-Bromotoluene (bp 185°)

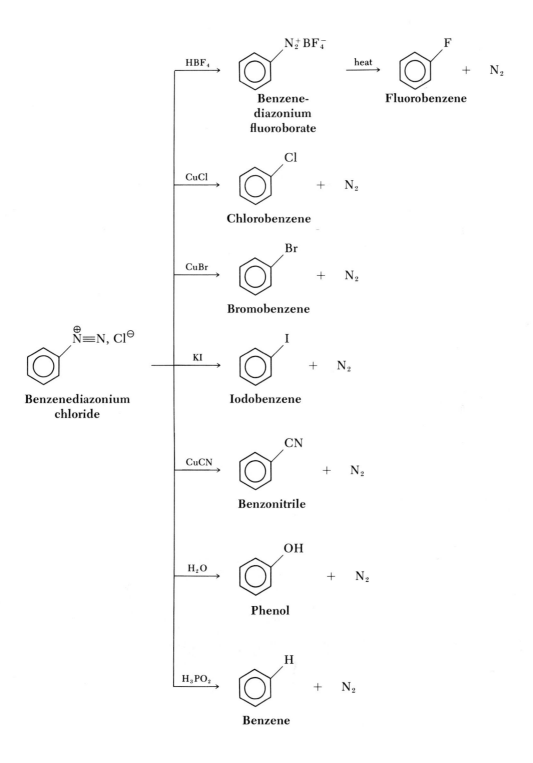

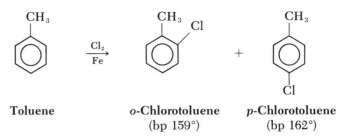

Thus first o-nitrotoluene and p-nitrotoluene are formed. Then because of the greater difference in their boiling points, these isomers can be separated by distillation; and the o-bromotoluene and p-bromotoluene can be formed from the diazonium salt as follows:

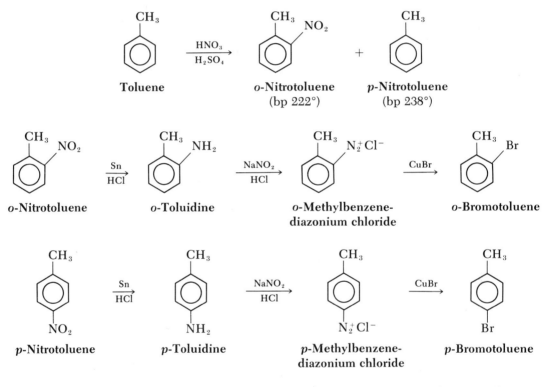

The importance of the replacement of the diazonium group by —CN lies not so much in the synthesis of the nitrile but in the hydrolysis of the nitrile to the carboxylic acid, which is illustrated as follows:

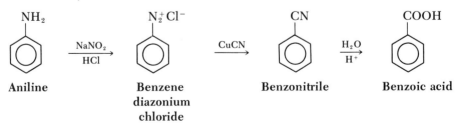

If a diazonium salt is treated with water, a phenol results.

Hypophosphorus acid (H_3PO_2) is a reducing agent that is used to cause the replacement of the diazonium group with —H.

The second type of reaction that diazonium salts undergo is coupling (substitution in an activated aromatic ring). A diazonium group can be substituted for hydrogen in another ring that has been highly activated by a group such as —OH, —NH$_2$, —NHR, or —NR$_2$. Following are some examples:

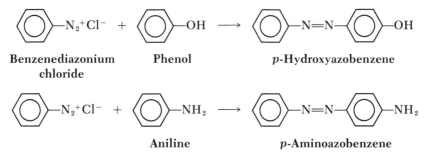

Benzenediazonium Phenol   *p*-Hydroxyazobenzene
chloride

Aniline   *p*-Aminoazobenzene

Substitution occurs in the position *para* to the activating group; but if the *para* position is blocked, the coupling will occur in the position *ortho* to the strongly activating group. For example:

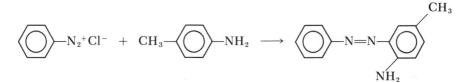

Benzenediazonium ***p*-Toluidine** **2-Amino-5-methylazobenzene**
chloride

The —N=N— group is called the *azo group,* and compounds containing this group are called *azo compounds.* They are strongly colored; depending on their structure, they are red, yellow, orange, blue, or green. Thus they are used as dyes and acid-base indicators:

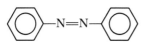
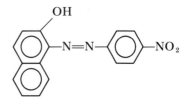

Azobenzene: **Para red: dye**
orange-red dye

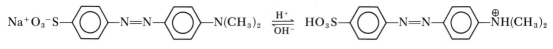

Basic solution: **Acidic solution:**
yellow   red

Methyl orange:
acid base indicator

Conversion into amides and sulfonamides

An amide is formed whenever an acid chloride, such as acetyl chloride, is treated with ammonia (Chapter 9):

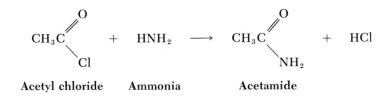

Acetyl chloride   Ammonia   Acetamide

Primary and secondary aliphatic and aromatic amines also react with acid chlorides to form amides; however, tertiary amines do not form amides, since they do not have a proton to lose:

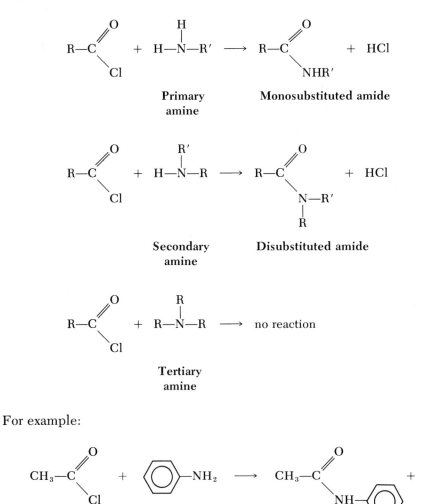

For example:

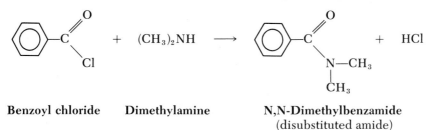

Benzoyl chloride　　**Dimethylamine**　　**N,N-Dimethylbenzamide**
(disubstituted amide)

As derivatives of amines, amides are used for identifying amines, just as 2,4-dinitrophenylhydrazones, oximes, and semicarbazones, as derivatives, are used for identifying aldehydes and ketones. The majority of the amides derived from amines are white, crystalline solids with characteristic melting points. Also, lack of a reaction indirectly characterizes an amine as tertiary.

In addition, these amides are useful to the organic chemist in syntheses. For example, aniline can be treated with acetyl chloride to form an amide. The resulting acetylated amino group acts as a blocking group for directing incoming groups to the *para* position of the ring and then is conveniently removed by hydrolysis. Thus the organic chemist is able to introduce substituents into the position *para* to the amino group:

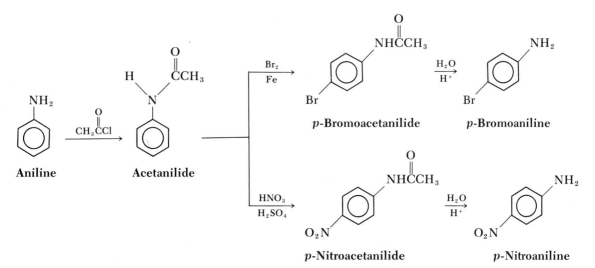

The need for a blocking group in electrophilic substitution reactions is discussed in the next section.

Amines also react with benzenesulfonyl chloride, a derivative of benzenesulfonic acid. The reaction forms a class of compounds called *sulfonamides*. Following are some examples:

$$\text{C}_6\text{H}_5\text{—SO}_2\text{Cl} \quad + \quad \text{RNH}_2 \quad \longrightarrow \quad \text{C}_6\text{H}_5\text{—SO}_2\text{N(H)—R} \quad + \quad \text{HCl}$$

Benzenesulfonyl　　　　　　　　　　**Monosubstituted**
chloride　　　　　　　　　　　　　**sulfonamide**

$$\text{Ph–SO}_2\text{Cl} \;+\; (\text{CH}_3)_2\text{NH} \;\longrightarrow\; \text{Ph–SO}_2\overset{\displaystyle\text{CH}_3}{\underset{}{\text{N}}}\text{–CH}_3 \;+\; \text{HCl}$$

Dimethylamine   N,N-Dimethylbenzenesulfonamide
(disubstituted sulfonamide)

The formation of sulfonamides is the basis of the *Hinsberg test,* which is used to distinguish primary, secondary, and tertiary amines. Primary and secondary amines form sulfonamides when shaken with benzenesulfonyl chloride in the presence of aqueous potassium hydroxide. However, the sulfonamides of primary and secondary amines are chemically different. Primary amines form sulfonamides that are soluble in potassium hydroxide, since the hydrogen atom remaining on the nitrogen is acidic; and the alkyl-sulfonamide is regenerated from the salt by acidification of the solution:

$$\text{Ph–SO}_2\text{Cl} + \text{RNH}_2 \xrightarrow{\text{KOH}} \underset{\underset{H \;\to\; Acidic}{|}}{\text{Ph–SO}_2\text{N–R}} \underset{\text{HCl}}{\overset{\text{KOH}}{\rightleftharpoons}} \underset{\underset{R}{|}}{\text{Ph–SO}_2\ddot{\text{N}}\overset{\ominus}{}\text{K}^{\oplus}}$$

Benzenesulfonyl **Primary**   **Alkylsulfonamide**
 chloride   **amine**   (soluble in base;
            insoluble in acid)

On the other hand, the sulfonamide formed from a secondary amine is insoluble in base, since it contains no acidic hydrogen:

$$\text{Ph–SO}_2\text{Cl} \;+\; \text{R}_2\text{NH} \xrightarrow{\text{KOH}} \text{Ph–SO}_2\text{NR}_2$$

Benzenesulfonyl   **Secondary**   **Dialkylsulfonamide**
 chloride   **amine**   (insoluble in base and acid)

Thus a primary amine and a secondary amine are distinguished in the Hinsberg test by the fact that a primary amine yields a clear solution, acidification of which precipitates the alkylsulfonamide, whereas a secondary amine yields a dialkylsulfonamide that is unaffected by either acid or base. Tertiary amines, since they have no hydrogen on nitrogen, do not react with benzenesulfonyl chloride:

$$\text{Ph–SO}_2\text{Cl} \;+\; \text{R}_3\text{N} \xrightarrow{\text{KOH}} \text{no reaction}$$

Benzenesulfonyl **Tertiary**
 chloride   **amine**

11-7. Describe how the Hinsberg test can distinguish between the following pairs of amines:

(a) Aniline and N-methylaniline

  Answer: Both amines are individually shaken with benzenesulfonyl chloride and KOH. Aniline (primary amine) yields a clear solution, acidification of which precipitates the sulfonamide. N-Methyl-aniline (secondary amine) forms a sulfonamide that is unaffected by either acid or base.

(b) *tert*-Butylamine and trimethylamine

(c) Diethylamine and triethylamine

Sulfonamides that are known for their antibacterial activity are called *sulfa drugs*. The parent compound, sulfanilamide, is prepared as follows:

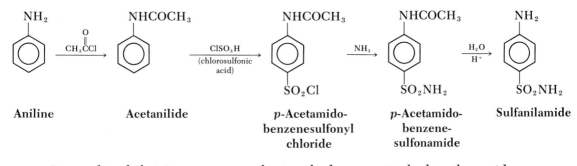

| Aniline | Acetanilide | *p*-Acetamido-benzenesulfonyl chloride | *p*-Acetamido-benzene-sulfonamide | Sulfanilamide |

It was found that in some cases the two hydrogens attached to the amide nitrogen could be replaced to produce a more active drug. Many of these substituted sulfanilamides are prepared from *p*-acetamidobenzenesulfonyl chloride according to the following equation:

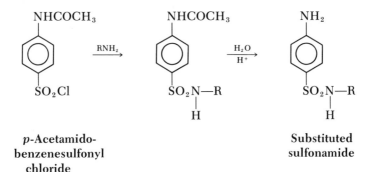

p-Acetamido-benzenesulfonyl chloride   Substituted sulfonamide

Some of the more effective sulfa drugs are as follows:

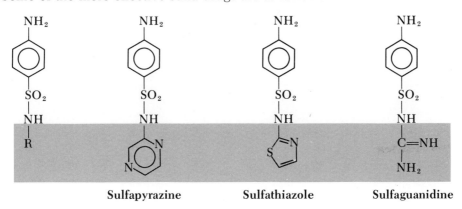

Sulfapyrazine   **Sulfathiazole**   **Sulfaguanidine**

The structure of sulfanilamide very closely resembles the structure of *p*-aminobenzoic acid, which is essential to the growth and reproduction of many microorganisms:

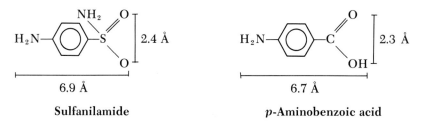

Sulfanilamide   ***p*-Aminobenzoic acid**

Therefore, the microorganism can and does mistake the sulfanilamide for the *p*-aminobenzoic acid and accidentally incorporates the wrong substance into its growth pattern. This process is irreversible. Since the growth system does not function properly, the microorganism cannot reproduce and ultimately dies.

The sulfa drugs are very effective against some types of pneumonia, blood poisoning, and gonorrhea but can cause toxic reactions in some people. Today, the sulfa drugs have been largely replaced by antibiotics such as penicillin and tetracycline:

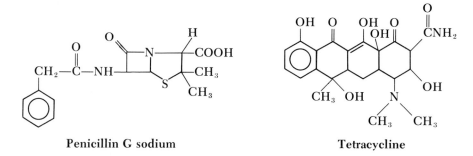

Penicillin G sodium   **Tetracycline**

Electrophilic substitution in aromatic amines

The groups —NH$_2$, —NHR, and —NR$_2$ are strong ring activators and direct incoming groups to the *ortho* and *para* positions. As a consequence, certain peculiarities exist in substitution in aromatic amines. With aniline as an example, these peculiarities are discussed as follows:

1. Bromination of aniline occurs at both *ortho* positions and at the *para* position to the amino group, and the ring is so activated that a Lewis acid catalyst is not needed:

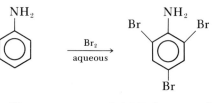

Aniline   **2,4,6-Tribromoaniline**

However, a single bromine atom can be introduced if the amino group is acetylated before bromination is carried out and then the resulting bromo-amide is hydrolyzed to give the bromoamine:

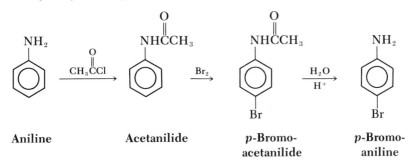

| Aniline | Acetanilide | *p*-Bromo-acetanilide | *p*-Bromo-aniline |

The acetylated amino group is sometimes referred to as a blocking or protecting group, since the incoming group is directed to the *para* position. The size of the acetylated amino group minimizes the attack of an incoming group at the *ortho* position. The incoming group, however, will go to one *ortho* position if the *para* position is blocked:

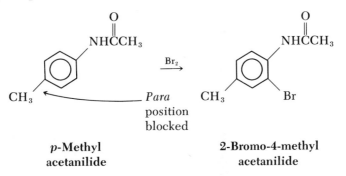

| *p*-Methyl acetanilide | 2-Bromo-4-methyl acetanilide |

The —NHCOCH$_3$ group, therefore, is an *ortho, para* director but is less activating than an —NH$_2$ group. The use of the acetal linkage in the selective reduction of the carbon–carbon double bond in the presence of a carbon–oxygen double bond is another example of the use of a blocking group (Chapter 10).

2. Nitration is also carried out with the acetylated amine. Nitric acid is fairly good as an oxidizing agent, and direct nitration of an aromatic ring leads to tarry oxidation products. Furthermore, nitration is carried out in acid solution and amines, being bases, form the —NH$_3^+$ group, which is a *meta* director. Hence, nitration of an aromatic amine leads mainly to the product with the nitro group in the *meta* position:

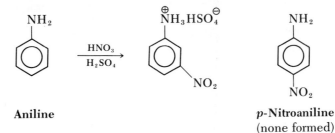

| Aniline | | *p*-Nitroaniline (none formed) |

Thus nitration of the acetylated amine is the best way to prepare the desired nitroaniline:

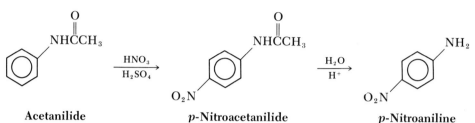

Acetanilide   *p*-Nitroacetanilide   *p*-Nitroaniline

3. Friedel-Crafts alkylation does not occur if the $—NH_2$, $—NHR$, or $—NR_2$ group is present in the aromatic ring. The strongly basic nitrogen ties up the Lewis acid needed for ionization of the alkyl halide (Chapter 4), and the formation of the positive charge on nitrogen probably contributes somewhat to preventing the reaction by deactivating the ring:

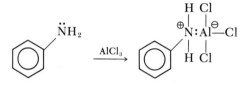

Heterocyclic amines

The nitrogen-containing compounds in which the nitrogen atom is a member of the ring system in addition to carbon atoms are called *heterocyclic amines*. Following are some important heterocyclic amines with five- and six-member ring systems:

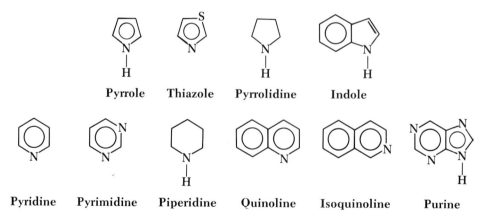

Pyrrole   Thiazole   Pyrrolidine   Indole

Pyridine   Pyrimidine   Piperidine   Quinoline   Isoquinoline   Purine

All these compounds are aromatic except pyrrolidine and piperidine, which are saturated heterocyclic amines and are the reduced forms of pyrrole and pyridine, respectively.

Heterocyclic amines with substituted groups are named as derivatives of the parent heterocyclic amines, and in the numbering of the ring, the nitrogen atoms are given the lowest possible numbers:

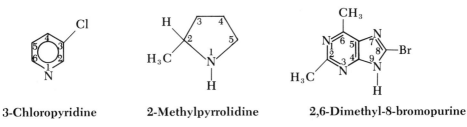

3-Chloropyridine   **2-Methylpyrrolidine**   **2,6-Dimethyl-8-bromopurine**

Many of these heterocyclic amines occur in nature. They have important roles in the biological and physiological activity in animals and plants. Most of the structures of these naturally occurring heterocyclic amines are complex.

The pyrrole ring structure occurs in heme. Heme, in the form of an oxygen-heme complex, enables hemoglobin to transport oxygen from the lungs to the body tissues.

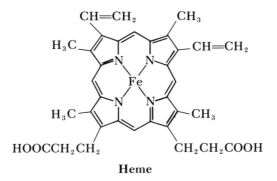

Heme

Some of the compounds in which the indole ring system is found are as follows:

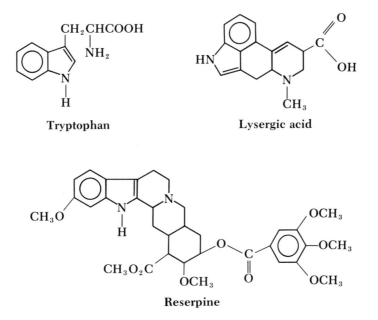

Tryptophan   **Lysergic acid**

Reserpine

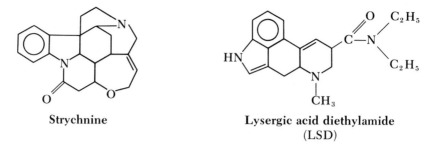

Strychnine   **Lysergic acid diethylamide**
                                              (LSD)

Many of the compounds containing the indole ring system have pronounced physiological activity. Lysergic acid is produced by a fungus called ergot and grows as a parasite on rye. Lysergic acid diethylamide (LSD), which does not occur naturally, is capable of producing hallucinations. Reserpine is used somewhat as a tranquilizer and in the control of high blood pressure. Strychnine is used as a cardiac stimulant. Tryptophan is an amino acid.

Some important compounds containing the pyridine ring system are as follows:

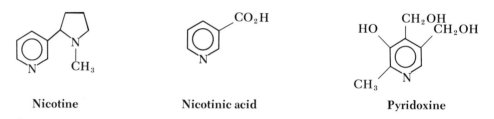

Nicotine   **Nicotinic acid**   **Pyridoxine**

Nicotine is a component of tobacco leaves. Nicotinic acid is a vitamin known as niacin. Niacin is used effectively in the treatment of pellagra, a disease caused by a deficiency of niacin. The disease occurs particularly among people whose diets consist mainly of corn, which is low in niacin. It is characterized by skin eruptions and gastrointestinal disturbances. Pyridoxine is vitamin B_6 (Chapter 14).

The ring system of piperidine is found in several naturally occurring compounds. Among these are coniine (p. 312), a component of hemlock, which poisoned Socrates, and cocaine, found in the leaves of the coca plant, which has medicinal use as a local anesthetic and an antimalarial drug:

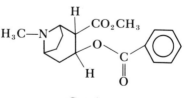

Cocaine

The isoquinoline ring system is found in papaverine, and the reduced isoquinoline ring is found in morphine and codeine:

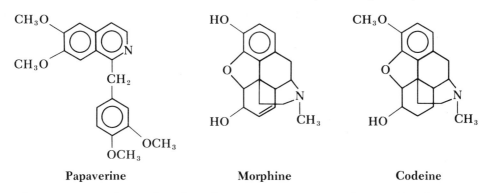

Papaverine	**Morphine**	**Codeine**

These compounds are found in the opium poppy. Morphine and codeine are used medicinally as pain relievers.

Caffeine, a natural purine product, is found in the leaves of the tea plant and in the coffee bean. It provides the stimulation derived from tea, coffee, and cola soft drinks.

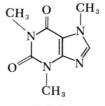

Caffeine

The thiazole ring is found in luciferin, the substance responsible for the luminescence of the firefly:

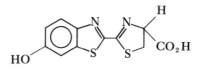

Luciferin

Penicillin G sodium whose structure is shown on p. 322, also contains the thiazole ring.

Chemistry of pyrrole and pyridine

Pyrrole is a heterocyclic amine with a five-member ring, and pyridine is a heterocyclic amine with a six-member ring.

The nitrogen atom of pyrrole has three sp^2 orbitals, each of which has one electron; and perpendicular to the plane of the sp^2 orbitals is a p orbital containing a pair of electrons. Thus nitrogen has the configuration $(sp^2)^1(sp^2)^1(sp^2)^1(p)^2$. Two of the sp^2 orbitals overlap with two sp^2 orbitals of carbon and the third an s orbital of hydrogen to form σ bonds. Since the two electrons in the p orbital lie in a plane perpendicular to the σ bonds, they can overlap with the p orbitals of the carbons. This overlapping results in a π electron cloud of six electrons:

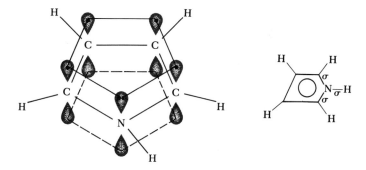

The nitrogen atom of pyridine has three sp^2 orbitals, two of which contain one electron and one of which contains two electrons, and one p orbital containing one electron. Thus the configuration of nitrogen is $(sp^2)^1(sp^2)^1(sp^2)^2(p)^1$. The two sp^2 orbitals that each contain one electron form bonds with two of the ring carbons. The sp^2 orbital containing two electrons lies in the same plane as the σ bonds. The p orbital of nitrogen lies in a plane perpendicular to the bonds and overlaps the p orbitals on the carbons to form a π electron cloud of six electrons:

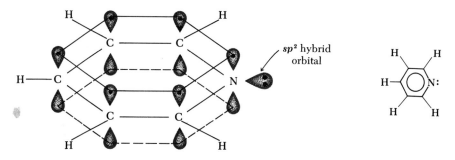

The structures of these molecules give rise to the properties of basicity and aromaticity. Pyrrole is a weakly basic amine because the two electrons in the p orbital of the nitrogen atom are incorporated into the π electron cloud and therefore are unavailable for sharing. The nitrogen atom in pyridine, on the other hand, has an unshared electron pair in an sp^2 orbital that is available for bond formation. The availability of the electron pair is illustrated by the reaction between pyridine and dilute hydrochloric acid:

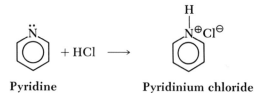

Even though pyridine possesses characteristics of a base, it is still a weaker base than ammonia as shown by the basicity constants:

	K_b
Ammonia	1.8×10^{-5}
Pyridine	2.0×10^{-9}
Pyrrole	1.0×10^{-14}

The aromatic character of benzene is attributed to a π electron cloud of six electrons. Since pyrrole and pyridine possess a π electron cloud of six electrons, they would be expected to have aromatic properties.

Like benzene, pyrrole and pyridine are planar. This planarity allows maximum overlap of the p orbitals. The bond angles in benzene are 120°; but the substitution of a nitrogen atom for a carbon atom reduces the bond angles in pyridine to an average value of slightly less than 120°, and in pyrrole the bond angles are about 108° probably because, in addition to substitution of a nitrogen atom for a carbon atom in the ring system, there is compression of the ring system to five atoms:

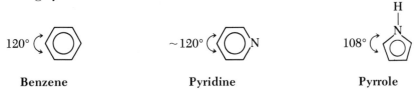

Benzene	**Pyridine**	**Pyrrole**

The characteristic reaction of an aromatic compound is electrophilic substitution. Following are some examples illustrating the preferred position of substitution in pyrrole and pyridine. Pyrrole undergoes substitution predominantly in the number 2 position, whereas pyridine undergoes substitution in the number 3 position:

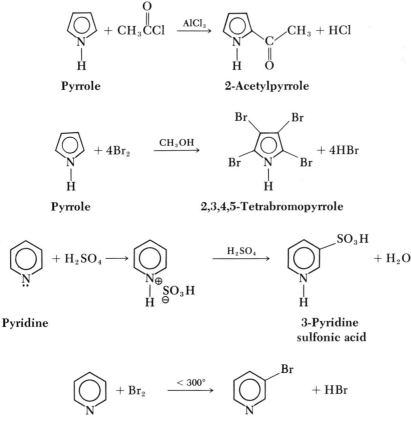

Drugs: use and abuse

Currently, there is much popular interest in amines because of their use and misuse as drugs. The physiologically active amines currently receiving attention are the hallucinogens, stimulants, and narcotic analgesics.

The *hallucinogens* are drugs that can provoke perceptual changes that have no basis in physical reality. Generally, vision is greatly altered; and hallucinations, increased energy, and panic can result from the use of an average amount of a hallucinogen. The amount used and the duration of the effect vary greatly with the drug. The use of a large amount can result in anxiety, psychosis, exhaustion, hallucinations, tremors, vomiting, and panic. There is the risk of "bad trips," and there have been reports of "flashbacks." Examples of hallucinogens are the synthetic drug LSD and the naturally occurring compound mescaline, a component of the peyote cactus. These substances have no medical use.

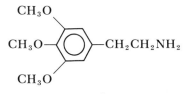

Mescaline

Stimulants are drugs that stimulate the body's central nervous system, resulting in increased alertness and reduction of fatigue. These include amphetamine (benzedrine), cocaine, caffeine, and nicotine.

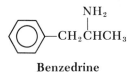

Benzedrine

These drugs are used primarily to treat mild depression, narcolepsy (overwhelming desire to sleep during normal waking hours), and obesity through their appetite control function. They are also used to combat fatigue during dangerous or prolonged tasks. For example, cocaine can bring about a sense of self-confidence and power and intense exhilaration, and nicotine use results in relaxation and constriction of blood vessels. In general, the repeated misuse of stimulants leads to insomnia, excitability, delusions, hallucinations, and psychosis. Since stimulants deplete physical and mental reserves, the heavy user often loses weight, becomes aggressively irritable, and sleeps very little. Outbursts of irrational, aggressive, and violent behavior can occur. Often death results from a combination of malnutrition and toxic exhaustion of the brain.

Narcotics are drugs that relieve pain and induce sleep. They include drugs made from opium, namely, morphine and codeine; heroin, a derivative of morphine and three times stronger than morphine; meperidine (Demerol); and methadone.

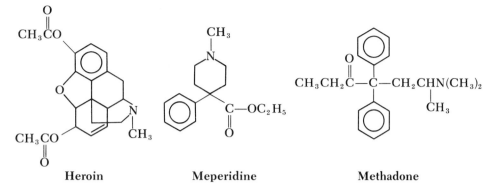

Heroin   Meperidine   Methadone

Heroin, which has no medicinal value, accounts for about 90% of the narcotic addiction problem in the United States. In 1972 there were from 200,000 to 400,000 heroin addicts in this country. Meperidine and morphine are prescribed as pain killers. Codeine is contained in many prescription cough remedies. The use of average amounts of narcotics leads to the short-term effects of relaxation, relief of pain and anxiety, decreased alertness, euphoria, and hallucination. Death can result from the use of large amounts. Continued abuse leads to lethargy, constipation, weight loss, temporary sterility and impotence, and withdrawal sickness. Withdrawl sickness may consist of shaking, heavy perspiration, abdominal pains, muscle aches, and violent jerks. Although the body develops a tolerance to the effects of narcotic drugs, users face a psychological and physical dependence on the drugs.

Summary

Organic compounds containing nitrogen include amines, diazonium salts, heterocyclic amines, nitro compounds, nitriles, and amides.

Amines are classified as primary, secondary, or tertiary, according to the number of groups attached to the nitrogen atom. The names of simple amines are formed by the names of alkyl groups attached to the nitrogen followed by the word -*amine*. In the names of more complicated amines, the prefix *amino*- or prefixes derived from it are used. The simplest aromatic amine is called aniline.

The methods of preparation of amines include reduction of nitro compounds and nitriles, reductive amination of aldehydes and ketones, Hofmann degradation of amides, and reduction of amides.

As organic bases, aliphatic amines are stronger bases than ammonia. Aromatic amines are the weakest bases, but their strength as bases can be increased by the substitution of electron-releasing groups for hydrogen in the *para* position.

The amino ($-NH_2$), alkyl amino ($-NHR$), and dialkylamino ($-NR_2$) groups are strong *ortho,para* directors.

Following is a summary of the reactions that amines undergo. So that electrophilic substitution can be illustrated, aromatic amines are used as the examples.

Reagent	With $C_6H_5NH_2$ (primary)	Product With C_6H_5NHR (secondary)	With $C_6H_5NR_2$ (tertiary)
HCl	$C_6H_5\overset{+}{N}H_3\ Cl^-$	$C_6H_5\overset{+}{N}H_2RCl^-$	$C_6H_5\overset{+}{N}HR_2\ Cl^-$
CH_3I (excess)	$C_6H_5\overset{+}{N}(CH_3)_3I^-$	$C_6H_5\overset{+}{N}R(CH_3)_2I^-$	$C_6H_5\overset{+}{N}R_2(CH_3)I^-$
R'COCl	C_6H_5NHCOR'	$C_6H_5N(R)COR'$	No reaction
$NaNO_2$, HCl	$C_6H_5N_2^+Cl^-$	$C_6H_5-N-N=O$ R	p-NOC$_6$H$_4$NR$_2$
Br_2	*Ortho* and *para* substitution	*Ortho* and *para* substitution	*Ortho* and *para* substitution
HNO_3, H_2SO_4	m-NO$_2$C$_6$H$_4$NH$_2$	m-NO$_2$C$_6$H$_4$NHR	m-NO$_2$C$_6$H$_4$NR$_2$
$C_6H_5SO_2Cl$	$C_6H_5NHSO_2C_6H_5$	$C_6H_5NSO_2C_6H_5$ R	No reaction

Heterocyclic amines contain a nitrogen atom as a member of the ring system. Many of these compounds occur naturally and possess biological and physiological activity. Pyrrole is a weaker base than pyridine. Both pyrrole and pyridine possess aromatic character.

Important terms

Amide
Azo group
Blocking group
Diazonium salt
Heterocyclic amine
Hinsberg test
Hofmann degradation
Nitrile
Nitro compound

Primary amine
Pyridine
Pyrrole
Quaternary ammonium salt
Sandmeyer reaction
Secondary amine
Sulfanilamide
Sulfonamide
Tertiary amine

Problems

11-8. Give a correct name for each of the following molecules:

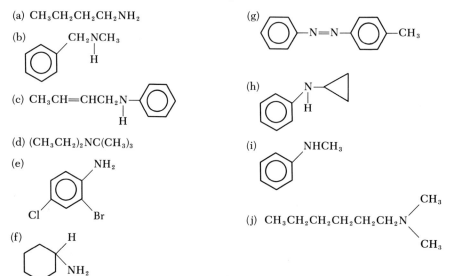

(a) $CH_3CH_2CH_2CH_2NH_2$

(b) (phenyl) CH_2NCH_3 | H

(c) $CH_3CH=CHCH_2N$—(phenyl) | H

(d) $(CH_3CH_2)_2NC(CH_3)_3$

(e) (ring with NH$_2$, Cl, Br)

(f) (cyclohexane ring with H and NH$_2$)

(g) (phenyl)—N=N—(phenyl)—CH$_3$

(h) (phenyl with N—cyclopropyl) | H

(i) (phenyl) NHCH$_3$

(j) $CH_3CH_2CH_2CH_2CH_2CH_2N$ with CH$_3$ and CH$_3$

11-9. Classify each of the amines in problem 11-8 as primary, secondary, or tertiary.

11-10. Draw the structure for each of the following compounds:

    (a) Ethylisopropylamine
    (b) 2-Nitrophenylmethylamine
    (c) 2-(N-*n*-Propylamino)hexane
    (d) *n*-Butyl-*sec*-butylamine
    (e) Cyclopentylamine

11-11. Using equations, give the product of the reaction of aniline with each of the following reagents:

    (a) Dilute H_2SO_4
    (b) Propionyl chloride (CH_3CH_2COCl)
    (c) Excess methyl iodide
    (d) Product of c with moist Ag_2O
    (e) 2-Pentanone ($CH_3COCH_2CH_2CH_3$); then H_2 and Ni
    (f) $NaNO_2 + HCl$

11-12. Give the product of the reaction of benzenediazonium chloride with each of the following reagents:

    (a) CuCl
    (b) CuCN
    (c) Product of b + H_2O/H^+
    (d) KI
    (e) H_3PO_2
    (f) H_2O
    (g) CuBr
    (h) HBF_4, heat
    (i) N,N-Dimethylaniline

11-13. Write the steps for each of the following syntheses:

    (a) *n*-Propylamine from ethyl bromide
    (b) *p*-Toluidine from benzene
    (c) *m*-Nitroaniline from benzoic acid
    (d) 1,2-Diaminoethane from ethylene
    (e) 1-Aminobutane from *n*-propyl alcohol
    (f) Tetramethylammonium bromide from methyl bromide
    (g) Di-*n*-butylamine from *n*-butanal
    (h) Fluorobenzene from benzene
    (i) *p*-Bromoaniline from aniline
    (j) *m*-Bromoaniline from nitrobenzene
    (k) Acetanilide from aniline
    (l) 2-Nitropyrrole from pyrrole

11-14. Write an equation for each of the following preparations:

    (a) *n*-Propylamine via reductive amination of aldehyde
    (b) Benzylamine via Hofmann degradation of amide
    (c) Iodobenzene via diazonium salt
    (d) *p*-Methoxyazobenzene via diazonium salt coupling
    (e) *p*-Nitroaniline via use of blocking group and electrophilic substitution
    (f) Coniine hydrochloride via acidification

11-15. Draw the structure for each of the following compounds:

    (a) Tertiary amine
    (b) Secondary aromatic amine
    (c) Benzenesulfonamide

(d) Heterocyclic amine containing six members in the ring
(e) Quaternary ammonium salt
(f) Diazonium salt
(g) Azo compound

11-16. Explain why each of the following statements is true:

(a) Cyclohexylamine is a stronger base than aniline.
(b) *p*-Methoxyaniline is a stronger base than aniline.

11-17. Give the product of each of the following reactions. Indicate if no reaction occurs.

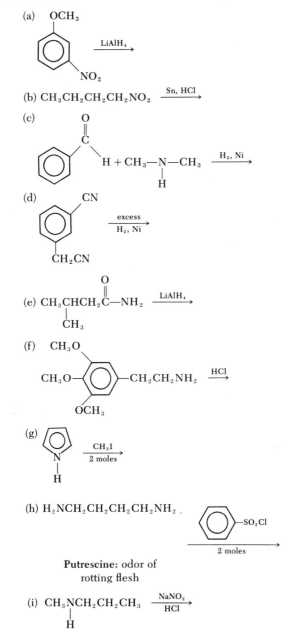

(a)

$$\xrightarrow{\text{LiAlH}_4}$$

(b) $CH_3CH_2CH_2CH_2NO_2 \xrightarrow{\text{Sn, HCl}}$

(c)

$$\xrightarrow{\text{H}_2, \text{Ni}}$$

(d)

$$\xrightarrow[\text{H}_2, \text{Ni}]{\text{excess}}$$

(e) $CH_3CHCH_2\overset{\overset{\displaystyle O}{\|}}{C}-NH_2 \xrightarrow{\text{LiAlH}_4}$

(f)

$$\xrightarrow{\text{HCl}}$$

(g)

$$\xrightarrow[\text{2 moles}]{\text{CH}_3\text{I}}$$

(h) $H_2NCH_2CH_2CH_2CH_2NH_2$.

$$\xrightarrow{\text{2 moles}}$$

Putrescine: odor of rotting flesh

(i) $CH_3NCH_2CH_2CH_3 \xrightarrow[\text{HCl}]{\text{NaNO}_2}$

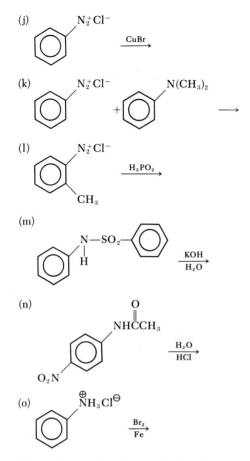

(j)

$N_2^+Cl^-$

$\xrightarrow{CuBr}$

(k)

$N_2^+Cl^-$ + $N(CH_3)_2$ $\longrightarrow$

(l)

$N_2^+Cl^-$

CH_3

$\xrightarrow{H_3PO_2}$

(m)

N—SO₂—

H

$\xrightarrow[H_2O]{KOH}$

(n)

O
‖
NHCCH₃

O_2N

$\xrightarrow[HCl]{H_2O}$

(o)

$\overset{\oplus}{N}H_3 Cl^{\ominus}$

$\xrightarrow[Fe]{Br_2}$

11-18. When adipic acid and hexamethylenediamine are heated together, a polyamide is formed, This polyamide is commercially known as Nylon 66, since the dicarboxylic acid and the diamine each contain six carbon atoms. Write the structure for a portion of this polyamide.

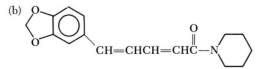

$$HO-\overset{\overset{O}{\|}}{C}-(CH_2)_4-\overset{\overset{O}{\|}}{C}-OH \ + \ H_2N-(CH_2)_6-NH_2 \ \xrightarrow{heat} \ H_2O \ + \ ?$$

Adipic acid   **Hexamethylenediamine**

11-19. Identify the heterocyclic amine system found in each of the following molecules:

(a) $CONHNH_2$

N

Isoniazid: used in treatment of tuberculosis

(b)   O

O

O

$CH{=}CHCH{=}CHC-N$

Piperine: Main constituent of black pepper

(c)

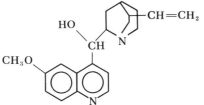

Quinine: found in bark of cinchona tree and used as antimalarial drug

(d)

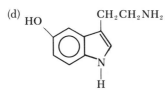

Serotonin: found in brain and functions by establishing stable pattern of mental activity

(e)

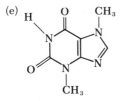

Theobromine: constituent of cocoa bean

11-20. An amine of the formula C_2H_7N gives a benzenesulfonamide that is insoluble in aqueous KOH. Draw a possible structure for the amine.

11-21. An amine of the formula C_7H_9N gives a benzenesulfonamide that is soluble in aqueous KOH but insoluble in aqueous HCl. Draw a possible structure for the amine.

11-22. A commercially available blond dye contains p-phenylendiamine (0.3%), p-N-methyl-aminophenol (0.5%), p-aminodiphenylamine (0.15%), o-aminophenol (0.15%), pyro-cathechol (0.25%), resorcinol (0.25%), and inert solvent (98.40%). Draw the structures of the compounds containing amino groups and classify each one as a primary, secondary, or tertiary amine.

Self-test

1. Give a correct name for each of the following molecules:

   (a) $(CH_3CH_2)_2NH$

   (b)

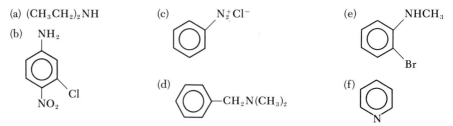

2. Draw the complete structure for each of the following compounds:

   (a) o-Nitroaniline   (d) Diisopropylamine
   (b) N-Ethylaniline   (e) 2-(N,N-Dimethylamino)pentane
   (c) Pyrrole   (f) Sulfanilamide

3. Give the structure of the organic product of each of the following reactions:

   (a) $CH_3CH_2NH_2 + HCHO$; then H_2, Ni $\longrightarrow$

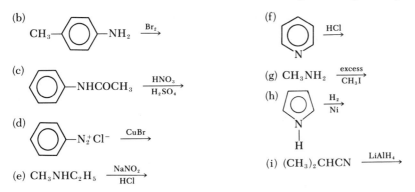

(e) CH$_3$NHC$_2$H$_5$ $\xrightarrow[\text{HCl}]{\text{NaNO}_2}$

(i) (CH$_3$)$_2$CHCN $\xrightarrow{\text{LiAlH}_4}$

4. Give the reagent and/or condition that will cause each step to occur as written in the following transformation:

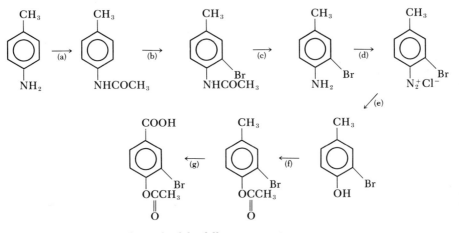

5. Give the correct answer for each of the following questions:

(a) What type of hybrid orbital does the unshared electron pair of nitrogen in an amine occupy?
(b) What class of organic compounds is obtained from the reduction of aromatic nitro compounds?
(c) If a carbonyl compound is treated with a secondary amine in reductive amination, what class of amine is formed?
(d) Which is the stronger nitrogen base: p-toluidine or aniline?
(e) If the reaction between an amine and sodium nitrite and HCl yields a yellow oil, what is the class of amine undergoing the reaction?
(f) Which reagent will convert benzenediazonium chloride to benzene: water or H$_3$PO$_2$?
(g) Sulfonamides are a class of compounds generally formed from the reaction of a primary or secondary amine with an acid. What is the name of the acid?
(h) Coniine is the component of hemlock. What is the name of the heterocyclic ring system found in coniine?
(i) Which is the stronger base: pyrrole or pyridine?
(j) What is the structure of the acylated amino blocking group?

Answers to self-test

1. (a) Diethylamine
   (b) 3-Chloro-4-nitroaniline
   (c) Benzenediazonium chloride
   (d) Benzyldimethylamine
   (e) o-Bromo-N-methylaniline or 2-bromo-N-methylaniline
   (f) Pyridine

2. (a)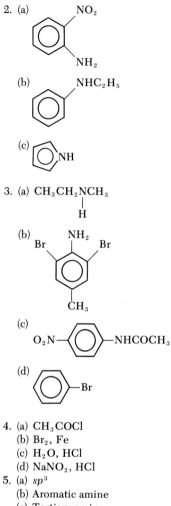

(b)

(c)

3. (a) $CH_3CH_2NCH_3$
$\quad\quad\quad\quad\quad |$
$\quad\quad\quad\quad\quad H$

(b)

(c)

(d)

4. (a) CH_3COCl
(b) Br_2, Fe
(c) H_2O, HCl
(d) $NaNO_2$, HCl

5. (a) sp^3
(b) Aromatic amine
(c) Tertiary amine
(d) p-Toluidine

$$\left(CH_3-\!\!\!\bigcirc\!\!\!-NH_2 \right)$$

(d)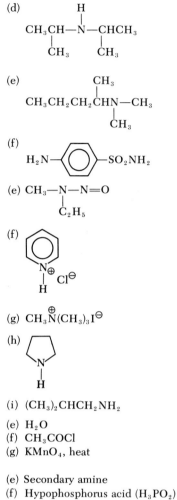

$$CH_3CH-N-CHCH_3$$
$$\quad\;\; |\quad\quad\;\; |$$
$$\quad\; CH_3\quad\; CH_3$$

(e)
$$\quad\quad\quad\quad\quad CH_3$$
$$CH_3CH_2CH_2CHN-CH_3$$
$$\quad\quad\quad\quad\quad\quad |$$
$$\quad\quad\quad\quad\quad\quad CH_3$$

(f)
$$H_2N-\!\!\!\bigcirc\!\!\!-SO_2NH_2$$

(e) $CH_3-N-N{=}O$
$\quad\quad\quad\quad |$
$\quad\quad\quad\; C_2H_5$

(f)

(g) $CH_3\overset{\oplus}{N}(CH_3)_3 I^\ominus$

(h)

(i) $(CH_3)_2CHCH_2NH_2$

(e) H_2O
(f) CH_3COCl
(g) $KMnO_4$, heat

(e) Secondary amine
(f) Hypophosphorus acid (H_3PO_2)
(g) Benzenesulfonic acid
(h) Piperidine or reduced pyridine
(i) Pyridine
(j) CH_3C-
$\quad\quad\;\; \|$
$\quad\quad\;\; O$

chapter 12

Identification of structure by classical and modern techniques

Organic chemists are continually preparing new compounds and isolating compounds that have existed for many years from plant and animal sources. Not only are the preparation and isolation of these compounds of scientific interest, but also many of these compounds are used to serve mankind in materials such as medicines, paints, varnishes, plastics, detergents, foods, and insecticides.

One of the first things an organic chemist does with a new compound is to determine its structure. This may involve simply determination of what functional groups it contains, or it may involve determination of the complete structure. The *structure of a molecule* is the order in which atoms of a molecule are bonded to each other and the nature of the bonds holding these atoms together.

A complete structure identification involves first determination of the elements present. The elements found in organic compounds, in addition to carbon and hydrogen, are oxygen, halogens, nitrogen, and sulfur.

Then the empirical formula is determined. The *empirical formula* gives the atomic ratios of the elements present. The molecular weight of the compound is then determined from the mass spectrum. Once the molecular weight is known, the molecular formula can be determined. The *molecular formula* gives the exact numbers and kinds of atoms that make up the molecule.

The next step is determination of the functional groups present in the compound. This tells the chemist to what family (such as alkene, ketone, or carboxylic acid) the compound belongs. If the unknown compound were a new compound, most likely some chemical reaction would be used to determine what functional groups are present. For example, if the unknown compound decolorized bromine, some type of unsaturation such as that found in alkenes or alkynes would be suspected of being present. Many of these simple chemical tests for functional groups are given in Table 12-1. In addition, spectral methods are used to determine the functional groups present in the unknown compound. The spectra most useful to the organic chemist are the infrared and the nuclear magnetic resonance.

The final step involves piecing together all the information obtained by these methods to give the total structure of the compound. If the unknown compound is suspected of being identical to a known compound, it is converted into another compound called a *derivative*. It is then determined whether or not the derivative is the same as that derived in the same chemical reaction with the known compound. (The technique of derivative preparation is described in Chapter 10.) Classically, the method of derivative preparation is used to confirm that a compound is identical to a known compound. Sometimes, particularly if the compound is a new compound, the assigned structure is confirmed by an independent synthesis of the compound so that there is no doubt about its structure.

Thus a complete structure determination involves the following steps:
1. Detection of elements present
2. Determination of molecular weight and molecular formula
3. Characterization of compound by chemical tests and spectra
4. Derivative preparation for confirmation if compound is already known or independent synthesis if it is new

The first three points are discussed in some detail in this chapter. Derivative preparation is discussed in Chapter 10. An independent synthesis is described in the application at the end of this chapter.

Detection of elements

The elements most commonly found in organic compounds, in addition to carbon and hydrogen, are oxygen, the halogens, nitrogen, and sulfur. In this study of organic compounds, the presence of carbon and hydrogen is assumed. The presence of hydrogen can be readily confirmed by nuclear magnetic resonance spectroscopy.

The detection of oxygen is commonly done by infrared spectroscopy. The detection of the halogens, nitrogen, and sulfur in an organic compound involves a sodium fusion reaction in which the organic compound is treated with hot molten sodium metal. The covalently bonded halogen (X), nitrogen, and sulfur are converted into halide, cyanide, and sulfide anions, respectively:

$$\text{organic compound} \atop \text{containing C, H, O, N, S, X} \quad \xrightarrow[\text{heat}]{\text{Na}} \quad CO_2 \;+\; H_2O \;+\; CN^- \;+\; S^{2-} \;+\; X^-$$

$$\textbf{Cyanide} \quad \textbf{Sulfide} \quad \textbf{Halide}$$

Halogen is identified by precipitation with silver nitrate as the insoluble silver halide (AgX):

$$X^- + AgNO_3 \longrightarrow AgX \downarrow + NO_3^-$$

$$\textbf{Silver} \atop \textbf{nitrate}$$

If the test is positive, tests for the specific halogens (which will not be described here) must then be performed. Nitrogen, which is converted to cyanide in the fusion process, is indicated by a blue cyano complex formed from ferrous ammonium sulfate:

$$CN^- \quad + \quad Fe(NH_4)_2(SO_4)_2 \quad \longrightarrow \quad Fe(CN)_6{}^{4-}$$

Ferrous ammonium sulfate	Cyano complex (blue)

Sulfur is detected as a dark brown precipitate of lead sulfide when the solution is treated with lead acetate:

$$S^{2-} \quad + \quad Pb(C_2H_3O_2)_2 \quad \longrightarrow \quad 2(C_2H_3O_2)^- \quad + \quad PbS \downarrow$$

Lead acetate	Lead sulfide (brown)

Empirical and molecular formulas

Quantitative elemental analysis is the next step in establishing the structure of a compound. The most common analyses are those for carbon, hydrogen, and nitrogen. The amount of oxygen is determined as the difference. The compound is completely combusted to yield carbon dioxide, water, and nitrogen. Classically, the water formed is adsorbed onto a drying agent such as calcium chloride or magnesium perchlorate, the carbon dioxide is adsorbed onto a strong base such as sodium hydroxide, and the nitrogen is determined by volume. In the more modern technique of gas chromatography, the combusted sample is recorded as a chromatogram with three peaks (water, carbon dioxide, and nitrogen). The area under each peak is a measure of the amount of each of the three components. The gas chromatographic method is very accurate and requires about one-fifth the amount of sample as the classical method.

Once the amount (in percentage) of each element is obtained, the empirical formula can be calculated. The empirical formula, which is the simplest formula for that compound, shows the relative number of each kind of atom in the molecule.

Let us consider the analysis of a sample of benzaldehyde as an example of determination of the empirical formula. The gas chromatographic method of quantitative elemental analysis shows that the sample of benzaldehyde contains 79.2% carbon, 5.7% hydrogen, and 15.1% oxygen. To calculate the empirical formula for benzaldehyde, we will assume that in 100 g of the sample, there are 79.2 g of carbon, 5.7 g of hydrogen, and 15.1 g of oxygen. Then the number of gram atoms of each element is obtained by division of each individual weight by the respective atomic weight:

$$\text{Gram-atoms of C:} \quad \frac{79.2}{12} = 6.6$$

$$\text{Gram-atoms of H:} \quad \frac{5.7}{1} = 5.7$$

$$\text{Gram-atoms of O:} \quad \frac{15.1}{16} = 0.94$$

Next we convert these amounts to smallest whole numbers:

$$C = \frac{6.6}{0.94} = 7 \qquad H = \frac{5.7}{0.94} = 6 \qquad O = \frac{0.94}{0.94} = 1$$

Thus the empirical formula for benzaldehyde is C_7H_6O.

Before the molecular formula can be determined, the molecular weight must be known. The *mass spectrum* of the compound gives the exact value of the molecular weight. Basically, the function of a mass spectrometer is the production of ions from a sample by electron bombardment and the separation of the various ions according to their mass-to-charge ratios (m/e). The mass spectrum is a record of the masses and abundances of the various ions produced from a sample. The heaviest ion in the mass spectrum, which is formed by the loss of a single electron from the molecule, is referred to as the molecular ion. The other ions, which are formed by cleavage of bonds in the molecular ion, are called fragment ions. Fragment ions have lower masses than molecular ions and are important to the organic chemist because they can be used as building blocks to reconstruct the molecular structure. The molecular ion corresponds to the molecular formula and thus the molecular weight of the compound. The molecular ion peak generally has the highest mass of any peak in the spectrum. For example, the mass spectrum of benzaldehyde is shown in Fig. 12-1. The mass spectrum indicates that the molecular weight of benzaldehyde is 106. Thus the molecular formula of benzaldehyde is exactly the same as the empirical formula: C_7H_6O.

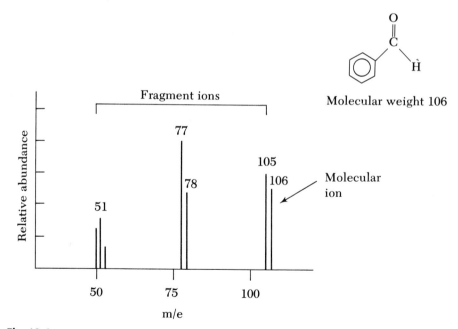

Fig. 12-1

Mass spectrum of benzaldehyde. Molecular ion appears at m/e value of 106. Molecular weight is 106.

12-1. The empirical formula for butyric acid is C_2H_4O, and its mass spectrum is shown as follows. What is the molecular formula of the compound?

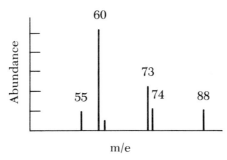

Chemical tests for functional groups

Occasionally, an unknown organic compound can be classified according to its solubility or insolubility in water (H_2O), dilute hydrochloric acid (HCl), dilute sodium hydroxide (NaOH), sodium bicarbonate ($NaHCO_3$), or concentrated sulfuric acid (H_2SO_4). Following are some general solubility rules:

1. Soluble in H_2O: alcohols, aldehydes, acids, amines, amides, esters, and ketones containing five or less carbons
2. Soluble in NaOH and $NaHCO_3$: acids containing more than five carbons
3. Soluble in NaOH; insoluble in $NaHCO_3$: phenols
4. Soluble in HCl: aliphatic amines containing more than five carbons (aromatic amines are not soluble in HCl)
5. Insoluble in H_2O; soluble in H_2SO_4: alkenes, alkynes, ethers and alcohols, aldehydes, ketones, and esters
6. Insoluble in H_2O, NaOH, HCl, and H_2SO_4: hydrocarbons and halogen-containing compounds

Functional groups can also be identified by a number of chemical reactions. Many of these classification reactions are discussed in other chapters, and they are summarized in Table 12-1.

12-2. Using the solubility rules and the functional group tests, identify the class of compound that fits each of the following descriptions:

    (a) Soluble in H_2O and dilute HCl   *Answer:* Aliphatic amine (five or less carbons)
    (b) Soluble in NaOH but not $NaHCO_3$ *Answer:* Phenols
    (c) Soluble in H_2SO_4 and decolorizes Br_2 in CCl_4
    (d) Soluble in H_2O and gives yellow precipitate of iodoform
    (e) Gives brown precipitate with $KMnO_4$ and white precipitate with ammonia solution of $AgNO_3$

In many cases, the presence or absence of functional groups is determined spectroscopically.

Table 12-1

Functional group classification tests

	Reagent	Observation	Indication
Bromine:	Br_2	Decolorization of bromine solution; no gas liberated	Unsaturation (C=C, C≡C)
Ammoniacal silver nitrate:	$Ag(NH_3)_2NO_3$	(a) White precipitate (b) Silver mirror	(a) Terminal C≡C (b) Aldehyde
Sodium hydroxide and iodine:	$NaOH + I_2$	Yellow precipitate	Methyl ketone ($-\overset{\displaystyle O}{\overset{\|}{C}}-CH_3$) or secondary alcohol of type $RCHOHCH_3$
Potassium permanganate:	$KMnO_4$ (cold, alkaline)	Brown precipitate (MnO_2)	Unsaturation (C=C, C≡C)
Sodium metal:	Na	Evolution of hydrogen gas	Carboxylic acid (RCOOH), phenol (ArOH), or alcohol (ROH)
Sodium hydroxide:	NaOH	Evolution of hydrogen gas	Carboxylic acid, phenol
Sodium bicarbonate:	$NaHCO_3$	Evolution of hydrogen gas, CO_2	Carboxylic acid
Dilute hydrochloric acid:	HCl	Solubility	Aliphatic amine (RNH_2)
Chloroform and aluminum chloride:	$CHCl_3$, $AlCl_3$	Change of color from orange to red	Aromatic hydrocarbon (ArR)
Chromium trioxide in dilute sulfuric acid:	CrO_3, H_2SO_4	Change of color from orange to blue-green	Primary or secondary alcohol (ROH)
2,4-Dinitrophenyl-hydrazine:	$O_2N-\langle\bigcirc\rangle\overset{\displaystyle NO_2}{-}NHNH_2$	Formation of precipitate, color from yellow to orange	Aldehyde (RCHO) or ketone (R_2CO)
Ferric chloride:	$FeCl_3$	Change of color of solution (green to blue, violet to red)	Phenol
Benzenesulfonyl chloride:	$\langle\bigcirc\rangle-SO_2Cl$	White precipitate	Primary or secondary amine
Acetyl chloride	$CH_3\overset{\|}{\underset{\displaystyle O}{C}}-Cl$	Sweet fruity odor	Primary, secondary, or tertiary alcohol

Physical methods of identification

When molecules interact with light waves, certain waves are absorbed. The absorption of light (or electromagnetic radiation as it is sometimes called) by a molecule provides specific information about the structure of the molecule.

Normally, molecules exist in the lowest energy (most stable) state, called the *ground state*. When a molecule absorbs light, its energy is raised to a higher energy (less stable) state, called an *excited state*. The difference in energy, ΔE, between the ground state and the excited state is equal to the amount of light absorbed, which is equal to the product of Planck's constant, h, and the frequency of the wave, ν. Thus:

$$E_{\text{absorbed}} = \Delta E = h\nu$$

The transition from one energy level to another is indicated by a vertical arrow, as shown in the following diagram:

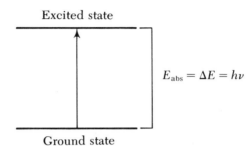

However, molecules are selective in absorbing light. Only certain levels can be attained, and not all light has the proper energy to cause a molecule to change from its ground state to an excited state. When a molecule absorbs light with energy equal to the difference between two energy levels, that is, when $\Delta E = h\nu$, an absorption spectrum is produced. Following is an example:

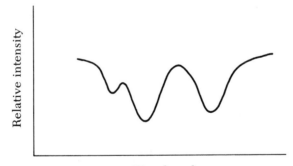

The instruments commonly used by the organic chemist can detect electromagnetic radiation of wavelengths ranging from 200 nanometers (nm) to 15 microns (μ). (A nanometer is one thousandth of a micron.) The regions of greatest interest to the organic chemist are the ultraviolet (UV) (200 to 400 nm), visible (400 to 800 nm), and infrared (IR) (2μ to 15μ) regions. In addition, the region between 10^{10} and 10^{11} nm, called the nuclear magnetic resonance (NMR) region, is of great importance. The relationships between these regions in the electromagnetic spectrum are shown in Fig. 12-2.

The mechanisms of absorption of energy are different in the UV, IR, and NMR regions. However, the fundamental process is the same: certain discrete amounts of energy are absorbed. Molecules possess electronic, vibrational, and rotational energy levels. The relationships between these energy levels in a molecule are shown in Fig. 12-3.

Absorption of light in the UV region is associated with transitions between electronic energy levels. Such transitions usually result in the promotion of electrons in a stable orbital (say e_0) to a higher-energy electronic level (say e_1). The energy associated with transitions between electronic levels corresponds to the absorption of light of wavelengths of 400 to 1000 nm.

Absorption of energy in the IR region is associated with transitions between vibrational and rotational energy levels. There are two reasons for

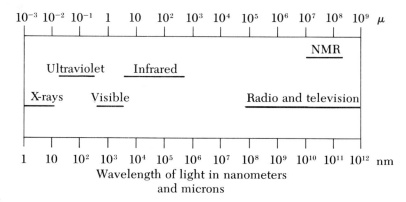

Fig. 12-2
Portion of electromagnetic spectrum. Energy increases from right to left.

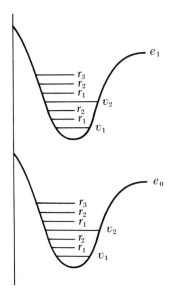

Fig. 12-3
Electronic (e), vibrational (v), and rotational (r) energy levels in a molecule.

such transitions. First, molecules are flexible, and the bond between two atoms can increase, that is, vibrate along its axis (stretch). This is commonly referred to as a change in dipole moment:

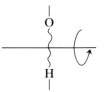

Vibrating bond

Second, the molecule can rotate about an axis through its center of gravity:

Transitions between rotational levels require less energy than those between vibrational levels.

Absorption of radio waves (another portion of the electromagnetic spectrum) in the NMR region is associated with the magnetic properties of certain nuclei when they are placed in a magnetic environment. Transitions between energy states in this case require only a small amount of energy, less than that required for either UV or IR absorptions.

Ultraviolet and visible spectroscopy

Normally, the UV-visible portion of the electromagnetic spectrum is considered to extend from 200 to 750 nm. The energy associated with the radiation in this region of the spectrum is such that the radiation causes transitions between electronic energy levels when it interacts with certain types of molecules. A typical spectrum is shown in Fig. 12-4.

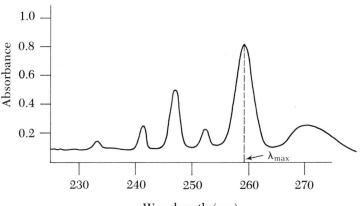

Fig. 12-4
UV spectrum of benzene in ethanol. λ_{max} is the wavelength of maximum absorbance for benzene.

Most frequently, the types of organic molecules that absorb or interact with this type of radiation are those that possess π bonds. For example, aromatic hydrocarbons, conjugated olefins, carbonyl compounds, and other compounds possessing double bonds interact with UV and visible radiations.

The type of information obtainable from a UV-visible spectrum is generally limited as far as identification of specific organic molecules is concerned. This is because the absorption bands are generally broad and similar compounds may have spectra that are so similar that they cannot be distinguished and because if the molecule contains no double bond or functional groups that interact with this type of radiation, there may be no spectrum at all. The greatest use of this region of the spectrum is made in the quantitative estimation of compounds that do absorb radiation in this region.

Today it is possible for all colleges and universities and even some high schools to have instruments that are very good for work in this protion of the spectrum. The instruments range from those in which the operator must pick out the specific wavelength that is going to be used to those that are completely automated and give a spectrum of the wavelength versus absorption over any region desired.

Infrared spectroscopy

IR radiation encompasses the region from 2μ to 15μ in the electromagnetic spectrum. However, IR spectroscopy is sometimes used in the range of 2μ to 50μ. The radiation in this portion of the spectrum is sensed as heat. It is of the right energy to interact with molecules to cause changes in the vibrational energy levels in the bonds holding two atoms together provided there is a change in the dipole moment. IR spectroscopy is the most generally useful and widely applied of the three spectroscopic methods discussed in this chapter. The instrumentation is within the reach of most small schools and universities. It is also useful because not only does each organic molecule, regardless of the functional group present, have its own unique IR spectrum, which makes its identification absolute, but also each functional group is observed in a different region of the spectrum. If the original spec-

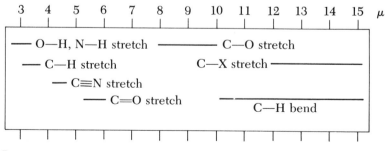

Fig. 12-5
IR absorption bands.

trum of the compound is not known, much data can still be obtained, since the functional groups present can still be identified. Generally, the spectrum of a compound is obtained and then examined for the presence of certain absorption bands, which are matched with known bands. Typical absorption bands are shown in Fig. 12-5 and tabulated as follows:

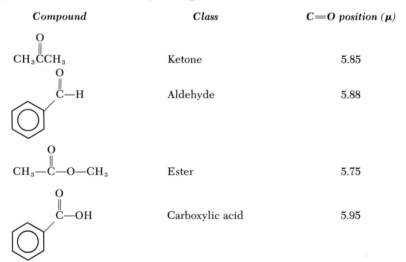

Group	*Region (μ)*
O—H, N—H	2.7-3.3
C—H (—C≡C—H, C=C, ⬡—H)	3.0-3.4
C—H (—C—H)	3.3-3.7
C≡C, C≡N	4.2-4.9
C=O (acids, aldehydes, anhydrides, ketones, amides, esters)	5.3-6.1
C=C (aliphatic, aromatic)	5.9-6.2
C—H (aromatic)	10.0-15.4

It is interesting that regardless of the rest of the molecule, a functional group present in two different compounds shows up in practically the same place in the spectra. For example, a carbonyl group, whether in a ketone, aldehyde, ester, or carboxylic acid, shows up in very much the same position even though the functional groups are quite different. The positions of absorption of the four carbonyl compounds are as follows:

Compound	*Class*	*C=O position (μ)*
CH_3CCH_3 (ketone, C=O)	Ketone	5.85
⬡—C—H (C=O)	Aldehyde	5.88
CH_3—C—O—CH_3 (C=O)	Ester	5.75
⬡—C—OH (C=O)	Carboxylic acid	5.95

It is easy for a spectroscopist to distinguish which of the carbonyl-containing compounds is present by examining the rest of the spectrum. For example, let us consider the IR spectrum for a carbonyl compound in Fig. 12-6. The spectrum shows the presence of a carbonyl group at about 5.9μ. We can eliminate acetone (CH_3COCH_3) and methyl acetate ($CH_3CO_2CH_3$), since neither of these compounds is aromatic and the spectrum definitely shows aromatic absorption between 10μ to 15μ. Therefore, the spectrum is that of either benzaldehyde (C_6H_5CHO) or benzoic acid (C_6H_5COOH). Benzoic acid has an oxygen–hydrogen (O–H) grouping, and the spectrum lacks any absorption in the 2.7μ to 3.3μ region (O–H region). Thus the spectrum must be that of benzaldehyde. This type of reasoning can be used to distinguish several compounds that contain the same functional group.

12-3. (a) Is the IR spectrum shown as follows that of benzyl alcohol ($C_6H_5CH_2OH$) or benzyl methyl ether ($C_6H_5CH_2OCH_3$)? Explain your choice.

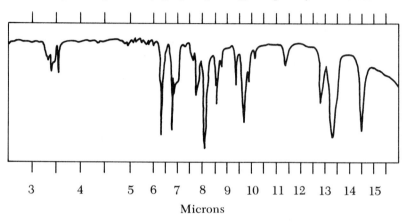

Microns

(b) Is the IR spectrum shown as follows that of benzophenone, benzonitrile (C_6H_5CN), or benzoic acid? Explain you choice.

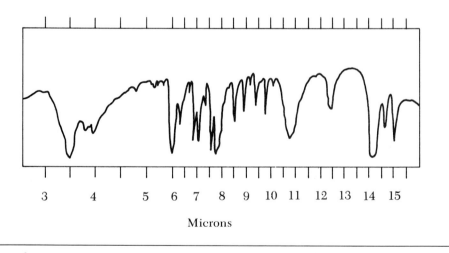

Microns

Nuclear magnetic resonance spectroscopy

In theory, NMR spectroscopy is the most complicated of all the spectroscopic methods. It is based on the fact that nuclei have associated with them spin quantum numbers that give them a magnetic moment and, in essence, make them act as if they were small magnets. Some nuclei possess spin numbers of 0; therefore, they do not act as magnets and are unaffected

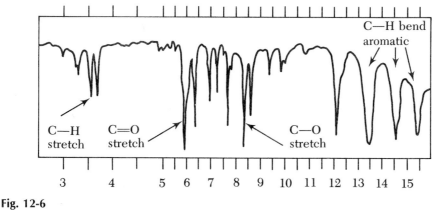

Fig. 12-6
IR spectrum of a carbonyl compound.

by a magnetic field. It is fortunate for the organic chemist that the nuclei of such common elements as ^{12}C, ^{16}O, and ^{32}S have spin quantum numbers of 0. If this were not the case, the spectra would be so complex that their usefulness would be limited. The nucleus most frequently investigated by NMR spectroscopy is the proton, although the nuclei of other elements such as ^{19}F and ^{31}P are often investigated.

The spin number of a proton is $\frac{1}{2}$; thus under the influence of a magnetic field the proton can align itself with the applied magnetic field ($\uparrow$) (the low-energy possibility) or against the applied field ($\downarrow$) (the high-energy possibility). This is illustrated as follows:

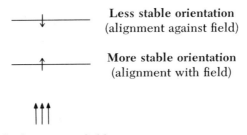

The energy required to make low-energy protons absorb energy and flip to become aligned against the field is in the region of radio frequency of about 60 megacycles (Mc). When the field strength is such that protons are caused to flip from the more stable alignment with the field to the less stable alignment against the field, energy is absorbed and a signal is recorded. If all protons absorbed energy at the same radio frequency with the same applied magnetic field, then all organic compounds would simply give one line for the absorption of the protons in the molecule. Fortunately, the chemical environments of the individual protons affect their magnetic environments; therefore, different protons in the same molecule absorb energy at different regions of the spectrum. In practice, the radio frequency is held constant,

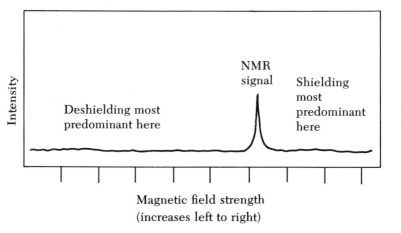

Intensity

Deshielding most predominant here

NMR signal

Shielding most predominant here

Fig. 12-7
NMR signal.

Magnetic field strength
(increases left to right)

and the magnetic field is varied over a small region with a value of about 14,042 gauss. The protons with the highest electron density about them (for example, those in saturated compounds) require the greatest magnetic field to make them absorb energy. The end of the spectrum, at which the absorption signal of this type of proton is found, is the upfield portion of the spectrum, and the protons are shielded. As the electron density about the protons decreases, the protons become deshielded, and the position of the absorption signal moves downfield. Deshielding can arise because of the electronegativity of neighboring atoms, the type of bonding between atoms, or the resonance effects within the molecule. A typical NMR signal is shown in Fig. 12-7. For routine work an NMR costs $10,000 to $35,000.

Three important pieces of information can be obtained from an NMR spectrum. The first is the positions of the peaks in the spectrum, which give the organic chemist some insight into the type of environment about the protons giving rise to the absorptions. This information is similar to the type of information gained from IR spectroscopy. However, it is limited in that the only signal seen is due to protons, so that the presence of other functional groups, such as carbonyl groups, can be inferred but cannot be positively confirmed.

Protons within a given molecule that absorb energy at the same field strength are *magnetically equivalent protons*. Each group of magnetically equivalent protons gives rise to a signal. For example, in methane all four protons are chemically and thus magnetically equivalent protons. Thus there is one signal for the four protons in methane. Following are some other examples:

CH_3Cl: all three protons equivalent; thus one NMR signal

CH_3CH_2OH: three sets of magnetically equivalent protons (CH_3—, —CH_2—, and OH); thus three NMR signals

H_3C—⬡—CH_3: two sets of magnetically equivalent protons (CH_3—, and all four ring protons); thus two NMR signals

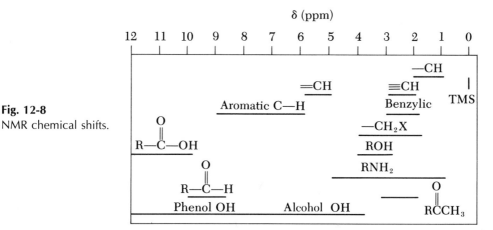

Fig. 12-8
NMR chemical shifts.

The exact positions of these signals, or *chemical shifts* as they are commonly called, are measured with a standard compound as a reference point. The compound used is *tetramethylsilane* (TMS):

$$\begin{array}{c} CH_3 \\ | \\ CH_3-Si-CH_3 \\ | \\ CH_3 \end{array}$$

In most cases, the chemical shifts for protons are downfield from TMS. The most commonly used scale is the *delta (δ) scale*, in which the position of the TMS signal is assumed to be 0 parts per million (ppm). The chemical shifts of most protons have δ values between 0 and 10 ppm. The chemical shifts of various protons relative to TMS are shown in Fig. 12-8 and tabulated as follows:

Type of proton	Chemical shift (ppm)
TMS	0
Aliphatic: —C—H	1-2
Olefinic: =C—H	5-6
Acetylenic: ≡C—H	2-3
Aromatic:	6-9
Benzylic: ⬡—CH₂—	2-3
Halide: XCH₂—	2-4
Alcoholic: ROH	3-4
Acid: RCOOH	10-12
Aldehyde: RCHO	9-10
Amino: NH₂	1-5
Phenolic: ⬡—OH	4-12
Keto: R—C—CH₃ (C=O)	2-3

The second piece of information, which is the most complicated for the chemist to interpret, is the exact number of hydrogens immediately adjacent to the proton giving rise to a particular signal. In practice, the signal from a particular proton is not always a single peak. Frequently, the signal from a proton or equivalent protons is a multiplet ranging from two to seven or possibly more peaks. The term applied to the phenomenon that brings about this splitting is *spin multiplicity*. It has been observed that if a certain proton has one proton as a neighbor, its signal appears as two peaks; if it has two equivalent protons as neighbors, its signal appears as three peaks; if it has three neighbors, its signal appears as four peaks; and so on. This observation is generalized as the $(n + 1)$ *rule*, which states that a signal from a proton or group of equivalent protons appears as $(n + 1)$ peaks where n is the number of equivalent hydrogens on adjacent carbons. The following diagram shows the distinction between equivalent and adjacent protons:

Adjacent protons

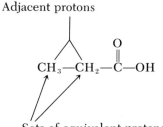

Sets of equivalent protons

The explanation for the $(n + 1)$ rule is somewhat involved and is beyond the scope of this text. The examples in Table 12-2 illustrate its usefulness. In general, protons bonded to either oxygen or nitrogen do not cause splitting patterns in adjacent protons.

12-4. What splitting pattern would you expect for each of the underlined protons?

(a) $CH_3CH_2OCH_2CH_3$ *Answer:* Triplet: adjacent carbon bears two protons; $2 + 1 = 3$.

(b) $\bigcirc$—OCH_3 *Answer:* Singlet: no adjacent carbon.

(c) $(CH_3)_2CHCl$

(d) $C_6H_5CH_2CH_3$

(e) $CH_3CH_2CCH_3$
              $\overset{\|}{O}$

The third and perhaps most important piece of information obtainable from NMR spectroscopy is the number of each type of proton that appears in a spectrum. Only one nucleus is being observed, that of hydrogen, and its sensitivity to detection is the same regardless of the environment. Thus two protons give twice the amount of signal of one proton, and three protons give

Table 12-2
Splitting patterns in some organic molecules

Compound	Structure and number of signals	Splitting pattern
Acetone	$\underset{\text{a}\qquad\text{a}}{CH_3\overset{\overset{\textstyle O}{\|}}{C}CH_3}$	None (singlet)
Propionic acid	$\underset{\text{a}\quad\text{b}\quad\text{c}}{CH_3CH_2\overset{\overset{\textstyle O}{\|}}{C}OH}$	(a) Triplet (2 adjacent + 1) (b) Quartet (3 adjacent + 1) (c) None (singlet)
Acetaldehyde	$\underset{\text{a}\quad\text{b}}{CH_3\overset{\overset{\textstyle O}{\|}}{C}{-}H}$	(a) Doublet (1 adjacent + 1) (b) Quartet (3 adjacent + 1)
p-Xylene	$CH_3\text{—}\underset{\text{b}}{\underset{\bigcirc}{\hexagon}}\text{—}CH_3$ a a	(a) Singlet (b) Singlet (adjacent protons equivalent)
Ethyl alcohol	$\underset{\text{a}\quad\text{b}\quad\text{c}}{CH_3CH_2OH}$	(a) Triplet (b) Quartet (c) Singlet

Table 12-3
Peak areas in some organic molecules

Compound	Structure and number of signals	Expected peak area
Propionic acid	$\underset{\text{a}\quad\text{b}\qquad\text{c}}{CH_3CH_2\underset{\underset{\textstyle O}{\|}}{C}{-}OH}$	a:b:c = 3:2:1
Acetaldehyde	$\underset{\text{a}\qquad\text{b}}{CH_3\underset{\underset{\textstyle O}{\|}}{C}{-}H}$	a:b = 3:1
p-Xylene	$CH_3\text{—}\underset{\text{b}}{\underset{\bigcirc}{\hexagon}}\text{—}CH_3$ a a	a:b = 6:4
Ethyl alcohol	$\underset{\text{a}\quad\text{b}\quad\text{c}}{CH_3CH_2OH}$	a:b:c = 3:2:1

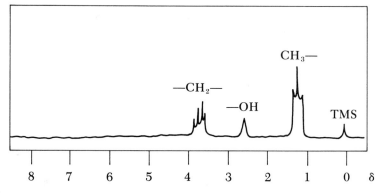

Fig. 12-9
NMR spectrum of ethyl alcohol (CH_3CH_2OH).

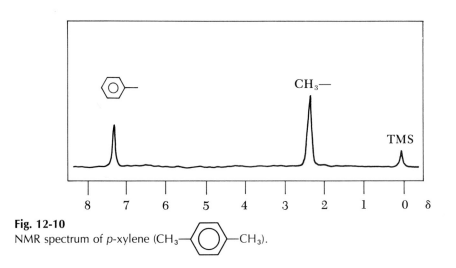

Fig. 12-10
NMR spectrum of *p*-xylene (CH_3—⬡—CH_3).

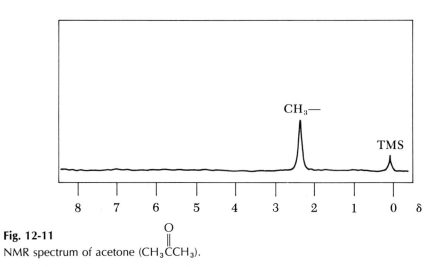

Fig. 12-11
NMR spectrum of acetone ($CH_3\overset{\overset{\displaystyle O}{\|}}{C}CH_3$).

three times the amount of signal of one proton. After the initial spectrum is obtained, the instrument can be set up to give a linear output or integration of the areas under the peaks. This linear output is then measured in an arbitrary fashion, and the relative heights of the different portions of the integration are related to one another. Thus the number of protons giving rise to each absorption can be calculated. Table 12-3 shows the expected peak areas in some representative organic molecules.

12-5. Predict the number of signals and the expected peak area for each proton in each of the following molecules:

(a) Acetic acid (CH_3COOH)   *Answer:* Two signals in CH_3COOH with
                                               a   b
                                 peak areas of $a:b = 3:1$.

(b) Propanal (CH_3CH_2CHO)   *Answer:* Three signals in CH_3CH_2CHO
                                               a   b   c
                                 with peak areas of $a:b:c = 3:2:1$.

(c) $CH_3CH_2NH_2$
(d)

$$CH_2-\overset{\overset{\displaystyle O}{\|}}{C}-CH_3$$

(e) Cyclohexane

With the positions of the peaks giving information about the chemical environment of the protons, the multiplicity of the peaks giving information regarding the number of protons on adjacent carbons, and the integration giving the number of each of the different types of protons, in most cases it is possible to identify an organic compound completely. If it is not possible to identify a compound from the NMR spectroscopic data, the addition of other spectroscopic data almost always makes it possible to identify the structure of a given compound. The NMR spectra of some typical organic compounds are shown in Figs. 12-9, 12-10, and 12-11.

Application of classical and spectral techniques
in identification of pheromone

Let us consider an example of the isolation of a compound from a natural source, identification of it by classical and spectral techniques, and confirmation of its structure by an independent synthesis. Chemists have discovered that one way of successfully controlling losses of crops and trees caused by insects depended on isolating, identifying, and synthesizing the sex pheromones found in these insects. Some examples of sex pheromones are the four compounds from the male boll weevil. These four active compounds, to which only the female boll weevil responds, include two alcohols and two aldehydes, the structures of which are as follows:

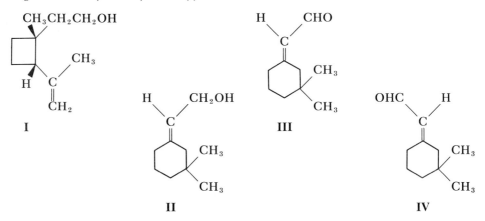

These four compounds were isolated by extraction of them from the fecal material of the male boll weevil with the solvent methylene chloride (CH_2Cl_2), concentration of the solvent extract, steam distillation of the concentrate, and fractionation of the distillate by column chromatography. Compounds I and II were collected individually and calculated to be present in concentrations of 0.76 ppm (41.6 mg) and 0.57 ppm (31.2 mg), respectively, in 54.7 kg of fecal material. Compounds III and IV were collected as a single fraction and were each found to be present in 0.06 ppm (3.3 mg) in 54.7 kg of fecal material.

All four compounds were then identified with IR, mass, and NMR spectra. We will consider only the identification of the structure of compound II. The IR spectrum of compound II had a band at $2.77\,\mu$, indicating the presence of an ÖH group; a band at $3.0\,\mu$ to $3.7\,\mu$, indicating the presence of C–H bonds; and a band at $5.9\,\mu$ to $6.2\,\mu$, indicating the presence of a C=C bond:

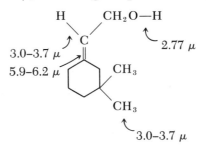

Some of the peaks of the mass spectrum were as follows: 154 (representing molecular ion and thus molecular weight), 139 (corresponding to loss of a methyl group, which is 15 mass units), 136 (corresponding to loss of a water molecule by dehydration of the alcohol), and 107 (corresponding to loss of an oxonium ion $[CH_2O^+H_2]$ and a methyl group).

The NMR spectrum had the following peaks: 5.53δ (triplet, =CH—); 4.05δ (doublet, RR′C=CHCH$_2$OH); 2.09δ (multiplet, two protons) overlapping 2.00δ (singlet, two protons), suggesting two methylenes adjacent to a double bond, with one unsplit; 1.83δ to 1.13δ (multiplet, broad, four protons); and 0.95δ (singlet, geminal dimethyls):

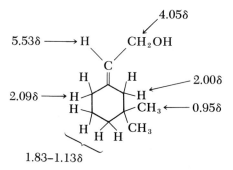

The *cis* configuration about the double bond was assigned by comparison of the NMR spectrum of this compound with the NMR spectra of the *cis* and *trans* ester precursors in the synthesis of compound II.

Compound II was also subjected to ozonolysis:

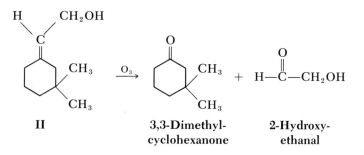

| | 3,3-Dimethyl- | 2-Hydroxy- |
| **II** | cyclohexanone | ethanal |

The resulting compounds were separated by gas chromatography and were identified by comparison of their spectra with those of the known compounds.

Compound II was synthesized by the addition of ethyl bromoacetate to 3,3-dimethylcyclohexanone (Reformatsky reaction, Chapter 9) in the presence of zinc to produce the β-hydroxy ester. Dehydration of the ester yielded a mixture of *cis* and *trans* unsaturated ester precursors, which were separated by gas chromatography and identified on the basis of NMR spectra. The *cis* unsaturated ester was then reduced to compound II with $LiAlH_4$. These steps are shown on p. 360.

The IR, NMR, and mass spectra and the biological activity of the synthesized compound II were found to be identical to those of the natural product.

Summary

This chapter deals with the identification of the structures of organic compounds by classical methods and modern spectral techniques. Structure is identified by a number of different steps. First, the kinds of elements present are determined. The next step is determination of the empirical formula followed by determination of the molecular weight (most likely from the unknown compound's mass spectrum) and the molecular formula. Once the molecular formula is obtained, the kinds of functional groups are determined. This is done by classical chemical reactions, usually involving some change

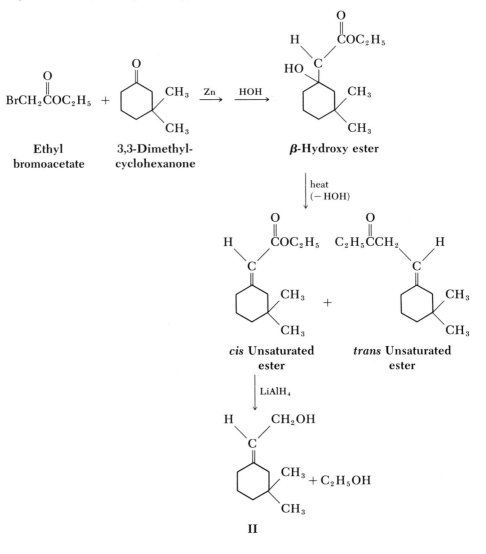

in color, or through modern techniques using the infrared spectrum or nuclear magnetic resonance spectrum of the compound. In the NMR spectrum, the position of the peak tells about the proton's chemical environment, the multiplicity of the peak shows the number of protons on adjacent carbons, and the integration or peak height indicates the number of the type of proton. Any one of these techniques may identify the unknown compound's structure, but coupled data almost always makes it possible to identify the structure. If the unknown compound is believed to be identical to a known compound, its structure is confirmed by derivative preparation or by IR and NMR spectra. If the compound is new, an independent synthesis is usually performed.

Important terms

Adjacent protons	Mass spectrum
Chemical shift	Molecular formula
Derivative	$(n + 1)$ rule
Empirical formula	Nuclear magnetic resonance spectrum
Equivalent protons	Splitting pattern
Excited state	Tetramethylsilane
Ground state	Ultraviolet spectrum
Infrared spectrum	

Problems

12-6. Identification of an unknown compound involves conversion of it through a known chemical reaction into a compound whose melting point is known with certainty. Amides $(RCONH_2)$, anilides $(RCONHC_6H_5)$, and p-nitrobenzyl esters $(RCOOCH_2C_6H_4p\text{-}NO_2)$ are derivatives of acids. Using the melting points of the acids as well as their derivatives in the following table, indicate which acid or acids each of the compounds a through d is most likely to be.

	Acid mp (°C)	Amide mp (°C)	Anilide mp (°C)	p-Nitro-benzyl ester mp (°C)
2-Butenoic acid	72	161	118	67
Phenylacetic acid	77	156	118	65
α-Hydroxybutyric acid	79	98	136	80
α-Hydroxyacetic acid	80	120	97	107
2-Iodopropanoic acid	82	101		
Iodoacetic acid	83	95	143	
p-Nitrophenylacetic acid	153	198	198	
2,5-Dichlorobenzoic acid	153	155		
m-Chlorobenzoic acid	154	134	122	107
m-Bromobenzoic acid	156	155	136	105
p-Chlorophenoxyacetic acid	158	133	125	

(a) Mp 76° to 77°; anilide, mp 118°; negative test with Br_2 in CCl_4
(b) Mp 80° to 82°; negative test with $AgNO_3$ for halogen
(c) Mp 152° to 153°; positive chlorine test; amide, mp 135°
(d) Mp 156° to 157°; positive halogen test; p-nitrobenzyl ester, mp 105°

12-7. Decide whether the data in column A applies to the compound in Column B or to that in Column C:

A	B	C
(a) Soluble in H_2O	CH_3CH_2COOH	$CH_3CH{=}CH(CH_2)_7COOH$

(b) IR spectrum:

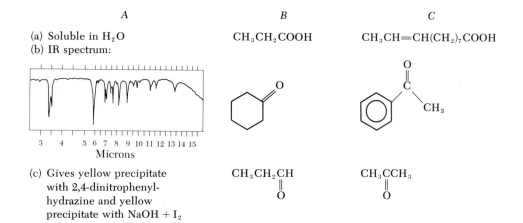

A	B	C
(c) Gives yellow precipitate with 2,4-dinitrophenyl-hydrazine and yellow precipitate with NaOH + I_2	CH_3CH_2CH (with $\parallel$ O)	CH_3CCH_3 (with $\parallel$ O)

(d) NMR spectrum: CH_3CH_3 CH_3CH_2Br

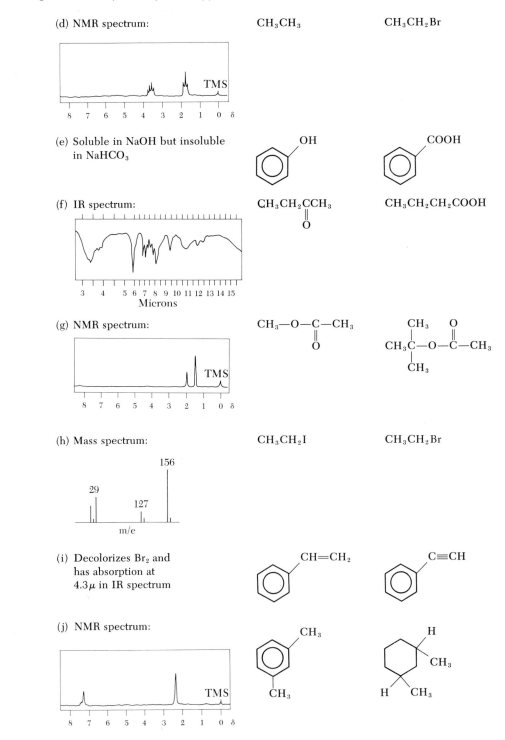

(e) Soluble in NaOH but insoluble in NaHCO$_3$

(f) IR spectrum: $CH_3CH_2CCH_3$ ‖ O $CH_3CH_2CH_2COOH$

(g) NMR spectrum: $CH_3—O—C—CH_3$ ‖ O

$$CH_3\overset{CH_3}{\underset{CH_3}{C}}—O—\overset{O}{C}—CH_3$$

(h) Mass spectrum: CH_3CH_2I CH_3CH_2Br

(i) Decolorizes Br$_2$ and has absorption at 4.3μ in IR spectrum

$CH=CH_2$ $C≡CH$

(j) NMR spectrum:

12-8. Using the following data, determine the structural formula of the compound:

Elemental analysis: 77.8% carbon, 7.4% hydrogen, 14.8% oxygen

Mass spectrum: molecular ion at 108

Tests: with 2,4-Dinitrophenylhydrazine, negative; with sodium hydroxide + iodine, negative; with acetyl chloride, sweet, fruity odor

IR spectrum:

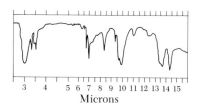

Microns

NMR spectrum:

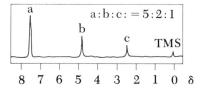

a:b:c: = 5:2:1

Self-test

1. What four steps are involved in the determination of the complete structure of an organic compound?
2. How are the elements in an organic compound detected?
3. How do empirical and molecular formulas differ? How is the molecular weight of an organic compound generally determined?
4. In what two ways are functional groups identified in an organic compound? (*Hint:* one way involves a physical method, and the other involves a chemical method).
5. Why is the type of information obtained from an ultraviolet-visible spectrum limited as far as the identification of specific organic molecules is concerned?
6. What type of information about an organic molecule can be inferred from an infrared spectrum?
7. Briefly describe the three pieces of information that can be obtained from a nuclear magnetic resonance spectrum.
8. Give the correct answer for each of the following questions:

   (a) An organic compound is fused with sodium. In what ionic form is the nitrogen present?
   (b) An organic compound is soluble in NaOH but not in $NaHCO_3$. Is the compound phenol or benzoic acid?
   (c) An organic compound decolorizes bromine, and no gas is liberated. Is the compound benzene or 1-hexene?
   (d) Which portion of the electromagnetic spectrum is of higher energy: the infrared or ultraviolet?
   (e) The ultraviolet region is associated with transitions between which types of levels available to molecules?
   (f) How is it possible to distinguish an aldehyde and a carboxylic acid on the basis of the infrared spectrum?
   (g) What is the name of the compound used as a standard in the nuclear magnetic resonance spectrum?
   (h) What is the splitting pattern for the methyl protons in ethyl alcohol?
   (i) How many signals are there in the nuclear magnetic resonance spectrum of propanoic acid?
   (j) Which of the three pieces of information useful to organic chemists in the nuclear magnetic resonance region is most useful in distinguishing p-xylene (H_3C—⬡—CH_3)

   and 1,3,5-trimethylbenzene?

Answers to self-test

1. A complete structure determination involves (1) detection of the elements present, (2) determination of the molecular weight and the molecular formula, (3) characterization of the compound by chemical tests and through spectra, and (4) either derivative preparation (for a known compound) or independent synthesis (for an unknown compound).

2. The presence of carbon and hydrogen is assumed. The other elements are detected by a sodium fusion reaction in which the organic compound is treated with hot molten sodium metal. All covalently bonded elements are converted into ionic species, which are detected through simple chemical tests.

3. The empirical formula gives the atomic ratios of the elements present. The molecular formula gives the exact numbers and kinds of atoms making up the molecule. The molecular weight of an organic compound is generally determined from its mass spectrum.

4. Functional groups in organic compounds are determined from chemical tests (such as those in Table 12-1) or from spectral data. The spectral data most commonly used are infrared spectroscopy and nuclear magnetic resonance spectroscopy.

5. The absorption bands are generally broad, and similar compounds may have spectra that are so similar that the two cannot be distinguished; or if the molecule contains no double bond or functional groups that interact with ultraviolet-visible radiation, there may be no spectrum at all.

6. In general, the presence or absence of various functional groups can be inferred.

7. The three useful pieces of information are (1) position of the peaks (type of environment about the proton giving rise to the signal), (2) the number of hydrogens immediately adjacent to the proton giving rise to a particular signal, and (3) the number of each type of proton that appears in the spectrum.

8. (a) Cyanide (CN^-)
   (b) Phenol
   (c) 1-Hexene
   (d) Ultraviolet
   (e) Electronic
   (f) Presence or absence of OH stretch at $\sim 3\mu$
   (g) Tetramethylsilane
   (h) Triplet: CH_3CH_2OH
   (i) Three: CH_3CH_2COOH
   (j) Number of protons giving rise to each signal

chapter 13

Organic chemistry of biomolecules

The organic chemist can synthesize complex structures from smaller molecules by modification of a known complex molecule prepared by other organic chemists or by isolation of the compound from living systems. The synthetic biochemical processes in man involve the conversion of relatively simple carbohydrates and amino acids, acetate, phosphate, and some inorganic materials such as sodium and chloride ions into the complex carbohydrates, proteins, lipids, and nucleic acids on which life depends. The human body is capable of producing these complex structures and using them as sources of energy. The chemistry of these organic molecules is discussed in this chapter; and the methods by which they are metabolized, that is, the reactions they undergo in the living systems, are discussed in Chapter 14.

The most abundant compound in all living systems is water. The three other major types of compounds found in living systems are carbohydrates, proteins, and lipids. Nucleic acids, which are present in minute quantities, are also important to living systems. As Table 13-1 shows, the relative amounts of these molecules vary widely in plants and animals.

Lipids

Lipids, particularly the triglycerides, are the body's best reserve energy supplies. They provide about twice as much energy as carbohydrates and

Table 13-1
Composition of substances from plants and animals

	Water	Carbohydrate	Protein	Lipid
Apple	85	14	0.4	0.5
Banana	75	23	1.2	0.2
Spinach	93	3	2.3	0.3
Potato	78	18	2.2	0.1
Fish	83	0.4	16	0.4
Honey	20	80	0	0
Pork chop	52	0	17	30.0

proteins per gram of substance for body processes. They also serve many other purposes. They are a heterogeneous class of widely distributed naturally occurring compounds that are insoluble in water but soluble in organic solvents. They include such types of substances as neutral fats, waxes, phospholipids, cerebrosides, steroids, and fat-soluble vitamins, each of which will be discussed in more detail.

Neutral fats

The neutral fats are esters of glycerol and fatty acids and are called *triglycerides*. They have the following general structure:

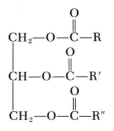

If one or all of the R groups are different, the fat is a mixed triglyceride. If all the groups are alike, the fat is a simple triglyceride. Following are some examples:

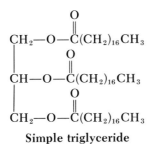

Simple triglyceride   **Mixed triglyceride**

Table 13-2
Names and structures of some fatty acids

Name	Number of carbons	Structure
Caprylic	8	$CH_3(CH_2)_6COOH$
Capric	10	$CH_3(CH_2)_8COOH$
Lauric	12	$CH_3(CH_2)_{10}COOH$
Myristic	14	$CH_3(CH_2)_{12}COOH$
Palmitic	16	$CH_3(CH_2)_{14}COOH$
Stearic	18	$CH_3(CH_2)_{16}COOH$
Oleic	18	$cis\text{-}CH_3(CH_2)_7CH{=}CH(CH_2)_7COOH$
Linoleic	18	$cis,cis\text{-}CH_3(CH_2)_4CH{=}CHCH_2CH{=}CH(CH_2)_7COOH$
Linolenic	18	$cis,cis,cis\text{-}CH_3CH_2CH{=}CHCH_2CH{=}CHCH_2CH{=}CH(CH_2)_7COOH$

The R groups in triglycerides are derived from fatty acids, which are straight-chain monocarboxylic acids. They are named fatty acids because they are found in fats. In general, fatty acids have an even number of carbon atoms, ranging from 2 to 20. Some are unsaturated, and the double bond is between carbons 9 and 10. Several of the unsaturated acids have more than one double bond; for the most part, these double bonds are at carbons 9, 12, and 15. Furthermore, the unsaturated acids are *cis* compounds. The names and structures of the more common fatty acids are given in Table 13-2.

13-1. Draw the structures of the triglycerides with the following components:

(a) Three lauric acids   *Answer:* $CH_2OOC(CH_2)_{10}CH_3$
$$CHOOC(CH_2)_{10}CH_3$$
$$CH_2OOC(CH_2)_{10}CH_3$$

(b) Three acid residues of 16 carbon atoms each
(c) One octanoic acid, one 14-carbon acid, and an unsaturated acid with a carbon–carbon double bond after carbon 9 only

---

The nature of a fatty acid determines the physical state of a neutral fat. Saturated fatty acids tend to make neutral fats solids at room temperature, whereas unsaturated fatty acids tend to make neutral fats liquid at room temperature. In other words, the more saturated fatty acids a neutral fat contains, the higher the melting point of the fat will be.

The neutral fats with low melting points are referred to as oils. Many of these oils, such as olive oil and other vegetable oils, are used in cooking; they are just as nutritious as butter, since they provide the unsaturated acids we need. The disadvantage of these oils is that their unsaturation makes them more susceptible to oxidative rancidity. Therefore, many oils are hydrogenated to produce fats of desired softness that are less prone toward rancidity. The resulting compounds are solid fats, such as shortening and oleomargarine. For example, the hydrogenation of the major triglyceride in olive oil is accomplished with a catalyst as follows:

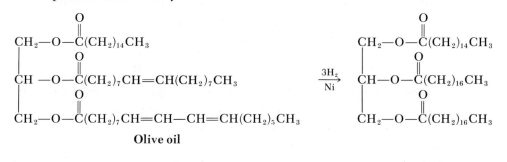

Olive oil

13-2. What is the structure of the product of hydrogenation of a triglyceride composed of two oleic acids and one linolenic acid residue. Assume that 5 moles of hydrogen is required to hydrogenate the triglyceride.

Answer: $CH_2O_2C(CH_2)_7CH=CH(CH_2)_7CH_3$
$CHO_2C(CH_2)_7CH=CH(CH_2)_7CH_3$
$CH_2O_2C(CH_2)_7CH=CHCH_2CH=CHCH_2CH=CHCH_2CH_3$

$$\xrightarrow[\text{Ni}]{5H_2}$$

$CH_2O_2C(CH_2)_{16}CH_3$
$CHO_2C(CH_2)_{16}CH_3$
$CH_2O_2C(CH_2)_{16}CH_3$

Reactions of neutral fats. The most important reactions of neutral fats are hydrolysis and, to a lesser extent, hydrogenation. One example of hydrolysis is *saponification,* the making of soap.

Triglycerides are also hydrolyzed in the digestive tract. This reaction is catalyzed by the enzyme lipase and yields glycerol and fatty acids. Because of the alkaline nature of the digestive tract, the fatty acids are converted to the metal ion salts of fatty acids.

Saponification is the hydrolysis of a neutral fat in the presence of a base, such as NaOH or KOH, that is, essentially, the hydrolysis of an ester in an alkaline medium. The products of this reaction are glycerol and a mixture of metal ion salts of fatty acids, which are called *soaps.* These fatty acid residues are straight-chain carboxylates that usually contain 8 to 18 carbons and may be saturated or unsaturated. A general saponification reaction is shown as follows:

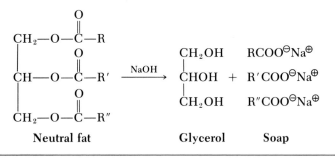

Neutral fat	**Glycerol**	**Soap**

13-3. What is the mixture of soaps formed in the saponification of olive oil?

Answer: The names or structures of the products are glycerol, sodium palmitate, $CH_3(CH_2)_7CH=CH(CH_2)_7COO^-Na^+$, and $CH_3(CH_2)_5(CH=CH)_2(CH_2)_7COO^-Na^+$.

Soaps vary in composition. Air can be whipped in to make a soap float. If it is a potassium salt, it is a soft soap. If it is formed from olive oil, it is Castile soap. Perfumes, dyes, and germicides can be added.

All soaps, however, have one structural feature in common. They contain a large nonpolar hydrocarbon end that is oil soluble and a polar end that is water soluble. This structure accounts for the cleansing action of soap. In the following molecule, the hydrocarbon chain is the nonpolar end and the carboxylate anion is the polar end:

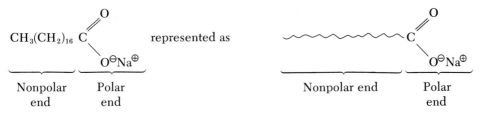

| Nonpolar end | Polar end | | Nonpolar end | Polar end |

The oils on our bodies (the dirt and grime) generally do not mix with water, so that separate oil and water layers are formed. The hydrocarbon end of the soap molecule dissolves in the oil droplets because both it and the oil layer are nonpolar. The polar end of the soap molecule projects into the water layer. Hundreds of soap molecules act in this fashion, and the oil globules are suspended in the water solution as shown in the following diagram and washed away:

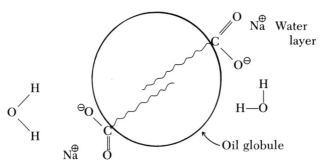

Hydrogenation of an unsaturated glyceride is usually controlled so that the resulting product is not completely saturated but yet is a solid at room temperature. However, under severe conditions of hydrogenation, an unsaturated triglyceride is reduced completely to a mixture of long-chain alcohols and glycerol. Following is an example:

$$
\begin{array}{l}
\text{CH}_2\text{—O—}\overset{\text{O}}{\overset{\|}{\text{C}}}(\text{CH}_2)_{10}\text{CH}_3 \\
\text{CH—O—}\overset{\text{O}}{\overset{\|}{\text{C}}}(\text{CH}_2)_{12}\text{CH}_3 \\
\text{CH}_2\text{—O—}\overset{\text{O}}{\overset{\|}{\text{C}}}(\text{CH}_2)_7\text{CH}=\text{CH}(\text{CH}_2)_7\text{CH}_3
\end{array}
\xrightarrow[\text{Ni}]{6\text{H}_2}
$$

$\text{CH}_3(\text{CH}_2)_{10}\text{CH}_2\text{OH}$ **Lauryl alcohol**

$\text{CH}_3(\text{CH}_2)_{12}\text{CH}_2\text{OH}$ **Myristyl alcohol**

$\text{CH}_3(\text{CH}_2)_{16}\text{CH}_2\text{OH}$ **Stearyl alcohol**

$$
\begin{array}{ccc}
\text{CH}_2\text{—CH—CH}_2 \\
| \quad\quad | \quad\quad | \\
\text{OH} \quad \text{OH} \quad \text{OH}
\end{array}
\quad \textbf{Glycerol}
$$

Coconut oil

The alcohols formed in the hydrogenation of unsaturated triglycerides can be sulfonated by treatment with sulfuric acid. The resulting alkyl hydrogen sulfates can then be neutralized with a base such as sodium hydroxide to yield the sulfates known as *detergents*. Following is an example of this process:

$$CH_3(CH_2)_{10}CH_2OH \xrightarrow{H_2SO_4} CH_3(CH_2)_{10}CH_2OSO_2OH \xrightarrow{NaOH} CH_3(CH_2)_{10}CH_2OSO_2O^{\ominus}Na^{\oplus}$$

Lauryl alcohol   **Lauryl hydrogen sulfate**   **Sodium lauryl sulfate**

Detergents are similar to soaps in cleansing ability but have the advantage of being more effective in hard water (water containing calcium and magnesium ions) because the calcium and magnesium sulfates of detergents are water soluble whereas the calcium and magnesium carboxylates of soaps are not.

Recently, sulfonates of aromatic hydrocarbons with substituted alkyl groups have been used as detergents. These detergents, which are commonly called *alkylbenzenesulfonates,* have the following general structure:

The importance of the nature of the side chain (R group) was discovered when detergents were passed into sewage systems. Apparently, the bacteria that feed on sewage cannot metabolize certain detergents. If the R group is branched, the action of bacteria on the detergent is blocked; but bacteria can destroy the nonbranched (normal) R groups. A detergent that contains normal alkyl groups is said to be biodegradable, since it can be destroyed by microorganisms. An example of a biodegradable detergent is sodium *n*-dodecylbenzene sulfonate:

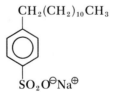

Waxes

Waxes are lipids, but they do not contain glycerol. They are esters of long-chain fatty acids with long-chain alcohols. Following are some typical examples of waxes:

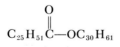

$\overset{\displaystyle O}{\overset{\displaystyle \|}{C_{25}H_{51}C}}$—$OC_{30}H_{61}$	$\overset{\displaystyle O}{\overset{\displaystyle \|}{C_{15}H_{31}C}}$—$OC_{30}H_{61}$	$\overset{\displaystyle O}{\overset{\displaystyle \|}{C_{15}H_{31}C}}$—$OC_{16}H_{33}$
Myricyl cerotate: carnauba wax, component of floor and car wax	**Myricyl palmitate:** beeswax	**Cetyl palmitate:** sperm whale head oil

Waxes are found on the surfaces of the feathers of birds and, since they are not wettable, help birds stay afloat. Waxes are used to coat fruit to protect it from moisture loss and invasion by foreign organisms. They are also used in making candles, ointments, salves, and lanolin.

Phospholipids

Phospholipids are esters of glycerol in which one of the OH groups of glycerol has been esterified by phosphoric acid. The general structure of phospholipids is as follows:

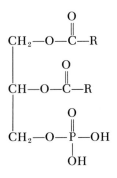

Following is a specific example of a phospholipid:

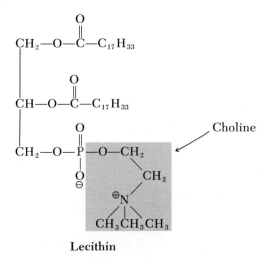

Lecithin

This phospholipid contains oleic acid and choline, a quaternary ammonium compound with a strong basicity about equal to that of KOH. Phospholipids containing fatty acids and choline are called *lecithins*. Lecithins are constituents of brain and nervous tissues and soybeans.

If the fatty acid on the central carbon in a lecithin is removed by hydrolysis, the resulting compound is a lysolecithin, which causes disintegration of red blood cells. The bite of poisonous snakes is fatal because the venom contains an enzyme capable of converting lecithins into lysolecithins. Following is an example:

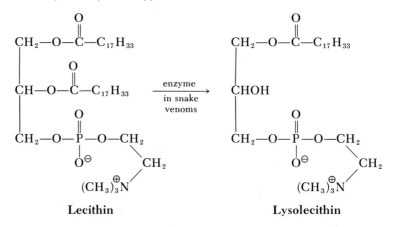

Lecithin   **Lysolecithin**

Another group of phospholipids are the *cephalins,* which are found in brain tissue and are involved in the blood-clotting process. Following is an example of a cephalin containing ethanolamine as the base:

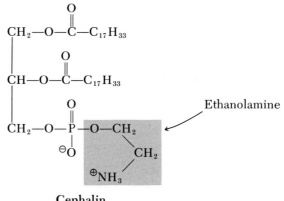

Cephalin

Cerebrosides

Cerebrosides are found in brain matter. The general structure is as follows:

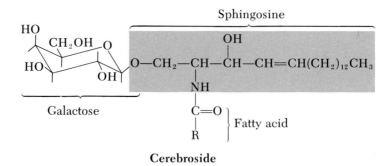

Cerebroside

When hydrolyzed, cerebrosides yield a straight-chain fatty acid, sphingosine, and galactose.

13-4. Sphingosine is a nitrogen-containing base (an amine). Draw the structure of sphingosine.   *Answer:*

$$CH_3(CH_2)_{12}CH{=}CHCHCHCH_2OH$$

with OH on the third-from-right carbon and NH_2 on the adjacent carbon.

Gaucher's disease is a hereditary condition characterized by enlargement of the liver and spleen and caused by the accumulation of large amounts of cerebrosides. Glucose replaces galactose in the cerebroside molecule; this results in a decrease in the breakdown of cerebrosides and hence an accumulation of them.

Steroids

Steroids are found in the nonsaponifiable fraction of animal tissue. (The saponifiable fraction is the triglycerides). Shown as follows are the steroid nucleus, contained by all steroids, and cholesterol, the most prevalent steroid in animal tissues:

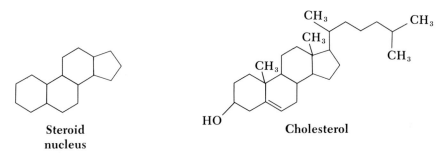

Steroid
nucleus

Cholesterol

Cholesterol is found abundantly in brain, nerve, and glandular tissues and is the chief component of gallstones. The steroid nucleus is found in a number of other biologically important lipids, including the bile salts, steroid hormones, and vitamin D.

Bile salts. These are found in bile, which is formed in the liver and secreted into the gallbladder. The bile salts are strong surface-active agents that help to disperse fat molecules so that they can be absorbed or more rapidly broken down by the action of enzymes. Bile salts are salts of the compound formed by the reaction between cholic acid and glycine, an amino acid, which is shown as follows:

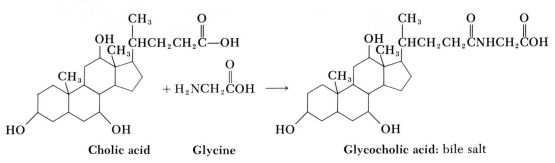

Cholic acid   Glycine   Glycocholic acid: bile salt

The bile salts aid in fat digestion and then are absorbed and returned to the liver, where they can be used again.

Steroid hormones. These include the male and female sex hormones and the adrenocortical hormones. The structural formulas of the sex hormones are shown on p. 157. The adrenocortical hormones are produced by the adrenal cortex under control of the pituitary gland. One of the most important of these hormones is cortisone, the structure of which is as follows:

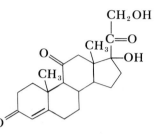

The release of cortisone and the other adrenocortical hormones is controlled by the pituitary gland. The pituitary secretes adrenocorticotropic hormone (ACTH), which stimulates the adrenal cortex and regulates the biosynthesis of the adrenocortical hormones.

Many steroids such as cortisone have been found to be useful as antiinflammatory agents in the treatment of rheumatoid arthritis. Addison's disease, which results from underactivity of the adrenal cortex, also responds to treatment with cortisone. Chemically, Addison's disease results in an electrolyte imbalance. The patient exhibits weakness, pigmentation of the skin and mucous membranes, loss of weight, and low blood pressure, and death can result.

Fat-soluble vitamins

Vitamins are a group of organic molecules that occur in food and are necessary for the health, growth, and reproduction of animal organisms. Vitamins are classified as water soluble (thiamine, pantothenic acid, riboflavin, niacin, ascorbic acid, pyridoxine, biotin, folic acid, and vitamin B_{12}) or as fat soluble (vitamins A, D, E, and K). The fat-soluble vitamins are lipids and thus are discussed in this section.

Vitamin A. It is found in fish livers, butter, eggs, and cream cheese. In human beings, vitamin A is also formed from carotenes in the intestinal wall and, to some extent, in the liver. Carotenes are found in green and yellow vegetables. Following are examples of a carotene and a form of vitamin A:

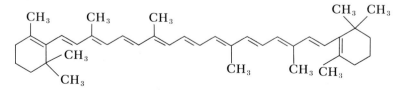

β-Carotene

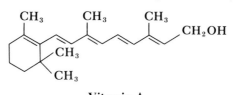

Vitamin A
(all-*trans*-retinol)

Vitamin A plays many roles in the animal body, but one of the most important roles is its function in visual processes. A deficiency of vitamin A results in night blindness, an inability to see in dim light, which essentially results from the lack of formation of rhodopsin.

The visual process involves rhodopsin, a visual pigment, which consists of a protein called opsin and 11-*cis*-retinal (the aldehyde of vitamin A). When light strikes rhodopsin, the opsin-11-*cis*-retinal complex dissociates and isomerization of 11-*cis*-retinal to all-*trans*-retinal occurs, yielding opsin and all-*trans*-retinal. This photochemical reaction makes vision possible. The structures of 11-*cis*-retinal and all-*trans*-retinal are as follows:

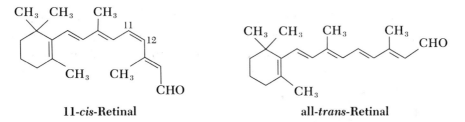

11-*cis*-Retinal   **all-*trans*-Retinal**

All-*trans*-retinal is then reduced to all-*trans*-retinol (vitamin A alcohol). Light also causes the association of 11-*cis*-retinal with opsin to regenerate rhodopsin. 11-*cis*-Retinal can also be regenerated either by isomerization of all-*trans*-retinal or by oxidation of 11-*cis*-retinol. The visual process is outlined as follows:

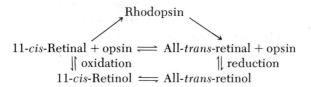

11-*cis*-retinal is also formed by the isomerization of all-*trans*-retinal in the liver.

Vitamin D. There are two important forms of vitamin D: ergocalciferol (Vitamin D_2) and cholecalciferol (vitamin D_3). Vitamin D_2 is formed when ergosterol, a steroid found in yeast and mold, is irradiated with ultraviolet light. Vitamin D_3 is formed when 7-dehydrocholesterol, a steroid found in the skins of animals, is irradiated with ultraviolet light. Vitamin D_3 is found abundantly in fish liver oils. The formation of these vitamins is shown as follows:

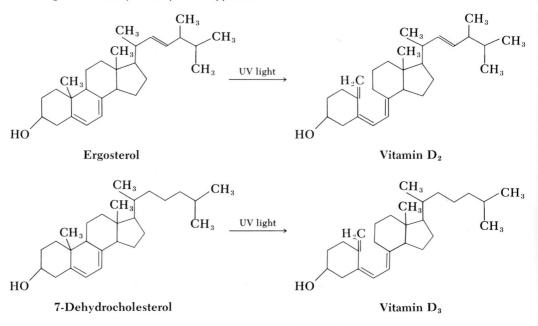

Ergosterol Vitamin D₂

7-Dehydrocholesterol Vitamin D₃

A deficiency of vitamin D causes a disease known as rickets, which is characterized by deformities of the spine and bowlegs, caused by softening and bending of the bones, and a failure of proper formation of tooth enamel.

Vitamin E. The active form of vitamin E is α-tocopherol, which has the following structure:

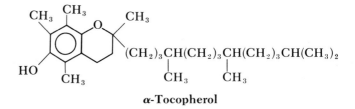

α-Tocopherol

Vitamin E is an antioxidant. It serves the animal organism by protecting body fats from destruction by chemical oxidation. It is also added to unsaturated fats to prevent rancidity.

It has been proposed that vitamin E is involved in slowing the aging process in man by deactivating peroxides formed in metabolism. If these peroxides were not deactivated, the triglycerides found in the living system would be oxidized, since they contain a high percentage of unsaturated fatty acids, whose unsaturated bonds are particularly susceptible to attack by the peroxides. The proposed scheme by which vitamin E retards aging is as follows. Oxygen, itself a free radical, reacts with the unsaturated fatty acid portion of the triglyceride molecule, forming radicals (ROO·). These radicals then react with the triglyceride by addition or hydrogen abstraction to form other radicals and ultimately break down the fat tissue. Vitamin E interferes with this last reaction by deactivating the ROO· radicals and forming a resonance-stabilized radical of vitamin E. This scheme is illustrated as follows:

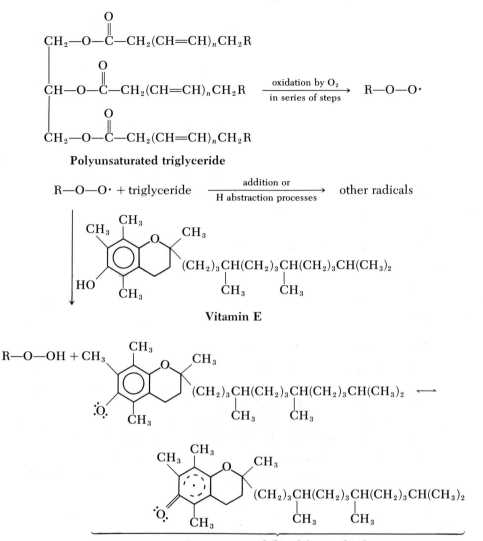

Polyunsaturated triglyceride

Vitamin E

Resonance-stabilized free radical

Scientists recognize that vitamin E is a requirement in human nutrition, but whether or not it can retard human aging to a significant extent is still being explored. They do agree, however, that the need for vitamin E depends on the amount of polyunsaturated triglycerides in the diet.

Wheat germ oil is the best source of vitamin E.

Vitamin K. There are several forms, one of which is a 1,4-naphthoquinone with the following structure:

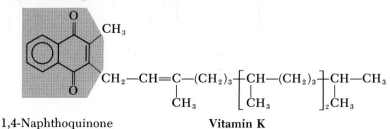

1,4-Naphthoquinone   **Vitamin K**

Vitamin K is essential in the synthesis of prothrombin by the liver and thus in the blood-clotting process. Basically, in the blood-clotting process, prothrombin is converted to thrombin, an enzyme, which then converts plasma fibrinogen to fibrin, which forms the blood clot. Thus a deficiency of the vitamin causes a reduction in the formation of prothrombin and an increase in the blood-clotting time.

The richest sources of vitamin K are alfalfa, cabbage, and spinach.

Carbohydrates

Carbohydrates include sugars, starches, and various forms of cellulose, such as wood, paper, and cotton. The carbohydrate glucose, which is a sugar, is a product of photosynthesis in plants. It forms the larger carbohydrate molecules of cellulose, which provides the structural support in plants, and starches. Starch molecules can be stored by the plant or by the animal that eats the plant and ultimately used as a source of glucose, which is used to form energy for the living system.

Some industrial processes utilize carbohydrates. For example, they are used as a source of ethyl alcohol in fermentation.

Chemical nature of carbohydrates

The simple carbohydrates are polyhydroxy aldehydes or polyhydroxy ketones. The more complicated carbohydrates are compounds that can be hydrolyzed to polyhydroxy aldehydes or ketones. Carbohydrates are classified as *monosaccharides*, which cannot be hydrolyzed to simpler compounds; *disaccharides*, which can be hydrolyzed to two monosaccharides; or *polysaccharides*, which can be hydrolyzed to many monosaccharides. Following are examples of a monosaccharide, disaccharide, and polysaccharide:

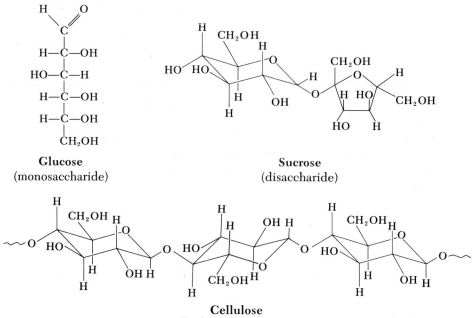

Glucose
(monosaccharide)

Sucrose
(disaccharide)

Cellulose
(polysaccharide segment)

Monosaccharides are classified further as follows. One containing an aldehyde group is called an *aldose,* and one containing a keto group is called a
ketose.

Aldehyde group

Keto group

Some specific examples are aldopentoses, aldoses that contain five carbons;
aldohexoses, which contain six carbons and an aldehyde group; and keto-
hexoses, which contain six carbons and a keto group.

13-5. Classify the following saccharides as aldoses or ketoses:

(a) CHO
   |
   CHOH
   |
   CHOH
   |
   CH$_2$OH

Answer: Aldotetrose

(b) H   O
    \\ //
     C
     |
H—C—OH
     |
   CH$_2$OH

Answer: Aldotriose

(c) CH$_2$OH
   |
   C=O
   |
   CHOH
   |
   CHOH
   |
   CH$_2$OH

(d)   CH$_2$OH
        |
        C=O
        |
   HO—CH
        |
    HC—OH
        |
    HC—OH
        |
    CH$_2$OH

Monosaccharides

Stereoisomerism. Let us consider the stereoisomerism of glyceraldehyde
as the basis of stereoisomerism of monosaccharides. The structure of glyceraldehyde is as follows:

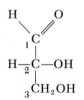

Glyceraldehyde is an aldotriose, a three-carbon monosaccharide containing
an aldehyde group. It is considered to be the simplest sugar, since it possesses
an aldehyde group and primary and secondary alcohol groups. Carbon 2 of
glyceraldehyde is chiral; that is, it is bonded to four different groups. Thus
glyceraldehyde exists as nonsuperimposable mirror images, or enantiomers
(Chapter 5), which are as follows:

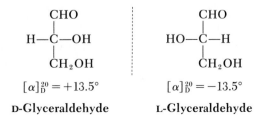

CHO
   |
H—C—OH
   |
  CH$_2$OH

$[\alpha]_D^{20} = +13.5°$

D-Glyceraldehyde

CHO
   |
HO—C—H
   |
  CH$_2$OH

$[\alpha]_D^{20} = -13.5°$

L-Glyceraldehyde

Since these are nonsuperimposable mirror images, they are optically active. A mixture of equal amounts of the two glyceraldehyde isomers (racemic modification) has no optical activity, since the rotation of one isomer cancels out the rotation of the other isomer.

The isomer of glyceraldehyde with a specific rotation of +13.5° is dextrorotatory and is designated by (+). The other isomer is levorotatory and is designated by (−). The configurations of atoms in other sugars are related to those of the glyceraldehyde isomers. Those sugars that are configurationally related to dextrorotatory glyceraldehyde are designated by D-, and those related to the levorotatory glyceraldehyde are designated by L-. Thus the designations D- and L- refer not to the direction of rotation but to the configurational relationship. Furthermore, the configurational relationship applies only to other monosaccharides with a formula that is oriented with the aldehyde at the top (or carbon-1 in the IUPAC system), the CH_2OH group at the bottom, and the groups on carbon atoms in between horizontal, that is, pointing toward you. The OH group on the next to the bottom carbon then must be on the right in the D- designation. It is possible for a sugar to have the D- designation but be levorotatory or to have the L- designation but be dextrorotatory. Following are some examples of sugars and their designations:

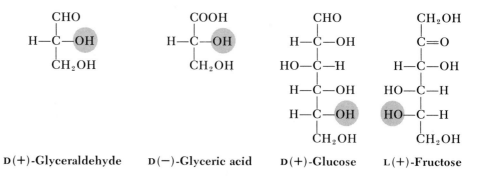

D(+)-Glyceraldehyde   D(−)-Glyceric acid   D(+)-Glucose   L(+)-Fructose

If there is more than one chiral carbon atom in a molecule, the number of isomers is 2^n were n is the number of different chiral carbons. As an example, consider the following aldohexose:

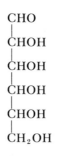

There are 2^4, or 16, stereoisomers—eight pairs of enantiomers. The configurations, names, and directions of rotation of the plane of polarized light of the eight D-stereoisomers are shown as follows:

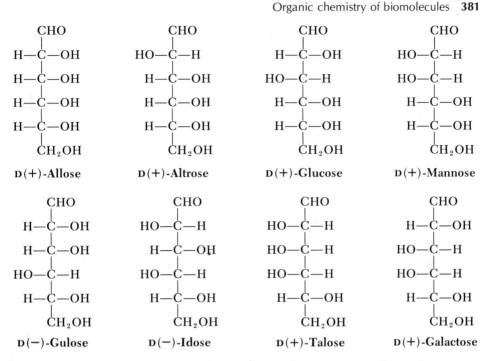

D(+)-Allose   D(+)-Altrose   D(+)-Glucose   D(+)-Mannose

D(−)-Gulose   D(−)-Idose   D(+)-Talose   D(+)-Galactose

The other eight stereoisomers have the L- designation and are the mirror images, enantiomers, of the D-stereoisomers. Of the 16 isomers, glucose is by far the most important.

Glucose: structure and configuration. Glucose is a polyhydroxy aldehyde and an aldohexose. It is the most abundant aldohexose in nature, and it plays

Table 13-3
Comparison of carbonyl reactions of aldehydes and glucose

Carbonyl compound	Reaction				
	Reduction	Oxidation	Tollens' reagent	With $C_6H_5NHNH_2$	With CH_3OH
Aldehyde $R-C\overset{O}{\underset{H}{\diagup\!\!\!\diagdown}}$	RCH_2OH	$RCOOH$	$RCOO^{\ominus}$ + Ag mirror	$RC=NNHC_6H_5$ $\mid$ H	2 moles H $\mid$ $R-C-OCH_3$ $\mid$ OCH_3
Glucose $\overset{H}{\underset{(CHOH)_4}{\overset{}{C}}}\overset{O}{\diagup\!\!\!\diagdown}$ $(CHOH)_4$ CH_2OH	CH_2OH $(CHOH)_4$ CH_2OH	$COO^{\ominus}$ $(CHOH)_4$ $COOH$	$COOH$ $(CHOH)_4$ + CH_2OH Ag mirror	$CH=NNHC_6H_5$ $C=NNHC_6H_5$ $(CHOH)_3$ CH_2OH	1 mole Two different monomethyl derivatives

a special role in biological processes. It is found in blood and muscle tissue, and it is stored in liver and muscle tissues in the form of glycogen or is converted to fat and stored. Oxidation of glucose provides greater than half of the energy required by the human body. It is fitting then to use glucose as a basis of a discussion of the chemistry of carbohydrates.

Glucose would be expected to undergo many of the reactions of aldehydes. The carbonyl reactions of an aldehyde and glucose are compared in Table 13-3. Aldehydes undergo reduction and oxidation, give a positive test with Tollens' reagent, form a phenylhydrazone, and form an acetal with 2 moles of methanol. Similarly, glucose undergoes reduction and oxidation and gives a phenylhydrazone characteristic of an aldehyde. However, it reacts with only 1 mole of methanol to form two different monomethyl acetals.

Reduction of glucose leads to an alcohol (D-sorbitol), and oxidation of glucose by HNO_3 leads to D-glucaric acid:

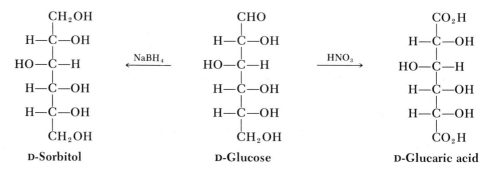

Glucose is oxidized by Tollens' reagent $[Ag(NH_3)_2NO_3]$ to a carboxylic acid salt and an observable silver precipitate that forms as a mirror on the walls of the test tube:

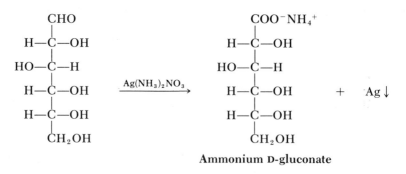

Glucose reacts with phenylhydrazine to give a phenylhydrazone characteristic of an aldehyde. However, the reaction does not stop at this stage. The hydroxyl group on the adjacent carbon is oxidized by a second molecule of phenylhydrazine, and the resulting carbonyl group reacts with a third molecule of the reagent to yield a phenylosazone. The reaction stops at this stage. The steps of this reaction are shown as follows:

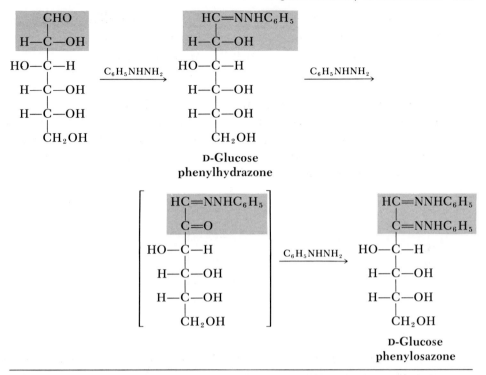

13-6. Mannose and glucose give phenylosazones with the same structure, since they differ in configuration about carbon 2 only. Monosaccharides that differ only in configuration about carbon 2 are said to be *epimers*. Show the reactions that mannose undergoes to form the same phenylosazone as glucose.

Answer:

```
        CHO              CHO
    H—C—OH       HO—C—H     ⟵  Epimeric carbon
   HO—C—H       HO—C—H
    H—C—OH       H—C—OH
    H—C—OH       H—C—OH
      CH₂OH          CH₂OH

   D-Glucose       D-Mannose
```

13-7. Draw the structure of the product of the reaction of D-mannose with each of the following reagents:

(a) HNO_3 *Answer:*

```
        CHO                    CO₂H
    HO—C—H                 HO—C—H
    HO—C—H      HNO₃       HO—C—H
     H—C—OH    ⟶           H—C—OH
     H—C—OH                 H—C—OH
       CH₂OH                  CO₂H
```

(b) $NaBH_4$
(c) $Ag(NH_3)_2NO_3$
(d) $C_6H_5NHNH_2$

Glucose reacts with 1 mole of methyl alcohol to form two different monomethyl derivatives, methyl α-D-glucoside and methyl β-D-glucoside:

$$C_6H_{12}O_6 \; + \; CH_3OH \; \xrightarrow{H^+} \; (C_6H_{11}O_5)OCH_3 \; + \; H_2O$$

Glucose   **Methyl D-glucoside**

This behavior, which differs from that of the carbonyl group in aldehydes, suggests that the carbonyl group in glucose is not free but is tied up in combination with one of the hydroxyl groups. Experiments have shown that the hydroxyl group on carbon 5 is tied up with the aldehyde group. Hence, glucose is a cyclic hemiacetal with the following structure:

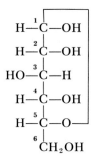

This hemiacetal formation explains why there are two different monomethyl glucosides. In the hemiacetal formation, carbon 1 becomes chiral; therefore, there are two forms of D-glucose, α-D-glucose and β-D-glucose. These two forms differ only in configuration about carbon 1, which is the hemiacetal carbon:

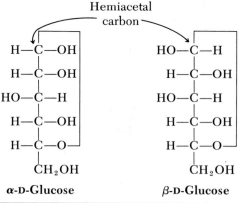

13-8. Draw the structures of methyl α-D-glucoside and methyl β-D-glucoside.

Answer: Methyl α-D-glucoside:

These formulas, which are called Fischer projection formulas, do not satisfactorily show the structure of the glucose molecules. Thus Haworth projection formulas were devised. The Haworth formulas are three-dimensional perspective structures showing the ring projecting out of the plane of the paper and the substituent groups above or below the plane of the ring. More recently, conformational formulas came into use. It is known that six-member rings are not planar but are puckered. The puckering results in a chair conformation for the ring. The Fischer, Haworth, and conformational formulas for α-D-glucose are as follows:

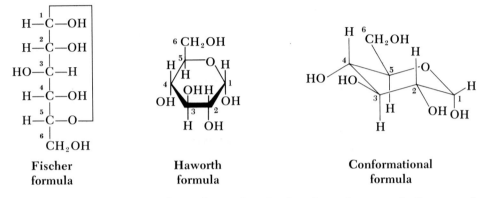

| Fischer formula | Haworth formula | Conformational formula |

A group that appears to the right in the Fischer formula extends downward in the Haworth and conformational formulas. For D-sugars, the OH on carbon 1 is down in the α-isomer and up in the β-isomer. Thus the Haworth and conformational formulas for β-D-glucose are as follows:

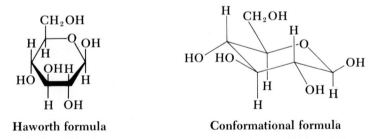

Haworth formula   Conformational formula

The fact that β-D-glucose is the only one of the 32 isomeric aldohexoses in which the larger hydroxyl and CH_2OH groups occupy equatorial positions in the conformational formula may explain why it is the most abundant monosaccharide in nature.

The hemiacetal of glucose has a six-member ring structure. The hemiacetals or hemiketals of other sugars may have five-member rings. The terms *furanose* and *pyranose* are used to denote five- and six-member rings in sugars. These terms indicate a relationship to the cyclic ethers, furan and pyran, which are shown as follows:

Pyran   **Furan**

Hence, the D-isomers of glucose are more correctly called α-D-glucopyranose and β-D-glucopyranose. The hemiketal form of D-fructose is a furanose; thus the isomers are called D-fructofuranose. α-D-Fructofuranose is shown as follows:

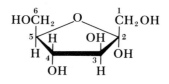

13-9. Write the Fischer, Haworth, and conformational formulas for D-mannose, and give a more correct name for D-mannose in terms of the hemiacetal ring.

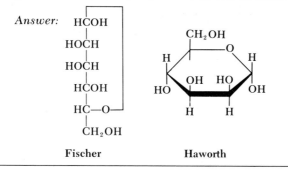

Fischer   Haworth

Mutarotation. When either α- or β-D-glucose is dissolved in water, an equilibrium mixture of both isomers is gradually formed. This can be observed with a polarimeter, since the observed specific rotation changes with time until a certain value is reached. The specific rotation of a solution of α-D-glucose decreases with time from +112° to an equilibrium value of +52.5°, whereas the specific rotation of a solution of β-D-glucose increases with time from +18.7° to an equilibrium value of +52.5°. This phenomenon is called *mutarotation.* It is commonly observed for reducing sugars, which are sugars that contain a carbonyl group in the form of a hemiacetal that is free to react. Hemiacetals exist in equilibrium with the aldehyde. Thus mutarotation results from interconversion of the α- and β-isomers to the equilibrium mixture as in the following example:

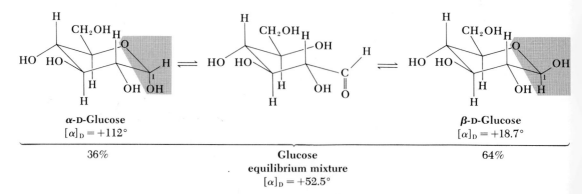

α-D-Glucose $[\alpha]_D = +112°$

β-D-Glucose $[\alpha]_D = +18.7°$

36%   Glucose   64%
equilibrium mixture
$[\alpha]_D = +52.5°$

At equilibrium, the mixture is 36% α-D-glucose and 64% β-D-glucose. The amount of free aldehyde present at equilibrium is only 0.024%. The β-isomer predominates because the hydroxyl group on carbon 1 is in the more stable equatorial position whereas it is in the axial position in the α-isomer.

Acetal formation: glycosides. When a hemiacetal is treated with an alcohol in the presence of a trace of acid, an acetal is formed:

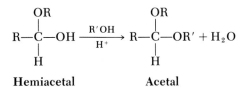

For example, β-D-glucose, which is a hemiacetal, undergoes an acid-catalyzed reaction with methanol to form an acetal called a *glycoside*:

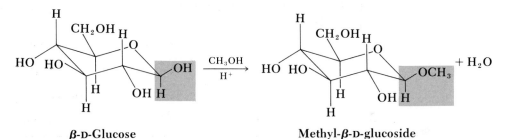

Since the α- and β-isomers of D-glucose are in equilibrium, the α-isomer is also formed:

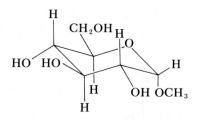

Methyl α-D-glucoside

Kiliani-Fischer synthesis. This is a technique of introducing a new carbon atom into a monosaccharide. The first step involves formation of a cyanohydrin by the reaction between an aldehyde (for example, D-glyceraldehyde) and HCN (p. 280). As shown on p. 388, the planar carbonyl group can be attacked from either side. Cyanohydrin formation results in formation of diastereomers. Even though attack by path *a* appears to give the enantiomer of attack by path *b*, the stereochemical relation of the two cyanohydrins is diastereomeric because the $CH_2OHCHOH$ groupings in the two isomers are not nonsuperimposable mirror images.

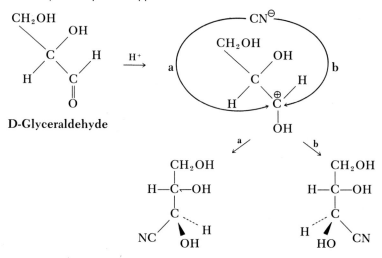

Diastereomeric cyanohydrins of D-glyceraldehyde

The cyanohydrins are not isolatable but are hydrolyzed by acid to the isomeric aldonic acids:

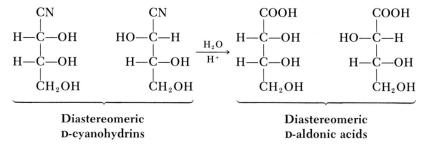

Separation of the diasteromeric D-aldonic acids is accomplished by appropriate methods, and each is converted to an internal, or cyclic, ester, which is called a *lactone*. The lactone is formed when the aldonic acid loses a molecule of water:

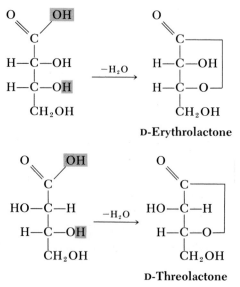

Each lactone is then reduced by sodium borohydride ($NaBH_4$) in aqueous solution to the corresponding aldose:

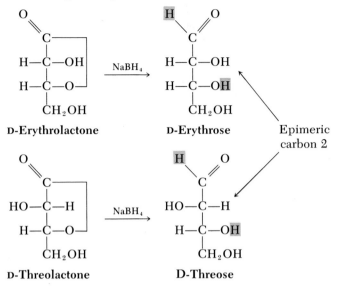

In this synthesis, the lactone (ester) is reduced to the aldehyde rather than the carboxylic acid because of the ease of reduction of esters compared to that of carboxylic acids. The reduction of a carboxylic acid is more difficult to carry out, and the final reduction product would most likely be the alcohol and not the aldehyde:

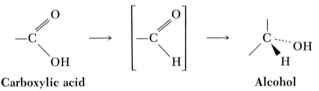

Table 13-4
Kiliani-Fischer synthesis of some aldoses

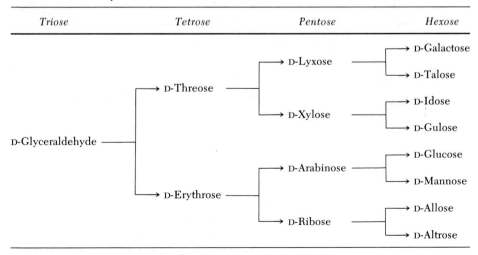

Triose	Tetrose	Pentose	Hexose
			D-Galactose
		D-Lyxose	D-Talose
	D-Threose		D-Idose
		D-Xylose	D-Gulose
D-Glyceraldehyde			D-Glucose
		D-Arabinose	D-Mannose
	D-Erythrose		D-Allose
		D-Ribose	D-Altrose

The reduction of the ester, on the other hand, can be more easily controlled and stopped at the aldehyde stage:

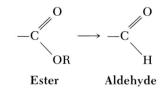

Ester   Aldehyde

Thus the Kiliani-Fischer synthesis converts an aldose into two diastereomeric aldoses, each containing one more carbon atom. These two aldoses differ only in the configuration about carbon 2 and thus are epimers. Various aldoses that can be made by the Kiliani-Fischer synthesis are shown in Table 13-4.

13-10. Write the equations for the reactions that illustrate how D-glucose and its epimer D-mannose would be prepared from D-glyceraldehyde in the Killiani-Fischer synthesis.   *Answer:* Use Table 13-4 and the reactions in this section to go from D-glyceraldehyde to D-erythrose to D-arabinose and then, finally, to D-glucose and D-mannose.

Fructose. The most common ketose is fructose, a 2-ketohexose. It has the following structure:

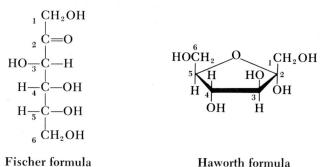

Fischer formula   Haworth formula

In the free state, fructose is found in fruits and honey and is the only sugar in human semen. In the combined state with glucose, it occurs in sucrose. This naturally occurring fructose is levorotatory; thus it is sometimes called levulose. It also gives the same osazone as D-glucose and therefore is D-fructose:

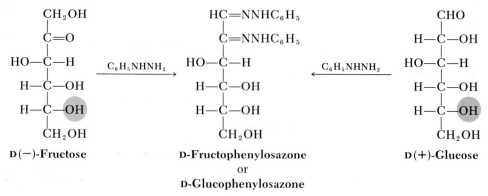

D(−)-Fructose   D-Fructophenylosazone   D(+)-Glucose

or

D-Glucophenylosazone

Fructose, like glucose, is a reducing sugar, since it gives positive tests with Tollens' reagent and Fehling's solution (Chapter 10). It can be distinguished from aldohexoses with bromine water, which oxidizes an aldose to an aldonic acid as in the following example but does not react with fructose:

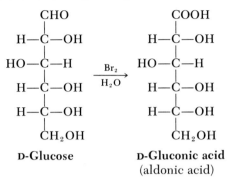

D-Glucose	D-Gluconic acid
	(aldonic acid)

Ribose. This is an aldopentose. It has the following structure:

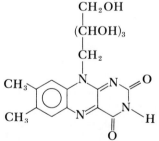

Fischer formula   Haworth formula

Ribose occurs in a number of biologically important compounds. It is part of the vitamin B_{12} molecule and of nucleosides and ribonucleic acids (RNA), which are cellular materials.

The alcohol related to ribose, ribitol, is part of riboflavin, or vitamin B_2, which is shown as follows:

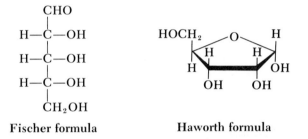

Amino sugars. These are monosaccharides in which an amino group is substituted for one of the hydroxyl groups. Following are some examples:

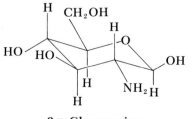

β-D-Glucosamine

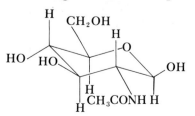

N-Acetyl-β-D-glucosamine

D-Glucosamine is found in human blood, brain tissue, lungs, adrenal glands, heart, and bile. It has an amino group in place of the hydroxyl group on carbon 2 of D-glucose. N-Acetyl-β-D-glucosamine is part of the polysaccharide chitin, found in the shells of crabs, lobsters, and shrimps.

Disaccharides

Disaccharides are composed of two monosaccharides. The union of the two monosaccharides involves the elimination of a molecule of water and the formation of a glycoside linkage. Thus disaccharides are glycosides. When the hydrogen on carbon 1 projects downward in the formula, the glycoside linkage is β; when the hydrogen projects upward, the linkage is α. The β linkage is illustrated below for cellobiose.

When hydrolyzed by acids or enzymes, disaccharides yield two molecules of monosaccharide. The monosaccharides produced differ with the disaccharides hydrolyzed. For example, enzymatic hydrolysis of cellobiose yields β-D-glucose:

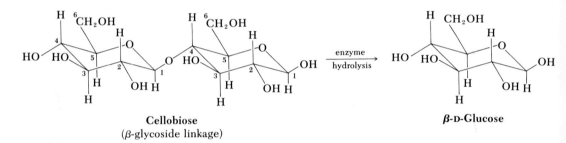

Cellobiose
(β-glycoside linkage)

β-D-Glucose

The monosaccharides formed in the hydrolysis of some other disaccharides are listed in Table 13-5.

Lactose. Human milk and cow's milk are about 5% lactose, which is commonly referred to as milk sugar.

The molecular formula of lactose is $C_{12}H_{22}O_{11}$, and the structural formula is as follows:

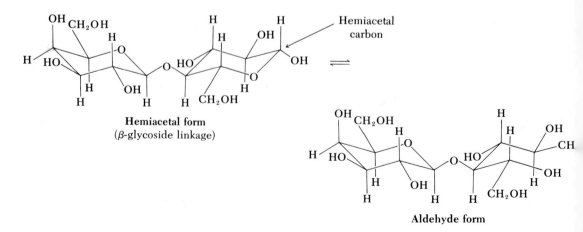

Hemiacetal form
(β-glycoside linkage)

Aldehyde form

Table 13-5
Monosaccharides formed in hydrolysis of some disaccharides

Disaccharide	Monosaccharides
Lactose	D-Glucose and D-galactose
Sucrose	D-Glucose and D-fructose
Maltose	D-Glucose (2 molecules)
Cellobiose	D-Glucose (2 molecules)

Lactose contains a molecule of D-galactose and of D-glucose. The glucose unit contains a hemiacetal group that is free to react. Thus lactose is a reducing sugar. In addition, lactose undergoes mutarotation and forms an osazone.

Sucrose. This is commonly called sugar and is obtained from sugarcane and sugar beets. It is found in nature in the saps and juices of fruits and vegetables.

The molecular formula of sucrose is $C_{12}H_{22}O_{11}$, and the structural formula is as follows:

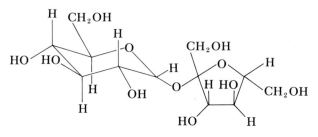

α-Glycoside linkage

Sucrose contains a molecule of glucose and of fructose. It does not reduce Tollens' reagent or Fehling's solution and thus is a nonreducing sugar, since the hemiacetal groups are blocked. Furthermore, it does not form an osazone and does not exhibit mutarotation.

When sucrose is hydrolyzed by dilute aqueous acid or by the action of invertase (an enzyme in yeast), equal amounts of D(+)-glucose and D(−)-fructose are formed. However, sucrose itself is dextrorotatory. The hydrolysis produces a change in the sign of rotation from positive to negative. This process is often called the *inversion of sucrose*. The resulting mixture of glucose and fructose, which is levorotatory, is called invert sugar. Invert sugar is the principal organic component of honey. Bees supply the enzyme invertase, which converts sucrose to invert sugar.

The sugars used by people as sweeteners are not equally sweet. However, the degree of sweetness is difficult to determine, since human tasters must be used. Table 13-6 gaves some idea about the sweetness of a variety of sugars and sugar substitutes discovered by man relative to sucrose.

Table 13-6

Sweetness of some sugars and sugar substitutes compared to sucrose

Sugar or sugar substitute		Relative sweetness
Name	Structure	
Lactose		0.15
Maltose		0.30
Glucose		0.70
Sucrose		1.0
Invert sugar		1.3
Fructose		1.7
Cyclamate		30
Sucaryl		51
Saccharin		200-700
Dulcin		350
2-*n*-Propoxy-5-nitroaniline		4000

Saccharin is probably the best known artificial sweetening agent. It is about 700 times as sweet as sucrose but has a bitter taste for some people. Dulcin is no longer in use, since it has recently been found to produce cancer in rats. 2-*n*-Propoxy-5-nitroaniline is toxic. Sodium cyclohexylsulfamate (Sucaryl) has no apparent aftertaste. Calcium cyclohexylsulfamate (cyclamate) has been removed from the market by the Food and Drug Administration, since it has been shown to cause cancer in some experimental animals.

Recently, a natural sweetener of soluble protein nature was isolated from the West African wild red berries classified as *Sioscoreophyllum cumminsii.* It is reported to be up to 3000 times sweeter than an equal weight of sucrose.

This substance, named monellin, is the first protein reported to elicit a sweet taste in human beings and is the sweetest natural product known.

A proposed theory based on hydrogen bonding of the cause of the sweet taste of substances is explained in Chapter 6.

Maltose and cellobiose. Maltose is the product of the incomplete hydrolysis of starch. Cellobiose is obtained from the hydrolysis of cellulose.

Both maltose and cellobiose have the molecular formula $C_{12}H_{22}O_{11}$. Each forms an osazone, is a reducing sugar, and undergoes mutarotation. Furthermore, they yield D-glucose when hydrolyzed. In cellobiose, the D-glucose units have a β-glycoside linkage; in mannose, they have an α-glycoside linkage. The structures of cellobiose and maltose are as follows:

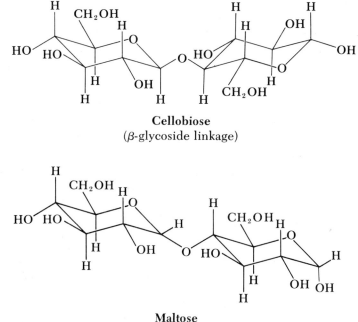

Cellobiose
(β-glycoside linkage)

Maltose
(α-glycoside linkage)

13-11. Which of the following compounds is a reducing sugar?

(a)

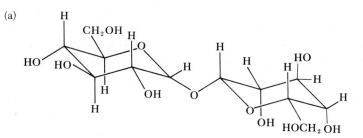

Answer: The glycoside linkage between the two glucose molecules is through the hemiacetal carbons. This sugar is nonreducing.

(b)

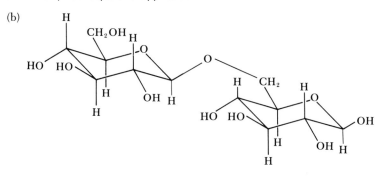

Polysaccharides

Polysaccharides are naturally occurring polymers of high molecular weight. They are made up of many monosaccharide units, which are joined together by glycoside linkages. The most important polysaccharides are cellulose and starch.

Cellulose. It constitutes about half the cellular material of trees and other plants and gives them there rigidity. It is a polymer of D-glucose and has a molecular weight ranging from about 50,000 to 400,000 units per molecule. Since one glucose molecule accounts for 180 units, there are 277 to 2200 glucose units per molecule. Cellulose has the following structural formula:

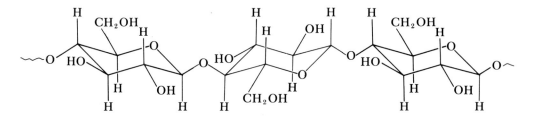

Cellulose is a raw material for a large quantity of commerical products. Cellulose can be extracted from wood or used in the form of cotton. In either case, the —OH groups undergo several reactions to form a variety of esters, such as cellulose nitrates, cellulose acetates, and xanthates.

Cellulose dinitrate contains two nitrate groups per glucose unit. This partially nitrated cellulose is pyroxylin, which when dissolved in ether and alcohol, forms a transparent film called collodion. Collodion is used as a covering over cuts and skin abrasions. A solution of pyroxylin in ketones and esters is used as a lacquer. When pyroxylin is gelatinized with camphor, a material called celluloid is formed. Although it has now been replaced, celluloid was one of the first commercial polymers to be used as photographic film. Nitration of cellulose so that there are three nitrate groups per glucose unit gives cellulose trinitrate, which is a powerful explosive known as guncotton used in smokeless powder as a propellant.

Cellulose diacetate is formed when cellulose is acetylated with acetic anhydride and sulfuric acid. When a solution of the diacetate in acetone is passed through a spinnerette into a stream of warm air to evaporate the acetone, a product called acetate rayon is obtained. However, if the acetone solution is rolled out on warm rollers into sheet form, the product is used in safety film.

When cellulose is treated with a sodium hydroxide solution and then with carbon disulfide, a xanthate is formed:

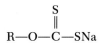

where R is cellulose residue. The xanthate solution is allowed to stand until it changes to a viscous yellow liquid called viscose. When the viscose is passed through a spinnerette into a dilute acid solution, the cellulose xanthate is decomposed and cellulose is regenerated in the form of viscose rayon, which has many of the properties of cotton and is used as a synthetic substitute for it. If the viscose solution is passed through a slit, cellulose is obtained as cellophane, a packaging material.

Starch. It is the form in which most plants store carbohydrates as a reserve food supply. It occurs in seeds, fruits, and the roots of plants in the form of granules.

Starch contains two components: a water-soluble fraction called *amylose* and a water-insoluble fraction called *amylopectin*. The ratio of amylopectin to amylose is about 4:1.

Amylose is a long-chain linear polymer with a molecular weight of about 40,000. Each chain contains about 200 D-glucose units which are joined together by α-1,4-glycoside linkages. The stucture of a portion of the amylose polymer is as follows:

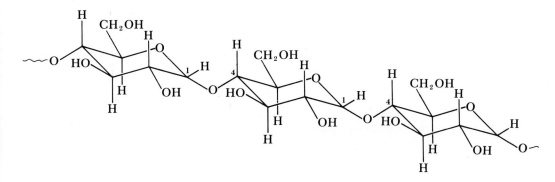

Amylopectin is a branched polymer. The structure consists of linear chains of about 25 D-glucose units with the chains joined by 1,6-glycoside linkages between carbon 1 on one chain and carbon 6 on the next chain. The structure of a portion of the amylopectin polymer is as follows:

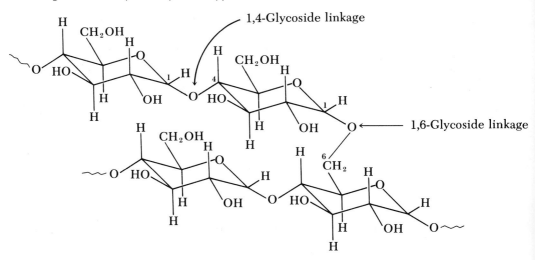

Glycogen. This is the reserve carbohydrate stored in the liver and muscles of animals. Its structure is very similar to that of amylopectin, but it is more highly branched and contains fewer (8 to 12) glucose residues per chain.

Heparin. This is a blood anticoagulant present in the liver, lungs, thymus, spleen, and blood. It is a sulfated polymer containing repeating units of D-glucuronic acid and D-glucosamine. The structure of a portion of the heparin polymer is as follows:

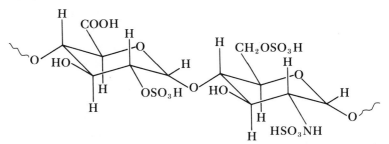

Chitin. This polysaccharide is present in the shells of crustaceans and insects. It is a linear polymer of N-acetyl-D-glucosamine. The structure of a portion of chitin is as follows:

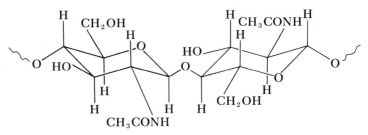

Proteins

Proteins are large complex polymers found in living systems. The monomeric units of protein molecules are α-amino acids, which are bonded by peptide (amide) linkages:

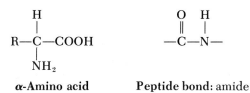

α-Amino acid   **Peptide bond:** amide

Proteins are important for a number of reasons. Cellular growth is intimately related to synthesis of them. The include such diverse compounds as the enzymes (the catalysts of many metabolic reactions), hormones (the regulators of many metabolic reactions), and antibodies. They are also important in our diet, since they are sources of amino acids, the building units necessary for cell growth.

Proteins are polymers with high molecular weights ranging from several thousand to several million. For example, insulin has a molecular weight of 5700; trypsin, 13,500; hemoglobin, 67,000; and urease, 370,000.

Amino acids

Kinds and structure. The amino acids commonly found in proteins, which are α-amino acids, are listed in Table 13-7.

13-12. Find the amino acid(s) in Table 13-7 that fit(s) each of the following descriptions:

    (a) Contains no chiral carbons   *Answer:* Glycine
    (b) α-Aminosuccinic acid   *Answer:* Aspartic acid
    (c) Amide   *Answer:* Asparagine, glutamine
    (d) 2-Amino-3-methylpentanoic acid
    (e) Isomer of 2-amino-3-methylpentanoic acid
    (f) 2-Aminopentanedioic acid
    (g) Thioether
    (h) Contains phenol hydroxyl group
    (i) Dicarboxylic acid
    (j) Thiol

All the amino acids except glycine contain at least one chiral carbon and are optically active. All the naturally occurring α-amino acids except glycine have the L configuration and are related to L-glyceraldehyde.

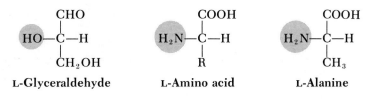

L-Glyceraldehyde   **L-Amino acid**   **L-Alanine**

In Table 13-7, the structures of the amino acids contain an amino group and a carboxyl group. Certain properties, such as high melting point, insolubility in organic solvents, and solubility in water, indicate that the amino acid molecule is actually best represented by the following dipolar structure, referred to as a *zwitterion:*

Table 13-7
Amino acids occurring in protein

Name	Abbreviation	Structure
Alanine	Ala	$CH_3CHCOOH$ $\quad\ \ |$ $\quad\ \ NH_2$
Arginine	Arg	$H_2NCNHCH_2CH_2CH_2CHCOOH$ $\quad\ \ \|\!\|\qquad\qquad\qquad\ \ |$ $\quad\ \ NH\qquad\qquad\qquad\ NH_2$
Asparagine	Asn	$H_2NCCH_2CHCOOH$ $\quad\ \ \|\!\|\qquad\ \ |$ $\quad\ \ O\quad\ \ NH_2$
Aspartic acid	Asp	$HOOCCH_2CHCOOH$ $\qquad\qquad\ |$ $\qquad\qquad NH_2$
Cysteine	CySH	$HSCH_2CHCOOH$ $\qquad\quad\ |$ $\qquad\quad NH_2$
Cystine	CyS-SCy	$HOOCCHCH_2SSCH_2CHCOOH$ $\qquad\ |\qquad\qquad\quad\ |$ $\qquad\ NH_2\qquad\qquad\ NH_2$
Glutamic acid	Glu	$HOOCCH_2CH_2CHCOOH$ $\qquad\qquad\qquad\ |$ $\qquad\qquad\qquad NH_2$
Glutamine	Gln	$H_2NCCH_2CH_2CHCOOH$ $\quad\ \ \|\!\|\qquad\qquad |$ $\quad\ \ O\qquad\qquad NH_2$
Glycine	Gly	CH_2COOH $|$ NH_2
Histidine	His	
Isoleucine	Ile	CH_3 $\qquad\qquad |$ $CH_3CH_2CHCHCOOH$ $\qquad\qquad\quad\ |$ $\qquad\qquad\quad\ NH_2$
Leucine	Leu	$(CH_3)_2CHCH_2CHCOOH$ $\qquad\qquad\qquad\ |$ $\qquad\qquad\qquad NH_2$
Lysine	Lys	$H_2NCH_2CH_2CH_2CH_2CHCOOH$ $\qquad\qquad\qquad\qquad\quad |$ $\qquad\qquad\qquad\qquad\quad NH_2$
Methionine	Met	$CH_3SCH_2CH_2CHCOOH$ $\qquad\qquad\qquad |$ $\qquad\qquad\qquad NH_2$
Phenylalanine	Phe	

Table 13-7
Amino acids occurring in protein — cont d

Name	Abbreviation	Structure
Proline	Pro	
Serine	Ser	$HOCH_2CHCOOH$ $\quad\quad\quad\; \overset{\displaystyle |}{NH_2}$
Threonine	Thr	HO $\;\;\;\; |$ $CH_3CHCHCOOH$ $\quad\quad\quad\quad\; \overset{\displaystyle |}{NH_2}$
Tryptophan	Try	$CH_2CHCOOH$ $\quad\quad\; \overset{\displaystyle |}{NH_2}$
Tyrosine	Tyr	$HO-$⟨◯⟩$-CH_2CHCOOH$ $\quad\quad\quad\quad\quad\;\; \overset{\displaystyle |}{NH_2}$
Valine	Val	$(CH_3)_2CHCHCOOH$ $\quad\quad\quad\quad\;\; \overset{\displaystyle |}{NH_2}$

$$R-CH-COO^{\ominus}$$
$$\overset{|}{\underset{\oplus}{N}H_3}$$

The addition of an acid to this dipolar ion (I) causes the formation of a carboxyl group (II), whereas the addition of base causes the formation of an amino group (III):

$$R-CH-COOH \;\overset{H^{\oplus}}{\rightleftharpoons}\; R-CH-COO^{\ominus} \;\overset{OH^{\ominus}}{\longrightarrow}\; R-CH-COO^{\ominus}$$
$$\overset{|}{_{\oplus}NH_3} \quad\quad\quad\quad \overset{|}{_{\oplus}NH_3} \quad\quad\quad\quad \overset{|}{NH_2}$$
$$\textbf{II} \quad\quad\quad\quad\quad \textbf{I} \quad\quad\quad\quad\quad \textbf{III}$$

Amino acid cation   **Zwitterion**   **Amino acid anion**
(acid solution)   (basic solution)

In an acidic solution (II), the amino acid has a positive charge and migrates toward the cathode if placed in an electric field. In a basic solution (III), the amino acid has a negative charge and migrates toward the anode if placed in an electric field. The zwitterion (I) has a net charge of zero and does not migrate in an electric field. The pH at which the amino acid has no net charge and is not attracted to either electrode is called the *isoelectric point*.

At the isoelectric point, amino acids show their lowest solubility in water. Because of the varied structures and numbers of acidic and basic groups in a protein molecule, each protein has a different isoelectric point. Thus a particular protein will precipitate from a solution when the pH of the solution is equal to its isoelectric point. This is the basis of the classical method of separating one protein from another.

The curdling of milk to produce sour milk is an example of the effect on the solubility of a protein caused by a change in pH. The principal protein found in milk is casein. The isoelectric point of casein, at which it is very slightly soluble in water, is 4.7. The pH of milk is about 6.5. Milk curdles, or sours, by enzymatic action of bacteria or by the addition of lemon juice. Both lactic acid, produced by bacterial action, and citric acid, the principal component of lemon juice, lower the pH of milk. This results in the formation of aggregates of molecules of casein held together by the attraction of positive

and negative charges. In other words, casein $\left(-\overset{|}{\underset{\overset{|}{\oplus}NH_3}{C}}\sim C\overset{\overset{\displaystyle O}{\diagup\diagup}}{\diagdown\underset{\ominus}{O}}\right)$, represented as

$\oplus\ominus$, forms aggregates, $\oplus\ominus\ \oplus\ominus\ \oplus\ominus\ \oplus\ominus$, at the isoelectric point. Thus precipitation occurs; that is, the milk curdles.

Amino acids in the zwitterion form also serve living systems in the role of buffers by neutralizing small amounts of acids or bases so that pH of the system remains constant. An amino acid at the isoelectric point can neutralize either an acid (H_3O^+, hydronium ion) or a base (OH^-) as follows:

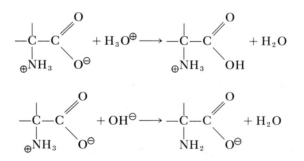

For example, proteins found in the blood are capable of acting as buffers to keep the pH in the range of 7.00 to 7.90.

Reactions. Amino acids contain both an amino group and a carboxyl group. Therefore, they would be expected to undergo the reactions of compounds containing amino groups and those containing carboxyl groups.

The carboxyl group can form a salt with a base. It can also form an ester, and the resulting ester can be reduced to an amino alcohol or can form an amide. These reactions are illustrated as follows with glycine as the starting material:

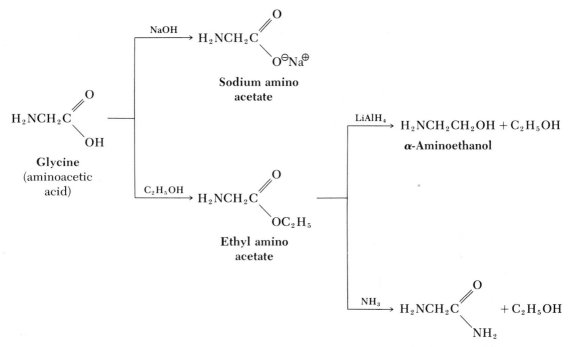

The amino group can form a salt with an acid, an amide with an acid chloride, or an N-alkylated amine with an alkyl halide. These reactions are illustrated as follows:

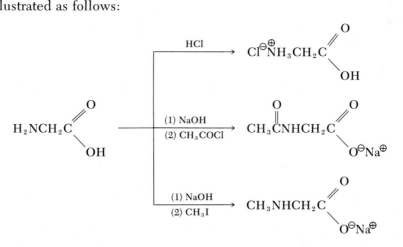

The amino group can also react with 2,4-dinitrofluorobenzene (DNFB) via a nucleophilic aromatic substitution mechanism. Following is an example:

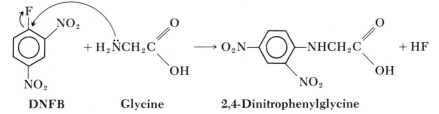

This reaction has been found to be useful as a method of determining the amino ends of peptides and is discussed in more detail later in this chapter.

13-13. Give the organic product expected from each of the following reactions:

(a)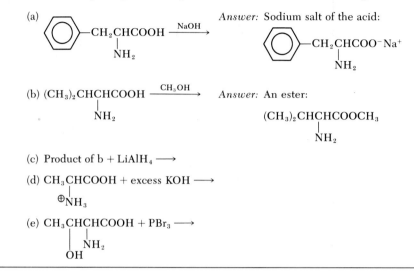

Answer: Sodium salt of the acid:

(b) $(CH_3)_2CHCHCOOH$ with NH_2 $\xrightarrow{CH_3OH}$ *Answer:* An ester:

$(CH_3)_2CHCHCOOCH_3$ with NH_2

(c) Product of b + LiAlH$_4$ $\longrightarrow$

(d) $CH_3CHCOOH$ with $\overset{\oplus}{N}H_3$ + excess KOH $\longrightarrow$

(e) $CH_3CHCHCOOH$ with OH and NH_2 + PBr$_3$ $\longrightarrow$

Peptides

One of the most important reactions of amino acids is the formation of peptides by the interaction between the carboxyl group of one amino acid and the amino group of a second amino acid. In the process, a molecule of water is eliminated, and a peptide linkage, an amide group, is formed. Following is an example of this reaction:

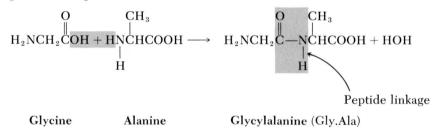

Peptide linkage

Glycine   Alanine   Glycylalanine (Gly.Ala)

Glycylalanine, which contains two amino acids is a dipeptide. A tripeptide contains three amino acids; a tetrapeptide, four amino acids; a pentapeptide, five amino acids; and so on. Finally, there are polypeptides. Following are some examples:

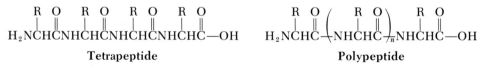

Tetrapeptide   Polypeptide

As can be seen from the general structures of the tetrapeptide and the polypeptide and the structure of glycylalanine, a peptide has a free amino group on one end of the peptide chain and a free carboxyl group on the other end. By convention, the N-terminal amino acid (with the free amino group) is written on the left end of the peptide chain, and the C-terminal amino acid (with the free carboxyl group) is written on the right end:

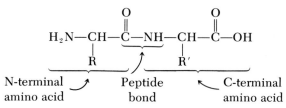

| N-terminal amino acid | Peptide bond | C-terminal amino acid |

Some of the more interesting examples of peptides are those formed in the hypothalamus in the brain. They stimulate the pituitary gland to release hormones that regulate the activities of other endocrine glands. One of these peptides is thyrotropin-releasing hormone (TRH), a tripeptide composed of glutamic acid, histidine, and proline. In TRH, glutamic acid exists as a cyclic amide (a lactam), and the carboxyl group of proline (the C-terminal amino acid) is replaced by an amide group (CONH$_2$). The structure of TRH is as follows:

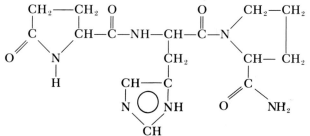

TRH causes the pituitary gland to release thyroid-stimulating hormone (TSH), which stimulates the thyroid gland in terms of growth and the production of thyroxine.

Oxytocin, an octapeptide, is a hormone formed in the pituitary gland that causes muscle contractions of the uterus. Its structure is as follows:

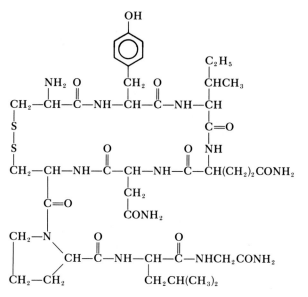

When written in the abbreviated form, the structure of oxytocin is as follows:

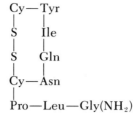

Another factor from the hypothalamus that effects the release of hormones is a decapeptide that indirectly regulates the maturation of the follicle and production of estrogen in the female. Its abbreviated structure is as follows:

Glu—His—Tyr—Ser—Tyr—Gly—Leu—Arg—Pro—Gly(NH₂)

In both oxytocin and this decapeptide, the C-terminal amino acid, glycine, is in the form of its amide, Gly(NH₂).

13-14. Draw the complete structure of the decapeptide just discussed.

Corticotropin, a hormone that stimulates the adrenal cortex to produce steroid hormones, contains 39 amino acids. The abbreviated structure of this compound is as follows:

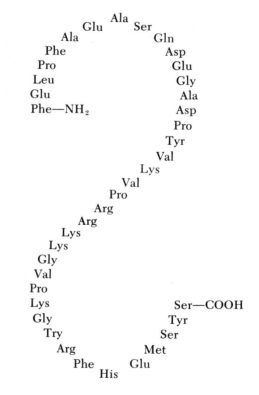

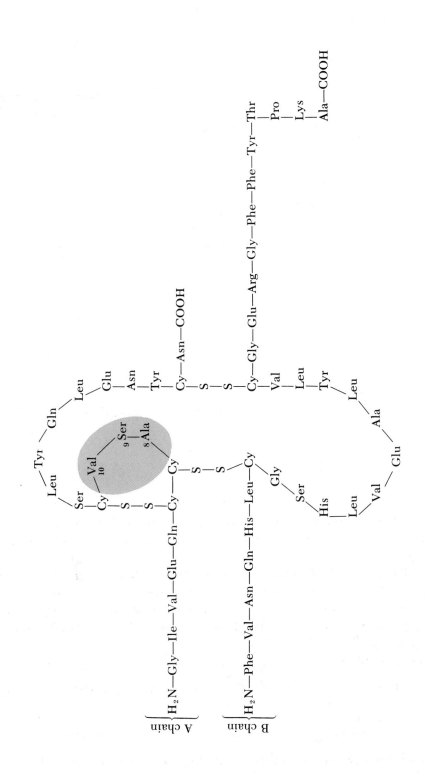

Insulin is a protein hormone produced in the pancreas. It is essential in the regulation of carbohydrate metabolism, which is discussed in Chapter 14. It contains two peptide chains: the A chain with 21 amino acids and the B chain with 30 amino acids. The A and B chains are held together by disulfide linkages of cystine. The amino acid sequence in beef insulin is shown on p. 407. The amino acids 8, 9, and 10 of the A chain vary from animal to animal. For example, in sheep the 8-9-10 amino acid sequence is alanine-glycine-valine, and in horses the 8-9-10 sequence is threonine-glycine-isoleucine. These differences are advantageous to the diabetic who may be allergic to one type of insulin. The allergic reaction can be minimized or even avoided completely when the animal source of insulin is changed.

13-15. (a) Draw the stucture for a pentapeptide, using the amino acids in Table 13-7.
　　　(b) Which amino acid of this pentapeptide is the N-terminal amino acid?
　　　(c) Which amino acid of this pentapeptide is the C-terminal amino acid?

13-16. Draw the two structures for the dipeptide that is made up of isoleucine and serine.

13-17. Draw a structure for each of the following compounds:

　　　(a) Alanylglycine

　　　(b) Asp.Val.Phe

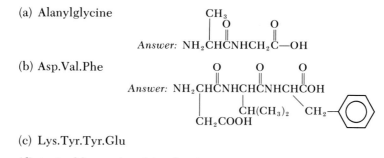

　　　(c) Lys.Tyr.Tyr.Glu

　　　(d) Arginylthreonylmethionylserine

Chemical synthesis of peptides. The purpose of synthesizing peptides is to confirm their structure and to make peptides identical with naturally occurring ones.

The method of peptide synthesis is to join together amino acids in a particular sequence to form chains of correct lengths. At first glance, this appears to be quite simple. For example, in synthesizing glycylalanine, it would seem reasonable to condense glycine with alanine as in the following equation:

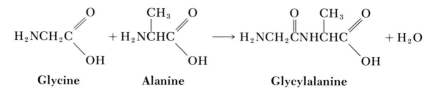

Unfortunately, several other completing reactions can occur: the formation of glycylglycine from two glycine molecules, the formation of alanylglycine from alanine and glycine, and the formation of alanylalanine from two alanine molecules. These reactions are as follows:

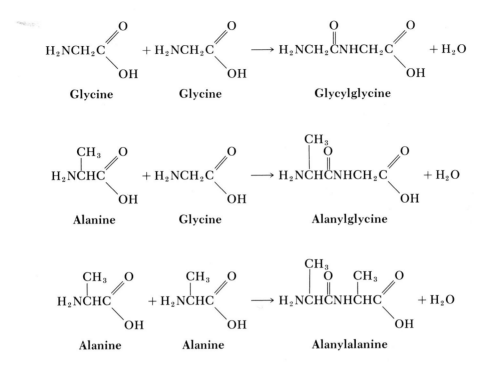

Glycine   Glycine   Glycylglycine

Alanine   Glycine   Alanylglycine

Alanine   Alanine   Alanylalanine

Thus this method of simply condensing the carboxyl group of one amino acid with the amino group of another amino acid leads to the formation of a mixture of dipeptides.

This difficulty has been solved by the following procedure. First, the amino acid is protected; that is, it is reacted with carbobenzoxy chloride

$\left(\text{\includegraphics-CH}_2\text{OCCl}\right)$, which protects the amino group by rendering it

unreactive. The protected amino group will be the peptide's N-terminal amino acid.

The second step involves condensation of the carbobenzoxy amino acid with the second amino acid containing the free amino group. This condensation is accomplished with dicyclohexylcarbodiimide, which acts as a dehydrating agent and as an activator of the carboxyl group in forming the peptide bond. Dicyclohexylcarbodiimide is converted to dicyclohexyl urea. This step results in formation of the peptide linkage.

The final step is the removal of the protecting carbobenzoxy group by hydrogenation with a palladium catalyst. This is accomplished without a breaking of the peptide linkage.

These steps in peptide synthesis of (1) protection of the amino group, (2) formation of the peptide bond, and (3) removal of the protecting group are illustrated as follows for the synthesis of glycylalanine:

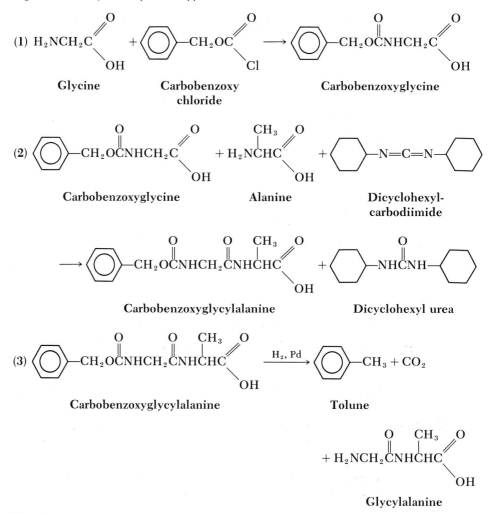

(1) H₂NCH₂C (Glycine) + Carbobenzoxy chloride → Carbobenzoxyglycine

(2) Carbobenzoxyglycine + Alanine + Dicyclohexylcarbodiimide →

→ Carbobenzoxyglycylalanine + Dicyclohexyl urea

(3) Carbobenzoxyglycylalanine $\xrightarrow{H_2,\ Pd}$ Tolune + CO₂

+ Glycylalanine

13-18. Write the equations for the synthesis of the dipeptide alanylglycine by the protecting-group method.

The dipeptide chain can be lengthened by the process of treating it with a third amino acid before removing the protecting group. For example, the reaction between carbobenzoxyglycylalanine and valine is as follows:

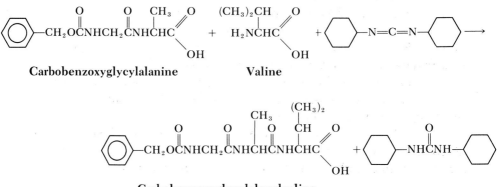

Carbobenzoxyglycylalanine + Valine + Dicyclohexylcarbodiimide →

Carbobenzoxyglycylalanylvaline

Solid-phase peptide synthesis. The method of synthesizing a peptide by the technique of combining amino acids with smaller peptides in a predetermined sequence is very tedious and time consuming, and more importantly, the yields are poor. Recently, a solid-phase peptide synthesis was developed by Merrifield. This method has the advantage of avoiding the intermediate isolation steps characteristic of the multistep synthesis previously described. The method is rapid, and the yields are good. The amino acid that will ultimately be the C-terminal amino acid containing the free carboxyl end of the peptide is bonded through its carboxyl end to an active group on a solid polymer. After synthesis is completed at the desired stage, the peptide is easily removed from the solid polymer. The steps involved in the Merrifield synthesis are summarized as follows:

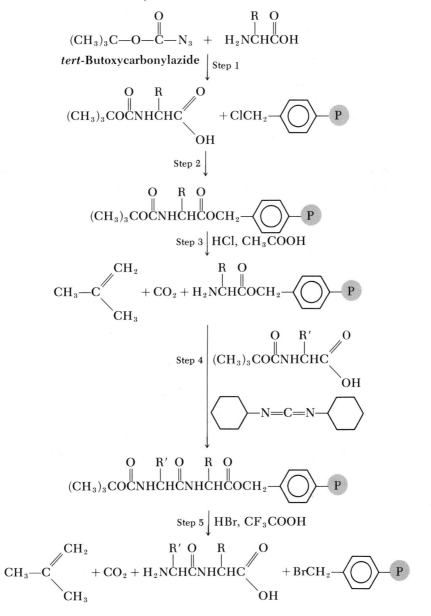

The amino group of the first amino acid is protected by reaction with *tert*-butoxycarbonylazide (step 1). The protected amino acid is then bonded to the active site on a solid polymer through the carboxyl group of the protected amino acid (step 2). The product of step 2 is not isolated, but the protecting group is washed away with hydrochloric acid and acetic acid (step 3). The peptide chain is elongated by treatment of a second protected amino acid with the peptide bonded to the polymer in the presence of dicyclohexyldiimide (step 4). After being washed, the peptide chain is extended further or is removed from the polymer surface with hydrobromic acid and anhydrous trifluoroacetic acid (step 5). The N-protecting group is also removed in step 5. The advantage of this method is that it can be automatically controlled and programmed so that the reagents can be added and the product washed at the appropriate times. Insulin (51 amino acids) and ribonuclease (124 amino acids) have been synthesized by this method.

Structure of proteins

The number and sequence of amino acids in a peptide chain determine the *primary structure* of the protein.

The long chain of a peptide is capable of twisting and turning into a unique pattern of coils, or helices, known as the *secondary structure* of the protein. This structure results from hydrogen bonding between the amide hydrogen of one amino acid in the chain and the carbonyl group of another amino acid in the same chain. It can be visualized with each chain coiled to form a right-handed helix similar to a spiral staircase:

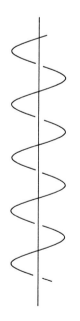

The helix is stabilized and held together by the hydrogen bonds:

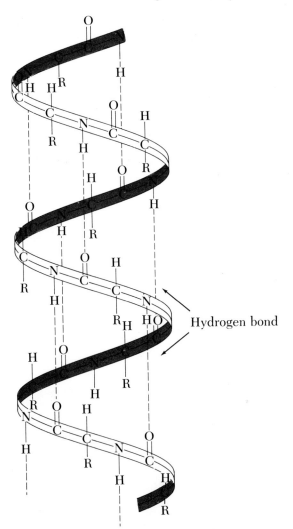

Another kind of secondary structure is formed by hydrogen bonding between adjacent chains. It is referred to as a pleated sheet. The characteristic feature of the pleated sheet structure is that the chains are antiparallel; that is, the adjacent chains head in opposite directions. Following is an example:

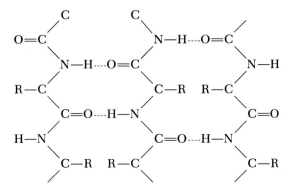

Higher levels of organizational structure of the peptide lead to a unique three-dimensional form associated with biological activity. These higher levels of organizational structure are the tertiary and quaternary structures of the peptide.

Determination of primary structure. Let us consider an example of the determination of the primary structure. A hexapeptide is isolated, and complete hydrolysis of a sample indicates that it is composed of the following six amino acids: valine, serine, tyrosine, isoleucine, alanine, and glycine. It is also determined that it contains a free amino group (N-terminal amino acid) at one end and a carboxylic acid group (C-terminal amino acid) at the other end.

One method of determining the sequence of amino acids involves the reaction of 2,4-dinitrofluorobenzene with the N-terminal amino acid. This reaction is illustrated as follows:

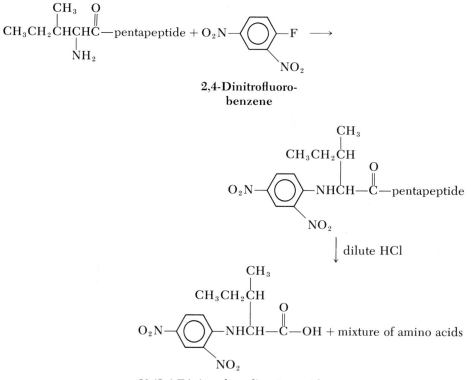

N-(2,4-Dinitrophenyl)amino acid

The N-(2,4-dinitrophenyl)amino acid is then separated from the other amino acids and identified. This method suffers from the disadvantage that the dilute HCl hydrolyzes not only the N-terminal amino acid peptide bond but also other peptide bonds in the protein. The sequence of amino acids is therefore not accurately determined.

A better method involves derivative preparation: the reaction of phenyl isothiocyanate with the N-terminal amino acid to form a thiourea, the hydrolysis of which liberates the N-terminal amino acid as a phenylthiohydantoin and leaves the remainder of the molecule intact. This reaction is illustrated as follows:

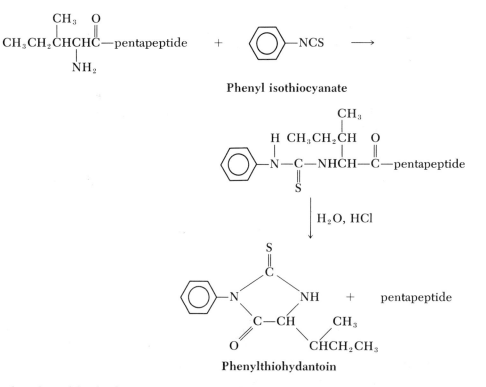

Phenyl isothiocyanate

Phenylthiohydantoin

The phenylthiohydantoin is separated from the pentapeptide and identified. The pentapeptide can then be treated with phenyl isothiocyanate to identify the new N-terminal amino acid. This technique can be repeated until the complete sequence of amino acids is determined.

The C-terminal amino acid is determined through the hydrolysis of the hexapeptide with the enzyme carboxypeptidase, which hydrolyzes the peptide bond adjacent to the free carboxyl group. Then the reaction with carboxypeptidase can be used to determine each new C-terminal amino acid.

As determined by these methods, isoleucine is the N-terminal amino acid and glycine is the C-terminal amino acid of the hexapeptide example:

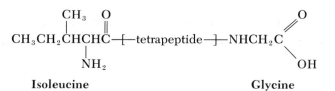

Isoleucine   **Glycine**

However, the use of phenyl isothiocyanate to determine each succeeding N-terminal amino acid and of carboxypeptidase to determine each succeeding C-terminal amino acid is limited. In practice, the hexapeptide is partially hydrolyzed to a mixture of dipeptides, tripeptides, tetrapeptides, and pentapeptides. Each of these fragments is analyzed for the N-terminal and C-terminal amino acids. For example, suppose the fragments have been identified as follows:

1. Ile.Ser.Ala
2.   Ser.Ala
3.   Ser.Ala.Val
4.   Val.Tyr
5.   Val.Tyr.Gly

By overlapping of the various fragments, the sequence of amino acid is determined to be Ile.Ser.Ala.Val.Tyr.Gly. Thus the complete structure of the hexapeptide is as follows:

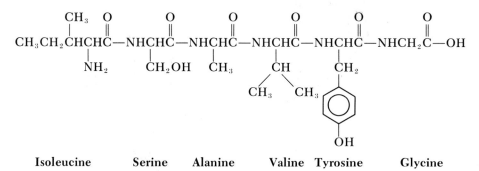

Isoleucine   Serine   Alanine   Valine   Tyrosine   Glycine

In summary, the primary structure of a protein is determined by partial hydrolysis of the protein into smaller peptide fragments followed by identification of each peptide fragment by the N-terminal amino acid and C-terminal amino acid determinations.

13-19. What is the sequence of amino acids in a peptide that when partially hydrolyzed gives the fragments His.Gly, Phe.Tyr, Asp.Phe, and Gly.Asp? Histidine is the N-terminal amino acid, and tyrosine is the C-terminal amino acid.

Answer: His-[tripeptide]-Tyr

N-terminal   C-terminal
amino acid   amino acid

His.Gly
Gly.Asp
Asp.Phe
Phe.Tyr
_____
His.Gly.Asp.Phe.Tyr

Nucleic acids

Nucleic acids are important to living systems because they are involved in the transmission of genetic information and in protein synthesis in the cell. They are polymers with nucleotides as the repeating, or monomeric, units. Each nucleotide is made up of a pentose sugar, a nitrogen-containing base, and phosphoric acid.

The bases are heterocyclic amines derived from purine and pyrimidine. A pyrimidine base consists of a heterocyclic ring containing two nitrogen atoms. The pyrimidine bases found in nucleic acids are cytosine, uracil, and thymine:

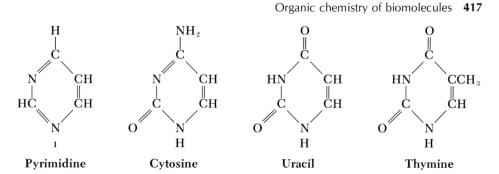

| Pyrimidine | Cytosine | Uracil | Thymine |

The purine bases are adenine and guanine:

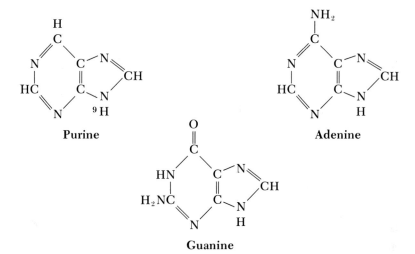

| Purine | Adenine |

Guanine

The pentose sugars in nucleic acids are D-ribose and D-2-deoxyribose, which occur in the furanose form:

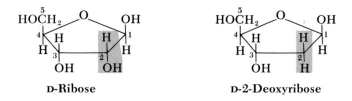

D-Ribose   **D-2-Deoxyribose**

The combination of a purine or pyrimidine base with D-ribose or D-2-deoxyribose results in a *nucleoside*:

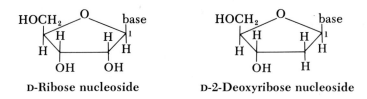

D-Ribose nucleoside   **D-2-Deoxyribose nucleoside**

The linkage is between nitrogen 1 in a pyrimidine base or nitrogen 9 in a purine base and carbon 1 of the pentose sugar. Following are two examples:

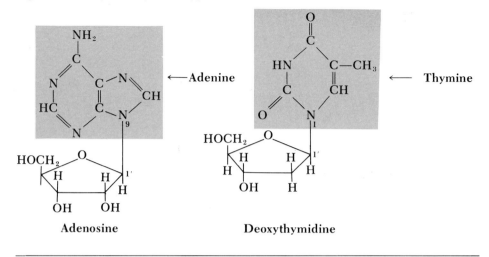

Adenosine   Deoxythymidine

13-20. Draw the structure of the nucleoside formed from D-ribose and cytosine.

Answer: Nucleoside is a sugar-base is D-ribose—cytosine

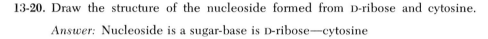

A *nucleotide*, which is a combination of a base, a sugar, and a phosphate, is formed when one of the hydroxyl groups of the sugar in a nucleoside is esterified by phosphoric acid. Following are two examples:

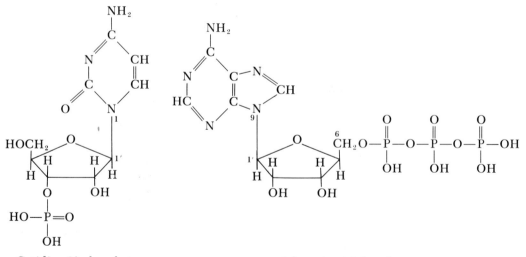

Cytidine-3′-phosphate   Adenosine triphosphate

Adenosine triphosphate (ATP) is a source of energy in metabolism. This energy arises from the phosphate bonds.

13-21. Draw the structure of the nucleotide formed from D-2-deoxyribose, thymine, and phosphoric acid.

Answer: Nucleotide=base-sugar–phosphate=thymine—D-2-deoxyribose—
phosphate

A *nucleic acid,* which is a polymer with a nucleotide repeating unit, is more complex than a nucleotide or a nucleoside:

$$n(\text{base—sugar—phosphate}) \longrightarrow (\text{base—sugar—phosphate})_n$$

Nucleotide	**Nucleic acid**

Following is a summary of the successive levels of structural complexity of nucleic acid components:

	base	base	base
	|	|	|
Nucleic acid:	⌒sugar—phosphate—sugar—phosphate—sugar—phosphate⌒		

	base
	|
Nucleotide:	sugar—phosphate

	base
	|
Nucleoside:	sugar

Purine or pyrimidine: base

Ribose or deoxyribose: sugar

Two important nucleic acids are deoxyribonucleic acid (DNA) and ribonucleic acid (RNA).

The primary structure of DNA consists of nucleotide chains with phosphate linkage between carbon 3' of one sugar molecule and carbon 5' of the next sugar molecule. The primary structure of a segment of a DNA molecule is shown as follows:

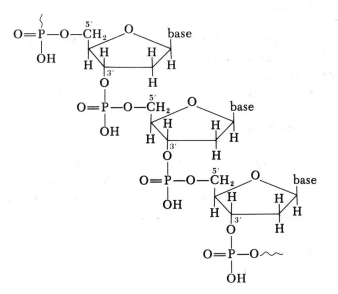

The nitrogen bases found in DNA are adenine, guanine, cytosine, and thymine. The number of purine bases (adenine and guanine) is equal to the number of pyrimidine bases (cytosine and thymine). Furthermore, the ratio of adenine to thymine is 1:1, and the ratio of cytosine to guanine is 1:1.

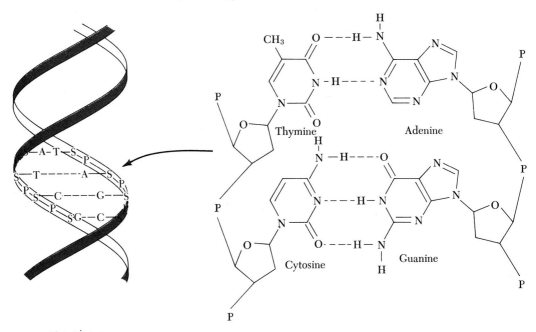

Fig. 13-1
Double helix of DNA illustrating base pairing. S, sugar; P, phosphate; A, adenine; T, thymine; C, cytosine; G, guanine.

13-22. Draw a segment of a DNA polymer composed of adenine, guanine, cytosine, and thymine.

Recent investigations have shown that DNA actually consists of two nucleotide chains coiled in a double helix. The nucleotide chains coil in opposite directions and are held in the helix by hydrogen bonding between adenine in one chain and thymine in the other chain and between cytosine in one chain and guanine in the second chain. The hydrogen bonding stabilizes the double helix. An example of the double helix of DNA is shown in Fig. 13-1.

The primary structure of RNA is similar to that of DNA in that it contains the phosphate ester linkages at carbons 3′ and 5′. However, RNA is different in several respects. Thymine is replaced by uracil. There is a large variation in base content, and there is no 1:1 ratio of purine bases to pyrimidine bases. Furthermore, the pentose sugar is D-ribose. Following is a comparison of the components of RNA and DNA:

RNA	*DNA*
Phosphoric acid	Phosphoric acid
D-Ribose	D-2-Deoxyribose
Adenine	Adenine
Guanine	Guanine
Cytosine	Cytosine
Uracil	Thymine

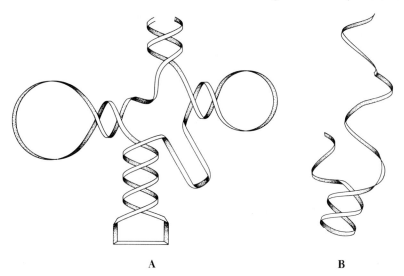

Fig. 13-2
Secondary structure of, **A,** tRNA and, **B,** rRNA.

The secondary structure of RNA consists of single, flexible nucleotide chains that fold on themselves to some extent and result in some helical content. In addition, the helical regions are separated by straight segments. The secondary structure is illustrated in Fig. 13-2.

There are several types of RNA: transfer RNA (tRNA), ribosomal RNA (rRNA), and messenger RNA (mRNA). The largest of these in terms of molecular size is rRNA. mRNA has a base composition complementary to DNA and transmits the genetic code of DNA to the protein synthesizing site. tRNA, the smallest of the three in terms of molecular size, functions by binding specific amino acids and carrying them to a specific site on mRNA so that they are then incorporated into the growing peptide chain. The biosynthesis of proteins in the cell, which is discussed in Chapter 14, involves a relationship between DNA and RNA.

Summary

Besides water, the most abundant compounds in living systems are lipids, carbohydrates, proteins, and nucleic acids.

Lipids are a heterogeneous class of naturally occurring compounds that are soluble in organic solvents. They include neutral fats, waxes, phospholipids, cerebrosides, steroids, and fat-soluble vitamins. Neutral fats, or triglycerides, are esters of glycerol and long-chain fatty acids. They can be saponified to give soaps or partially hydrogenated so that they have a degree of softness and the tendency toward rancidity is reduced. A triglyceride can be completely hydrogenated to give long-chain alcohols, which can be separated and used to make detergents.

Carbohydrates are polyhydroxy aldehydes or ketones and are classified as monosaccharides, disaccharides, or polysaccharides. Glucose is the most abundant and perhaps the most important monosaccharide. It undergoes many of the reactions characteristic of aldehydes. However, when it is treated with methanol, it forms two different monomethyl derivatives. It exists in an α-form and a β-form. Since its carbonyl group is in the form of a hemiacetal that is free to react, it is a reducing sugar and thus undergoes mutarotation. Mutarotation results from interconversion of the α- and β-isomers to an equilibrium mixture. Monosaccharides can be made larger by the Kiliani-Fischer synthesis, which involves the formation of a cyanohydrin. An aldose is converted into two diastereomeric aldoses, each containing one more carbon atom. These diastereomers differ in configuration about carbon 2 and are called epimers. The most common disaccharide is sucrose, which contains a molecule of glucose and of fructose and is a nonreducing sugar. Two important polysaccharides are cellulose and starch. Cellulose forms the skeletal structure of plants, and starch is a reserve source of sugar.

Proteins are naturally occurring polymers composed of amino acids and characterized by the peptide linkage, which is an amide group. Amino acids undergo many of the reactions characteristic of the amino group and the carboxyl group. Peptides can be synthesized by condensation of one amino acid with another whose amino group has been protected by carbobenzoxy chloride. In the Merrifield solid-state synthesis, another method of peptide synthesis, an amino acid that will be the C-terminal amino acid of the peptide is bonded to a polymer through its carboxyl group. The advantage of this synthesis is that all intermediate peptides do not have to be isolated. The primary structure of a protein is determined by the number and sequence of amino acids. It can be determined by partial hydrolysis of a protein to peptides. Each peptide is identified by the N-terminal and C-terminal amino acid determinations. The secondary structure is the helix, which results from hydrogen bonding between amino acids in the same chain. Tertiary and quaternary structures of proteins are also important.

Nucleic acids contain a purine or pyrimidine base, a ribose or deoxyribose sugar, and phosphate. They are important because they are involved in the transmission of genetic information and in protein synthesis in the cell. Two important nucleic acids are DNA and RNA.

Important terms

Aldose	Fat-soluble vitamin
Amino acid	Furanose
Bile salt	Glycoside
Carbobenzoxy chloride	Invert sugar
Carbohydrate	Isoelectric point
Cerebroside	Ketose
C-terminal amino acid	Kiliani-Fischer synthesis
Detergent	Lipid
Double helix	Merrifield synthesis
Epimer	Monosaccharide

Mutarotation	Purine
Neutral fat	Pyranose
N-terminal amino acid	Pyrimidine
Nucleic acid	Reducing sugar
Nucleoside	Saponification
Nucleotide	Secondary structure of protein
Peptide bond	Soap
Phenyl isothiocyanate	Steroid
Phenylthiohydantoin	Steroid hormone
Phospholipid	Wax
Polysaccharide	Zwitterion
Primary structure of protein	

Problems

13-23. Why are proteins, lipids, and carbohydrates important to living systems?

13-24. Draw structures for the following pairs of molecules and explain how they differ:

   (a) Fat and fatty acid
   (b) Cholesterol and cortisone
   (c) Saturated triglyceride and polyunsaturated triglyceride
   (d) D(+)-Glucose and D(+)-mannose
   (e) α-D-Glucose and β-D-glucose
   (f) D-Glyceraldehyde and D-erythrose
   (g) Maltose and cellobiose
   (h) Cellobiose and chitin
   (i) Leucine and isoleucine
   (j) D-Alanine and L-alanine
   (k) Glycylalanine and alanylglycine
   (l) DNA and RNA
   (m) 9-Methylpurine and 1-methylpyrimidine
   (n) Ribose and deoxyribose
   (o) Nucleoside and nucleotide
   (p) ATP and AMP (adenosine monophosphate)
   (q) Nucleotide and nucleic acid
   (r) Tripeptide and pentapeptide

13-25. Write the structural formula for a triglyceride that contains three different fatty acids and is classified as polyunsaturated.

13-26. What are the conditions for saponification of the triglyceride in problem 13-25? Identify the triglyceride by naming the products of saponification.

13-27. Draw all the geometric isomers of linoleic acid.

13-28. Discuss the stereochemistry of cholesterol. How many chiral carbons does it have?

13-29. Using a diagram, discuss how a soap cleans.

13-30. What are the similarities and differences between the compounds of each of the following pairs?

   (a) Neutral fats and waxes   (c) Lecithins and cerebrosides
   (b) Phospholipids and steroids   (d) Soap and detergent

13-31. Write a structural formula for β-D-(+)-glucopyranose, and explain the significance of β-, -D-, -(+)-, -gluco-, -pyran-, and -ose in reference to the formula?

13-32. With an appropriate equation, illustrate the concept of mutarotation for glucose. Is this reaction an epimerization? Which form of glucose is present in the highest concentration? Explain your choice.

13-33. For glucose, draw the structure for each of the following:
(a) Structural isomer   (c) Diastereomer
(b) Enantiomer   (d) Epimer

13-34. Write the equation for the formation of a disaccharide from two glucose molecules. Is the bonding in this disaccharide typical of starch or cellulose? Is this disaccharide an α- or a β-glycoside? Is it a reducing sugar?

13-35. The conformational formula of β-maltose (4-O[α-D-glucopyranosyl]-β-D-glucopyranose) is as follows:

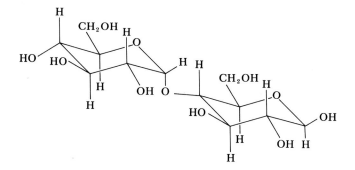

Draw the appropriate conformational formula of the structure of β-maltose modified to represent each of the following compounds:
(a) Amylose   (d) Sucrose
(b) Cellulose   (e) α-Maltose
(c) Lactose

13-36. Consider the following peptide:

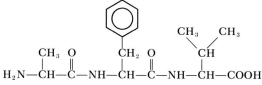

(a) Classify it on the basis of the number of amino acids.
(b) What is the name of it?
(c) Draw the structural formulas of the amino acids formed when it is hydrolyzed.

13-37. The synthesis of peptides with a specific primary structure is a difficult task. Explain the difficulties, and suggest a solution to the problem.

13-38. Write the complete structural formulas for two different monomers in a typical DNA molecule. Do the same for a typical RNA molecule.

13-39. Vasopressin is an octapeptide hormone. It has the following abbreviated formula:

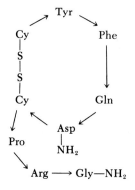

Draw the complete structure of vasopressin. How many peptide linkages are there?

13-40. What is meant by the term *base pair* in nucleic acid chemistry?

13-41. Write structural formulas to show hydrogen bonding between the compounds of each of the following pairs:
(a) Adenine and thymine
(b) Guanine and cytosine
(c) Adenine and uracil

13-42. Normal adult hemoglobin consists of 574 amino acids distributed among four polypeptide chains. Sickle cell anemia is a hereditary disease in which the red blood cells possess a characteristic sickle shape. The hemoglobin in such cells is not able to carry the necessary amount of oxygen throughout the body. In this abnormal hemoglobin, the glutamic acid residue of normal hemoglobin is substituted by valine. Partial hydrolysis of abnormal hemoglobin gives an octapeptide, among other peptides. In this octapeptide, valine is the N-terminal amino acid and lysine is the C-terminal amino acid. Partial hydrolysis of the octapeptide forms the following fragments (by convention, the N-terminal amino acid is on the left): Val.His.Leu, Pro.Val.Glu.Lys, Leu.Thr.Pro, and His.Leu.Thr. What is the amino acid sequence in abnormal hemoglobin?

Self-test

1. Draw a complete structure for each of the following compounds:
(a) Triglyceride
(b) Detergent
(c) Cholesterol
(d) Disaccharide
(e) Conformational formula for β-glucose
(f) Aldotetrose
(g) Osazone
(h) Alanine
(i) Dipeptide
(j) Form of glycine in solution of pH 5
(k) Purine base
(l) Nucleoside

2. Give the structure of the organic product of each of the following reactions:

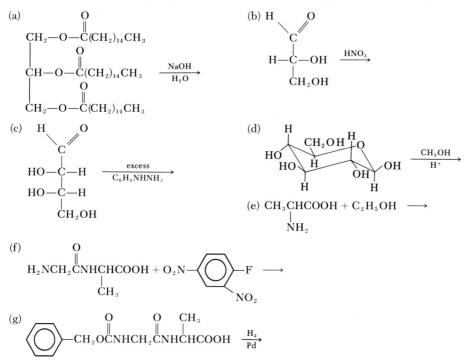

3. Briefly discuss the advantages of the Merrifield solid-phase synthesis of peptides.
4. Give the correct answer for each of the following questions:
   (a) What name is given to the mixture of sodium salts of fatty acids formed in saponification?
   (b) What vitamin serves the body as a possible antioxidant?
   (c) What name is given to a monosaccharide containing a keto group?
   (d) What is the name of the monosaccharide to which all other monosaccharides are related configurationally?
   (e) Is glucose best represented as an acetal or a hemiacetal?
   (f) Fructose exists as a furanose. How many atoms does this ring structure of fructose contain?
   (g) What is the stereochemical relationship of the two cyanohydrins formed in the Kiliani-Fischer synthesis?
   (h) If maltose and cellobiose give only D-glucose on hydrolysis, how do they differ structurally?
   (i) What structural feature is characteristic of a peptide?
   (j) What is the name of the pH level at which an amino acid has a net charge of zero and does not migrate in an electric field?
   (k) Draw the structural formula of a blocking reagent for the N-terminal amino acid in peptide synthesis.
   (l) What is the name given to the type of compound composed of phosphate, ribose, and a purine base?
   (m) What is the primary structure of a protein?
   (n) With what base is adenine paired in DNA?
   (o) In RNA what base replaces the thymine of DNA?

Answers to self-test

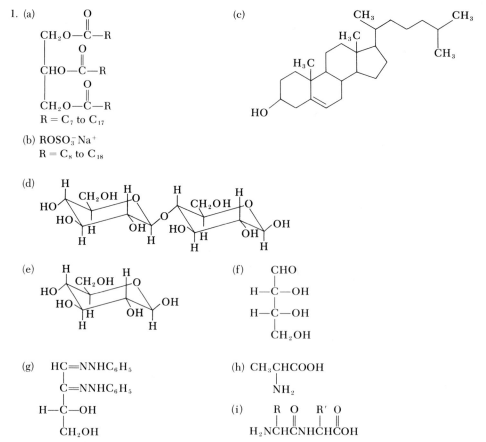

1. (a)

$$CH_2O-\overset{\overset{O}{\|}}{C}-R$$
$$CHO-\overset{\overset{O}{\|}}{C}-R$$
$$CH_2O-\overset{\overset{O}{\|}}{C}-R$$
$$R = C_7 \text{ to } C_{17}$$

(b) $ROSO_3^- Na^+$
$R = C_8$ to C_{18}

(c)

(d)

(e)

(f)
$$CHO$$
$$H-C-OH$$
$$H-C-OH$$
$$CH_2OH$$

(g) $HC=NNHC_6H_5$
$$C=NNHC_6H_5$$
$$H-C-OH$$
$$CH_2OH$$

(h) $CH_3CHCOOH$
$$NH_2$$

(i)
$$\underset{H_2NCHCNHCHCOH}{\overset{R \quad O \quad\quad R' \ O}{\quad\ \ \|\quad\quad\ \ \|}}$$

(j) $H_3\overset{\oplus}{N}CH_2COOH$

(l) One example of nucleoside, sugar-base unit:

(k)

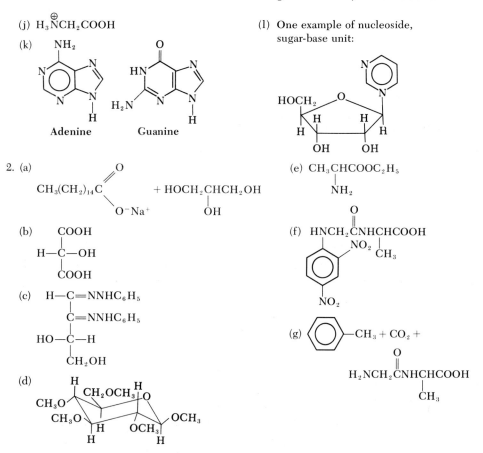

Adenine   Guanine

2. (a) $CH_3(CH_2)_{14}C\overset{O}{\diagdown}_{O^-Na^+}$ + $HOCH_2CHCH_2OH$
                                                          $\underset{OH}{|}$

(b)
$$\begin{array}{c} COOH \\ | \\ H-C-OH \\ | \\ COOH \end{array}$$

(c)
$$\begin{array}{c} H-C=NNHC_6H_5 \\ | \\ C=NNHC_6H_5 \\ | \\ HO-C-H \\ | \\ CH_2OH \end{array}$$

(d)

(e) $CH_3CHCOOC_2H_5$
        $\underset{NH_2}{|}$

(f) $HNCH_2\overset{O}{\overset{||}{C}}NHCHCOOH$

(g)

3. Merrifield's approach to peptide synthesis avoids the intermediate isolation steps characteristic of multistep syntheses. The amino acid that will ultimately contain the free carboxyl end of the peptide is bonded through its carboxyl end to an active group on a solid polymer. After synthesis is completed at the desired stage, the peptide is easily removed from the solid polymer.

4. (a) Soaps
   (b) Vitamin E
   (c) Ketose
   (d) D- or L-Glyceraldehyde
   (e) Hemiacetal
   (f) Five (four carbon atoms and one oxygen atom)
   (g) Diastereomers
   (h) In glycoside linkage
   (i) Amide $\left(-\overset{O}{\underset{||}{C}}-NH-\right)$

   (j) Isoelectric point
   (k) Carbobenzoxy chloride
   (l) Nucleotide
   (m) Number and sequence of amino acids
   (n) Thymine
   (o) Uracil

chapter 14

Organic chemistry of living systems

In this chapter we will study the chemical reactions that involve lipids, carbohydrates, proteins, and nucleic acids in living systems. Lipids, carbohydrates, proteins, and nucleic acids are the major organic components of living systems and are the foods that cells utilize to sustain life. The reactions carried out by cells are collectively referred to as *metabolism.* The area of chemistry that deals with these reactions of living systems is called *biochemistry.*

The cell

The cell can be considered to be a chemical industry. It is capable of performing a large number of specific and complex reactions rapidly, efficiently, and indefinitely. It can regulate its energy requirement and can capture and utilize a considerable amount of the energy released in metabolic processes.

Cells differ in size, structure, and appearance, depending on their function. However, all cells are complex aqueous solutions with the following

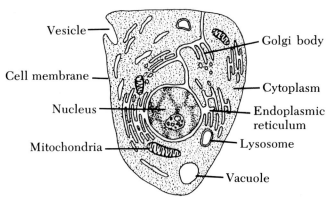

Fig. 14-1
Important features of a typical animal cell.

Table 14-1

Biochemical function and organic composition of various cellular units

Unit	Biochemical function	Composition
Cell membrane	Selective transport of ions and molecules into and out of cell	Lipids; proteins; carbohydrates
Nucleus	Site for transmission and regulation of hereditary characteristics	Mainly DNA; some RNA
Mitochondria	Cell's power plant; formation of ATP by oxidation, Krebs' cycle	Enzymes and substrates; very small amounts of RNA; DNA
Endoplasmic reticulum and ribosomes	Site of protein synthesis	RNA; protein
Golgi apparatus	Transport and excretion of materials within cells	RNA; protein
Lysosomes	Hydrolysis of foreign material and clearance of dead cells from tissue	Proteins, mainly hydrolytic enzymes
Vacuoles and vesicles	Transport of material into cell and removal of foreign material from cell	
Cytoplasm	Many pathways, such as glycolysis	Complex

features in common: cell membrane, nucleus, mitochondria, endoplasmic reticulum, ribosomes, lysosomes, cytoplasm, vacuoles and vesicles, and Golgi apparatus. These features are shown for a typical animal cell in Fig. 14-1. The composition and function of the various cellular units are summarized in Table 14-1.

Enzymes: catalysts in biological systems

Enzymes are proteins of very high molecular weights formed by living cells with the highly specific function of catalyzing biochemical reactions by lowering the transition state energy so that the reactions can be promoted under the conditions in the cells. Often high temperatures and high pressures are required for inorganic catalysts to be effective, as in the platinum- or nickel-catalyzed addition of hydrogen to an alkene (Chapter 3). Enzymes, however, catalyze reactions in the cell at body temperatures and at the pH of body fluids. Since they are highly specific, a large number of them are found in living systems. The various classes of enzymes are listed in Table 14-2.

The nomenclature of enzymes is complex. In general, their names are formed by the addition of the ending -*ase* to the root of the name of the substrate on which the enzyme acts. For example, a lipase is an enzyme that hydrolyzes lipids; and sucrase promotes the hydrolysis of sucrose. Other names are based on the activity of the enzyme. For example, oxidases promote oxidation, and decarboxylases catalyze the removal of carbon dioxide. Still other names identify the substrate and the nature of the reaction. For example, the name lactic acid dehydrogenase identifies lactic acid as the substrate and dehydrogenation as the reaction it promotes.

Table 14-2
Classification of enzymes according to specificity

Type	Function	General example	Specific example and catalytic function	Reaction
Oxidoreductase	Physiological oxidation processes in body	Dehydrogenase, oxidase, peroxidase	Succinic dehydrogenase: dehydrogenation of succinic acid to fumaric acid	$^{\ominus}OOC-CH_2-CH_2-COO^{\ominus} \rightarrow$ fumarate ($^{\ominus}OOC-CH=CH-COO^{\ominus}$)
Transferase	Transfer of chemical groups from one compound to another	Transaminase, methyltransferase, transacylase	Creatine kinase: transfer of phosphate group to creatine	$H_2N-C(=NH)-N(CH_3)-CH_2COOH \rightarrow HOPNHC(=NH)NCH_2COOH,\ CH_3$
Hydrolase	Hydrolysis of molecules	Amylase, lipase, protease	Glucose-6-phosphatase: hydrolysis of glucose-6-phosphate to glucose and phosphate	glucose-6-phosphate $\rightarrow$ glucose $+ PO_4^{3-}$
Lyase	Nonhydrolytic removal of chemical groups from compounds	Decarboxylase	Histidine decarboxylase: decarboxylation of histidine to histamine	histidine ($CH_2CHCOOH,\ NH_2$) $\rightarrow$ histamine ($CH_2CH_2NH_2$) $+ CO_2$
Isomerase	Catalysis of different types of isomerization	Cis-trans isomerase, racemase, intramolecular transferase, epimerase	Phosphoribuloepimerase: isomerization of D-ribulose-5-phosphate to D-xylulose-5-phosphate	D-ribulose-5-phosphate $\rightarrow$ D-xylulose-5-phosphate
Ligase	Catalysis of bond formation with aid of ATP	RNA ligase	Aminoacyl-tRNA synthetase: catalysis of L-amino acid + tRNA to aminoacyl-tRNA	$H_2NCH(CH_3)COOH + tRNA + ATP \rightarrow H_2NCH(CH_3)C(=O)tRNA + AMP$

Mode of enzyme catalysis in metabolic reactions

Enzymes catalyze specific metabolic reactions, which are those chemical changes occurring in living systems that cause the formation and destruction of substances. For example, succinic dehydrogenase (SDH) is capable of dehydrogenating succinic acid through a *trans* elimination reaction to yield fumaric acid; however, although all four hydrogens are structurally equivalent, maleic acid, the isomer formed through a *cis* elimination, is not produced:

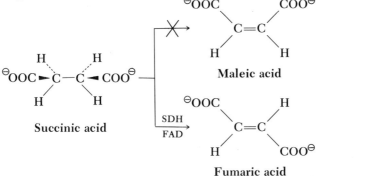

Maleic acid

Succinic acid

Fumaric acid

Cis elimination

Trans elimination

SDH recognizes only one kind or set of equivalent hydrogens:

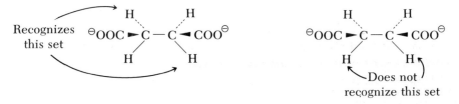

Recognizes this set

Does not recognize this set

This occurs because the enzyme molecule possesses an active binding site through which it and succinic acid (the substrate) form a complex (the enzyme-substrate complex), which then takes part in the chemical reaction. One explanation for this specificity of the enzyme is the *lock-and-key mechanism*. Basically, the surface shapes of the enzyme and the substrate fit together just as a key fits into a lock. In other words, the structure or specificity of the enzyme is such that it can fit and act on only one kind of substrate. The enzyme-substrate complex that is formed actually undergoes the chemical reaction. This is similar to the hydrogenation reaction of an alkene occurring on the surface of a platinum catalyst (Chapter 3). After the reaction takes place, the enzyme and modified substrate, or product, dissociate. This is illustrated as follows:

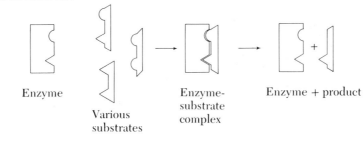

Enzyme

Various substrates

Enzyme-substrate complex

Enzyme + product

However, there is a limitation to the lock-and-key mechanism. It implies that the structure of the enzyme is fixed, yet current evidence indicates that the structure of the enzyme is actually flexible. According to the *induced-fit theory*, the substrate causes a shift in the conformation of the enzyme so that the enzyme assumes the proper orientation of its active binding sites. This is illustrated as follows:

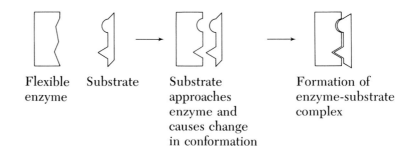

Flexible   Substrate   Substrate   Formation of
enzyme   approaches   enzyme-substrate
                            enzyme and   complex
                            causes change
                            in conformation

In some cases, the enzyme may have a coenzyme (discussed later) that facilitates the binding of the enzyme to the substrate. This is the case with SDH. Its coenzyme is flavin adenine dinucleotide (FAD). Electrical forces of attraction between the enzyme and the substrate also assist in the formation of the enzyme-substrate complex. In SDH, these electrical forces of attraction are provided by a sulfhydryl group (—SH). In the enzyme-substrate complex of the hypothetical mechanism of dehydrogenation of succinic acid, a sulfur ion abstracts a proton from the succinic acid and forms an —SH group. Simultaneously, the pair of electrons forms a carbon–carbon double bond, and a hydride ion is transferred to the coenzyme. The second hydrogen that appears in $FADH_2$ probably comes from the sulfhydryl group. This mechanism is illustrated as follows:

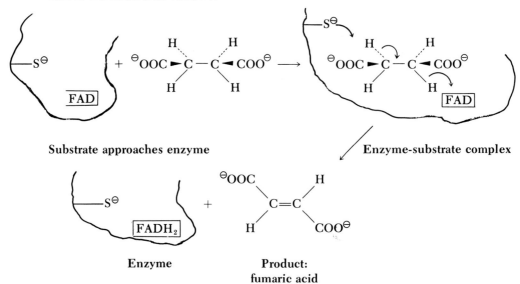

Substrate approaches enzyme   Enzyme-substrate complex

Enzyme   Product:
                             fumaric acid

Carboxypeptidase A is an enzyme that either cleaves peptides at the carboxyl end of a polypeptide substrate or cleaves esters:

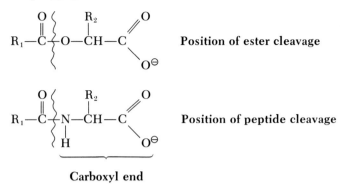

Carboxypeptidase contains over 300 amino acid residues. The active sites in the enzyme are the metal ion Zn^{2+}, glutamine (Glu-270), and arginine (Arg^+-145). The proposed mechanism for the action of carboxypeptidase A on esters is illustrated on p. 434.* According to this mechanism, the ester (or peptide) bond in the substrate undergoes nucleophilic attack by the carboxylate group of Glu-270. (Arg^+-145 holds the carboxylate group of the ester in the enzyme cavity.) The zinc ion, which is shown with a single positive charge because the other positive charge is neutralized by a carboxylate ligand from the enzyme, at the active site of the enzyme orients the carbonyl group of the substrate by polarizing it, thus facilitating attack by the carboxylate group (step 1). The tetrahedral intermediate that is formed breaks down to give an anhydride with the accompanying formation of an α-hydroxy acid anion from the C-terminal portion of the substrate (step 2). Then the anhydride species decomposes, regenerating the enzyme (step 3).

Proenzymes and coenzymes

Enzymes are synthesized by living systems. Although some enzymes contain a highly active site, they do not act on the tissues or glands that produce them because they are secreted in an inactive form. The inactive form of an enzyme is the precursor of an active enzyme in the body and is referred to as a *proenzyme*. It remains inactive until it reaches the reaction site. For example, pepsinogen, a proenzyme, is secreted into the stomach, where the stomach acid (hydrogen ions) converts it into pepsin, the active form of the enzyme.

Enzymes occasionally lose their activity. It was discovered that this is because some enzymes need a nonprotein material called a *coenzyme* for catalysis.

It is now known that many vitamins or their derivatives serve as coenzymes in metabolic reactions. Following are some examples. Thiamine, or vitamin B_1, contains both a pyrimidine ring and a thiazole ring. It occurs as a free molecule in cereal grains but as the coenzyme thiamine pyrophosphate

---

*Redrawn from Kaiser, E. T., and Kaiser, B. L.: Accounts of Chemical Research 5:219, 1972.

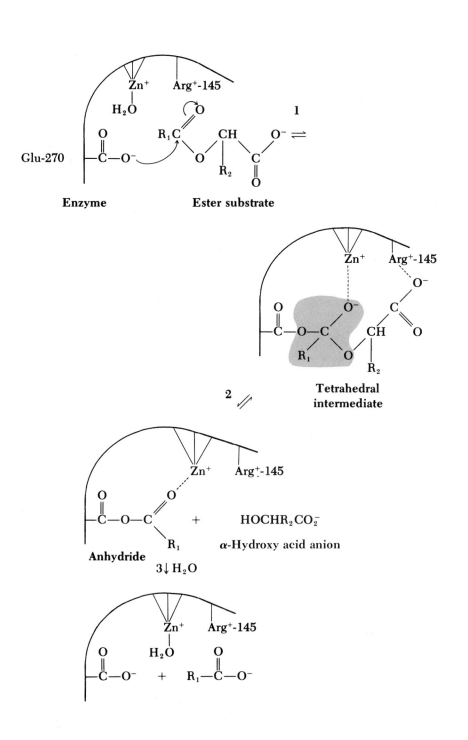

(TPP), also known as cocarboxylase, in yeast and meat. The structure of TPP is as follows:

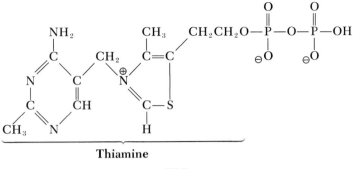

Thiamine

TPP

TPP functions in the oxidative decarboxylation of pyruvic acid to acetyl–coenzyme A. It reacts with pyruvic acid to give active acetaldehyde-TPP complex and carbon dioxide. The thiazole ring is the active site of this catalytic reaction:

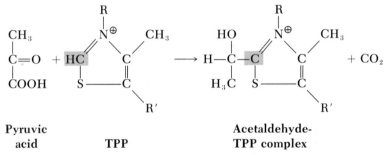

| Pyruvic acid | TPP | Acetaldehyde-TPP complex |

In this reaction and in those that follow, R is the remainder of the co-enzyme molecule, and the portion of the molecule undergoing the change is shaded.

Riboflavin, or vitamin B_2, is composed of a ribotol (an alcohol from the pentose ribose) and flavin (a pigment). A deficiency of vitamin B_2 causes inflammation of the cornea and sores and cracks at the corners of the mouth. The vitamin occurs in foods as part of two coenzymes: flavin mononucleotide (FMN) and flavin adenine dinucleotide (FAD), whose structures are as follows:

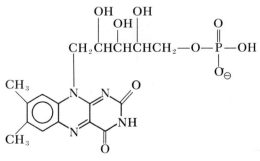

FMN

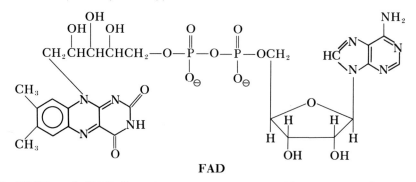

FAD

Both FMN and FAD function as coenzymes with enzymes that catalyze oxidation-reduction reactions. For example, FAD serves as a coenzyme with succinic dehydrogenase, which oxidizes succinic acid to fumaric acid. The flavin portion of the coenzyme is the active site in the reaction at the top of p. 437.

Another vitamin that functions as a coenzyme is niacin (nicotinamide). Lack of niacin in the diet results in pellagra, a disease characterized by skin lesions. Niacin contains the active site in two coenzymes: nicotinamide adenine dinucleotide (NAD^+) and nicotinamide adenine dinucleotide phosphate ($NADP^+$), whose structures are as follows:

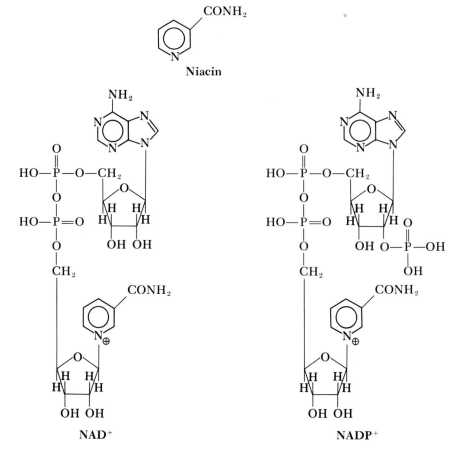

NAD$^+$ **NADP$^+$**

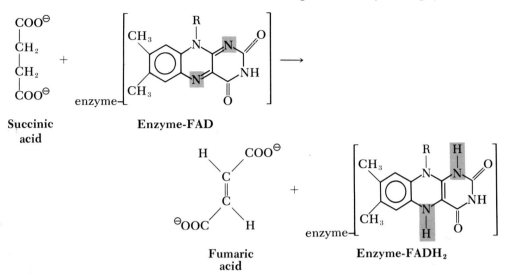

Succinic acid   Enzyme-FAD

Fumaric acid   Enzyme-FADH₂

Lactic acid dehydrogenase catalyzes the oxidation of lactic acid to pyruvic acid. In this oxidation reaction, NAD serves as the coenzyme and is reduced to NADH. Nicotinamide contains the active site:

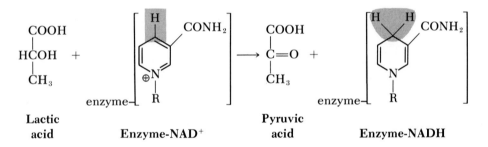

Lactic acid   Enzyme-NAD⁺   Pyruvic acid   Enzyme-NADH

The derivatives of pyridoxine, or vitamin B_6, function as coenzymes. The structures of pyridoxal, the aldehyde, and pyridoxal phosphate, its derivative, are as follows:

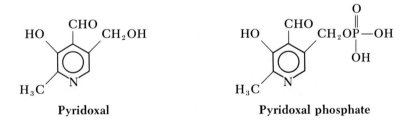

Pyridoxal   Pyridoxal phosphate

Pyridoxal phosphate is the major coenzyme involved in the metabolism of amino acids. It is a coenzyme in transamination, decarboxylation, and racemization of amino acids.

A final example of a coenzyme is a derivative of pantothenic acid, which has the following structure:

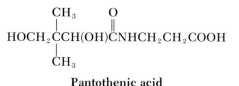

Pantothenic acid

Pantothenic acid forms part of coenzyme A (CoA), which has the following structure:

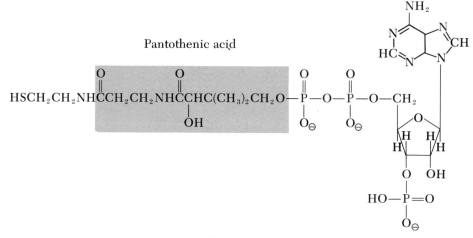

Coenzyme A

The reactive site in CoA is the sulfhydryl (SH) group. CoA functions mainly as acyl—CoA in acylation reactions, synthesis of fats, and the synthesis of steroids. Acetyl—CoA, which has the following structure, is a source of acetate for the Krebs' cycle:

$$CH_3\overset{O}{\underset{\|}{C}}-SCoA$$

Other vitamins that function as coenzymes are vitamin B_{12}, folic acid, biotin, and lipoic acid.

Enzyme inhibition

The inhibition of an enzyme renders it inactive because of the formation of a complex between the inhibitor and the active site of the enzyme, which then is not available to the substrate. Sulfanilamide is an example of an enzyme inhibitor. It inhibits the growth of microorganisms, since it can be mistaken for *p*-aminobenzoic acid, which is necessary for the growth of the microorganism (Chapter 11).

Another example is the inhibition of acetylcholinesterase. Acetylcholine (an ester of choline) plays a vital role in the transmission of nerve impulses in the body. It is combined with other substances in the nerve cell and is released when the cell is stimulated. Then it stimulates an adjacent nerve cell to release acetylcholine, which transmits the nerve impulses to the next nerve cell. These reactions are extremely fast (30 to 50 microseconds). After the transmission of the nerve impulse is complete, the acetylcholine released is deactivated by hydrolysis. This reaction is catalyzed by the enzyme acetyl-cholinesterase (enzyme—OH):

$$(CH_3)_3\overset{\oplus}{N}CH_2CH_2O\overset{\overset{\displaystyle O}{\|}}{C}CH_3 \xrightarrow{\text{enzyme—OH}} (CH_3)_3\overset{\oplus}{N}CH_2CH_2OH + CH_3COOH$$

Acetylcholine   Choline   Acetic acid

Any chemical that interferes with the deactivation process can cause paralysis or death. Nerve gases, insecticides, and some snake venoms interfere with the deactivation process by acting on the enzyme cholinesterase and inhibiting the hydrolysis of acetycholine to choline. Thus these compounds cause a continuous transmission of the impulses that control vital functions; this in turn causes paralysis of muscles, glands, and organs that are controlled by nerve cell impulses. Some of these chemicals are as follows:

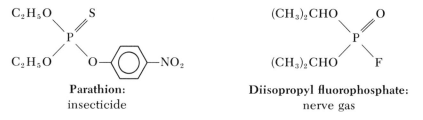

Parathion: Diisopropyl fluorophosphate:
insecticide   nerve gas

The chemical equation of the deactivation of the enzyme can be represented as follows:

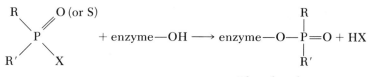

Phosphoryl enzyme

The phosphoryl enzyme is relatively stable and cannot be decomposed by water. Therefore, water is a poor antidote. A possible antidote is 2-pyridine aldoxime methyl iodide, which rapidly removes the organophosphate group from the enzyme at essentially nontoxic levels:

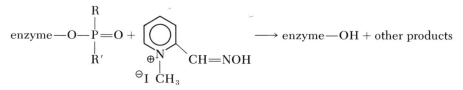

Energetics of biochemical reactions

To better understand the enzyme-catalyzed reactions in metabolic processes we will examine the energy relationships that are involved. The chemical changes brought about by metabolic processes depend on the

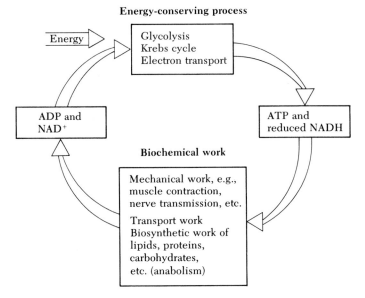

Fig. 14-2
A biochemical energy cycle.

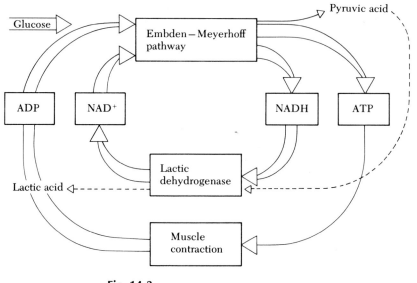

Fig. 14-3
A biochemical energy cycle for glucose.

energy available to the specific reactions. There are two types of metabolic processes: catabolism, the breakdown of substances, and anabolism, in which biomolecules such as polymers are synthesized from smaller molecules. Many of the reactions in these processes depend on the energy available from oxidation-reduction reactions. The chemical energy available from an exothermic reaction is conserved in the form of adenosine triphosphate (ATP) and reduced coenzymes and is used for energy-requiring biosynthetic reactions. Fig. 14-2 shows the relationship between these anabolic and catabolic processes.

A complex organic molecule such as glucose contains much potential energy; thus degradation of it releases much energy. This chemical energy is used in the synthesis of ATP, which then is used to perform mechanical, transport, and biosynthetic work. ATP functions as a carrier of energy-rich phosphate groups and NADH as a carrier of energy-rich electrons from catabolic reactions to energy-requiring anabolic reactions. In the energy cycle shown in Fig. 14-3, energy is supplied by the degradation of glucose to pyruvic acid. The chemical energy made available is in the form of NADH and ATP. The NADH is subsequently used in the energy-requiring process of reducing pyruvic acid to lactic acid. The energy conserved in the form of ATP, on the other hand, is used to do the mechanical work of muscle contraction.

There are several energy-rich compounds, including adenosine monophosphate (AMP), adenosine diphosphate (ADP), and ATP; however, the most important one is ATP. The structures of these compounds are as follows:

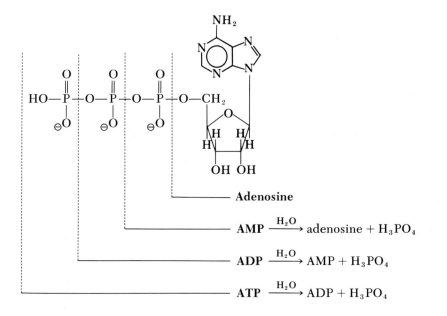

ATP possesses a structural characteristic that helps to explain the release of energy. Two of the phosphate linkages are anhydride bonds, which are easily hydrolyzed because there is a high repulsion energy between the negative oxygen atoms:

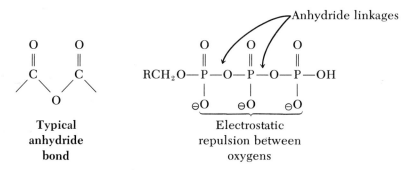

Typical anhydride bond	Electrostatic repulsion between oxygens

Following are other examples of high-energy compounds, some of which contain phosphorus:

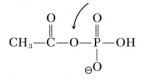

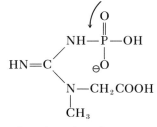

Acetyl phosphate	**Creatine phosphate**

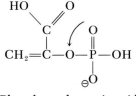

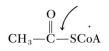

Phosphoenolpyruvic acid	**Acetyl–coenzyme A**

The high-energy bonds in these compounds are indicated by arrows.

Biochemical reactions forming ATP

One of the reactions occurring in glucose metabolism is the oxidation of glyceraldehyde-3-phosphate to 3-phosphoglyceric acid:

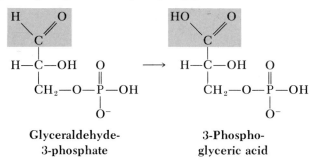

Glyceraldehyde-3-phosphate	3-Phospho-glyceric acid

When this oxidation of an aldehyde to a carboxylic acid is performed in a laboratory, the energy released is lost as heat. In living systems, the oxidation proceeds in two steps, and the energy released is transformed to ATP, which is used to drive other energy-requiring reactions. In the first step, phosphate is incorporated into glyceraldehyde-3-phosphate to give 1,3-di-phosphoglyceric acid, an anhydride. This conversion is catalyzed by NAD^+, which accepts electrons from the substrate, glyceraldehyde-3-phosphate, and is reduced to NADH:

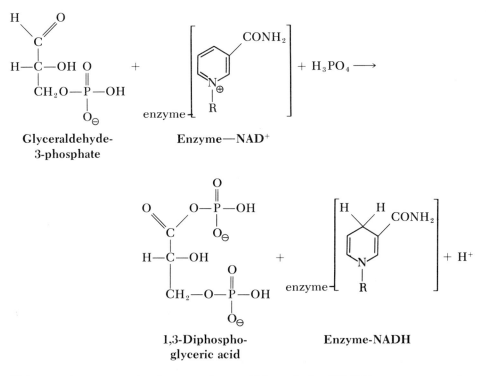

This reaction occurs in the cytoplasm. Although the NADH cannot enter the mitochondria, where it could be oxidized to provide energy for ATP synthesis, this energy can be transferred to the electron transport system via phospho-glycerol or β-hydroxybutyric acid. The mitochondria possess enzymes so that electrons from NADH can be transferred through a series of steps to oxygen. The result of this electron transfer is the formation of ATP. As long as oxygen is available and most tissues are well supplied with it, large amounts of NADH can be oxidized to NAD^+ with the subsequent formation of ATP. This pathway by which energy is obtained in the form of ATP is the *electron-transport system* and is called *oxidative phosphorylation*. A scheme depicting electron flow is illustrated here.

Substrate $\rightleftharpoons$ NADH $\xrightarrow{\text{ADP ATP}}$ FADH$_2$ $\rightleftharpoons$ Cyt b $\xrightarrow{\text{ADP ATP}}$ Cyt c $\xrightarrow{\text{ADP ATP}}$ Cyt a $\longleftrightarrow$ O$_2$

The cytochromes (cyt) are oxidative enzymes. The overall reaction for the electron-transport oxidation of NADH is as follows:

$$NADH + H^+ + 3\,\textcircled{P} + 3ADP + \tfrac{1}{2}O_2 \longrightarrow NAD^+ + 3ATP + H_2O$$

where $\textcircled{P}$ is the monophosphate group $—H_2PO_3$ in solution.

A second way in which ATP is formed is illustrated by the second step in the oxidation of glyceraldehyde-3-phosphate to 3-phosphogylceric acid. In this step, a phosphate group of the anhydride 1,3-diphosphoglyceric acid is transferred to ADP, and the released energy is conserved in ATP:

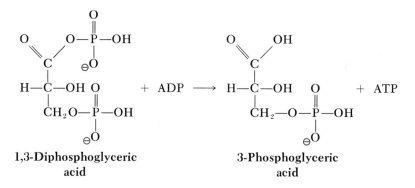

1,3-Diphosphoglyceric   3-Phosphoglyceric
         acid   acid

The energy in the form of ATP is subsequently used by the cell to drive other endothermic processes. This method of forming ATP involving reactions other than an electron-transport system is called *substrate-level phosphorylation*.

The two equations describing the NAD^+-enzyme-catalyzed conversion of glyceraldehyde-3-phosphate to 3-phosphoglyceric acid and ATP can be written as follows in the short-hand notation used by biochemists to describe the events occurring in metabolic reactions:

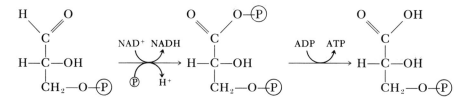

The normal horizontal arrows indicate the conversion of reactants to products. The curved arrows over the horizontal arrows indicate the reduction of NAD^+ to NADH and the conversion of ADP to ATP, respectively. The curved arrow under the first horizontal arrow represents the introduction of the phosphate group $-H_2PO_3$ into the product.

A third way in which ATP is formed is through the conversion of light energy into chemical energy in plants during photosynthesis, which takes place in the chloroplasts. This process, which can be represented as follows, involves the phosphorylation of ADP, and is called *photophosphorylation*:

$$2NAD^+ + 2H_2O + 4ADP + 4\text{\textcircled{P}} \longrightarrow O_2 + 4ATP + 2NADH + 2H^{\oplus}$$

ATP as energy source

In order for a biochemical endothermic reaction to occur, energy must be supplied from some other energy-releasing reaction. The chemical energy must be transferred from one reaction to the other through a common intermediate. As an example of this energy-transferring process, consider the conversion of 3-phosphoglyceric acid to glyceraldehyde-3-phosphate, which is part of a series of reactions through which glucose is synthesized from pyruvic acid:

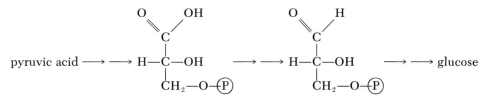

This particular biochemical transformation uses NADH and ATP as energy sources. The sequence of reactions that occur is as follows:

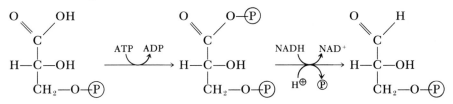

In the first reaction, ATP reacts with 3-phosphoglyceric acid to form ADP and 1,3-diphosphoglyceric acid, which is an acid anhydride and is more easily reduced to the aldehyde group of glyceraldehyde-3-phosphate. The energy released by the hydrolysis of ATP is transferred to the 1,3-diphophoglyceric acid. The second step is promoted by the oxidation of coenzyme NADH to NAD$^+$.

Metabolism

In general, carbohydrates, lipids, proteins, and nucleic acids in foods cannot be absorbed and utilized by the body until they are broken down into smaller molecules. Digestion is the process by which the body changes these complex carbohydrates, lipids, proteins, and nucleic acids into simpler molecules. Digestion hydrolyzes carbohydrates to form monosaccharides, lipids to form fatty acids and glycerol, and proteins to form amino acids. These hydrolysis reactions occur by the action of enzymes collectively referred to as hydrolases. Monosaccharides, fatty acids, and amino acids are then absorbed into the bloodstream and transported to various organs in the body, where they are stored until they are utilized by the body. In this section, the methods by which the body utilizes monosaccharides, fatty acids, and amino acids are described.

Carbohydrate metabolism

Carbohydrates are major sources of energy for living systems. Glucose, produced by hydrolysis of carbohydrates, is carried by the blood to the liver, where it is stored in the form of glycogen. Glycogen is a complex polysaccharide similar in structure to amylopectin (p. 397). Glucose is obtained from it when the level of blood glucose falls below normal. The formation of glycogen from glucose is *glycogenesis,* and the reverse process is *glycogenolysis.*

Glycogenesis: formation of glycogen. The hormone insulin is responsible for lowering the blood glucose level by promoting its uptake by tissue and its conversion into glycogen. It also regulates the oxidation of glucose by the tissues. It acts by controlling the phosphorylation of glucose to glucose-6-phosphate by the action of the enzyme glucokinase and promoting utilization of glucose. The condition in which insulin is absent and there is a decrease in the utilization of glucose is known as *diabetes mellitus.*

The steps in the formation of glycogen from glucose are as follows:

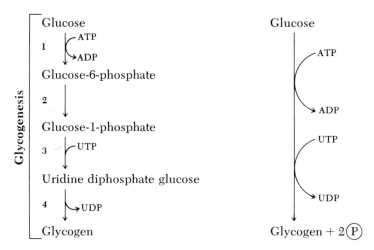

The energy for the conversion of glucose to glucose-6-phosphate (step 1) is supplied by ATP, and the transformation is catalyzed by glucokinase. Since the reaction occurs between an alcohol and an acid, it is commonly referred to as esterification:

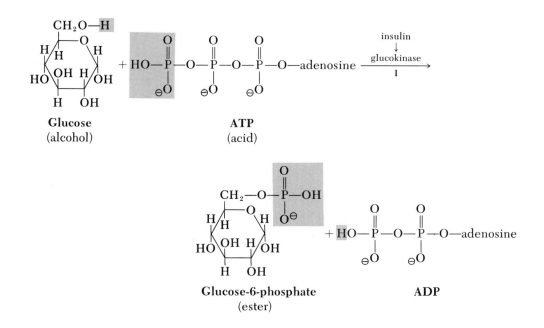

The enzyme phosphoglucomutase then aids the isomerization of glucose-6-phosphate to glucose-1-phosphate (step 2):

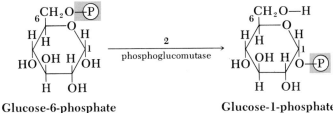

Glucose-6-phosphate   Glucose-1-phosphate

In step 3, glucose-1-phosphate reacts with uridine triphosphate (UTP) to form the nucleotide uridine diphosphate glucose (UDPG). This reaction can be considered to be the formation of an anhydride of UTP whose function is to activate the glucose molecule in glycogen formation in step 4:

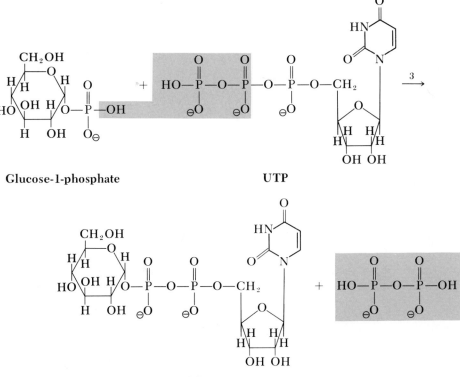

Glucose-1-phosphate   UTP

UDPG

In the final step (step 4), the activated glucose molecules in the form of the UDPG are then joined together through glycoside linkages to form glycogen. This reaction occurs in the presence of UDPG glycogen glucosyl transferase:

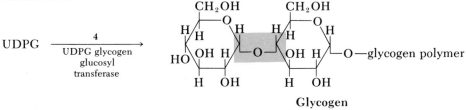

Glycogen

Glycogenolysis: formation of glucose. If the level of glucose in the blood drops because of fasting or increased muscle action, glycogen is broken down in the liver to form glucose, which is passed into the bloodstream to maintain the normal blood level of glucose. Glycogenolysis is initiated by the enzyme phosphorylase, which is found in the liver. The reactions occurring can be summarized as follows:

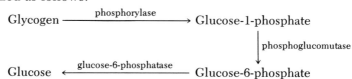

Metabolism of glucose. In living systems this involves the oxidation of glucose to carbon dioxide and water and the liberation of energy. The major intermediate in the metabolism of glucose is glucose-6-phosphate, which is formed in the liver from glucose. This transformation is under the control of insulin. Alternatively, it may be formed from glycogen by the action of phosphorylase. Glucose-6-phosphate metabolism follows first the Embden-Meyerhof pathway, which is essential because it leads to the formation of pyruvic acid, and then the Krebs' cycle, which utilizes the pyruvic acid formed in the Embden-Meyerhof pathway. The Embden-Meyerhof pathway is anaerobic; that is, the reactions take place in the cell in the absence of oxygen. The Krebs' cycle is aerobic; that is, it operates in the presence of oxygen. The Krebs' cycle transforms more energy to a useful form than the Embden-Meyerhof pathway.

Embden-Meyerhof pathway. The steps leading from the transformation of glucose to glucose-6-phosphate to pyruvic acid are as follows. The overall transformation is energy releasing with the energy being conserved in the form of 8 moles of ATP. (See opposite page.)

In step 1, glucose-6-phosphate is isomerized to fructose-6-phosphate:

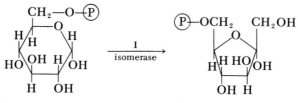

Step 2 is an endothermic reaction utilizing ATP as an energy source. Carbon 1 of fructose-6-phosphate is phosphorylated, that is, esterified, forming fructose-1,6-diphosphate:

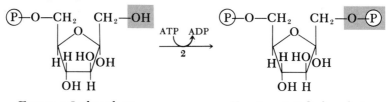

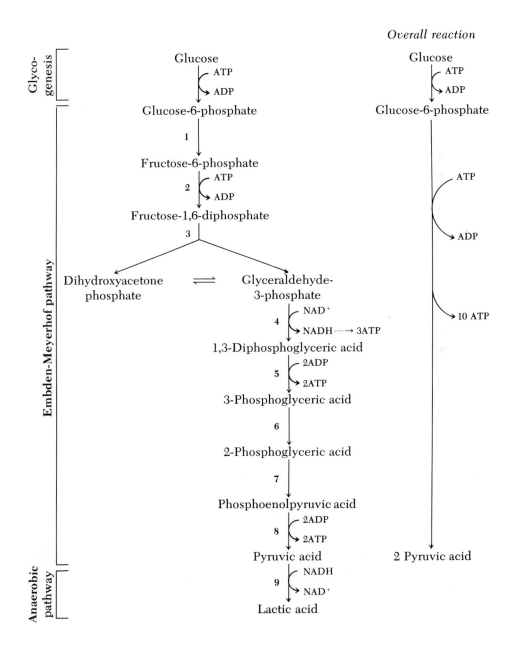

Step 3 is the cleavage of fructose-1,6-diphosphate to form two triose monophosphates:

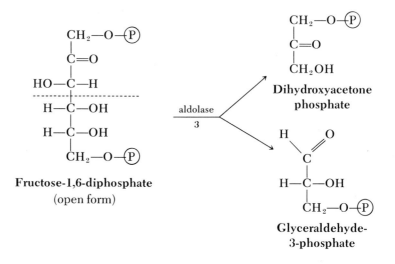

Step 4 is the conversion of glyceraldehyde-3-phosphate into 1,3-diphosphoglyceric acid. This step involves oxidation and the subsequent esterification of the resulting acid in which phosphoric acid is the source of the phosphate group:

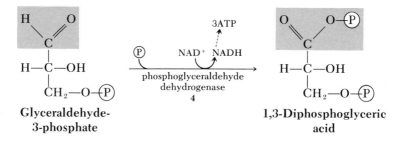

The oxidation portion of this reaction requires the coenzyme NAD^+, which is reduced to NADH. In cells that operate in the absence of oxygen, NAD^+ is regenerated in step 9 by the reduction of pyruvic acid to lactic acid. However, in the presence of oxygen, enzymes from the mitochondria oxidize NADH to NAD^+. This process is an example of the electron-transport system. The oxidation of 1 mole of NADH by molecular oxygen results in the formation of 3 moles of ATP as shown by the following equation:

$$NADH + H^+ + \tfrac{1}{2}O_2 + 3ADP + 3H_3PO_4 \longrightarrow NAD^+ + 3ATP + 4H_2O$$

Step 5 is an energy-releasing reaction resulting in the formation of ATP: 1 mole per monophosphate molecule or 2 moles per glucose molecule. The phosphate group on carbon 1 of 1,3-diphosphoglyceric acid is transferred rather than that group on carbon 3, since it is a high-energy anhydride linkage:

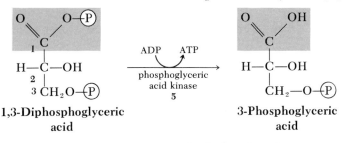

1,3-Diphosphoglyceric
acid

3-Phosphoglyceric
acid

Step 6 is an isomerization reaction in which the phosphate group on carbon 3 of 3-phosphoglyceric acid is transferred to carbon 2:

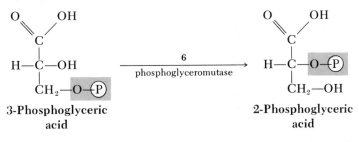

3-Phosphoglyceric
acid

2-Phosphoglyceric
acid

Step 7 is the dehydration of 2-phosphoglyceric acid to form phosphenol-pyruvic acid:

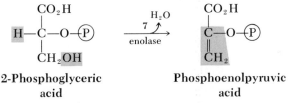

2-Phosphoglyceric
acid

Phosphoenolpyruvic
acid

In step 8, the transfer of phosphate from phosphoenolpyruvic acid gives rise to the intermediate enol tautomer of pyruvic acid, which is in equilibrium with the more stable keto tautomer. The reaction of phosphoenolpyruvic acid results in the formation of 2 moles of ATP per glucose molecule:

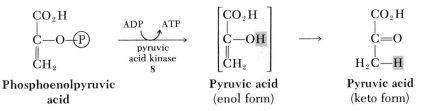

Phosphoenolpyruvic
acid

Pyruvic acid
(enol form)

Pyruvic acid
(keto form)

Step 9, which is the reduction of pyruvic acid to lactic acid, requires NADH as a coenzyme:

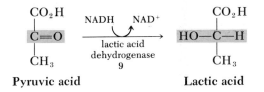

Pyruvic acid

Lactic acid

Step 9 is important only under anaerobic conditions.

The requirements for and the liberation of ATP in the Embden-Meyerhof pathway under aerobic conditions are summarized as follows:

Pathway	Process	Moles ATP per mole glucose
Glycogenesis	Glucose to glucose-6-phosphate	1 mole consumed
Embden-Meyerhof	Step 2	1 mole consumed
	Step 4 (electron transport)	6 moles liberated
	Step 5	2 moles liberated
	Step 8	2 moles liberated
	Net gain	8 moles liberated

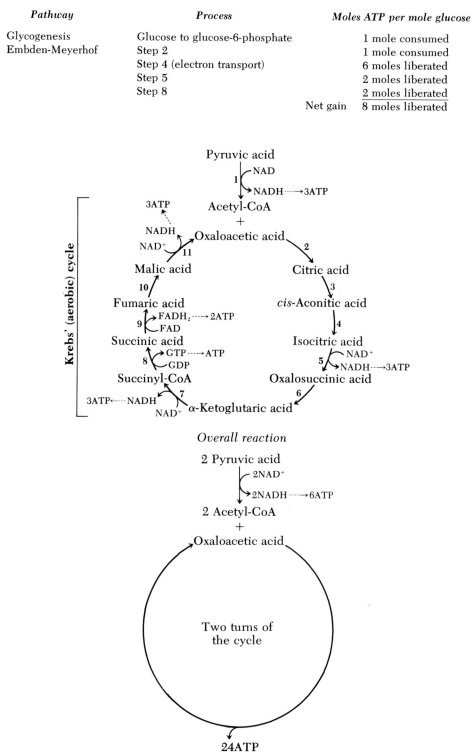

Krebs' cycle. In this cycle, first, pyruvic acid is decarboxylated to carbon dioxide and an acetyl group, which combines with coenzyme A. Next, the acetyl group combines with oxaloacetic acid. The cycle then proceeds through ten steps back to oxaloacetic acid. These ten steps take place under the influence of five coenzymes and occur in the presence of oxygen. The end products of pyruvic acid oxidation are carbon dioxide and water. Compared to the Embden-Meyerhof pathway, this cycle liberates a large amount of energy. The energy is conserved in the form of ATP, which is used for muscular work and for other energy requirements. The Krebs' cycle, which is shown on the opposite page, liberates 30 moles of ATP per mole of glucose.

Step 1 is the decarboxylation of pyruvic acid to carbon dioxide and an acetyl group, which combines with CoA to form acetyl-CoA:

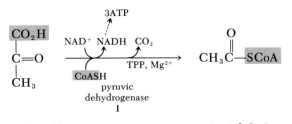

Pyruvic acid   Acetyl-CoA

This reaction involves the coenzyme NAD^+, which is reduced to NADH. Processes in the mitochondria then catalyze the oxidation of NADH to NAD^+ with the formation of 3 moles of ATP via the electron-transport system as shown by the following equation:

$$NADH + H^+ + \tfrac{1}{2}O_2 + 3ADP + 3H_3PO_4 \longrightarrow NAD^+ + 3ATP + 4H_2O$$

Thus the NAD^+ can be reused again in the cycle.

Step 2 is the condensation of acetyl-CoA and oxaloacetic acid to form citric acid. CoA is regenerated and can react with additional pyruvic acid to form acetyl-CoA. This step is an aldol type of condensation:

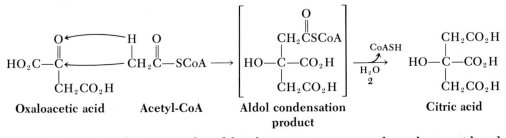

Oxaloacetic acid   Acetyl-CoA   Aldol condensation   Citric acid
                                          product

Steps 3 and 4 are catalyzed by the same enzyme and can be considered overall to be the isomerization of citric acid to isocitric acid:

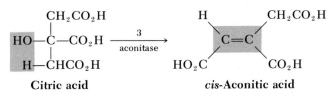

Citric acid   *cis*-Aconitic acid

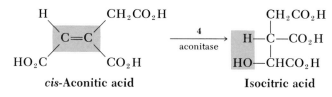

cis-Aconitic acid   Isocitric acid

Steps 5 and 6 are catalyzed by the same enzyme. However, step 5, which is the oxidation of a hydroxyl group to a keto group, requires NAD^+ as a coenzyme:

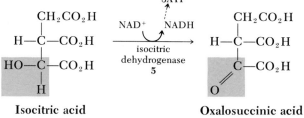

Isocitric acid   Oxalosuccinic acid

NADH, the reduced form of NAD^+, is oxidized with the formation of 3 moles of ATP via the electron-transport system as shown by the following equation:

$$NADH + 4H^+ + \tfrac{1}{2}O_2 + 3ADP + 3H_3PO_4 \longrightarrow NAD^+ + 3ATP + 4H_2O$$

Step 6 is the decarboxylation of oxalosuccinic acid to form α-ketoglutaric acid:

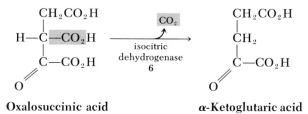

Oxalosuccinic acid α-Ketoglutaric acid

Step 7 is the conversion of α-ketoglutaric acid to succinyl-CoA. This step is a decarboxylation followed by thioester formation. The coenzyme NAD^+ is reduced to NADH, which is then oxidized to NAD^+ with the formation of 3 moles of ATP:

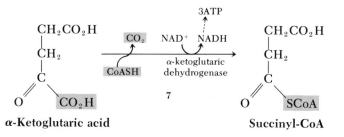

α-Ketoglutaric acid   Succinyl-CoA

Step 8 involves the hydrolysis of the energy-rich thioester bond of succinyl-CoA. The energy liberated from this transfer is conserved through the reaction of guanosine diphosphate (GDP) with phosphate in the medium to form guanosine triphosphate (GTP), which can be used as an energy source in biosynthetic work or used to make ATP:

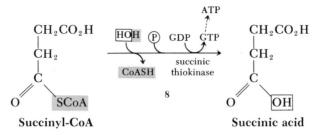

Succinyl-CoA   **Succinic acid**

Step 9 is catalyzed by succinic dehydrogenase, which causes the removal of two hydrogens (dehydrogenation) from succinic acid:

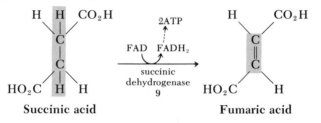

Succinic acid   **Fumaric acid**

The coenzyme FAD is reduced to $FADH_2$ in the process, and the subsequent oxidation of $FADH_2$ via the electron-transport system produces 2 moles of ATP as shown by the following equation:

$$FADH_2 + \tfrac{1}{2}O_2 + 2ADP + 2H_3PO_4 \longrightarrow FAD + 2ATP + 3H_2O$$

In step 10, fumarase catalyzes the addition of water to fumaric acid to form malic acid:

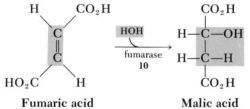

Fumaric acid   **Malic acid**

Step 11 is the oxidation of a hydroxyl group to a keto group:

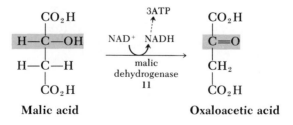

Malic acid   **Oxaloacetic acid**

The coenzyme NAD^+ is reduced to NADH in this reaction. Then NADH is oxidized to NAD^+ with the formation of 3 moles of ATP. The oxaloacetic acid formed reacts with acetyl-CoA, and the cycle is repeated. Each turn of the cycle liberates 2 moles of carbon dioxide, one in each of steps 6 and 7.

Table 14-3
ATP yield from oxidation of glucose

Pathway	Reaction	Coenzyme	Yield ATP per mole glucose
	Glucose ⟶ glucose-6-phosphate		−1
Embden-Meyerhof	Fructuse-6-phosphate ⟶ fructose-1,6-diphosphate		−1
	Glyceraldehyde-3-phosphate ⟶ 1,3-diphosphoglyceric acid	NAD^+	6
	1,3-Diphosphoglyceric acid ⟶ 3-phosphoglyceric acid		2
	Phosphoenolpyruvic acid ⟶ pyruvic acid		2
Krebs'	Pyruvic acid ⟶ acety-CoA	NAD^+	6
	Isocitric acid ⟶ oxalo-succinic acid	NAD^+	6
	α-Ketoglutaric acid ⟶ succinly-CoA	NAD^+	6
	Succinly-CoA ⟶ succinic acid		2
	Succinic acid ⟶ fumaric acid	FAD	4
	Malic acid ⟶ oxaloacetic acid	NAD^+	6
		Net yield	38

Since 1 mole of glucose forms 2 moles of pyruvic acid, the Krebs' cycle yields the following amount of energy in the form of ATP per mole of glucose:

Process	Moles ATP per mole glucose
Step 1	6 moles liberated
Step 5	6 moles liberated
Step 7	6 moles liberated
Step 8	2 moles liberated
Step 9	4 moles liberated
Step 11	6 moles liberated
Net gain	30 moles liberated

• • •

Table 14-3 summarizes the yield of ATP from the oxidation of glucose through the Embden-Meyerhof pathway and Krebs' cycle.

Lipid metabolism

The metabolism of lipids includes the metabolism of neutral fats, phospholipids, and steroids. The structures of these compounds are discussed in Chapter 13. Neutral fats are reserve sources of energy for living systems. They are transported by the blood and stored in fat depots under the skin as adipose tissue. Some fat is also stored around the vital organs, such as the kidneys and heart, to help to protect them against injury. Much is also stored as fat globules. Phospholipids and steroids are not stored in fat depots but

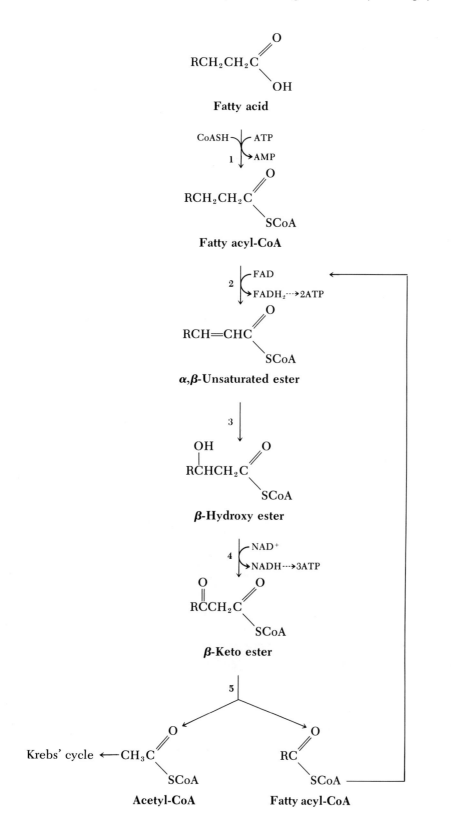

are essential components of tissues. They play a role in fat transport and in many cellular metabolic reactions.

Oxidation of fatty acids. Neutral fats can be broken down into glycerol and fatty acids by hydrolysis. The fatty acids then are released to the tissues to be used for energy. The fatty acids obtained from the hydrolysis of neutral fats are completely oxidized to carbon dioxide and water. The oxidation process also releases a large amount of energy. The oxidation of fatty acids is the major source of energy in lipid metabolism. The glycerol obtained in the hydrolysis is converted to glyceraldehyde and is metabolized by the Embden-Meyerhof pathway.

The oxidation of fatty acids, which occurs in the mitochondria, requires the participation of several enzymes and coenzymes. The process produces acetyl-CoA, which enters the Krebs' cycle and is converted to carbon dioxide, water, and energy. The reactions occurring in the oxidation of fatty acids are shown on p. 457. The process releases energy, which is conserved in the form of ATP.

The organic chemistry of these steps is discussed as follows with stearic acid as the example.

Step 1 involves the esterification of stearic acid to form a CoA derivative. This reaction is catalyzed by thiokinase and consumes 1 mole of ATP:

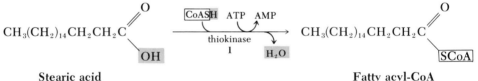

| Stearic acid | Fatty acyl-CoA |

In step 2, the stearic acid CoA derivative is dehydrogenated by a dehydrogenase with FAD as the coenzyme to form an α,β-unsaturated ester:

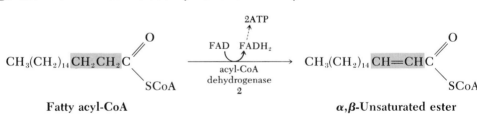

| **Fatty acyl-CoA** | α,β-**Unsaturated ester** |

FADH$_2$ is oxidized to FAD with the formation of 2 moles of ATP as shown by the following equation:

$$\text{FADH}_2 + \tfrac{1}{2}\text{O}_2 + 2\text{ADP} + 2\text{H}_3\text{PO}_4 \longrightarrow \text{FAD} + 2\text{ATP} + 3\text{H}_2\text{O}$$

Step 3 involves the hydration of the α,β-unsaturated ester by an acyl-CoA hydrase to give a β-hydroxy ester:

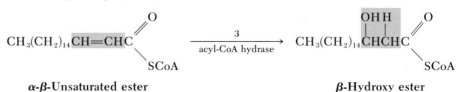

| α-β-**Unsaturated ester** | β-**Hydroxy ester** |

Step 4, which is the oxidation of a β-hydroxyl group to a β-keto group, is catalyzed by a dehydrogenase with NAD as the coenzyme:

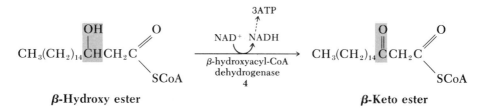

β-Hydroxy ester   **β-Keto ester**

NADH is oxidized to NAD^+ with the formation of 3 moles of ATP as shown by the following equation:

$$NADH + H^+ + \tfrac{1}{2}O_2 + 3ADP + 3H_3PO_4 \longrightarrow NAD^+ + 3ATP + 4H_2O$$

In step 5, the β-keto ester is cleaved to form acetyl-CoA and a new fatty acid derivative with two less carbons than the initial fatty acid:

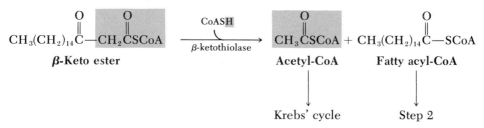

β-Keto ester   **Acetyl-CoA**   **Fatty acyl-CoA**

Krebs' cycle   Step 2

The acetyl-CoA enters the Krebs' cycle and is converted to carbon dioxide, water, and energy in the form of 12 ATP molecules. The new fatty acid CoA derivative reenters the cycle and forms acetyl-CoA and another fatty acid derivative containing two less carbons. This cycle continues until the product is two molecules of acetyl-CoA. In the case of stearic acid, eight turns of the cycle are required, and nine acetyl-CoA molecules enter the Krebs' cycle.

The amount of energy liberated in the oxidation of stearic acid is as follows:

Process	Coenzyme	Moles ATP	Turns of cycle	Net moles ATP
Step 1		1 mole consumed		-1
Step 2	FAD	2 moles liberated	8	16
Step 4	NAD^+	3 moles liberated	8	24
Krebs' cycle		12 moles liberated	9	108
				147

Thus the complete oxidation of 1 mole of stearic acid forms 147 moles of ATP. The formation of 108 moles of ATP is the result of the oxidation of 9 moles of acetyl-CoA in the Krebs' cycle.

Thus food fat is a very effective source of energy.

Starvation and diabetes: impaired metabolism. Both starvation and diabetes are conditions resulting from a restriction of carbohydrate metabolism with subsequent increase in fat metabolism to supply the body's energy needs and an increase in formation of ketone bodies, which include acetoacetic acid, β-hydroxybutyric acid, and acetone.

Starvation begins when the carbohydrate storage is depleted. The fat storage is then mobilized, and proteins are degraded to furnish fats and carbohydrates. Some of the amino acids are metabolized to produce aceto-acetyl-CoA. Acetoacetyl-CoA may then produce the ketone bodies faster than the tissues can utilize them:

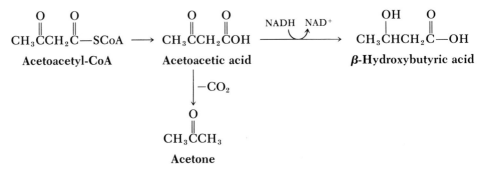

Although the blood level of glucose is elevated in diabetes, the glucose is not available for use by the cells. The fat storage is then mobilized. Hence, the production of ketone bodies is greater than utilization, and they begin to accumulate. Insulin restores the level of ketone bodies to normal.

Protein metabolism

The metabolism of proteins basically involves two processes: anabolism and catabolism. Anabolism is the process through which the body synthesizes amino acids or protein molecules by combining amino acids in specific sequences and arrangements. Catabolism, on the other hand, is the process by which proteins are enzymatically hydrolyzed to yield amino acids, which then undergo specific metabolic reactions.

The amino acids in living systems are absorbed after hydrolysis of protein during digestion. Once amino acids are absorbed into the bloodstream, they are transported to the liver. Some of these amino acids, especially those required for use by the liver, are removed from the blood, and others are synthesized by the liver and added to the bloodstream. Proteins are not stored as carbohydrates and lipids are. The amino acids are utilized for the synthesis of new proteins for tissues, hormones, and enzymes and are even used as sources of energy.

Catabolism. This involves the general reactions specific to the metabolism of each amino acid after the breakdown of proteins.

Some of the important metabolic reactions of amino acids are:

Oxidative deamination. This is a reaction in which the amino group is removed from an amino acid. It is catalyzed by an oxidase with FAD as the coenzyme. The products are a keto acid and, of course, ammonia. Following is an example:

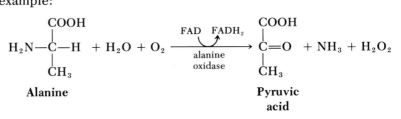

The product of this reaction, pyruvic acid, can then enter the Krebs' cycle.

The carbon structures of amino acids can be converted to glucose and stored as glycogen or can be metabolized in the Embden-Meyerhof pathway or Krebs' cycle and used as energy sources.

Transamination. This is a process by which an amino group is transferred from an amino acid to a keto acid, producing another amino acid and another keto acid. The new keto acid (formed from the original amino acid) may be metabolized in the Krebs' cycle. Following is an example of transamination:

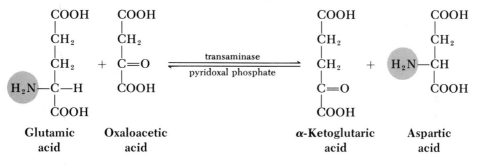

| Glutamic acid | Oxaloacetic acid | | α-Ketoglutaric acid | Aspartic acid |

The reaction is reversible, and the overall reaction requires pyridoxal phosphate (vitamin B_6) as the coenzyme.

Decarboxylation. This is a reaction in which amino acids lose a molecule of carbon dioxide to form amines. For example, histidine undergoes decarboxylation to form histamine:

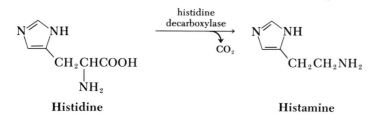

Histidine is known to relax and dilate blood vessels, and antihistamines are known to counteract these effects.

Another example is the decarboxylation of ornithine, first found in birds, to form putrescine, a foul-smelling amine with the characteristic odor of rotting flesh:

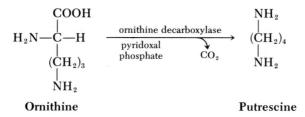

Another example is the decarboxylation of L-dopa to form dopamine. L-Dopa is an amino acid formed by the ring oxidation of the amino acid tyrosine in the presence of tyrosinase. These reactions are as follows:

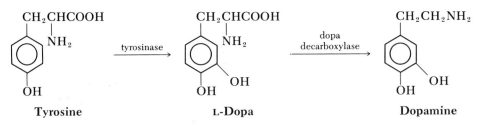

Tyrosine   L-Dopa   Dopamine

L-Dopa has been shown to be effective in the treatment of Parkinson's disease. For some unexplained reason, the brain of the individual with Parkinson's disease is unable to make L-dopa even though the enzyme tyrosinase is present and, hence, is unable to make dopamine. Dopamine is an intermediate in the formation of epinephrine (Adrenalin), which is a vasoconstrictor that is released into the bloodstream when an individual is frightened. It increases cardiac output and elevates the blood glucose level.

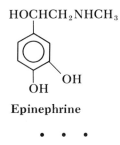

Epinephrine

• • •

Ring oxidation, transamination, and decarboxylation are found in the normal and abnormal metabolism of phenylalanine. The equations involved in phenylalanine metabolism are as follows:

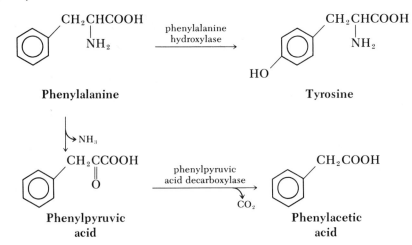

In normal metabolism, phenylalanine is converted to tyrosine in the presence of the enzyme phenylalanine hydroxylase. The abnormal metabolism of phenylalanine is the basis of a hereditary disease known as phenylketonuria (PKU), which results in mental retardation. The individual with PKU lacks the phenylalanine hydroxylase needed to convert phenylalanine to tyrosine;

thus phenylalanine, phenylpyruvic acid, and phenylacetic acid accumulate.

Urea cycle. The ammonia formed in deamination and the carbon dioxide formed in decarboxylation combine to form carbamyl phosphate and are eliminated as urea. The urea is formed in the liver, carried by the blood to the kidney, and excreted in the urine. It is the main nitrogenous end product of protein metabolism. It is synthesized from carbamyl phosphate through a series of reactions referred to as the urea cycle:

$$NH_3 + CO_2 \xrightarrow[\;\;\;\;\;]{2ATP \;\; 2ADP \;\;\; \textcircled{P}} H_2N-\overset{\overset{\displaystyle O}{\|}}{C}-O-\overset{\overset{\displaystyle O}{\|}}{\underset{\underset{\displaystyle O^{\ominus}}{|}}{P}}-OH \xrightarrow[\text{aspartic acid, ATP}]{\text{urea cycle}} H_2N-\overset{\overset{\displaystyle O}{\|}}{C}-NH_2$$

Carbamyl
phosphate   Urea

Anabolism: protein biosynthesis. There are two principal steps involved in the synthesis of a protein:

1. *Transcription.* In this process, the sequencing information of DNA is transferred to coded RNA.
2. *Translation.* In this process, which is the actual protein synthesis, the sequencing information of the coded RNA directs the order of insertion of specific amino acids.

DNA is often referred to as the keeper of the code of life or a template, since all proteins are formed according to precise instructions from it. DNA gives its instructions in the nucleus of the cell. It uncoils its helical strands to produce two strands (Fig. 14-4). As these strands are forming, nucleotides from the environment pair with complementary nucleotides in the original DNA strand to form a new molecule. In synthesis of RNA in the DNA template, the pairing to bases in the nucleotide occurs by base pairing of cytosine (C) to guanine (G) and adenine (A) to uracil (U). The new molecule formed is an RNA molecule and is a complementary copy of the original DNA molecule except in two respects: thymine is replaced by uracil, and deoxyribose is replaced by ribose. Once the RNA is formed, the DNA helix coils up again.

The two principal kinds of RNA formed are messenger RNA (mRNA) and transfer RNA (tRNA), previously called soluble RNA. The mRNA, which is a longer strand, is responsible for transferring instructions from DNA to the site of protein synthesis in the cell; it is protein specific, and there is a different message for each different protein. The tRNA, which is a shorter strand, is responsible for carrying an amino acid to the site of protein synthesis in the cytoplasm; it is amino-acid specific. Ribosomal RNA (or rRNA) is also formed, but its role in protein synthesis is less clear.

After the DNA molecule uncoils and mRNA is formed from the template (Fig. 14-4), the translation process begins. The mRNA molecule contains sets of three bases called *codons*, each of which is a code for a particular amino acid. The sequence of codons is complementary to the sequence of bases in the template of DNA. This is illustrated as follows for an octapeptide

Recoil DNA strand   Uncoil DNA strand

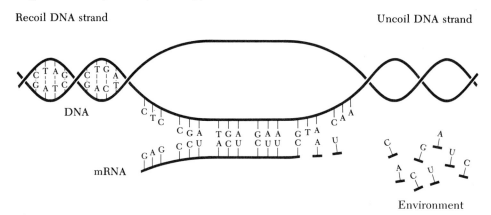

DNA

mRNA

Environment

Fig. 14-4

Transcription of genetic code present in DNA into mRNA for a normal hemoglobin octa-peptide sequence. The code letters in mRNA are the same as those in DNA (with U replacing T), but they are complementary to the letters in the DNA code that was transcribed. The code consists of sets of triplets.

sequence in normal hemoglobin, which is a protein containing four peptide chains, each consisting of nearly 150 amino acids:

DNA:⌇CAA—GTA—GAA—TGA—GGA—CTC—CTC—TTT⌇

mRNA:⌇GUU—CAU—CUU—ACU—CCU—GAG—GAG—AAA⌇

The complementary mRNA conveys the information coded in DNA to the tRNA molecules in the ribosomes in the cytoplasm. Each tRNA molecule is combined with an activated amino acid. Then the amino acids are assembled to produce the protein called for by the DNA. Thus this segment of the genetic information is translated into a specific octapeptide sequence in hemoglobin: H_2N—Val.His.Leu.Thr.Pro.Glu.Glu.Lys—.

Protein synthesis involves three phases: initiation, elongation, and termination.

Initiation begins with activation of the amino acids. This is necessary in order for the carboxyl group to be more reactive toward peptide bond formation. The carboxyl group is more reactive toward esterification if it is first converted to an acid anhydride or acid chloride (Chapter 9):

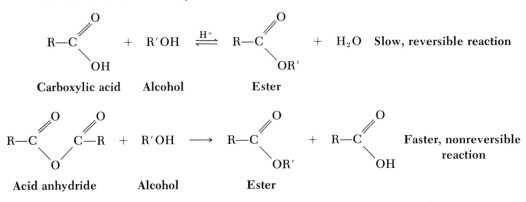

In a similar fashion, living systems convert an amino acid in the presence of ATP and an enzyme specific for that amino acid into an anhydride, which is more reactive:

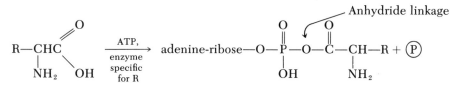

Thus the amino acid is activated. Then the anhydride is used to esterify the tRNA specific for the amino acid to produce aminoacyl-tRNA:

Aminoacyl-tRNA

The amino acid is transferred and bound to a site on mRNA in the form of the aminoacyl—tRNA. Each tRNA molecule contains an anticodon, a triplet of nucleotides whose sequence is different for each amino acid, that is complementary to a specific triplet (codon) on the mRNA. The codon-anticodon arrangement for an octapeptide sequence in hemoglobin to be synthesized is as follows:

DNA:⌇CAA—GTA—GAA—TGA—GGA—CTC—CTC—TTT⌇
mRNA:⌇GUU—CAU—CUU—ACU—CCU—GAG—GAG—AAA⌇
tRNA: CAA GUA GAA UGA GGA CUC CUC UUU

Prior to protein synthesis, mRNA moves into the cytoplasm and interacts with the ribosome to form an mRNA-ribosome complex. Different aminoacyl-tRNA molecules in the environment come into contact with the mRNA-ribosome complex. If the anticodon of the tRNA is complementary to the codon on the mRNA, the aminoacyl group binds to the ribosome to form an aminoacyl-tRNA-mRNA-ribosome complex. The next codon on mRNA is then exposed and binds with the complementary anticodon of another tRNA. Next, the carboxyl group of the first amino acid participates in the formation of an amide bond with the free amino group of the next incoming aminoacyl-tRNA molecule. The first amino acid is then free of its tRNA, which moves away from the complex. Thus a dipeptidyl-tRNA-mRNA-ribosome complex is formed. The ribosome then moves down the mRNA and positions the third codon so that it can react with the anticodon of another aminoacyl-tRNA molecule. This sequence of events, which is depicted in Fig. 14-5 for an octapeptide sequence in hemoglobin, is repeated until a message on the mRNA directs the termination of the synthesis, and the completed protein is released.

The alteration of a single base in the DNA molecule can change the amino acid specified by a particular codon in the mRNA, and the new amino acid can drastically change the characteristics of the synthesized protein. This substitution of one amino acid for another occurs in the abnormal hemoglobin found in sickle cell anemia. At low oxygen concentrations, the abnormal

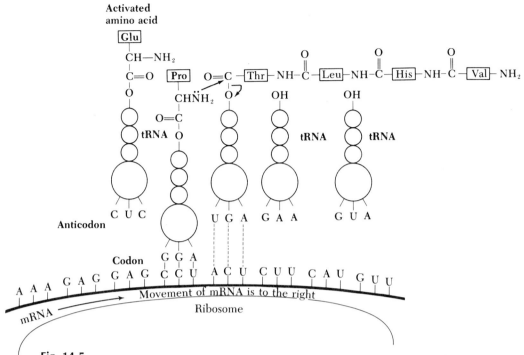

Fig. 14-5

Translation process in protein synthesis of a normal hemoglobin octapeptide sequence.

hemoglobin forms crystals, and the transport of oxygen by the hemoglobin is impaired. Thus the red blood cells of the individual with this condition assume a sickle shape. It is now known that sickle cell hemoglobin is due to the substitution of valine for glutamic acid as the sixth amino acid in the octapeptide sequence that has been discussed:

Normal hemoglobin:

H_2N—Val—His—Leu—Thr—Pro—Glu—Glu—Lys—

Sickle cell hemoglobin:

H_2N—Val—His—Leu—Thr—Pro—Val—Glu—Lys—

Glutamic acid has a polar carboxyl group, which helps to maintain the solubility of the normal hemoglobin in an aqueous medium; whereas valine, a nonpolar amino acid, promotes crystallization of the abnormal hemoglobin:

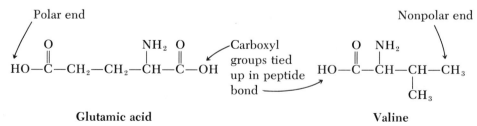

The substitution of one amino acid for another probably occurs through the changing of a single base in the DNA molecule that is responsible for the coding of normal hemoglobin. The genetic code in DNA, codon in mRNA, and anticodon in tRNA for glutamic acid and valine are as follows:

	Glutamic acid	*Valine*
DNA genetic code	CTC	CAC
mRNA codon	GAG	GUG
tRNA anticodon	CUC	CAC

Thus the substitution of the base adenine for thymine in DNA can result in sickle cell anemia. The orientation of the activated valine instead of glutamic acid toward mRNA would appear as follows for sickle cell hemoglobin:

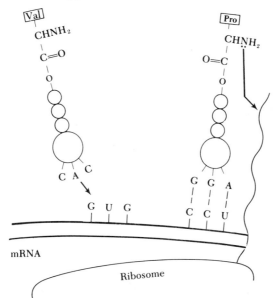

Relationship of metabolism of carbohydrates, lipids, and proteins

The relationship of the metabolism of carbohydrates, lipids, and proteins is summarized as follows:

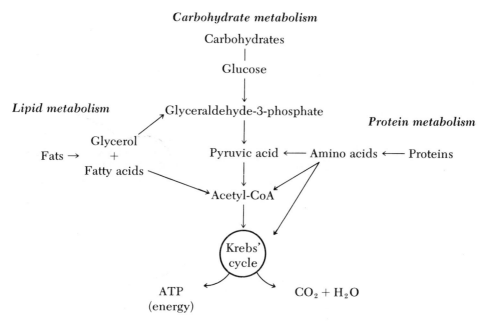

Summary

The cell is a chemical industry capable of performing a number of specific and complex metabolic reactions. Enzymes catalyze these reactions. They have specific functions and are characterized by an active binding site. The various classes of enzymes are listed in Table 14-2. The structure of an enzyme is flexible. It is now believed that its mode of operation involves a shift in its conformation by the substrate so that the enzyme can assume the proper orientation of its active binding sites. Many enzymes are synthesized in an inactive form called a proenzyme. When they are activated, the reaction site is formed. In some enzyme systems, catalysis depends on a protein and a coenzyme. Many vitamins serve as coenzymes.

Many metabolic reactions release energy, which is transformed into ATP, a high-energy compound. ATP is formed through the electron-transport system and substrate-level phosphorylation. It can then be used to drive other metabolic reactions.

Glucose is metabolized by its oxidation to carbon dioxide and water. This metabolism proceeds through the Embden-Meyerhof pathway, which liberates 8 moles of ATP, and the Krebs' cycle, which liberates 30 moles of ATP.

The hydrolysis of neutral fats produces glycerol and fatty acids. The oxidation of fatty acids releases more energy per mole than the metabolism of glucose.

The metabolism of proteins involves catabolism (breakdown of proteins and amino acids) and anabolism (biosynthesis of proteins and amino acids). Protein biosynthesis involves the teamwork of DNA and RNA. The information in DNA is transcribed into the structure in RNA, which is translated into the structure of the protein.

Important terms

Adenosine triphosphate (ATP)	Glycogenolysis
Anabolism	Induced-fit theory
Anticodon	Krebs' cycle
Catabolism	Lock-and-key mechanism
Cell	Metabolic reaction
Cell nucleus	Mitochondria
Codon	Oxidative phosphorylation
Coenzyme	Proenzyme
Deamination	Ribosomes
Electron-transport system	Substrate-level phosphorylation
Embden-Meyerhof pathway	Transamination
Enzyme	Transcription
Glycogenesis	Translation

Problems

14-1. Each of the following reactions is catalyzed by an enzyme. Name a general enzyme type that will catalyze each reaction. Select your answer from Table 14-2.

(a)

$$CH_3COOH \xrightarrow[\text{ATP} \quad \text{ADP}]{} CH_3\overset{\overset{\displaystyle O}{\|}}{C}-OPO_3H_2$$

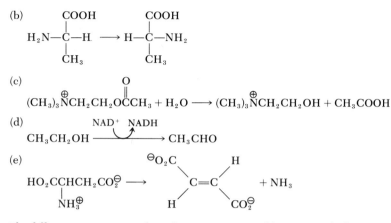

14-2. The following enzymes catalyze the reactions in problem 14-1. Which enzyme catalyzes which reaction?

(a) Alcohol dehydrogenase
(b) ATP acetate phosphotransferase
(c) Cholinesterase
(d) Aspartate ammonia lyase
(e) Alanine racemase

14-3. Explain the current theory of the mechanism of enzyme-catalyzed reactions.

14-4. What is the difference between a proenzyme and a coenzyme?

14-5. Using one of the vitamins known to be a coenzyme, explain how coenzymes work in metabolic reactions.

14-6. Name two ways in which ATP is formed.

14-7. What is the difference between glycogenesis and glycogenolysis?

14-8. What is the structural difference between UTP and ATP?

14-9. Answer the following questions in regard to the Embden-Meyerhof pathway (p. 448):

(a) What is the coenzyme?
(b) How is the reduced form of this coenzyme oxidized?
(c) Which step is a dehydration?
(d) Which step is an oxidation?
(e) Which step consumes a portion of a high-energy compound?
(f) Which step forms ATP on the substrate level?

14-10. In the Krebs' cycle (p. 453), which reaction fits each of the following descriptions?

(a) Hydration
(b) Decarboxylation
(c) Oxidation by NAD
(d) Oxidation by flavoprotein
(e) Dehydration
(f) Hydrolysis of thioester
(g) Aldol condensation

14-11. Explain how the electron-transport system functions in conjunction with the Krebs' cycle.

14-12. How is the Krebs' cycle related to the Embden-Meyerhof pathway?

14-13. Which of the 38 moles of ATP formed from the oxidation of glucose to carbon dioxide and water are formed in the following reactions?

(a) Glycolysis: substrate level
(b) Krebs' cycle: substrate level
(c) Electron-transport system

14-14. The oxidation of a fat produces more energy than the oxidation of glucose. Calculate the number of moles of ATP formed when hexanoic acid is completely oxidized according to the steps shown on p. 457. (This comparison is noteworthy, since both hexanoic acid and glucose contain six carbons.)

14-15. How many moles of ATP are produced in the complete oxidation of palmitic acid to carbon dioxide and water?

14-16. Name three important catabolic reactions of amino acids, and give an example of each.

14-17. Compare DNA, tRNA, and mRNA according to biological function.

14-18. Describe the process of transcription of the information in DNA?

14-19. What is the genetic code, and how is it translated?

14-20. What is the relationship between an activated amino acid and an anhydride?

14-21. The codon for alanine is GCU.

(a) What is the sequence of bases in DNA producing this codon?
(b) What is the anticodon?

14-22. Explain how the alteration of a single base in DNA can change the amino acid specified by a particular codon. What are the consequences of such a substitution of amino acids?

Self-test

1. Explain briefly how enzyme-catalyzed reactions generally occur.
2. FAD serves as a coenzyme for succinic dehydrogenase, which oxidizes succinic acid to fumaric acid. The flavin portion of the coenzyme is part of the active site for the reaction. Indicate the changes occurring in the active site.

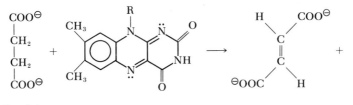

| Succinic acid | FAD-enzyme | Fumaric acid | FADH$_2$-coenzyme |

3. With respect to metabolic processes, what is the difference between anabolism and catabolism?
4. A structural characteristic of an energy-rich compound is the easily hydrolyzed anhydride linkage. Draw the structural formula of ATP, and show how it fulfills this structural characteristic.
5. How does phosphorylation in the electron-transport system differ from phosphorylation on the substrate level as a means of formation of ATP?
6. The Krebs' cycle is an aerobic pathway for the conversion of acetyl groups from pyruvic acid to carbon dioxide and water.

(a) What is an aerobic pathway?
(b) Is the conversion an energy-releasing process?
(c) How is a portion of the energy conserved?
(d) How does the energy conserved in the Krebs' cycle compare with that conserved in the Embden-Meyerhof pathway?

7. Briefly explain the difference between transcription and translation by describing the sequence of events occurring in each.

8. Using terms such as reduction, oxidation, deamination, decarboxylation, hydrolysis, hydration, or dehydrogenation, identify the process occurring in each of the following biochemical reactions:

(a)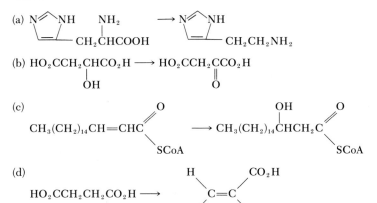

(b) $HO_2CCH_2CHCO_2H \longrightarrow HO_2CCH_2CCO_2H$
      $\qquad\qquad\quad |$ $\qquad\qquad\qquad \parallel$
      $\qquad\qquad\quad OH$ $\qquad\qquad\qquad O$

(c)

$$CH_3(CH_2)_{14}CH{=}CHC\overset{O}{\underset{SCoA}{\diagup}} \longrightarrow CH_3(CH_2)_{14}\overset{OH}{\underset{}{CHCH_2C}}\overset{O}{\underset{SCoA}{\diagup}}$$

(d)

$$HO_2CCH_2CH_2CO_2H \longrightarrow \overset{H}{\underset{HO_2C}{\diagdown}}C{=}C\overset{CO_2H}{\underset{H}{\diagup}}$$

(e) $ATP + H_2O \longrightarrow ADP + H_3PO_4$

9. Write the correct answer for each of the following questions:

   (a) What part of the cell has the function of transmitting and regulating hereditary characteristics?
   (b) Enzymes generally need some kind of nonprotein material for catalysis. What is this material called?
   (c) What is an example of a vitamin that serves as a coenzyme?
   (d) One of the steps in the Embden-Meyerhof pathway is the conversion of 1,3-diphosphoglyceric acid to 3-phosphoglyceric acid. The reaction occurs in the presence of ADP. What becomes of the phosphate that is released?
   (e) What is the name of the process in which glycogen is converted to glucose?
   (f) Is the conversion of glucose-6-phosphate to glucose-1-phosphate in glycogenesis described as oxidative phosphorylation, hydrolysis, or isomerization?
   (g) What is the main nitrogenous end product of the catabolism of protein?
   (h) Of the two kinds of RNA, which is amino-acid specific?
   (i) If a sequence of bases in DNA is TGA, what is the complementary sequence of bases in mRNA?
   (j) Living systems convert an amino acid into a more reactive form for transfer to the site of protein synthesis. In terms of structure, what is the more reactive form called?

Answers to self-test

1. The substrate approaches the flexible enzyme and causes the enzyme to assume a proper orientation of its active binding sites. An enzyme-substrate complex is formed, and the specific reaction takes place with this complex. After the reaction has occurred, the enzyme-substrate complex dissociates into the enzyme and product.

2.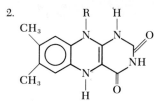

3. Anabolism is the synthesis of molecules; catabolism is the breakdown of substances.

4.

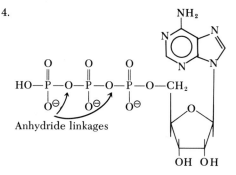

Anhydride linkages

5. The electron-transport system is a series of steps through which electrons are transferred to oxygen by the oxidation of NADH and FADH$_2$ with the formation of up to 3 moles of ATP. On the substrate-level, ATP is formed as a result of interconversion of a substrate other than an electron-transport material.

6. (a) It is a pathway that operates in the presence of oxygen.
   (b) Yes.
   (c) A portion of the energy is stored in ATP.
   (d) A larger amount of energy is stored in the form of ATP in the Krebs' cycle.

7. Transcription is the process in which the sequencing information of DNA is transferred to coded RNA. Translation is the process in which the sequencing information of the coded RNA directs the order of insertion of specific amino acids; it is the actual protein synthesis.

8. (a) Decarboxylation
   (b) Oxidation
   (c) Hydration
   (d) Dehydrogenation
   (e) Hydrolysis

9. (a) Nucleus
   (b) Coenzyme
   (c) Vitamin B$_1$, vitamin B$_2$, or niacin
   (d) Incorporated into ATP
   (e) Glycogenolysis
   (f) Isomerization
   (g) Urea
   (h) tRNA
   (i) ACU
   (j) Anhydride (aminoacyl-tRNA)

Selected answers

Chapter 1 Some fundamental concepts

1-1. (c) $\cdot \ddot{N} \cdot + 3H \cdot \longrightarrow H : \ddot{N} : H$
$\phantom{1-1. (c) \cdot \ddot{N} \cdot + 3H \cdot \longrightarrow H : }\ddot{H}$

(d) $\cdot \dot{\underset{\cdot}{C}} \cdot + 2 : \ddot{F} \cdot + 2H \cdot \longrightarrow H : \overset{\displaystyle :\ddot{F}:}{\underset{\displaystyle :\ddot{F}:}{C}} : H$

1-2. (d) sp^3; carbon is bonded to four other atoms
(e) sp^2; carbon is bonded to three other atoms
(f) sp^3; carbon is bonded to four other atoms
(g) sp; carbon is bonded to two other atoms
(h) sp^3; carbon is bonded to four other atoms

1-3. (d) 180°; orbitals of carbon are sp hybridized
(e) 120°; orbitals of carbon are sp^2 hybridized
(f) 120°; orbitals of carbon are sp^2 hybridized
(g) 180°; orbitals of carbon are sp hybridized
(h) 109.5°; orbitals of carbon are sp^3 hybridized

1-4. (a) Carboxylic acid
(c) Amine
(e) Ether
(g) Amide
(i) Aldehyde

1-5. (a)

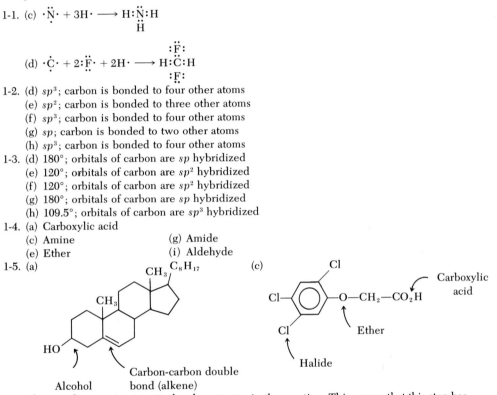

1-7. The rate-determining step is the slowest step in the reaction. This means that this step has the largest energy of activation. Step 3 has the largest energy of activation (E_{A_3}) and is the rate determining step.

1-8. The rate expression is rate $= k\, [CH_2{=}CH_2][H^+]$.

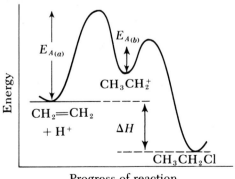

Progress of reaction

1-10. Nitrogen: $1s^2 2[(sp^3)^2(sp^3)^1(sp^3)^1(sp^3)^1]$
Oxygen: $1s^2 2[(sp^3)^2(sp^3)^2(sp^3)^1(sp^3)^1]$

1-11. (a) sp^2 (e) sp^3
    (c) sp (g) sp^2

The hybridization is determined by counting the number of atoms directly bonded to the circled carbon. See Table 1-3 in text.

Chapter 2 Hydrocarbons: structure, nomenclature, and preparation

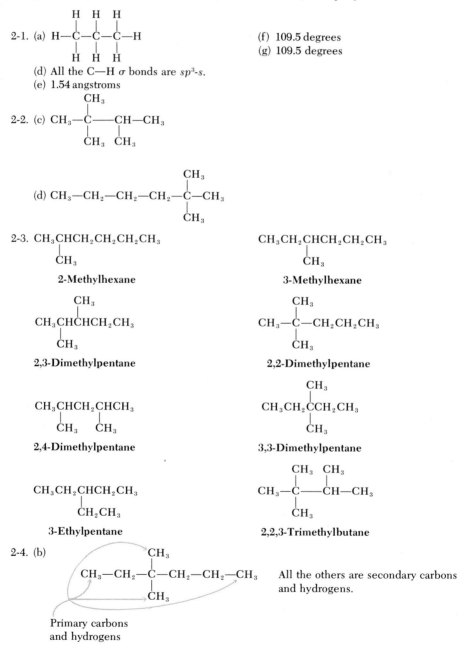

2-1. (a)

$$\begin{array}{ccc} H & H & H \\ | & | & | \\ H-C-C-C-H \\ | & | & | \\ H & H & H \end{array}$$

(f) 109.5 degrees
(g) 109.5 degrees

(d) All the C—H σ bonds are sp^3-s.
(e) 1.54 angstroms

2-2. (c)

$$CH_3-\underset{\underset{CH_3}{|}}{\overset{\overset{CH_3}{|}}{C}}----\underset{\underset{CH_3}{|}}{CH}-CH_3$$

(d)

$$CH_3-CH_2-CH_2-CH_2-\underset{\underset{CH_3}{|}}{\overset{\overset{CH_3}{|}}{C}}-CH_3$$

2-3.

$CH_3CHCH_2CH_2CH_2CH_3$
    |
    CH_3

2-Methylhexane

$CH_3CH_2CHCH_2CH_2CH_3$
         |
         CH_3

3-Methylhexane

      CH_3
      |
$CH_3CHCHCH_2CH_3$
    |
    CH_3

2,3-Dimethylpentane

     CH_3
     |
$CH_3-C-CH_2CH_2CH_3$
     |
     CH_3

2,2-Dimethylpentane

$CH_3CHCH_2CHCH_3$
    | |
    CH_3 CH_3

2,4-Dimethylpentane

        CH_3
        |
$CH_3CH_2CCH_2CH_3$
        |
        CH_3

3,3-Dimethylpentane

$CH_3CH_2CHCH_2CH_3$
        |
        CH_2CH_3

3-Ethylpentane

     CH_3 CH_3
     | |
$CH_3-C----CH-CH_3$
     |
     CH_3

2,2,3-Trimethylbutane

2-4. (b)

           CH_3
           |
$CH_3-CH_2-C-CH_2-CH_2-CH_3$
           |
           CH_3

All the others are secondary carbons and hydrogens.

Primary carbons and hydrogens

(c) Tertiary carbons and hydrogens   Primary carbons and hydrogens

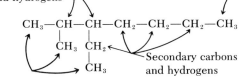

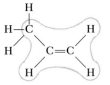

$$CH_3—CH—CH—CH_2—CH_2—CH_2—CH_3$$

Secondary carbons and hydrogens

Primary carbons and hydrogens

2-5. (e) Side-by-side overlap of two p orbitals
    (f) 120 degrees
    (g) The encircled atoms lie in the same plane.

2-6. (e) No, because there are no sp^2 carbon atoms
    (f) One sp-sp and two p-p or one σ and two π.
    (g) 180 degrees
    (h) The encircled atoms lie along the same straight line.

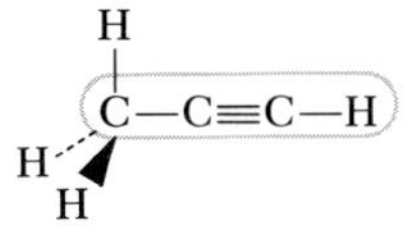

2-7. The isomeric hexenes (C_6H_{12}):

$$CH_3CH_2CH_2CH_2CH{=}CH_2$$

1-Hexene

$$CH_3CH_2CH_2CH{=}CHCH_3$$

2-Hexene

$$CH_3CH_2CH{=}CHCH_2CH_3$$

3-Hexene

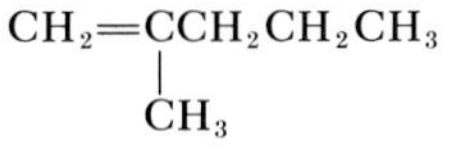

$$CH_2{=}CCH_2CH_2CH_3$$
           $|$
         CH_3

2-Methyl-1-pentene

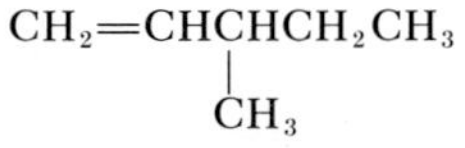

$$CH_2{=}CHCHCH_2CH_3$$
         $|$
        CH_3

3-Methyl-1-pentene

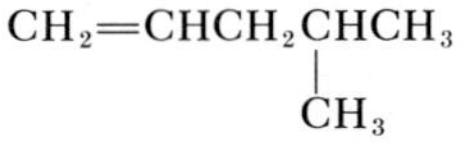

$$CH_2{=}CHCH_2CHCH_3$$
            $|$
          CH_3

4-Methyl-1-pentene

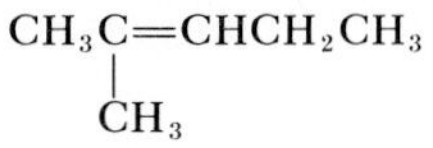

$$CH_3C{=}CHCH_2CH_3$$
      $|$
    CH_3

2-Methyl-2-pentene

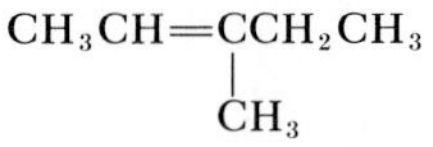

$$CH_3CH{=}CCH_2CH_3$$
         $|$
       CH_3

3-Methyl-2-pentene

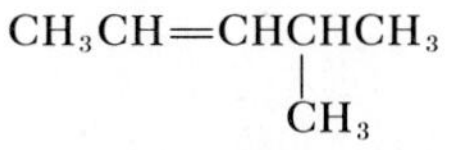

$$CH_3CH{=}CHCHCH_3$$
          $|$
        CH_3

4-Methyl-2-pentene

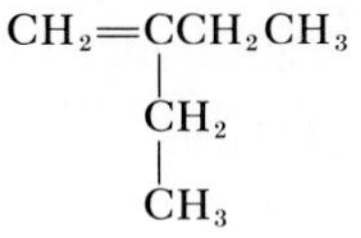

$$CH_2{=}CCH_2CH_3$$
      $|$
    CH_2
    $|$
    CH_3

2-Ethyl-1-butene

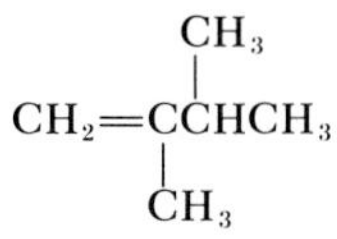

         CH_3
        $|$
$$CH_2{=}CCHCH_3$$
        $|$
      CH_3

2,3-Dimethyl-1-butene

$$CH_3C{=}CCH_3$$
     $|$ $|$
   CH_3 CH_3

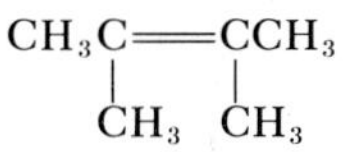

2,3-Dimethyl-2-butene

The isomeric hexynes:

$CH{\equiv}CCH_2CH_2CH_2CH_3$

1-Hexyne

$CH_3C{\equiv}CCH_2CH_2CH_3$

2-Hexyne

$CH_3CH_2C{\equiv}CCH_2CH_3$

3-Hexyne

$HC{\equiv}CCHCH_2CH_3$
|
CH_3

3-Methyl-1-pentyne

$HC{\equiv}CCH_2CHCH_3$
|
CH_3

4-Methyl-1-pentyne

$CH_3C{\equiv}CCHCH_3$
|
CH_3

4-Methyl-2-pentyne

$$CH_3$$
|
$CH{\equiv}C-C-CH_3$
|
CH_3

3,3-Dimethyl-1-butyne

2-8. Cycloalkanes (C_nH_{2n}) are isomeric with alkenes (C_nH_{2n}).
Cycloalkenes (C_nH_{2n-2}) are isomeric with alkynes (C_nH_{2n-2}).

2-9. The isomeric cyclo-C_6H_{12}:

Cyclohexane   Methylcyclopentane   1,1-Dimethylcyclobutane   1,2-Dimethylcyclobutane

1,3-Dimethylcyclobutane   Ethylcyclobutane   1,2,3-Trimethylcyclopropane   1,1,2-Trimethyl-cyclopropane

2-10. (b)

$$CH_3 \qquad CH_3$$
| |
$CH_3-C-CH_2I \longrightarrow CH_3-C{=}CH_2$
|
H

2-11. The energy of B and D is greater than that of C and E, respectively.

2-12. (c) $CH_3\overset{\oplus}{C}HCH_2$
|
CH_3

(d)

H

2-13. (c)

$$CH_3 \qquad\qquad CH_3 \qquad\qquad CH_3$$
| | |
$CH_3CHCH_2CH_2OH \xrightarrow[\text{heat}]{H^+} CH_3CHCH{=}CH_2 + CH_3C{=}CHCH_3$

(major product)

(d)

$$CH_3 \qquad\qquad CH_3 \qquad\qquad CH_3$$
| | |
$CH_3CCH_2CH_3 \xrightarrow[\text{heat}]{H^+} CH_2{=}CCH_2CH_3 + CH_3C{=}CHCH_3$
|
OH

(major product)

2-14. (a) 3-Methylnonane   (e) 2-Methyl-6-isopropylnonane
(c) 2-Ethyl-3-methyl-4-chloro-1,5-heptadiene

2-15. (a) Any of the isomers shown in Answer 2-7.

(c)

CH₃—C(CH₃)(CH₃)—CH₃, 2,2-Dimethylpropane

(g)

CH₃C=CCH₃ (with CH₃, CH₃ substituents), 2,3-Dimethyl-2-butene

(h)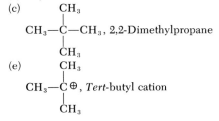
, *Trans*-2-pentene

(e)

CH₃—C⊕(CH₃)(CH₃), *Tert*-butyl cation

2-17. (a) Three
   (c) One, the H adjacent to bond 4.
   (e) 120 degrees, 180 degrees

2-18. (a) $CH_3CH_2CH_2CH=CH_2$

   1-Pentene

   (d)

   Cyclohexene

   (e) $CH_3CH=CHCH_2CH=CH_2$ + $CH_3CH=CHCH=CHCH_3$

   1,4-Hexadiene   **2,4-Hexadiene**
                         (conjugated diene)
                         (major product)

   (g) 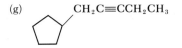 CH₂C≡CCH₂CH₃

   1-Cyclopentyl-2-pentyne

2-19. (a) $CH_3C\equiv CCH_2Cl$

   (g) $CH_3CHCH_2CHCH_2CH_2CH_3$
       with CH_3 substituent

   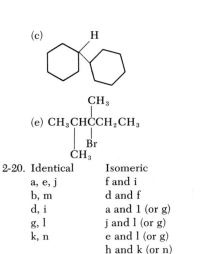

   (c)

   H

   (i)

   Cl

   $CH_3C=CHCCH_3$
   with Cl, Cl

   CH₃
   (e) $CH_3CHCCH_2CH_3$
       Br
       CH₃

2-20. Identical   Isomeric   Homologs
      a, e, j   f and i   c and h
      b, m   d and f
      d, i   a and 1 (or g)
      g, l   j and 1 (or g)
      k, n   e and 1 (or g)
                   h and k (or n)

2-21. (a) Heptane
   (b) $CH_3(CH_2)_5CH_3$
   (e) $CH_3CH_2CH_2CH_2CH_2CH_3$ *or* $CH_3CH_2CH_2CH_2CH_2CH_2CH_2CH_3$

      Hexane   **Octane**

2-22. $CH_3 \boxed{CH\!=\!C} CH_2CH_3 > CH_3CH_2 \boxed{C} CH_2CH_3 > \boxed{CH_2\!=\!CH} CHCH_2CH_3$

with CH_3 below the first boxed group, $\overset{\|}{C}H_2$ below the middle, and CH_3 below the last.

Chapter 3 Hydrocarbons: reactions

3-1. (c)

$CH_3CH_2CH_2CH_2CH_3 \xrightarrow[\text{UV light}]{Cl_2} CH_3CH_2CH_2CH_2CH_2Cl + CH_3CH_2CH_2CHCH_3 + CH_3CH_2CHCH_2CH_3$

with Cl below the second and third products.

1-Chloropentane

2-Chloropentane
(preferred product)

3-Chloropentane
(preferred product)

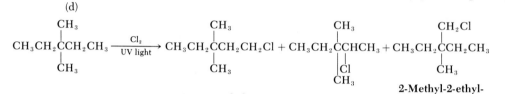

(d)

$CH_3CH_2\overset{CH_3}{\underset{CH_3}{C}}CH_2CH_3 \xrightarrow[\text{UV light}]{Cl_2} CH_3CH_2\overset{CH_3}{\underset{CH_3}{C}}CH_2CH_2Cl + CH_3CH_2\overset{CH_3}{\underset{Cl\,CH_3}{C}}CHCH_3 + CH_3CH_2\overset{CH_2Cl}{\underset{CH_3}{C}}CH_2CH_3$

3,3-Dimethyl-1-
chloropentane

3,3-Dimethyl-2-
chloropentane
(preferred product)

2-Methyl-2-ethyl-
1-chlorobutane

(e)

$CH_3\overset{CH_3}{\underset{CH_3}{C}}CH_2CH_3 \xrightarrow[\text{UV light}]{Cl_2} CH_3\overset{CH_3}{\underset{CH_3}{C}}CH_2CH_2Cl \;+\; CH_3\overset{CH_3}{\underset{Cl\,CH_3}{C}}CHCH_3 \;+\; ClCH_2\overset{CH_3}{\underset{CH_3}{C}}CH_2CH_3$

3,3-Dimethyl-1-
chlorobutane

2,2-Dimethyl-3-
chlorobutane
(preferred product)

2,2-Dimethyl-1-
chlorobutane

(f)

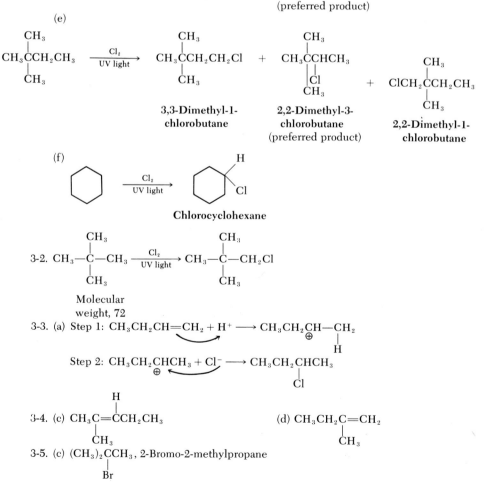

Chlorocyclohexane

3-2. $CH_3\!-\!\overset{CH_3}{\underset{CH_3}{C}}\!-\!CH_3 \xrightarrow[\text{UV light}]{Cl_2} CH_3\!-\!\overset{CH_3}{\underset{CH_3}{C}}\!-\!CH_2Cl$

Molecular
weight, 72

3-3. (a) Step 1: $CH_3CH_2CH\!=\!CH_2 + H^+ \longrightarrow CH_3CH_2\overset{\oplus}{C}H\!-\!\underset{H}{C}H_2$

Step 2: $CH_3CH_2\overset{\oplus}{C}HCH_3 + Cl^- \longrightarrow CH_3CH_2\underset{Cl}{C}HCH_3$

3-4. (c) $CH_3\overset{H}{C}\!=\!\overset{}{C}CH_2CH_3$ (d) $CH_3CH_2\overset{}{C}\!=\!CH_2$

with CH_3 below the C in (c) and CH_3 below the C in (d).

3-5. (c) $(CH_3)_2\overset{}{C}CH_3$, 2-Bromo-2-methylpropane

with Br below.

(d) $\underset{\underset{CH_3}{|}}{\overset{\overset{Br\ \ Br}{|\ \ \ |}}{CH_2CCH_2CH_2CH_3}}$, 1,2-Dibromo-2-methylpentane

(e) $\underset{\underset{CH_3}{|}}{CH_3CH_2CHCH_2Br}$, 1-Bromo-2-methylbutane

3-8. Yes. 1-Hexyne is a terminal alkyne and will give a white precipitate with $AgNO_3$ in alcohol. 1-Hexene will not.

3-9. (a) $\underset{\underset{CH_3}{|}}{CH_3CH_2CHCH_2CH_3} \xrightarrow[\text{UV light}]{Br_2} \underset{\underset{CH_3}{|}}{CH_3CH_2CHCH_2CH_2Br} + \underset{\underset{CH_3}{|}}{\overset{\overset{Br}{|}}{CH_3CH_2CHCHCH_3}}$

1-Bromo-3-methyl-   **2-Bromo-3-methyl-**
pentane   **pentane**

$$\underset{\underset{CH_3}{|}}{\overset{\overset{Br}{|}}{CH_3CH_2CCH_2CH_3}} \quad + \quad \underset{\underset{CH_2Br}{|}}{CH_3CH_2CHCH_2CH_3}$$

3-Methyl-3-bromopentane   **2-Ethyl-1-bromo-**
(preferred product)   **butane**

(c) $CH_3CH=CHCH_2CH_3 \xrightarrow[Ni]{H_2} CH_3CH_2CH_2CH_2CH_3$

Pentane

(e) $CH_3C{\equiv}CCH_3 + AgNO_3 \xrightarrow{\text{alcohol}} \text{No reaction}$

(g) $CH_3CH_2C{\equiv}CH + HBr \longrightarrow \underset{\underset{Br}{|}}{CH_3CH_2C=CH_2}$

2-Bromo-1-butene

(i) $\underset{\underset{CH_3}{|}}{CH_3C=CH_2} \quad + \quad KMnO_4 \quad \xrightarrow{\text{cold}} \quad \underset{\underset{CH_3}{|}}{\overset{\overset{OH}{|}}{CH_3C-CH_2OH}}$

2-Methyl-1,2-propanediol

(k)

(m) $\underset{\underset{CH_3}{|}}{HC{\equiv}CCH_2CHCH_3} \xrightarrow{2HCl} \underset{\underset{Cl}{|}\ \ \underset{CH_3}{|}}{\overset{\overset{Cl}{|}}{CH_3CCH_2CHCH_3}}$

2,2-Dichloro-4-methylpentane

3-10. Since A gives no white precipitate with $AgNO_3$ in alcohol, A cannot contain a terminal carbon-carbon triple bond. Several possibilities exist: $CH_3C{\equiv}CCH_3$, $CH_2=CH-CH=CH_2$, $CH_2=C=CHCH_3$. All three compounds will consume two moles of H_2 to form butane.

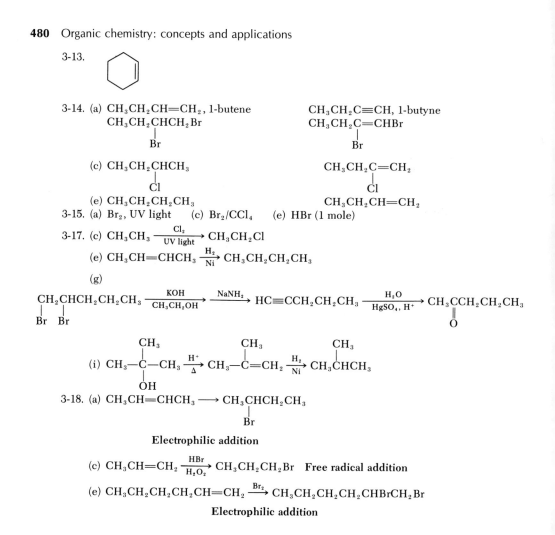

3-13.

3-14. (a) $CH_3CH_2CH=CH_2$, 1-butene $CH_3CH_2C\equiv CH$, 1-butyne

$CH_3CH_2CHCH_2Br$ $CH_3CH_2C=CHBr$
|
Br |
                                Br

    (c) $CH_3CH_2CHCH_3$ $CH_3CH_2C=CH_2$
              | |
             Cl   Cl

    (e) $CH_3CH_2CH_2CH_3$ $CH_3CH_2CH=CH_2$

3-15. (a) Br_2, UV light   (c) Br_2/CCl_4 (e) HBr (1 mole)

3-17. (c) $CH_3CH_3 \xrightarrow[\text{UV light}]{Cl_2} CH_3CH_2Cl$

    (e) $CH_3CH=CHCH_3 \xrightarrow[\text{Ni}]{H_2} CH_3CH_2CH_2CH_3$

    (g)

$CH_2CHCH_2CH_2CH_3 \xrightarrow[\text{CH}_3\text{CH}_2\text{OH}]{KOH} \xrightarrow{NaNH_2} HC\equiv CCH_2CH_2CH_3 \xrightarrow[\text{HgSO}_4,\ \text{H}^+]{H_2O} CH_3CCH_2CH_2CH_3$
| |
Br Br   ‖
                                                                                                      O

                    CH_3 CH_3 CH_3
                     | | |
    (i) $CH_3-C-CH_3 \xrightarrow[\Delta]{H^+} CH_3-C=CH_2 \xrightarrow[\text{Ni}]{H_2} CH_3CHCH_3$
                     |
                     OH

3-18. (a) $CH_3CH=CHCH_3 \longrightarrow CH_3CHCH_2CH_3$
                                                       |
                                                       Br

        Electrophilic addition

    (c) $CH_3CH=CH_2 \xrightarrow[\text{H}_2\text{O}_2]{HBr} CH_3CH_2CH_2Br$ **Free radical addition**

    (e) $CH_3CH_2CH_2CH_2CH=CH_2 \xrightarrow{Br_2} CH_3CH_2CH_2CH_2CHBrCH_2Br$

                     Electrophilic addition

Chapter 4   Aromatic hydrocarbons

4-1. (c) Positive with 1,3-hexadiyne only

    (d) For benzene: cyclohexane; for 1,3-hexadiyne: hexane

    (f) For benzene: one; for 1,3-hexadiyne: three; the three isomers are $XC\equiv CC\equiv CCH_2CH_3$, $HC\equiv CC\equiv CCHXCH_3$, and $HC\equiv CC\equiv CCH_2CH_2X$, where X is Br.

    (g) For benzene: six; for 1,3-hexadiyne: eight

    (h) For benzene: six; for 1,3-hexadiyne: four

    1,3-Hexadiyne could not be benzene. Benzene contains a cyclic cloud of six π electrons, gives only one monosubstitution product, forms cyclohexane when completely hydrogenated, and does not decolorize bromine/CCl_4 or $KMnO_4$.

4-2. (c)   (d)   (e)

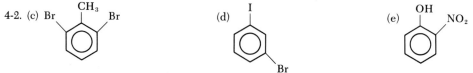

4-3. The position or location of the chloro group is not indicated.

4-4. (b) *p*-Bromoaniline or 4-bromoaniline

    (d) 4,5-Diaminotoluene

    (e) 2,4,5-Tribromobenzenesulfonic acid

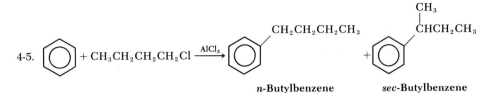

4-5.

The initially formed *n*-butyl cation rearranges to a more stable *sec*-butyl cation.

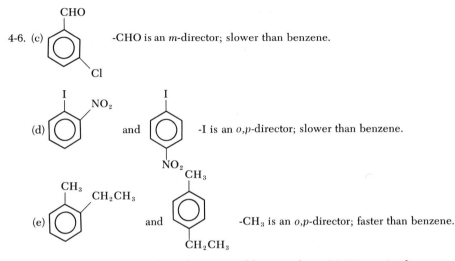

4-6. (c) -CHO is an *m*-director; slower than benzene.

(d) and -I is an *o,p*-director; slower than benzene.

(e) and -CH$_3$ is an *o,p*-director; faster than benzene.

4-7. The correct product is *m*-bromobenzoic acid because the —COOH is an *m*-director.

4-8. The resonance structure in the answer to the problem in the text shows that in the *o* and *p* resonance structures like charges appear on adjacent atoms. These structures are very unstable and do not occur in the case of *m*-attack. Therefore, —NH$_3^+$ is an *m*-director.

4-9. (c) The two substituents reenforce the indicated position.

(d) The —NH$_2$ group is a stronger *o,p*-director.

(e) COOH The two substituents (—COOH is an *m*-director and —Br is an *o,p*-director) reenforce the indicated positions.

4-10. (c)

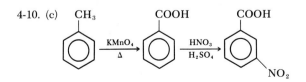

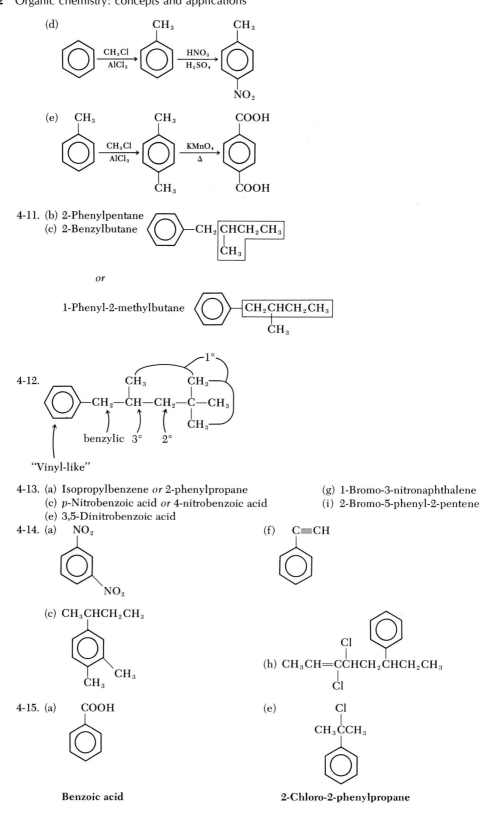

(d)

$$CH_3Cl \xrightarrow{AlCl_3} \xrightarrow{HNO_3 \atop H_2SO_4}$$

(e)

$$CH_3Cl \xrightarrow{AlCl_3} \xrightarrow{KMnO_4 \atop \Delta}$$

4-11. (b) 2-Phenylpentane
(c) 2-Benzylbutane   —CH₂CHCH₂CH₃ | CH₃

or

1-Phenyl-2-methylbutane   —CH₂CHCH₂CH₃ | CH₃

4-12.

—CH₂—CH—CH₂—C—CH₃

benzylic   3°   2°

"Vinyl-like"

4-13. (a) Isopropylbenzene *or* 2-phenylpropane   (g) 1-Bromo-3-nitronaphthalene
(c) *p*-Nitrobenzoic acid *or* 4-nitrobenzoic acid   (i) 2-Bromo-5-phenyl-2-pentene
(e) 3,5-Dinitrobenzoic acid

4-14. (a)   NO₂   (f)   C≡CH

NO₂

(c) CH₃CHCH₂CH₃

CH₃

(h) CH₃CH=CCHCH₂CHCH₂CH₃   with Cl and Cl

4-15. (a)   COOH   (e)   Cl
                                                            CH₃CCH₃

Benzoic acid   **2-Chloro-2-phenylpropane**

(c) CH$_3$CHCH$_3$ CH$_3$CHCH$_3$

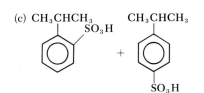

o- and *p*-**Isopropyl-
benzenesulfonic acid**

4-16. (a)   CH$_3$CHCH$_3$ CH$_3$CHCH$_3$ $\xrightarrow{H^+}$ CH$_3$CHCH$_3$ $\xrightarrow{-H_2O}$

Isopropylbenzene

4-18. (a)   CH$_3$ The two substituents reenforce the indicated position.

NO$_2$

(c)   CHO   The two substituents reenforce the indicated position.

CN

(e)   NH$_2$ All three substituents reenforce the indicated position.

CH$_3$
NO$_2$

4-19. (a) **Z** is an *o, p*-director because the product mixture in nitration, chlorination, or bromination is predominantly *o, p*.
 (b) **Z** is a deactivator because halogenation occurs less slowly than in benzene.
 (c) **Z** determines the kind of orientation. The group already on the ring determines the orientation, not the incoming group.
 (d) **Z** is probably —F, —Cl, —Br or —I.

4-21. (a)   NO$_2$ (c)

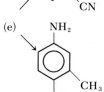

4-22. (a)

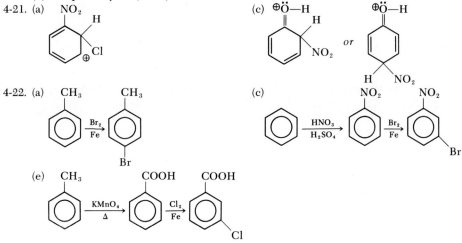

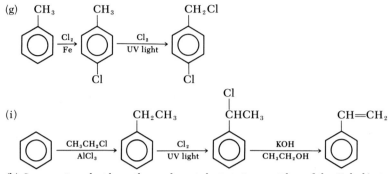

4-23. (b) Same as in a, but here the analogy is better since neither of the Kekulé structures has any real existence. Thus benzene is a hybrid of two structures that have no existence. Similarly, a rhinoceros is described as a hybrid of a dragon and a unicorn, both of which are mythical animals.

Chapter 5   Stereochemistry

5-1. $CH_3CH_2CH_2CH_2CH_3$

Hexane

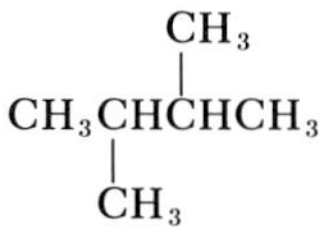

$$CH_3\underset{\underset{CH_3}{|}}{CH}\underset{\underset{CH_3}{|}}{CH}CH_3$$

2,3-Dimethylbutane

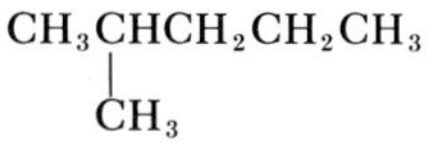

$$CH_3\underset{\underset{CH_3}{|}}{CH}CH_2CH_2CH_3$$

2-Methylpentane

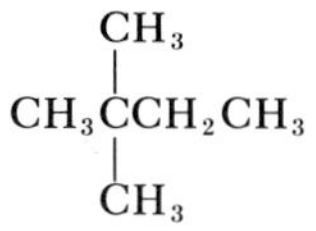

$$CH_3\underset{\underset{CH_3}{|}}{\overset{\overset{CH_3}{|}}{C}}CH_2CH_3$$

2,2-Dimethylbutane

$$CH_3CH_2\underset{\underset{CH_3}{|}}{CH}CH_2CH_3$$

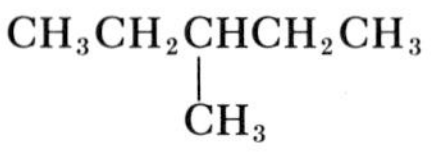

3-Methylpentane

5-2. $CH_3CH_2CH_2CH_2CH_2Cl$

1-Chloropentane

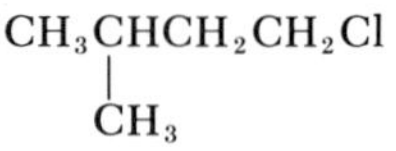

$$CH_3\underset{\underset{CH_3}{|}}{CH}CH_2CH_2Cl$$

1-Chloro-3-methylbutane

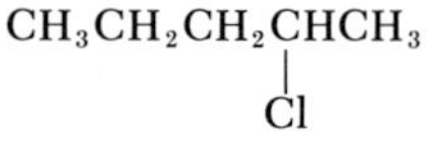

$$CH_3CH_2CH_2\underset{\underset{Cl}{|}}{CH}CH_3$$

2-Chloropentane

$$CH_3\underset{\underset{CH_3}{|}}{\overset{\overset{Cl}{|}}{CH}}CHCH_3$$

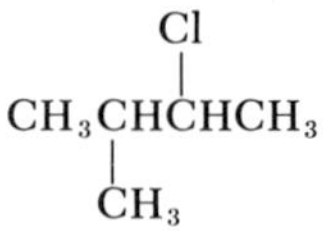

2-Chloro-3-methylbutane

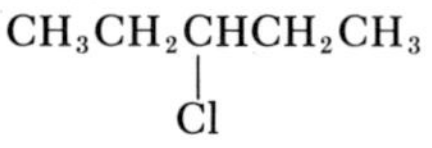

$$CH_3CH_2\underset{\underset{Cl}{|}}{CH}CH_2CH_3$$

3-Chloropentane

$$CH_3\underset{\underset{CH_3}{|}}{\overset{\overset{Cl}{|}}{C}}CH_2CH_3$$

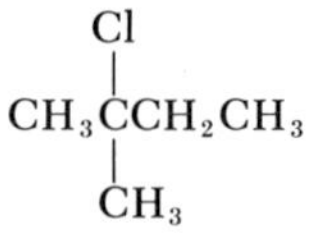

2-Chloro-2-methylbutane

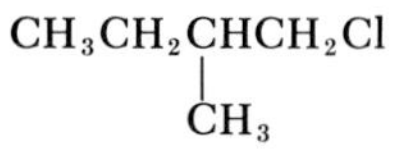

$$CH_3CH_2\underset{\underset{CH_3}{|}}{CH}CH_2Cl$$

1-Chloro-2-methylbutane

5-3. (d)

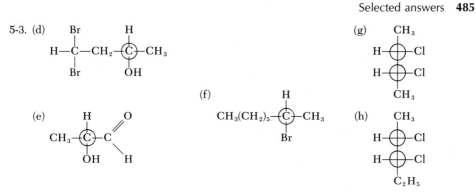

H—C—CH$_2$—C—CH$_3$ with Br, H on top, Br, OH below

(e)

CH$_3$—C—C with H, O on top and OH, H below

(f)

CH$_3$(CH$_2$)$_5$—C—CH$_3$ with H on top, Br below

(g)

CH$_3$ / H—Cl / H—Cl / CH$_3$

(h)

CH$_3$ / H—Cl / H—Cl / C$_2$H$_5$

5-4. (c) Optically active because the molecule has a nonsuperimposable mirror image
(d) Optically inactive because the molecule has a superimposable mirror image
(e) Optically active because the molecule has a nonsuperimposable mirror image

5-5. (b) The mirror image is

CH$_3$
HO—C—H
HO—C—H
HO—C—H
CH$_3$

(e) *Meso* compound
(f)

CH$_3$
HO——H
H——OH
CH$_3$

(g) Optically active because the molecule has a nonsuperimposable mirror image.

5-6. Once the planar free radical is formed, attack by a chlorine molecule is equally probable on either side. Hence, attack leads to the formation of equal amounts of enantiomers

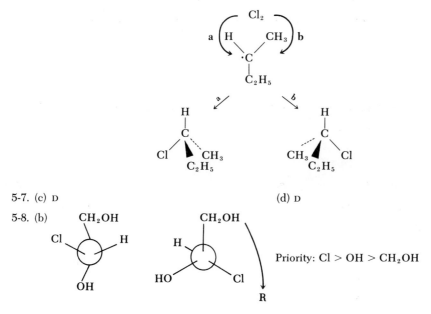

5-7. (c) D (d) D

5-8. (b)

CH$_2$OH with Cl, H and OH

CH$_2$OH with H, HO, Cl

Priority: Cl > OH > CH$_2$OH

R

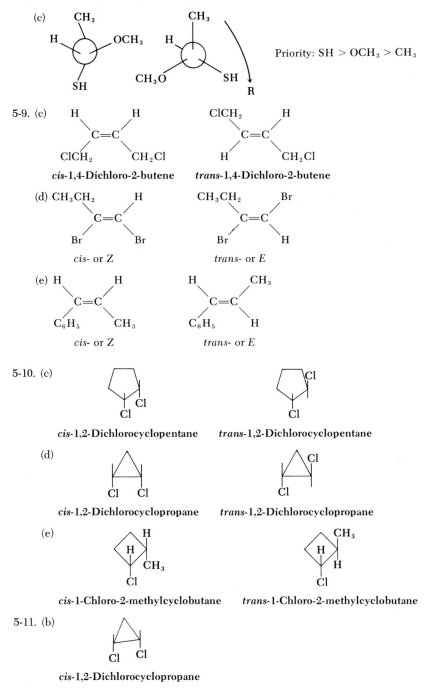

(c)

Priority: SH > OCH₃ > CH₃

R

5-9. (c)

cis-1,4-Dichloro-2-butene   trans-1,4-Dichloro-2-butene

(d) CH₃CH₂

cis- or Z   trans- or E

(e)

cis- or Z   trans- or E

5-10. (c)

cis-1,2-Dichlorocyclopentane   trans-1,2-Dichlorocyclopentane

(d)

cis-1,2-Dichlorocyclopropane   trans-1,2-Dichlorocyclopropane

(e)

cis-1-Chloro-2-methylcyclobutane   trans-1-Chloro-2-methylcyclobutane

5-11. (b)

cis-1,2-Dichlorocyclopropane

(e) The *cis*-isomer

(f)

1,1-Dichlorocyclopropane

(g) Diastereomers

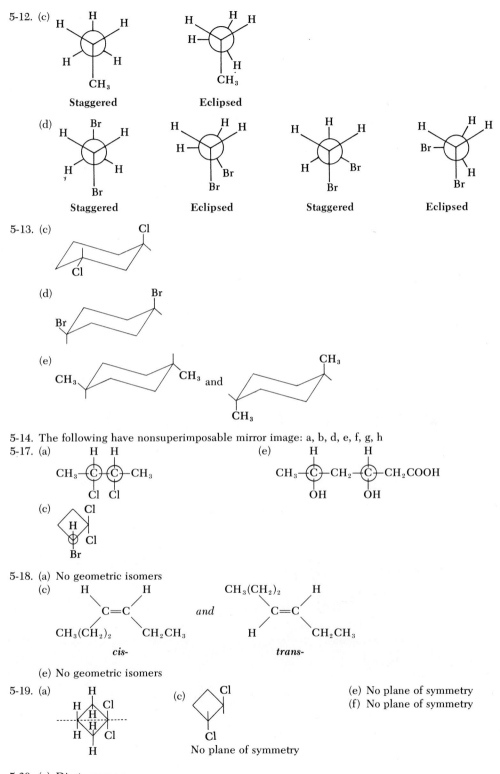

5-12. (c)

Staggered   Eclipsed

(d)

Staggered   Eclipsed   Staggered   Eclipsed

5-13. (c)

(d)

(e)

CH₃ and CH₃

5-14. The following have nonsuperimposable mirror image: a, b, d, e, f, g, h

5-17. (a)

CH_3—C—C—CH_3

(e)

CH_3—C—CH_2—C—CH_2COOH

(c)

5-18. (a) No geometric isomers

(c)

$CH_3(CH_2)_2$ CH_2CH_3

and

$CH_3(CH_2)_2$ CH_2CH_3

cis-   trans-

(e) No geometric isomers

5-19. (a)

(c)

No plane of symmetry

(e) No plane of symmetry
(f) No plane of symmetry

5-20. (a) Diastereomers
(c) Diastereomers

> (e) 1, 3, and 5
> (g) 1, 3, and 5
> (i) Inactive; the compounds have equal optical rotations but in opposite directions.
5-22. (a) *trans*
5-24. (a) Identical
> (c) *meso* compounds, identical
> (e) Diastereomers
> (g) Enantiomers
> (i) Enantiomers

Chapter 6 Alcohols and phenols

6-1. (c) 2,4-Dinitrophenol
> (d) 3-Cyano-5-bromophenol   *or*   5-cyano-3-bromophenol
> (e) *p*-Hydroxybenzyl alcohol   *or*   4-hydroxybenzyl alcohol
> (f) 3-Methylcyclopentanol

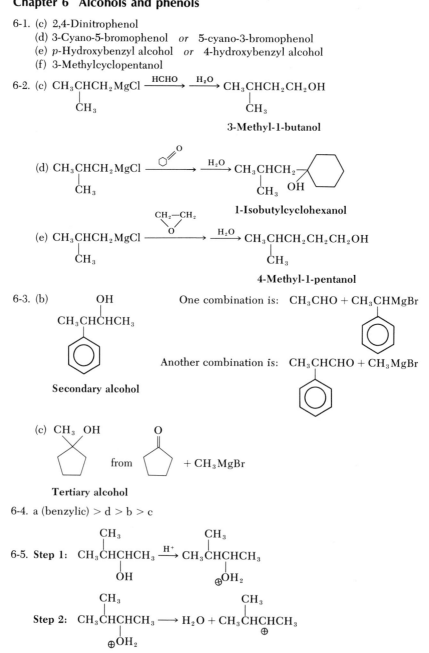

6-2. (c) CH_3CHCH_2MgCl $\xrightarrow{HCHO}$ $\xrightarrow{H_2O}$ $CH_3CHCH_2CH_2OH$
$\quad\quad\quad\quad$ | $\quad\quad\quad\quad\quad\quad\quad\quad\quad\quad\quad\quad\quad\quad$ |
$\quad\quad\quad\quad$ CH_3 $\quad\quad\quad\quad\quad\quad\quad\quad\quad\quad\quad\quad\quad$ CH_3

3-Methyl-1-butanol

(d) CH_3CHCH_2MgCl $\longrightarrow$ $\xrightarrow{H_2O}$ CH_3CHCH_2

1-Isobutylcyclohexanol

(e) CH_3CHCH_2MgCl $\longrightarrow$ $\xrightarrow{H_2O}$ $CH_3CHCH_2CH_2CH_2OH$

4-Methyl-1-pentanol

6-3. (b) $CH_3CHCHCH_3$ (OH)   One combination is: $CH_3CHO + CH_3CHMgBr$

Secondary alcohol

Another combination is: $CH_3CHCHO + CH_3MgBr$

(c) CH_3 OH   from   + CH_3MgBr

Tertiary alcohol

6-4. a (benzylic) > d > b > c

6-5. **Step 1:** $CH_3\overset{CH_3}{\underset{OH}{CHCHCH_3}}$ $\xrightarrow{H^+}$ $CH_3\overset{CH_3}{\underset{\oplus OH_2}{CHCHCH_3}}$

Step 2: $CH_3\overset{CH_3}{\underset{\oplus OH_2}{CHCHCH_3}}$ $\longrightarrow$ $H_2O + CH_3\overset{CH_3}{\underset{\oplus}{CHCHCH_3}}$

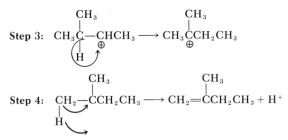

Step 3: $CH_3\underset{\underset{H}{|}}{\overset{\overset{CH_3}{|}}{C}}-\overset{\oplus}{C}HCH_3 \longrightarrow CH_3\overset{\overset{CH_3}{|}}{\underset{\oplus}{C}}CH_2CH_3$

Step 4: $CH_2-\overset{\overset{CH_3}{|}}{\underset{\oplus}{C}}CH_2CH_3 \longrightarrow CH_2{=}\overset{\overset{CH_3}{|}}{C}CH_2CH_3 + H^+$

6-6. *p*-Nitrophenol is the stronger acid because the negative charge on the phenolic oxygen can be delocalized over a greater number of atoms including the nitro oxygen:

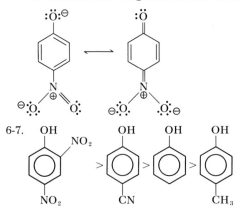

6-7.

6-8. $CH_3CH_2CH_2CH_2CH_2OH$

1-Pentanol
n-Pentyl alcohol

$CH_3CH_2\underset{\underset{CH_3}{|}}{C}HCH_2OH$

2-Methyl-1-butanol

$CH_3CH_2CH_2\underset{\underset{OH}{|}}{C}HCH_3$

2-Pentanol

$CH_3CH_2\overset{\overset{OH}{|}}{\underset{\underset{CH_3}{|}}{C}}CH_3$

2-Methyl-2-butanol

$CH_3CH_2\underset{\underset{OH}{|}}{C}HCH_2CH_3$

3-Pentanol

$CH_3\underset{\underset{OH}{|}}{C}H\overset{\overset{CH_3}{|}}{C}HCH_3$

3-Methyl-2-butanol

$CH_3-\overset{\overset{CH_3}{|}}{\underset{\underset{CH_3}{|}}{C}}-CH_2OH$

2,2-Dimethyl-1-propanol

6-10. (a) *m*-Chlorophenol or 3-chlorophenol
(c) 2-Methyl-4-chlorophenol
(e) 4-Bromo-1-heptyn-3-ol

(g) *p*-Phenylphenol or 4-phenylphenol
(i) 3-Methyl-1-pentanol

6-11. (a) $CH_3\overset{\overset{OH}{|}}{C}H\overset{\overset{Br}{|}}{C}HCHCH_2CH_2CH_2CH_2CH_3$

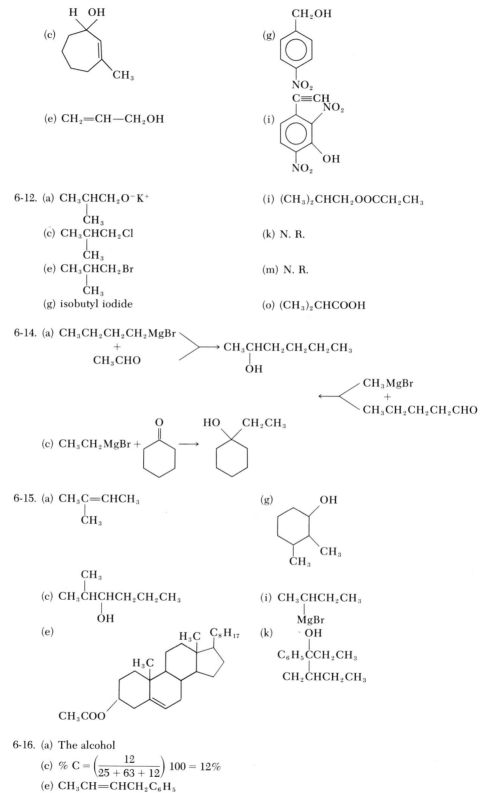

(c)

(g)

(e) $CH_2=CH-CH_2OH$

(i)

6-12. (a) $CH_3CHCH_2O^-K^+$
       |
       CH_3

(i) $(CH_3)_2CHCH_2OOCCH_2CH_3$

(c) CH_3CHCH_2Cl
       |
       CH_3

(k) N. R.

(e) CH_3CHCH_2Br
       |
       CH_3

(m) N. R.

(g) isobutyl iodide

(o) $(CH_3)_2CHCOOH$

6-14. (a) $CH_3CH_2CH_2CH_2MgBr$
       +
       CH_3CHO
       $\longrightarrow$ $CH_3CHCH_2CH_2CH_2CH_3$
                              |
                              OH
       $\longleftarrow$ CH_3MgBr
                         +
                         $CH_3CH_2CH_2CH_2CHO$

(c) $CH_3CH_2MgBr +$

6-15. (a) $CH_3C=CHCH_3$
       |
       CH_3

(g)   OH

CH_3
CH_3

(c) $CH_3CHCHCH_2CH_2CH_3$
       | |
       CH_3 OH

(i) $CH_3CHCH_2CH_3$
       |
       $MgBr$

(e)

H_3C C_8H_{17}
H_3C

(k) OH
       |
$C_6H_5CCH_2CH_3$
       |
$CH_2CHCH_2CH_3$

CH_3COO

6-16. (a) The alcohol

(c) $\% \ C = \left(\dfrac{12}{25 + 63 + 12}\right) 100 = 12\%$

(e) $CH_3CH=CHCH_2C_6H_5$

Chapter 7   Organic halides

7-1. (c) Allyl
    (d) Alkyl
    (e) Alkyl
    (f) Vinyl

7-2. (c) 4-Chloro-2-pentene
    (d) *trans*-1,2-Dibromocyclohexane
    (e) 2,2-Dimethyl-1-iodopropane
    (f) 1-Chloro-2-phenylethene
    (g) 3-Phenyl-1-bromopropane

7-3. (c) Electrophilic substitution

7-4. (c) $CH_3C{\equiv}CCH_3$; alkyne
    (d) $C_6H_5CH_2\overset{\oplus}{N}H_3Br^{\ominus}$; amine salt
    (e) CH_3CH_2I; alkyl iodide

(g) Alkyl
(h) Aromatic
(i) Allyl or benzyl
(j) Benzyl
(h) 2-Chlorobenzoic acid or
    o-Chlorobenzoic acid
(i) 3-Phenyl-3-bromo-1-propene
(j) Triphenylmethyl iodide

(d) Nucleophilic substitution
(f) CH_3NH_2; amine
(g) $CH_3(CH_2)_2CH_2SCH_2CH_3$; thioether

7-5. $\underset{\underset{\displaystyle CH_3}{|}}{CH_3CH_2CH}{-}Br + OH^{\ominus} \longrightarrow CH_3CH{=}CHCH_3 + CH_3CH_2CH{=}CH_2$
                               (major product)

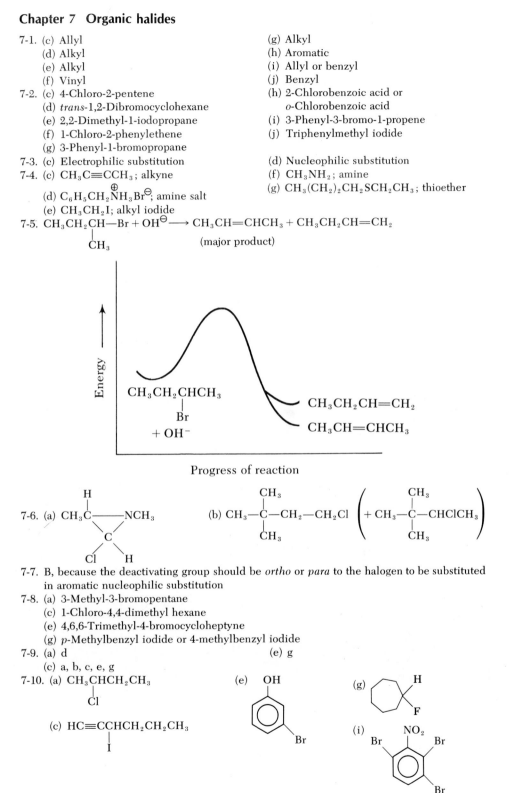

7-7. B, because the deactivating group should be *ortho* or *para* to the halogen to be substituted in aromatic nucleophilic substitution

7-8. (a) 3-Methyl-3-bromopentane
    (c) 1-Chloro-4,4-dimethyl hexane
    (e) 4,6,6-Trimethyl-4-bromocycloheptyne
    (g) *p*-Methylbenzyl iodide or 4-methylbenzyl iodide

7-9. (a) d   (e) g
    (c) a, b, c, e, g

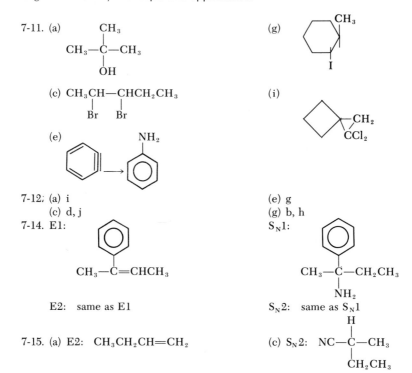

7-11. (a)

$$CH_3-\underset{\underset{OH}{|}}{\overset{\overset{CH_3}{|}}{C}}-CH_3$$

(c) $CH_3CH-CHCH_2CH_3$ with Br on each middle carbon

(e) (benzene) $\rightarrow$ (aniline with NH_2)

(g) (cyclohexane ring with CH_3 and I)

(i) (cyclobutane ring with $-CH_2$ and CCl_2)

7-12. (a) i (e) g
  (c) d, j (g) b, h
7-14. E1:

$$CH_3-\overset{(phenyl)}{\underset{}{C}}=CHCH_3$$

S_N1:

$$CH_3-\underset{\underset{NH_2}{|}}{\overset{(phenyl)}{C}}-CH_2CH_3$$

E2: same as E1

S_N2: same as S_N1

7-15. (a) E2: $CH_3CH_2CH=CH_2$

(c) S_N2:

$$NC-\underset{\underset{CH_2CH_3}{|}}{\overset{\overset{H}{|}}{C}}-CH_3$$

Chapter 8   Ethers and epoxides

8-1. (b) sp^2-sp^3 and sp^3-sp^3 (reading left to right)
8-2. (c) $CH_3OCH=CH_2$

(d) (phenyl)$-CH_2OCH_2CH_3$

(e) (cyclohexyl)$-O-$(cyclohexyl)

8-3.

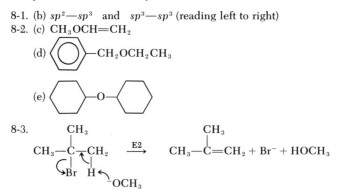

$$CH_3-\underset{}{C}-CH_2 \xrightarrow{\text{E2}} CH_3-\overset{\overset{CH_3}{|}}{C}=CH_2 + Br^- + HOCH_3$$

8-4. NaOH converts the phenol to stronger nucleophile, phenoxide ion.
8-5. (c) $C_2H_5Br + (CH_3)_2CHCH_2O^-K^+$ or $(CH_3)_2CHCH_2Br + C_2H_5O^-K^+$

(d) (phenyl)$-CH_2Br + O_2N-$(phenyl)$-O^-K^+$

8-6. (benzene ring with OCH_3, NO_2, CN)   The two substituents reenforce each other.

8-7. $\underset{\underset{O}{\diagdown\diagup}}{CH_2-CH_2} + CH_3CH_2OH \xrightarrow{H^+} CH_3CH_2OCH_2CH_2OH$

8-8. (a) Methyl ether
(c) Cyclopentyl methyl ether
(e) Ethylene oxide

(g) Isobutyl ethyl ether
(i) 4-Ethoxy-5-methyl-1-hexyne

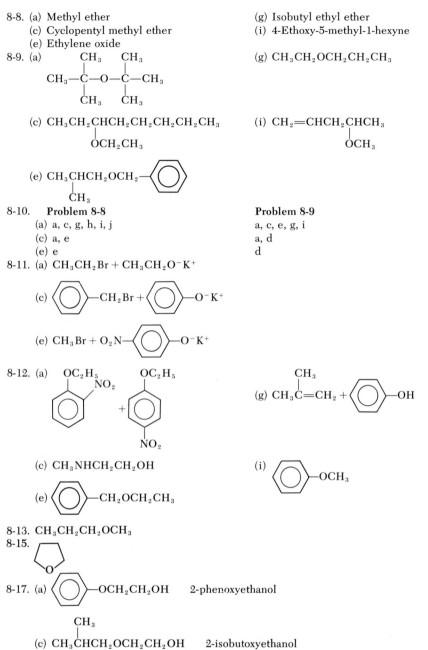

8-9. (a)

$$CH_3-\underset{\underset{CH_3}{|}}{\overset{\overset{CH_3}{|}}{C}}-O-\underset{\underset{CH_3}{|}}{\overset{\overset{CH_3}{|}}{C}}-CH_3$$

(g) $CH_3CH_2OCH_2CH_2CH_3$

(c) $CH_3CH_2CHCH_2CH_2CH_2CH_3$
      $\quad\quad\quad\;\; |$
      $\quad\quad\; OCH_2CH_3$

(i) $CH_2{=}CHCH_2CHCH_3$
      $\quad\quad\quad\quad\quad\;\; |$
      $\quad\quad\quad\quad\quad OCH_3$

(e) $CH_3CHCH_2OCH_2{-}$
      $\quad\; |$
      $\quad CH_3$

8-10. **Problem 8-8**
(a) a, c, g, h, i, j
(c) a, e
(e) e

Problem 8-9
a, c, e, g, i
a, d
d

8-11. (a) $CH_3CH_2Br + CH_3CH_2O^-K^+$

(c) $-CH_2Br + -O^-K^+$

(e) $CH_3Br + O_2N--O^-K^+$

8-12. (a) OC_2H_5 OC_2H_5
              NO_2
       $+$
                          NO_2

(g) $CH_3\underset{\underset{}{\overset{\overset{CH_3}{|}}{C}}}{=}CH_2 + -OH$

(c) $CH_3NHCH_2CH_2OH$

(i) $-OCH_3$

(e) $-CH_2OCH_2CH_3$

8-13. $CH_3CH_2CH_2OCH_3$
8-15.

8-17. (a) $-OCH_2CH_2OH$ 2-phenoxyethanol

      $\quad\quad\quad CH_3$
      $\quad\quad\quad\; |$
(c) $CH_3CHCH_2OCH_2CH_2OH$ 2-isobutoxyethanol

(e) $CH_3CH_2OCH_2CH_2OCH_2CH_2OCH_2CH_2OH$

Chapter 9 Carboxylic acid, carboxylic acid derivatives, and dicarboxylic acids

9-1. (c) CH_3CHCH_2COOH; 3-methylbutanoic acid
      $\quad\quad\; |$
      $\quad\quad CH_3$

(d) $HOOCCH_2CH_2COOH$; butanedioic acid

(e)

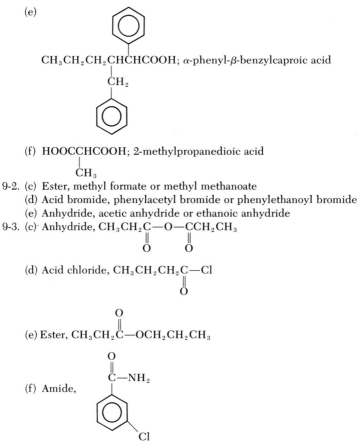

$CH_3CH_2CH_2CHCHCOOH$; α-phenyl-β-benzylcaproic acid

(f) $HOOCCHCOOH$; 2-methylpropanedioic acid

9-2. (c) Ester, methyl formate or methyl methanoate
(d) Acid bromide, phenylacetyl bromide or phenylethanoyl bromide
(e) Anhydride, acetic anhydride or ethanoic anhydride

9-3. (c) Anhydride, $CH_3CH_2C—O—CCH_2CH_3$

(d) Acid chloride, $CH_3CH_2CH_2C—Cl$

(e) Ester, $CH_3CH_2C—OCH_2CH_2CH_3$

(f) Amide,

9-4. The first step in synthesis (b) will not occur. Bromobenzene is an unreactive aromatic halide and will not undergo nucleophilic substitution.

9-5. (b) CH_3CHCH_2COOH

(c) —COOH

(d) —COOH, COOH

9-6. (d) —COOH

(e) —COOH

(f) Cl——COOH

9-7. $CH_3CH_2C—OCH_3$ The bulky *tert*-butyl group in the acid portion of the second ester will cause step 2 to occur very slowly.

9-8. (c) HO, $CH_2COC_2H_5$, CH_3, CH_3

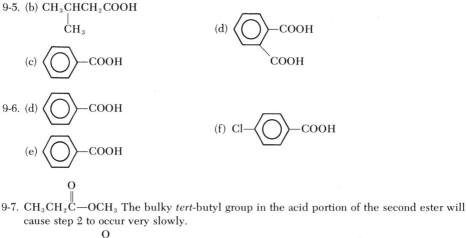

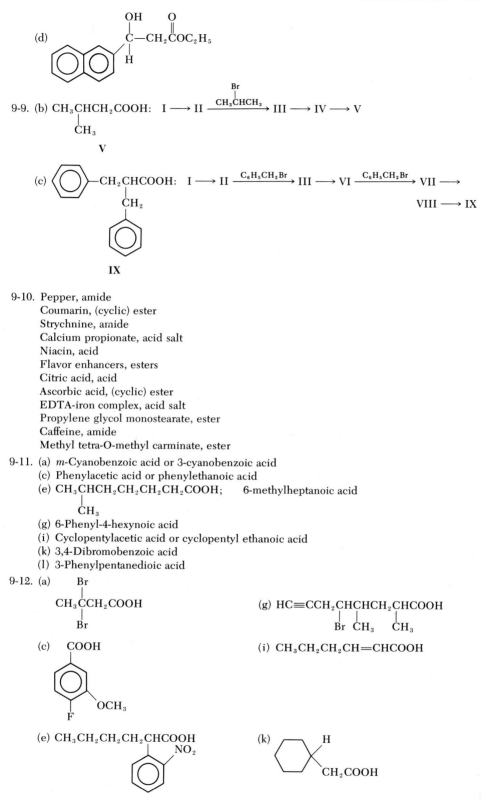

(d)

9-9. (b) CH_3CHCH_2COOH: I $\longrightarrow$ II $\xrightarrow{CH_3\overset{Br}{\underset{}{C}HCH_3}}$ III $\longrightarrow$ IV $\longrightarrow$ V

V

(c) $\langle \bigcirc \rangle - CH_2CHCOOH$: I $\longrightarrow$ II $\xrightarrow{C_6H_5CH_2Br}$ III $\longrightarrow$ VI $\xrightarrow{C_6H_5CH_2Br}$ VII $\longrightarrow$

VIII $\longrightarrow$ IX

IX

9-10. Pepper, amide
Coumarin, (cyclic) ester
Strychnine, amide
Calcium propionate, acid salt
Niacin, acid
Flavor enhancers, esters
Citric acid, acid
Ascorbic acid, (cyclic) ester
EDTA-iron complex, acid salt
Propylene glycol monostearate, ester
Caffeine, amide
Methyl tetra-O-methyl carminate, ester

9-11. (a) *m*-Cyanobenzoic acid or 3-cyanobenzoic acid
(c) Phenylacetic acid or phenylethanoic acid
(e) $CH_3CHCH_2CH_2CH_2CH_2COOH$;   6-methylheptanoic acid
(g) 6-Phenyl-4-hexynoic acid
(i) Cyclopentylacetic acid or cyclopentyl ethanoic acid
(k) 3,4-Dibromobenzoic acid
(l) 3-Phenylpentanedioic acid

9-12. (a) $CH_3\overset{Br}{\underset{Br}{C}}CH_2COOH$

(g) $HC{\equiv}CCH_2\overset{}{\underset{Br}{C}}HCH\overset{}{\underset{CH_3}{C}}HCH_2\overset{}{\underset{CH_3}{C}}HCOOH$

(c)   COOH ... OCH₃, F

(i) $CH_3CH_2CH_2CH{=}CHCOOH$

(e) $CH_3CH_2CH_2CH_2CHCOOH$, NO₂

(k)   H ... CH₂COOH

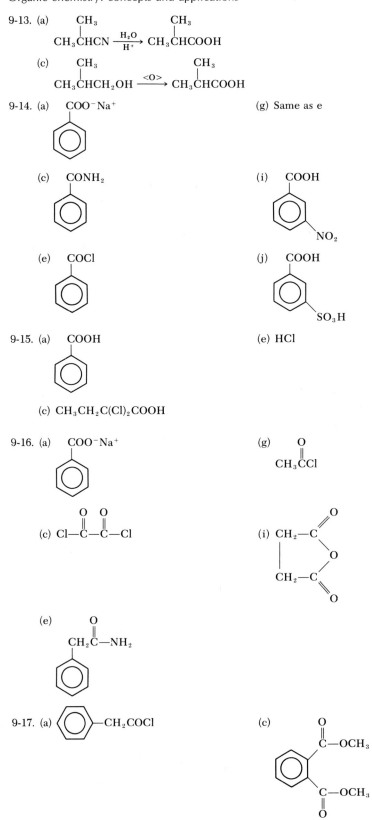

9-13. (a)

$$CH_3\overset{\overset{\displaystyle CH_3}{|}}{C}HCN \xrightarrow[H^+]{H_2O} CH_3\overset{\overset{\displaystyle CH_3}{|}}{C}HCOOH$$

(c)

$$CH_3\overset{\overset{\displaystyle CH_3}{|}}{C}HCH_2OH \xrightarrow{<O>} CH_3\overset{\overset{\displaystyle CH_3}{|}}{C}HCOOH$$

9-14. (a) COO^-Na^+

(g) Same as e

(c) $CONH_2$

(i) $COOH$, NO_2

(e) $COCl$

(j) $COOH$, SO_3H

9-15. (a) $COOH$

(e) HCl

(c) $CH_3CH_2C(Cl)_2COOH$

9-16. (a) COO^-Na^+

(g) $CH_3\overset{\overset{\displaystyle O}{\|}}{C}Cl$

(c) $Cl-\overset{\overset{\displaystyle O}{\|}}{C}-\overset{\overset{\displaystyle O}{\|}}{C}-Cl$

(i) CH_2-C, CH_2-C

(e) $CH_2\overset{\overset{\displaystyle O}{\|}}{C}-NH_2$

9-17. (a) $-CH_2COCl$

(c) $C-OCH_3$, $C-OCH_3$

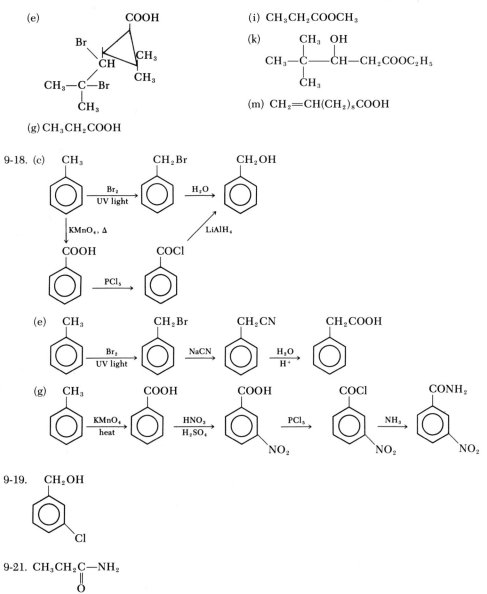

(e) COOH

Br

CH

CH₃

CH₃

CH₃—C—Br

CH₃

(g) CH_3CH_2COOH

(i) $CH_3CH_2COOCH_3$

(k)

$$CH_3—\overset{\overset{\displaystyle CH_3}{|}}{\underset{\underset{\displaystyle CH_3}{|}}{C}}—\overset{\overset{\displaystyle OH}{|}}{CH}—CH_2COOC_2H_5$$

(m) $CH_2{=}CH(CH_2)_8COOH$

9-18. (c) CH₃   CH₂Br   CH₂OH

$\xrightarrow[\text{UV light}]{Br_2}$ $\xrightarrow{H_2O}$

$\downarrow$ KMnO₄, Δ   / LiAlH₄

COOH   COCl

$\xrightarrow{PCl_5}$

(e) CH₃   CH₂Br   CH₂CN   CH₂COOH

$\xrightarrow[\text{UV light}]{Br_2}$ $\xrightarrow{NaCN}$ $\xrightarrow[H^+]{H_2O}$

(g) CH₃   COOH   COOH   COCl   CONH₂

$\xrightarrow[\text{heat}]{KMnO_4}$ $\xrightarrow[H_2SO_4]{HNO_3}$ $\xrightarrow{PCl_5}$ $\xrightarrow{NH_3}$

NO₂   NO₂   NO₂

9-19. CH₂OH

Cl

9-21. $CH_3CH_2\overset{\displaystyle O}{\underset{\displaystyle \|}{C}}{-}NH_2$

9-23. Butyric acid, $CH_3CH_2CH_2COOH$

Butyric anhydride, $CH_3(CH_2)_2\overset{\overset{\displaystyle O}{\|}}{C}{-}O{-}\overset{\overset{\displaystyle O}{\|}}{C}(CH_2)_2CH_3$

Methyl butyrate, $CH_3(CH_2)_2\overset{\overset{\displaystyle O}{\|}}{C}{-}OCH_3$

Butyramide, $CH_3(CH_2)_2\overset{\overset{\displaystyle O}{\|}}{C}NH_2$

Butyryl chloride, $CH_3(CH_2)_2\overset{\overset{\displaystyle O}{\|}}{C}{-}Cl$

Sodium butyrate, $CH_3(CH_2)_2\overset{\overset{\displaystyle O}{\|}}{C}-O^-Na^+$

9-24. (a) 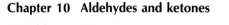 CH_3O- ⬡ $-\underset{\underset{\displaystyle Br}{|}}{\overset{\overset{\displaystyle Br}{|}}{C}}H\overset{}{C}H\overset{\overset{\displaystyle O}{\|}}{C}OCH_2CH_2OC_2H_5$

(c) CH_3O ⬡ $-CH=CH\overset{\overset{\displaystyle O}{\|}}{C}-OH + HOCH_2CH_2OC_2H_5$

Chapter 10 Aldehydes and ketones

10-2. (c) $CH_3-\overset{\overset{\displaystyle O}{\|}}{C}-CH_2-$ ⬡ , 1-phenyl-2-propanone

(d)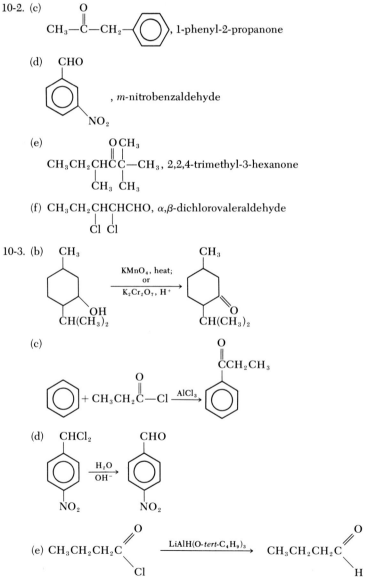

, m-nitrobenzaldehyde

(e) $CH_3CH_2\underset{\underset{\displaystyle CH_3}{|}}{C}H\overset{\overset{\displaystyle O}{\|}}{C}\underset{\underset{\displaystyle CH_3}{|}}{\overset{\overset{\displaystyle OCH_3}{|}}{C}}-CH_3$, 2,2,4-trimethyl-3-hexanone

(f) $CH_3CH_2\underset{\underset{\displaystyle Cl}{|}}{C}H\underset{\underset{\displaystyle Cl}{|}}{C}HCHO$, α,β-dichlorovaleraldehyde

10-3. (b)

$\xrightarrow[\text{K}_2\text{Cr}_2\text{O}_7,\ \text{H}^+]{\overset{\text{KMnO}_4,\ \text{heat};}{\text{or}}}$

(c) ⬡ $+ CH_3CH_2\overset{\overset{\displaystyle O}{\|}}{C}-Cl \xrightarrow{\text{AlCl}_3}$

(d)

$\xrightarrow[\text{OH}^-]{\text{H}_2\text{O}}$

(e) $CH_3CH_2CH_2\overset{\overset{\displaystyle O}{\|}}{\underset{\underset{\displaystyle Cl}{\diagdown}}{C}}$ $\xrightarrow{\text{LiAlH(O-}\textit{tert}\text{-C}_4\text{H}_9)_3}$ $CH_3CH_2CH_2\overset{\overset{\displaystyle O}{\|}}{\underset{\underset{\displaystyle H}{\diagdown}}{C}}$

10-4. (b)

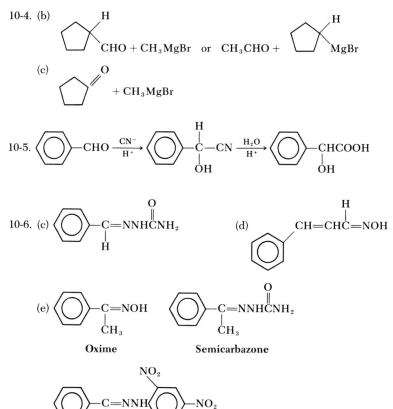

(c)

10-5.

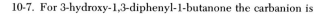

10-6. (c)

(d)

(e)

Oxime   **Semicarbazone**

2,4-Dinitrophenylhydrazone

10-7. For 3-hydroxy-1,3-diphenyl-1-butanone the carbanion is

10-10. (a) *n*-Propyl phenyl ketone
         or 1-phenyl-1-butanone
     (c) Heptanal

     (e) Cyclopentanone

     (g) Isopropyl *n*-butyl ketone or
         2-methyl-3-heptanone
     (i) 4,4-Dimethylpentanal or
         γ,γ-dimethylvaleraldehyde
     (j) Phenyl *p*-nitrophenyl ketone or
         p-nitrobenzophenone

10-11. (a) $CH_3CH_2CHCH_2CHCHO$
                    | |
                   CH_3 CH_3

     (c) $CH_3CH_2CH_2CH_2CHCH=CHCHO$
                              |
                             C_2H_5

     (e) $CH_3CH_2CH_2CHO$

     (g)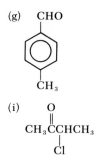

     (i) $CH_3\overset{O}{\overset{\|}{C}}CHCH_3$
                      |
                     Cl

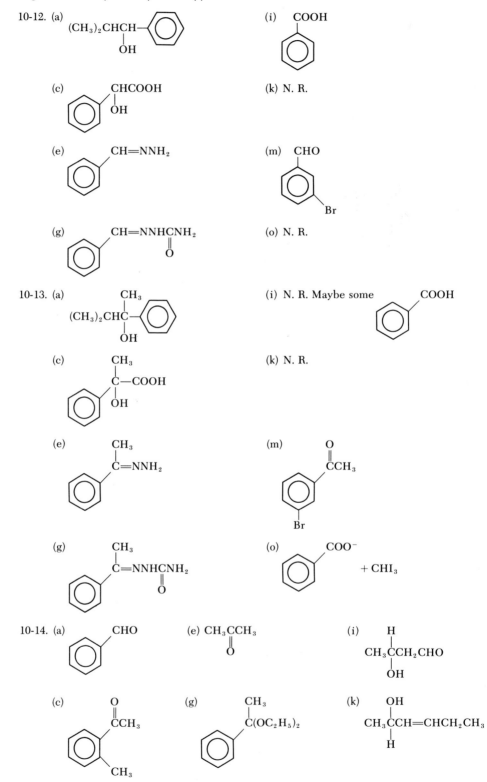

10-12. (a) $(CH_3)_2CHCH-C_6H_5$ with OH

(c) phenyl–CHCOOH with OH

(e) phenyl–CH=NNH_2

(g) phenyl–CH=NNHCNH_2 with O

(i) COOH (benzoic acid)

(k) N. R.

(m) CHO with Br (meta)

(o) N. R.

10-13. (a) $(CH_3)_2CHC(CH_3)-C_6H_5$ with OH

(c) phenyl with CH_3 / C—COOH / OH

(e) phenyl with CH_3 / C=NNH_2

(g) phenyl with CH_3 / C=NNHCNH_2 / O

(i) N. R. Maybe some COOH (benzoic acid)

(k) N. R.

(m) phenyl with O / CCH_3 and Br (meta)

(o) phenyl—COO^- + CHI_3

10-14. (a) phenyl–CHO

(c) phenyl with O / CCH_3 and CH_3 (ortho)

or the *m* or *p*-derivative

(e) CH_3CCH_3 with O

(g) phenyl with CH_3 / C(OC_2H_5)_2

(i) CH_3CCH_2CHO with H (top) and OH (bottom)

(k) $CH_3CCH=CHCH_2CH_3$ with OH (top) and H (bottom)

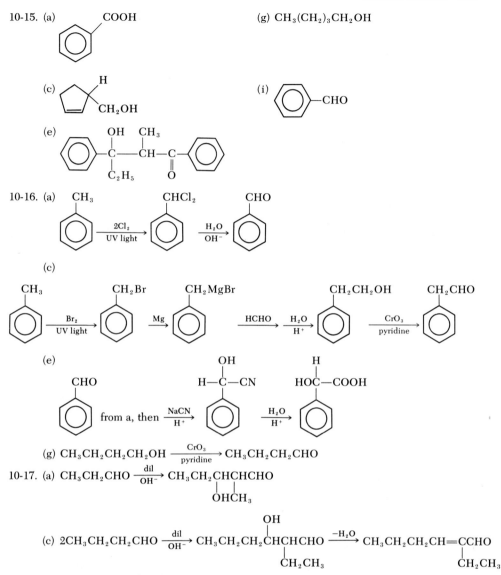

10-15. (a) COOH

(g) $CH_3(CH_2)_3CH_2OH$

(c) [structure with H and CH_2OH on cyclopentene ring]

(i) [benzene]—CHO

(e) [structure: phenyl—C(OH)(C$_2$H$_5$)—CH(CH$_3$)—C(=O)—phenyl]

10-16. (a) CH_3 [benzene] $\xrightarrow[\text{UV light}]{2Cl_2}$ $CHCl_2$ [benzene] $\xrightarrow[OH^-]{H_2O}$ CHO [benzene]

(c)

CH_3 [benzene] $\xrightarrow[\text{UV light}]{Br_2}$ CH_2Br [benzene] $\xrightarrow{Mg}$ CH_2MgBr [benzene] $\xrightarrow[H^+]{HCHO \quad H_2O}$ CH_2CH_2OH [benzene] $\xrightarrow[\text{pyridine}]{CrO_3}$ CH_2CHO [benzene]

(e)

CHO [benzene] from a, then $\xrightarrow[H^+]{NaCN}$ [benzene with] $H-\overset{OH}{\underset{}{C}}-CN$ $\xrightarrow[H^+]{H_2O}$ [benzene with] $HO\overset{H}{\underset{}{C}}-COOH$

(g) $CH_3CH_2CH_2CH_2OH \xrightarrow[\text{pyridine}]{CrO_3} CH_3CH_2CH_2CHO$

10-17. (a) $CH_3CH_2CHO \xrightarrow[OH^-]{\text{dil}} CH_3CH_2\underset{\underset{OHCH_3}{|\quad|}}{CH}CHCHO$

(c) $2CH_3CH_2CH_2CHO \xrightarrow[OH^-]{\text{dil}} CH_3CH_2CH_2\underset{\underset{CH_2CH_3}{|}}{\overset{\overset{OH}{|}}{CH}}CHCHO \xrightarrow{-H_2O} CH_3CH_2CH_2CH=\underset{\underset{CH_2CH_3}{|}}{C}CHO$

10-18. (a) Add Tollens' reagent; acetaldehyde will be oxidized and a silver mirror will be formed on the sides of the test tube.

(c) Fehling's test will be positive for propionaldehyde and will be indicated by a red precipitate.

(e) Iodoform test will be positive with acetophenone.

10-19. (a) Acetophenone

(c) Propionaldehyde

(e) Hexanal

Chapter 11 Organic nitrogen compounds

11-2. $CH_3CH_2CH_2CH_2CH_2NH_2$

$CH_3CH_2CH_2\underset{\underset{NH_2}{|}}{CH}CH_3$

1-Aminopentane

2-Aminopentane

$CH_3CH_2CHCH_2CH_3$
|
NH_2

3-Aminopentane

$CH_3CH_2CHCH_2NH_2$
|
CH_3

2-Methyl-1-aminobutane

NH_2
|
$CH_3CH_2CCH_3$
|
CH_3

2-Methyl-2-aminobutane

NH_2
|
$CH_3CHCHCH_3$
|
CH_3

2-Amino-3-methylbutane

$H_2NCH_2CH_2CHCH_3$
|
CH_3

1-Amino-3-methylbutane

CH_3
|
$CH_3CCH_2NH_2$
|
CH_3

2,2-Dimethyl-1-aminopropane

11-3. $CH_3CH_2CH_2CH_2NHCH_3$

methyl *n*-butylamine
1-N-Methylaminobutane
Secondary

$CH_3CH_2CH_2NHCH_2CH_3$

Ethyl *n*-propylamine
1-N-Ethylaminopropane
Secondary

CH_3—N—$CH_2CH_2CH_3$
|
CH_3

Dimethyl *n*-propylamine
1-N,N-Dimethylaminopropane
Tertiary

$CH_3CH_2NCH_2CH_3$
|
CH_3

Methyl diethylamine
Tertiary

$CH_3CHNHCH_2CH_3$
|
CH_3

Ethyl isopropylamine
Secondary

$CH_3CHCH_2NHCH_3$
|
CH_3

Methyl isobutylamine
Secondary

CH_3
|
CH_3NCHCH_3
|
CH_3

Dimethyl isopropylamine
Tertiary

$CH_3CH_2CHNHCH_3$
|
CH_3

Methyl *sec*-butylamine
Secondary

CH_3
|
CH_3C—$NHCH_3$
|
CH_3

Methyl *tert*-butylamine
Secondary

11-4. (a)
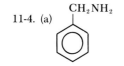

(c)

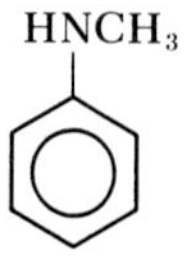

N-Methylaniline

(b)

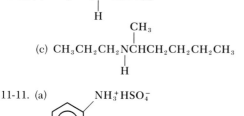

NH$_2$ CH$_3$

(d)

NH$_2$

CH$_3$

o-**Toluidine**
m- and *p*-toluidine
are possible structures

11-5. Route B: $CH_3CH_2CH_2CH_2OH \xrightarrow[\text{pyridine}]{CrO_3} CH_3CH_2CH_2CHO \xrightarrow{NH_3} \xrightarrow[\text{Ni}]{H_2}$
$CH_3CH_2CH_2CH_2NH_2$

Route C: $CH_3CH_2CH_2CH_2CH_2OH \xrightarrow{KMnO_4} CH_3(CH_2)_3COOH \xrightarrow{PCl_3} CH_3(CH_2)_3COCl$
$\xrightarrow{NH_3} CH_3(CH_2)_3CONH_2 \xrightarrow{NaOBr} CH_3(CH_2)_2CH_2NH_2$

Route D: $CH_3CH_2CH_2CH_2OH \xrightarrow{KMnO_4} CH_3CH_2CH_2COOH \xrightarrow{PCl_3} CH_3CH_2CH_2COCl$
$\xrightarrow{NH_3} CH_3(CH_2)_2CONH_2 \xrightarrow{LiAlH_4} CH_3(CH_2)_2CH_2NH_2$

11-6. (c) Cyclohexylamine
(d)

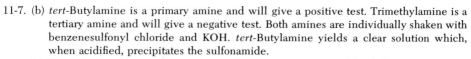

NH$_2$

(f)

NH$_2$

CH$_3$O

(g)

NH$_2$

H$_3$C   NO$_2$

11-7. (b) *tert*-Butylamine is a primary amine and will give a positive test. Trimethylamine is a
tertiary amine and will give a negative test. Both amines are individually shaken with
benzenesulfonyl chloride and KOH. *tert*-Butylamine yields a clear solution which,
when acidified, precipitates the sulfonamide.
(c) Diethylamine, a secondary amine, will give a positive test. Triethylamine, a tertiary
amine, will give a negative test.

11-8. (a) *n*-Butylamine or
1-aminobutane
(c) 1-N-Phenylamino-2-butene

(e) 2-Bromo-4-chloroaniline

(g) *p*-Methylazobenzene

(i) N-Methylaniline or
phenyl methylamine
(j) 1-N,N-Dimethylaminohexane

11-9. (a) Primary   (c) Secondary   (e) Primary   (i) Secondary

11-10.(a) $CH_3CH_2NCH(CH_3)_2$
    $\overset{|}{H}$

(c) $CH_3CH_2CH_2\overset{\overset{\textstyle CH_3}{|}}{N}CHCH_2CH_2CH_2CH_3$
    $\overset{|}{H}$

(e)

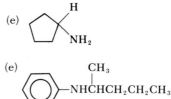

H

NH$_2$

11-11. (a)

NH$_3^+$HSO$_4^-$

(c)

$\overset{+}{N}$(CH$_3$)$_3$I$^-$

(e)

CH$_3$
—NHCHCH$_2$CH$_2$CH$_3$

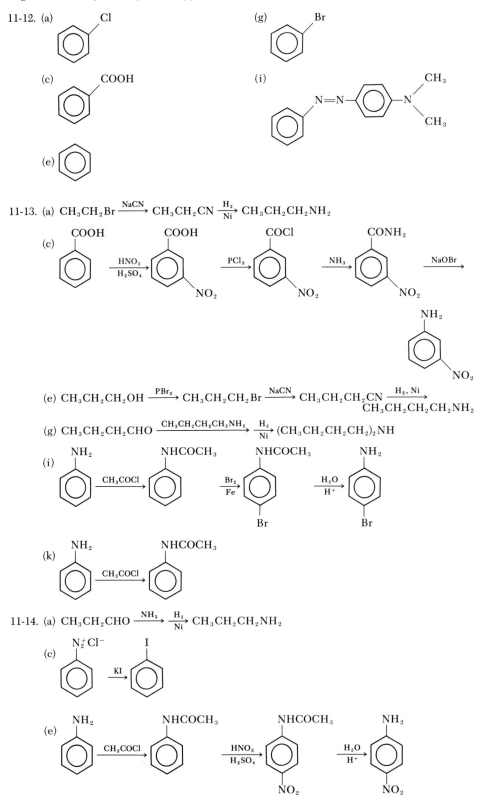

11-12. (a) Cl (benzene ring)

(g) Br (benzene ring)

(c) COOH (benzene ring)

(i) N=N— (benzene ring) —N(CH₃)₂

(e) (benzene ring)

11-13. (a) $CH_3CH_2Br \xrightarrow{NaCN} CH_3CH_2CN \xrightarrow[Ni]{H_2} CH_3CH_2CH_2NH_2$

(c) COOH $\xrightarrow[H_2SO_4]{HNO_3}$ COOH/NO₂ $\xrightarrow{PCl_3}$ COCl/NO₂ $\xrightarrow{NH_3}$ CONH₂/NO₂ $\xrightarrow{NaOBr}$ NH₂/NO₂

(e) $CH_3CH_2CH_2OH \xrightarrow{PBr_3} CH_3CH_2CH_2Br \xrightarrow{NaCN} CH_3CH_2CH_2CN \xrightarrow{H_2, Ni} CH_3CH_2CH_2CH_2NH_2$

(g) $CH_3CH_2CH_2CHO \xrightarrow[\text{}]{CH_3CH_2CH_2CH_2NH_2} \xrightarrow[Ni]{H_2} (CH_3CH_2CH_2CH_2)_2NH$

(i) NH₂ $\xrightarrow{CH_3COCl}$ NHCOCH₃ $\xrightarrow[Fe]{Br_2}$ NHCOCH₃/Br $\xrightarrow[H^+]{H_2O}$ NH₂/Br

(k) NH₂ $\xrightarrow{CH_3COCl}$ NHCOCH₃

11-14. (a) $CH_3CH_2CHO \xrightarrow{NH_3} \xrightarrow[Ni]{H_2} CH_3CH_2CH_2NH_2$

(c) N₂⁺Cl⁻ $\xrightarrow{KI}$ I

(e) NH₂ $\xrightarrow{CH_3COCl}$ NHCOCH₃ $\xrightarrow[H_2SO_4]{HNO_3}$ NHCOCH₃/NO₂ $\xrightarrow[H^+]{H_2O}$ NH₂/NO₂

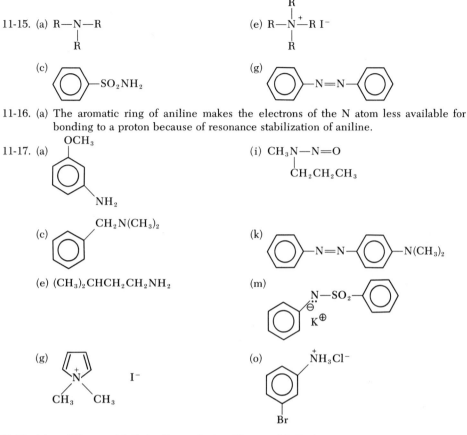

11-15. (a) R—N—R   (e) R—N$^+$—R I$^-$ (with R substituents)

(c) benzene ring—SO$_2$NH$_2$

(g) benzene ring—N=N—benzene ring

11-16. (a) The aromatic ring of aniline makes the electrons of the N atom less available for bonding to a proton because of resonance stabilization of aniline.

11-17. (a) (OCH$_3$ substituted benzene with NH$_2$)

(i) CH$_3$N—N=O with CH$_2$CH$_2$CH$_3$

(c) benzene with CH$_2$N(CH$_3$)$_2$

(k) benzene—N=N—benzene—N(CH$_3$)$_2$

(e) (CH$_3$)$_2$CHCH$_2$CH$_2$NH$_2$

(m) (benzene ring with N—SO$_2$—benzene, K$^\oplus$)

(g) (pyrrolium ring with N, CH$_3$ CH$_3$) I$^-$

(o) (benzene with $^+$NH$_3$Cl$^-$ and Br)

11-19. (a) pyridine   (c) Quinoline and piperidine   (e) Purine
11-20. CH$_3$NHCH$_3$

Chapter 12 Identification of structure by classical and modern techniques

12-1. The molecular weight is 88. The empirical weight of C_2H_4O is 44. Therefore, the molecular formula of butyric acid is $C_4H_8O_2$.
12-2. (c) Alkenes, alkynes
     (d) Secondary alcohols of the type $RCHOHCH_3$ or methyl ketones
     (e) Terminal alkyne
12-3. (a) Benzyl methyl ether. The absence of absorption at 2.7-3.3μ eliminates the alcohol.
     (b) Benzophenone. The lack of absorption at 4.2-4.9μ (C≡N) eliminates the benzonitrile, whereas the absence of absorption at 2.7-3.3μ (OH) eliminates benzoic acid.
12-4. (c) Septet; $n+1=6+1=7$
     (d) Singlet; protons on aromatic ring are equivalent
     (e) Quartet; $n+1=3+1=4$
12-5. (c) CH$_3$CH$_2$NH$_2$ a:b:c = 3:2:2
       a   b   c
     (d) (benzene ring with CH$_2$CCH$_3$ and C=O)   a:b:c = 5:2:3
             b   c
        a
     (e) (cyclohexane ring) All 12 protons are equivalent.

12-6. (a) Phenylacetic acid

(c) *m*-chlorobenzoic acid

12-7. (a) B

(c) C (test for methyl ketone)

(e) B (phenols are insoluble in $NaHCO_3$)

(g) C (for CH_3OCCH_3 the two signals would be of equal intensity)

$$\overset{\|}{O}$$

(i) C (absorption at 4.3μ is C≡C)

12-8.

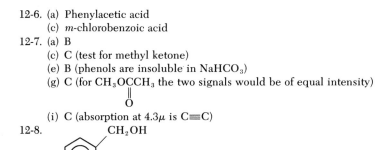

Chapter 13 The organic chemistry of biomolecules

13-1. (b)

$$\begin{array}{l} CH_2-O-\overset{\overset{\textstyle O}{\|}}{C}-(CH_2)_{14}CH_3 \\ \overset{\overset{\textstyle O}{\|}}{CH}-O-\overset{\overset{\textstyle O}{\|}}{C}-(CH_2)_{14}CH_3 \\ CH_2-O-\overset{\|}{C}-(CH_2)_{14}CH_3 \end{array}$$

(c)

$$\begin{array}{l} CH_2-O-\overset{\overset{\textstyle O}{\|}}{C}-(CH_2)_{6}CH_3 \\ \overset{\overset{\textstyle O}{\|}}{CH}-O-\overset{\overset{\textstyle O}{\|}}{C}-(CH_2)_{12}CH_3 \\ CH_2-O-\overset{\|}{C}-(CH_2)_{7}CH=CH(CH_2)_{7}CH_3 \end{array}$$

13-5. (c) Ketopentose

(d) Ketohexose

13-7.

(b)
$$\begin{array}{c} CH_2OH \\ HO-C-H \\ HO-C-H \\ H-C-OH \\ H-C-OH \\ CH_2OH \end{array}$$

(c)
$$\begin{array}{c} COOH \\ HO-C-H \\ HO-C-H \\ H-C-OH \\ H-C-OH \\ CH_2OH \end{array}$$

(d)
$$\begin{array}{c} CH=NNHC_6H_5 \\ C=NNHC_6H_5 \\ HO-C-H \\ H-C-OH \\ H-C-OH \\ CH_2OH \end{array}$$

13-8. Methyl β-D-glucoside

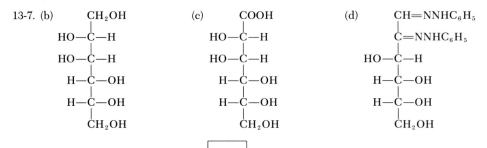

13-9. Conformational

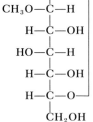

α-D-Mannopyranose

13-11. (b) Reducing sugar

13-12. (d) $CH_3CH_2CHCHCOOH$, isoleucine

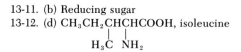

(e) Leucine
(f) Glutamic acid
(g) Methionine

(h) Tyrosine
(i) Glutamic acid, aspartic acid
(j) Cysteine

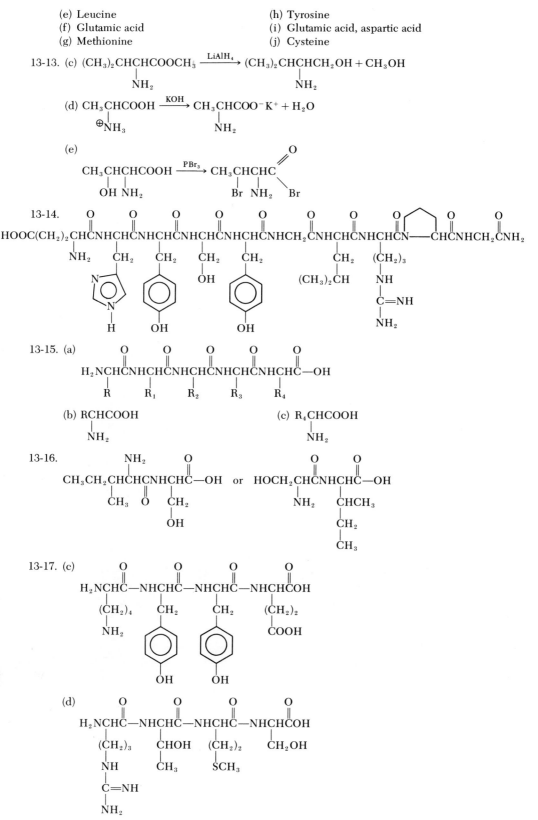

13-13. (c) $(CH_3)_2CHCHCOOCH_3 \xrightarrow{\text{LiAlH}_4} (CH_3)_2CHCHCH_2OH + CH_3OH$
            |NH₂ ... |NH₂

(d) $CH_3CHCOOH \xrightarrow{\text{KOH}} CH_3CHCOO^-K^+ + H_2O$
       $^{\oplus}NH_3$ ... NH_2

(e)
$CH_3CHCHCOOH \xrightarrow{\text{PBr}_3} CH_3CHCHC\overset{O}{\underset{Br}{\parallel}}$
   OH NH₂ ... Br NH₂

13-14.
$HOOC(CH_2)_2CHCNHCHCNHCHCNHCHCNHCHCNHCH_2CNHCHCNHCHCN\text{—}CHCNHCH_2CNH_2$

13-15. (a)
$H_2NCHCNHCHCNHCHCNHCHCNHCHC\text{—}OH$
   R   R₁   R₂   R₃   R₄

(b) $RCHCOOH$
       NH₂

(c) $R_4CHCOOH$
       NH₂

13-16.
$CH_3CH_2CHCHCNHCHC\text{—}OH$ or $HOCH_2CHCNHCHC\text{—}OH$

13-17. (c)
$H_2NCHC\text{—}NHCHC\text{—}NHCHC\text{—}NHCHCOH$
   (CH₂)₄   CH₂   CH₂   (CH₂)₂
   NH₂   ⬡OH   ⬡OH   COOH

(d)
$H_2NCHC\text{—}NHCHC\text{—}NHCHC\text{—}NHCHCOH$
   (CH₂)₃   CHOH (CH₂)₂ CH₂OH
   NH   CH₃   SCH₃
   C=NH
   NH₂

13-18.

13-20.

13-21.

13-22.

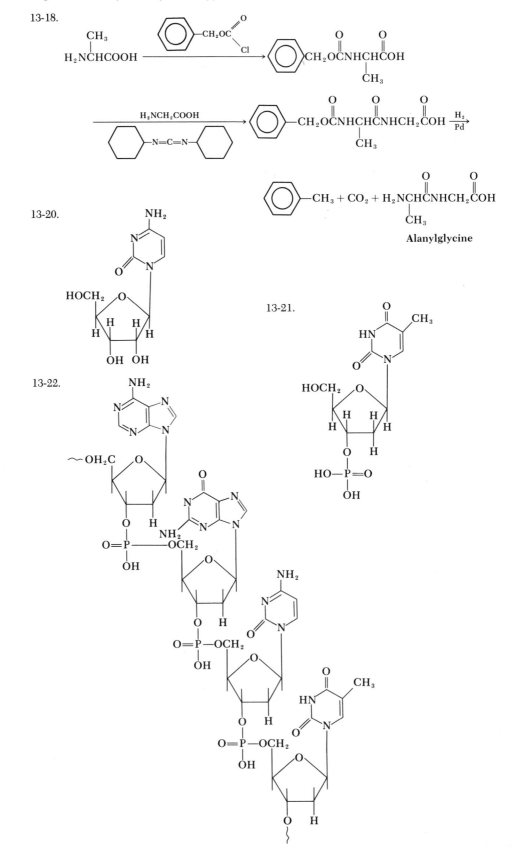

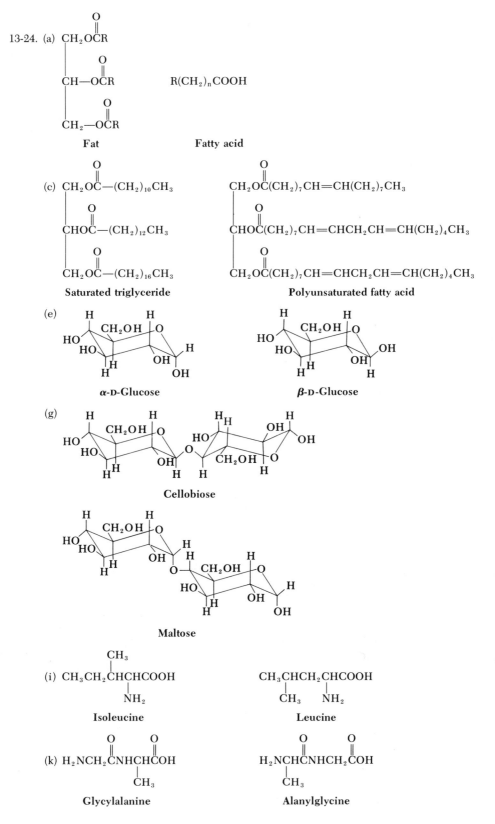

13-24. (a)

Fat   Fatty acid

(c) Saturated triglyceride

Polyunsaturated fatty acid

(e) α-D-Glucose

β-D-Glucose

(g) Cellobiose

Maltose

(i) Isoleucine

Leucine

(k) Glycylalanine

Alanylglycine

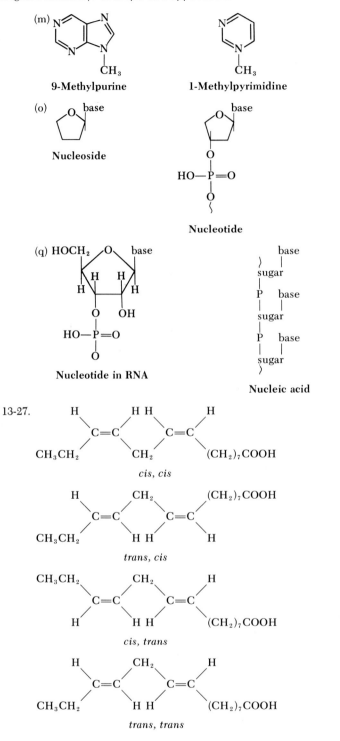

(m) 9-Methylpurine

1-Methylpyrimidine

(o) Nucleoside

Nucleotide

(q) Nucleotide in RNA

Nucleic acid

13-27.

cis, cis

trans, cis

cis, trans

trans, trans

13-30. (a) Neutral fats are esters of glycerol and fatty acids. Waxes are esters of fatty acids and long-chain alcohols.

(c) Lecithins are phospholipids that contain a base. Cerebrosides contain a straight chain fatty acid, a sugar, and a base (e.g. sphingosine).

13-33.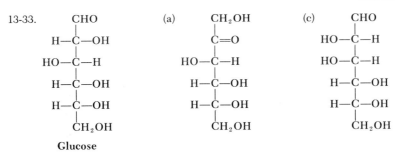

Glucose

13-35. (a) See text page 397.
(c) See text page 392.

(e)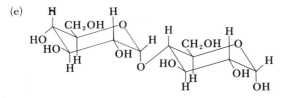

13-36. (a) Tripeptide
(b) Alanyltyrosylleucine
13-41. (a) See Fig. 13-1 page 420.
13-42. Val-his-leu-thr-pro-val-glu-lys

Chapter 14 The chemistry of the living system

14-1. (a) Transferase   (c) Hydrolase   (e) Lyase
14-2. (a) d   (c) c   (e) b
14-9. (a) Niacin from NAD^+
    (c) Step 7: dehydration of 2-phosphoglyceric acid to phosphenol pyruvic acid
    (e) Step 2: fructose-6-phosphate to fructose-1,6-diphosphate
14-10. (a) Step 10: hydration of fumaric acid
    (c) Step 11: oxidation of malic acid
    (e) Step 3: dehydration of citric acid
    (g) Step 2: condensation of acetyl coenzyme A and oxaloacetic acid
14-13. From Table 14-3.
    (a) 4   (c) 34
14-14. Step 1:   hexanoic acid $\longrightarrow$ hexanoyl —SCoA   − 2 ATP
    Steps 2-5: hexanoyl —SCoA $\longrightarrow$ 3 acetyl SCoA
        a. Step 2: 2FADH $\longrightarrow$ 2FAD ($2 \times 2 = 4$)   + 4 ATP
        b. Step 4: 2NADH $\longrightarrow$ 2NAD$^+$ ($2 \times 3 = 6$)   + 6 ATP
    Krebs: 3 acetyl SCoA $\longrightarrow$ $6CO_2 + 3H_2O$ + 3CoASH ($3 \times 12 = 36$) $\underline{+36\ ATP}$
                                                                    $+44\ ATP$

14-15. Step 1:   palmitic acid $\longrightarrow$ palmitoyl SCoA
    Steps 2-5: palmitoyl SCoA $\longrightarrow$ 8 acetyl SCoA   − 1 ATP
        a. Step 2: 7FADH $\longrightarrow$ 2FAD ($7 \times 2 = 14$)   + 14 ATP
        b. Step 4: 7NADH $\longrightarrow$ 2NAD$^+$ ($7 \times 3 = 21$)   + 21 ATP
    Krebs: 8 acetyl SCoA $\longrightarrow$ $16CO_2 + 8H_2O$ + 8CoASH ($8 \times 12 = 96$) $\underline{+ 96\ ATP}$
                                                                    $+130\ ATP$

14-21. (a) CGA   (b) CGA

Index